KJ

1, 2, 3, 4, 5, 11, 15, 17

ENGINEERING GEOLOGY AND GEOTECHNICS

PRINCIPLES OF
ENGINEERING GEOLOGY
AND GEOTECHNICS

Geology, Soil and Rock Mechanics, and Other Earth Sciences as Used in Civil Engineering

DIMITRI P. KRYNINE
Late Consulting Foundation Engineer; Formerly on the Faculty of Yale University and University of California

WILLIAM R. JUDD
Engineering Geologist, Chief Engineer's Staff, U.S. Bureau of Reclamation; Geotechnical Consultant

A
McGRAW-HILL
CLASSIC
TEXTBOOK
REISSUE

McGRAW-HILL BOOK COMPANY

New York St. Louis San Francisco Auckland Bogotá Caracas
Colorado Springs Hamburg Lisbon London Madrid Mexico Milan
Montreal New Delhi Oklahoma City Panama Paris San Juan
São Paulo Singapore Sydney Tokyo Toronto

PRINCIPLES OF ENGINEERING GEOLOGY AND GEOTECHNICS

Library of Congress Catalog Card Number 56-9631

252627282930 VBAVBA 89321098

ISBN 0-07-035560-6

PREFACE

Application of earth sciences to the solution of civil engineering problems, or *geotechnics*, constitutes the subject of this book. The point of view is that of the engineer, and the earth sciences, particularly geology, have been brought into the engineering pattern only when they have direct bearing upon the problems under discussion. The authors' goal is to present only those basic geotechnical principles that can provide a solid basis for the solution of problems connected with the natural environment of an engineering structure, particularly the surrounding ground. Case histories have been used only for the elucidation of principles.

The book has been designed as a textbook for civil engineering and advanced geology students and as a reference work for practicing civil engineers and engineering geologists. Those portions of the book that can be omitted at the first reading and all examples are printed in small type. The book has been made more compact, and thus more convenient for use, by condensing descriptions of laboratory and field test procedures; its mathematics are reduced to a minimum by omitting the derivation of formulas. Detailed information on an item of interest may be located, however, by using the list of references appended to each chapter.

Although the book is not formally subdivided into two major parts, the first eight chapters contain geotechnical information of a general character applicable to any structure. The remainder of the book, except Chap. 19, contains applications of this general information to specific kinds of structures. This concept may be useful in subdividing the material between two semesters when the book is used for teaching purposes.

Like any new branch of human knowledge, geotechnics advances from day to day. The authors endeavored to keep the manuscript as up to date as possible. The general silhouette of geotechnics was sufficiently clear cut when the book was written; however, new theories and improved techniques constantly appear in the current technical press. Monthly issues of the *Proceedings of the American Society of Civil Engineers* separates, publications of the British Institution of Civil Engineers, quarterly issues of the English journal *Géotechnique*, and pertinent papers of the *Proceedings of the Annual Meeting of the Highway Research Board* all may be referred to in this regard. New issues of geological journals also

v

occasionally present individual papers and articles of considerable geotechnical interest.

Acknowledgments. Numerous books, articles in technical and geological press, and United States government publications were helpful to the authors in the preparation of this text. Portions of the text also are based not only on published but also on unpublished material and practices of the U.S. Bureau of Reclamation through the kind permission of L. N. McClellan, Assistant Commissioner and Chief Engineer, to whom the authors wish to express their appreciation. Essential advice of general character was received from Dr. W. H. Irwin, Chief Geologist, and substantial material help in the preparation of individual portions of the book was obtained from various members of the Denver staff of the Bureau. An engineer especially interested in the book was Dart Wantland, already well-known as a contributor to the Jakosky book "Exploration Geophysics." Mr. Wantland furnished most of the examples of geophysical investigations used throughout the book and collaborated in preparing the geophysics sections of Chap. 6. Ground-water specialist Thomas P. Ahrens contributed the discussion of field methods of determining the permeability of crustal materials and read over Chap. 5. The airphoto interpretation and surveying methods in Chap. 7 were reviewed by photogrammetrist Wm. H. Hatfield. The counsel of Dr. R. C. Mielenz and the late Merle E. King proved useful in the preparation of the mineralogical discussions in Chaps. 1 and 8. Whitney M. Borland and Carl R. Miller, sedimentation specialists, assisted the authors in bringing the material in a portion of Chap. 12 as up to date as current practice permits.

Many persons and organizations outside the Bureau of Reclamation also graciously contributed their time. Dr. Paul D. Krynine of Pennsylvania State University greatly assisted in the final drafting of Chap. 1, and his advice was helpful at various stages of the manuscript's preparation. Dean K. B. Woods of Purdue University contributed general criticism of much benefit. The portion of Chap. 6 on soil investigations was reviewed and amplified by Richard J. Woodward of Woodward, Clyde & Associates, Consulting Civil Engineers of Oakland, California. The Denver partner of this firm, Dr. James L. Sherard, constructively criticized the draft of Chap. 16. The original manuscript of Chap. 18 was improved according to comments by Dr. Perry Byerly, professor at the University of California at Berkeley and California State Seismologist; John E. Rinne of San Francisco, structural engineer and specialist in aseismic design; and William K. Cloud of the San Francisco office, U.S. Coast and Geodetic Survey. Henry Degenkolb, San Francisco structural engineer, was helpful in lending some material on earthquakes.

The portion of Chap. 8 on explosives was approved by the Technical

Service, Explosive Department, E. I. du Pont de Nemours & Company. The description of the bore hole camera in Chap. 6 was approved by Engineering Research Associates Division of Remington Rand Company. The legal phraseology and context of Chap. 19 were reviewed by Simon Quiat, specialist in contract law; presentations of the citations in that chapter were reviewed by the American Law Book Company of Brooklyn, New York.

The source of illustrations, other than the authors' files, receives proper recognition on the pertinent captions. Last, but not least, the authors' appreciation goes to illustrator James Vitaliano for his work in the preparation of some difficult drawings.

The authors wish to protect their willing aides by warning the reader that although these experts were responsible for certain information, the authors assume full responsibility for the presentation of this information as it appears in the book.

DIMITRI P. KRYNINE
WILLIAM R. JUDD

CONTENTS

ix

INTRODUCTION: EARTH SCIENCES IN CIVIL ENGINEERING

When the St. Francis Dam in Southern California failed in 1928 with a loss of many lives and damages in millions of dollars, the civil engineering profession awoke to the idea that the careful design of a structure itself is not all that is required for its safety. Before that regrettable event, not all engineers could see clearly that the design of a structure should be preceded by a careful study of its environment, particularly the foundation material on which the structure was to be placed. After the failure of St. Francis Dam, the need of environment exploration with proper interpretation of the results was understood by all. It became advisable and, in serious cases, compulsory to consult a geologist about the characteristics of the natural materials at the emplacement of the proposed structure. In their turn, the geologists who were asked for assistance had to become acquainted with the engineering requirements of those materials and to obtain at least some notion of the work and behavior of engineering structures. Thus, a new type of specialist—the engineering geologist—appeared. The best known engineering geologist in the United States was Dr. Charles P. Berkey (1867–1955), who counseled engineers in the design and construction of numerous dams, tunnels, and other important structures.[1]*

Geology vs. Earth Sciences. Engineering geologists have been primarily geologists. Geology has been defined as "the science which treats of the origin, history and structure of the Earth, as recorded in the rocks, together with the forces and processes operating to modify the rocks."[2] According to this definition all studies of the earth's *atmosphere*, *hydrosphere* (all water at the earth's surface), and *lithosphere* (rocks, soils, and other constituents of the globe proper) are studied in geology. In actual practice there are limitations to this definition. Examination of existing geology texts shows that the principal subject of discussion are rocks and minerals (including subsurface water) and forces acting on them; the ultimate purpose of the examination is "to explore the long geologic record contained in the rocks."[3] Commendable results have been

---

* Superior numerals in the text correspond to the numbered references at the end of each chapter.

1

obtained in this field by the geologists when one considers their methods of study—primarily observation, inference, and reasoning and, in exceptional cases, direct subsurface exploration to limited depths. Certain studies of the earth, however, require the use of other means, e.g., elaborate instrumentation or mathematical analysis and advanced physics and chemistry beyond the knowledge of the average geologist. Accordingly, several sciences or branches of human knowledge have been developed to enable the study of some portions or features of the earth without being part of geology proper. In modern scientific literature and in some college curricula the complex of sciences concerned with the study of the earth is known as earth sciences. Geology is, of course, the most prominent in the family of earth sciences.

Engineering Geology vs. Geology. A distinction should be made between geology and engineering geology. Engineering geology as it now stands may be defined as a branch of human knowledge that uses geological information combined with practice and experience to assist the engineer in the solution of problems in which such knowledge may be applicable.[8] Engineering geology differs from geology proper primarily in scope. In the construction of engineering structures, depths of over 300 ft are practically never exceeded. Thus, civil engineering activities, and hence possible need of geological consultation, are concentrated in a relatively thin layer as compared with that portion of the earth's crust (perhaps 10 or 20 miles thick) considered accessible for geological studies. Ordinary geological studies generally involve large areas; the result is that the information given on geological maps may be much too general for engineering purposes and does not show local geological conditions in the detail which the engineer requires. Prior to the design of an important structure, engineering geological investigations are generally concentrated at the site of the structure and consist of direct subsurface exploration by boring holes in the ground or making excavations, as explained in Chap. 6. The results are analyzed, and geological reports submitted. The engineer generally wants to know whether a given constituent of the earth's crust—rock or soil, either in its natural state (in place or otherwise) or in a processed condition—fits in with his construction program; if it does not, then he desires to know if and how it can be made to fit. Such information should be presented to the engineer in clear and simple terms. Complex classifications of rocks and the history of the earth are of little professional value to an engineer, though as a layman he may be greatly interested.

Engineering Geology vs. Geotechnics. As the field of geology applied to civil engineering enlarged, it gradually became clear that the geologist's discoveries and deductions would have to be translated into practical terms and applications. This translation required that the geologist have substantial, rather than superficial, notions of engineering. The geólogist in responsible charge should have sufficient knowledge of design and construction problems to assist in the solution of such problems.[4] Concurrently, the geologists working with engineers understood that geological information alone was insufficient. One striking example of this insufficiency is the lack of any geological information on the design of structure foundations on soils. Only very heavy structures such as masonry dams, some large bridge piers, and special underground plants are carried to rock; the rest of the structures, in fact the majority, are founded on soil materials if rock is not present or conveniently close to the surface. Whereas no hints as to the behavior of loaded soils can be found in geology, the engineers not only have sufficient practice along these lines but have elaborated the theoretical bases of a new earth science—soil mechanics—helpful in the design and construction of earth structures in general. Experienced geologists have recommended the study of soil mechanics to their colleagues,[5] indicating that "a working knowledge of the principles of soil mechanics is an important adjunct to the knowledge of a geologist engaged in applying his science to the practical construction problems."[6] In this book the elements of soil mechanics are given in Chap. 4.

The advent of soil mechanics aroused a great interest in geology among engineers, since they realized that the efficiency of soil mechanics may be further increased by combining it with geology. A considerable contingent of engineers is studying geology and trying to use it in practical work.

Thus, engineering geology reinforced with useful information from other earth sciences and adequate notions of engineering is gradually being transformed into a new branch of human knowledge—*geotechnics*. This term is in common use in several countries but has not been, as yet, officially accepted in the United States, although many people here use the term. It is believed that this transitional period should be ended with the general recognition and proper use of the term "geotechnics" by all concerned. The present book has been written with this belief.

Earth Sciences Used in Geotechnics. Knowledge of basic properties of rocks and minerals constituting the earth's crust is a prerequisite to geotechnical studies. Detailed study of rocks is the object of *petrology* and its branch petrography, or systematic description of rocks. The term "petrology" is more general and involves also the properties and origin of the rocks. Chapter 1 and, to some extent, Chap. 2 cover the elements of petrology required for engineering purposes. Chapter 3 introduces a portion of *geomorphology*, the description and origin of the shape (landforms) in which the constituents of the earth's crust occur on its surface. Only landforms made of soils are described.

Pedology, known also as "soil science," is a study of the uppermost layers of the earth's crust formed of soils, and the study is primarily confined to those soils important in agronomy.   It was founded originally by a group of Russian geologists and agronomers.   Fresh ideas of soil formation were advanced, and new soil classifications elaborated.   Though probably adequate for agronomical purposes, these classifications often are objectionable from the geologic or engineering point of view.   All in all, however, pedology based on physics, chemistry, and bacteriology grew, with the active participation of American scientists, into a science completely independent of geology.   Pedologic knowledge is applicable in irrigation and drainage projects and in road and highway engineering, and the geotechnical specialist should keep his eye on the progress of this science, particularly in its application to the study of soil moisture.   *Soil mechanics,* another earth science already mentioned, was originated by engineers and counts many prominent names in the ranks of its workers, the best known worker in this field being Karl Terzaghi.   Soil mechanics is based on physics and mechanics, particularly on the study of stresses and strains in soils.   Methods of soil mechanics are gradually penetrating into geology; this penetration is typified by the material on "rock mechanics" that appears in Chap. 2.   It should be emphasized that soil mechanics, besides being an earth science, is a basic engineering science.[7]   The same is true of rock mechanics.

Besides rocks and soils, the earth's crust contains subsurface water, which may strongly influence the behavior of engineering structures. *Hydrology* is a science that studies both surface and subsurface water; however, the study of subsurface water sometimes is termed *geohydrology* and is discussed in Chap. 5.   An important earth science is *geophysics,* the application of the methods of physics to the study of mass properties of rocks and soils.   Its application to subsurface investigation is spread throughout the book.   A branch of geophysics—*seismology,* the study of the earth's vibrations—is presented in Chap. 18 in so far as its application to "aseismic," or earthquake-proof, design is concerned.

Sedimentation, as the cause of the formation of sedimentary rocks and oil reservoirs, is intensively studied in general and petroleum geology.   In geotechnics, the term "sedimentation" is used in the sense of the filling and obstruction by the settling "sediment" of rivers, canals, harbors, and water reservoirs behind dams.   This kind of sedimentation, or "deposition," is discussed in Chap. 12 and partly in Chap. 16.

The reader should realize that not all ramifications of geology and not all earth sciences are necessarily utilized in geotechnics.   For example, *paleontology* and *paleobotany,* which deal with the study of the former fauna and flora of the earth now found in the form of fossils, seldom are helpful in the solution of geotechnical problems.   Conversely, however, it should be recognized that at the present time full advantage has not been taken of *meteorology*, the science of the atmosphere dealing with wind, rain, storms, hurricanes, and little known phenomena in the stratosphere. It is possible that unexpected results may be obtained from the meteorologic studies when ultramodern instrumentation is applied. Finally, geography should also be considered as an earth science. The terms "physical geography" and "physiography" have practically the same meaning as "geomorphology" as far as geotechnics is concerned.

Historical Geology. Unlike most materials, the older the rock, the stronger it is and the better are its foundation properties. Engineers who have worked both on the East and the West Coast of the United States know that Eastern rocks are stronger than the Western, except, of course, in those cases when the foundation properties of older rocks have been

Geological Time Scale (Abridged)

Era and life	Period	Epoch	Approximate age of rocks, millions of years
Cenozoic: mammals and modern flora	Quaternary	Recent Pleistocene	0–1
	Tertiary	Pliocene Miocene Oligocene Eocene	1–60
Mesozoic: reptiles	Cretaceous Jurassic Triassic		60–200
Paleozoic: amphibians, fishes, and higher invertebrates	Permian		
	Carboniferous	Pennsylvanian Mississippian	200–500
	Devonian Silurian Ordovician Cambrian		
Proterozoic: lower invertebrates	Pre-Cambrian		500–3,000
Archeozoic: primordial life or none			

seriously altered by movements of the earth's crust, particularly by mountain-making processes (orogeny). Also, an experienced engineer does not expect to find the peat deposits on the West Coast comparable in thickness to some Eastern ones. These and similar facts, together with the recent history of a locality, e.g., changes in the course of a river, may be important from the geotechnical point of view, whereas geological changes that occurred millions of years ago are immaterial (unless active or causing recent changes). Though the study of *historical geology* as a science is of little practical value to the engineer, it is advisable for him to be acquainted with the terms of the "geochronological table" or "geological time scale" as shown on page 5.

As seen from this table, geological time is subdivided into eras, periods, and epochs. The age of the earth is probably greater than the estimated maximum age of the rocks.

REFERENCES

1. Application of Geology to Engineering Practice, "Berkey Volume," Geological Society of America, New York, 1950. Throughout the present book, this volume is referred to simply as "Berkey Volume."
2. Rice, C. M.: "Dictionary of Geological Terms," J. W. Edwards, Publisher, Inc., Ann Arbor, Mich., 1950.
3. Longwell, C. R., A. Knopf, and R. F. Flint: "Physical Geology," 2d ed., John Wiley & Sons, Inc., New York, 1948. A 3d ed. of this book is currently available.
4. Burwell, E. B., Jr., and G. D. Roberts: The Geologist in the Engineering Organization, "Berkey Volume," 1950.
5. Taber, Stephen: *Science*, June 27, 1952.
6. Burwell, E. B., Jr.: Geology in Dam Construction: Pt I, "Berkey Volume," p. 33, 1950.
7. Casagrande, Arthur: General Report, Third Conference on Soil Mechanics and Foundation Engineering, Zurich, August, 1953.
8. Happ, S. C.: Engineering Geology Reference List, *Bull. GSA*, vol. 66, pp. 993–1030, August, 1955.

CHAPTER 1

ROCKS AND MINERALS

An engineering structure consists of a *superstructure* and a *substructure*. The latter transmits the weight of the structure and superimposed loads to the underlying part of the earth's crust or foundation. A more precise term in this case is *foundation material*. The term "foundation" also is applied to the substructure or to its lower part. The term "footing" is used for those parts of the substructure in direct contact with the underlying foundation material. To design and build the superstructure, it is necessary to know the properties of the materials to be used. Similarly, to appraise the foundation material for the design of the substructure, it is necessary to know the properties of the rocks and soils constituting the foundation material. This knowledge is especially important for the engineering geologist, but also it is the engineer's responsibility to understand the behavior of rocks and soils forming the foundation materials. The knowledge of the properties of rocks and soils also is of great value when the entire engineering structure is to be built chiefly of natural materials, e.g., earth dams.

Geologists often apply the term "rock" to all constituents of the earth crust. In this book, however, the engineering usage of subdividing these constituents into *rocks* and *soils* is followed. To the engineer and engineering geologist (or geotechnician), hard and compact natural materials of the earth crust are rocks and their derivatives are soils.

MINERALS

1.1. Identification of Minerals. Rocks are composed of minerals. A *mineral* is a natural inorganic substance of a definite structure and chemical composition. The appearance of a well-*crystallized* mineral is representative of its atomic structure. However, in the "cryptocrystalline" or in the rounded minerals, crystalline appearance does not show correct internal structure. The same is true of amorphous (shapeless) minerals.

In general engineering practice, rocks and minerals are identified by *megascopic* methods, i.e., only with the naked eye, a low-power hand lens, or a magnifying glass.   Complex rocks and minerals may require the use of *analytical* methods for proper identification.   The first approach in such an analysis is to saw a very thin section from the specimen by means of a diamond saw.   This thin section of rock is then glued to a microscope glass slide and ground down until the rock section becomes transparent to light.   The resulting *thin section* is then studied under a microscope, usually with the aid of polarized light.   Besides such petrographic studies of thin sections, other analytical methods are X-ray analysis (particularly useful for fine-grained minerals such as clay minerals) and chemical analyses of varying degrees of complexity, ranging from simple tests with hydrochloric acid, HCl (Sec. 1.3), through silicate analysis and into spectrographic methods.   In certain cases, particularly in the study of metallic ores, the *blowpipe analysis* is used.   A mineral fragment is acted upon by a blowpipe jet formed by forcing air from a blowpipe through a gas flame of a laboratory burner.   The reaction of the mineral to the effect of the jet and the color of the flame are observed.[1]

1.2. Physical Properties of Minerals.[2,3]   The most important properties of rock-making minerals are the following:

Color and Streak.   The normal eye easily recognizes the color of a mineral; however, for standard colors used in rock identification, standard color charts are available.[4]   In order not to be misled by a weathered surface or by a secondary coloring from impurities such as iron, the investigator should expose a fresh surface by chipping off a corner of the specimen. It must be realized in using color for identification, that some minerals exhibit a wide range of colors.

If several specimens of one rock type are collected in the same region and color comparisons are to be made, a small collection of rock chips of varying colors should be prepared.   These can be glued to a piece of cardboard or plywood and, properly labeled, can be used as a natural color chart.

If a mineral is rubbed against an unglazed porcelain or china plate, a colored or colorless (white) *streak* of minute mineral particles is left on the plate.   The color of the streak often is characteristic of the mineral.

Hardness.   The hardness of a mineral (symbol H) is expressed by its number in Mohs scale of hardness (Table 1.1).   Each mineral listed in that scale can scratch all minerals of smaller numbers but in turn can be scratched only by minerals of a higher number than its own.   For instance, quartz ($H = 7$) can scratch all members of Mohs scale from 1 to 6, but not those above 7.   Approximate identification methods are indicated in Table 1.1.

When a mineral is scratched with a knife, it is necessary to observe carefully whether the white mark left on the mineral surface is actually an indentation in the surface (a scratch) or is merely a mark left by a streak of particles from the steel in the knife. The same is true for other scratching objects.

For example, if an unknown mineral is scratched by a piece of orthoclase and happens to be quartz, a white streak left on the unknown mineral will be merely fine particles of orthoclase rather than an actual scratch on the quartz surface.

Cleavage and Fracture. If struck with a sharp object, a mineral breaks along a definite crystallographic plane (cleavage plane), which is parallel to a crystal face. The cleavage face is usually a perfectly smooth plane surface that appears to be polished. It can be detected by holding the mineral so that light will reflect from the cleavage faces as from a mirror. If a mineral has more than one cleavage, the angle between the two

TABLE 1.1. Mohs Scale of Hardness

Standard mineral	Hardness H	Approximate identification method
Talc............	1	Will mark cloth
Gypsum.........	2	Can be scratched by a fingernail
Calcite..........	3	Can be scratched by a copper penny
Fluorite.........	4	
Apatite.........	5	Can be scratched by a pocket knife
Orthoclase.......	6	Will scratch window glass
Quartz....,....	7	Cannot be scratched by a steel file
Topaz..........	8	
Corundum......	9	Will scratch most metals, but not diamond
Diamond........	10	Will scratch any substance except other diamonds

Example. If the mineral can be scratched by a penny but not by a fingernail, its hardness is about 3.

cleavage faces can be determined roughly by rotating the mineral in reflected light and estimating the angle between the two reflected light rays. Some minerals have what is known as *difficult* or *hard* cleavage. In these minerals, it is difficult to determine the cleavage face without the aid of a microscope.

Fracture in a mineral also can be induced by a sharp blow; however, the resultant surface is irregular and bears no relation to the crystal faces of the mineral. Often, only an expert can distinguish between fracture and true cleavage. Fractures commonly are described as *conchoidal*—similar to a smooth convex or concave surface; *uneven*—a rough, irregular surface with angular and rounded projections; *splintery*—a self-descriptive term; and *hackly*—a jagged surface resembling the end of a steel rod that has been broken in compression.

Tenacity. The ability of a mineral to withstand crushing, tearing, or bending is its *tenacity.* It is commonly measured as follows: *brittle*— shatters and powders easily; *malleable*—can be hammered into thin sheets;

sectile—can be cut into thin shavings by a knife; *ductile*—can be drawn into a wirelike form; *flexible* (a better term being "inelastic")—can be bent but does not return to its original shape after the bending force is released; and *elastic*—can be bent but returns to its original shape after the bending force is released.

Crystal Form.   Except for the amorphous minerals, every mineral has a definite crystal form bounded by several or many crystal faces (planes) and belongs to a definite crystallographic system.   The latter is charac-terized by crystallographic axes (*a*, *b*, *c*) which often are the mutually perpendicular axes of symmetry of the crystal.   A few cases in which the engineer is concerned with mineral crystal forms are discussed at various points in this book.

Specific Gravity (G or sp. gr.).   The specific gravity of a mineral or rock is the ratio between the weight of a given volume of the mineral or rock and the weight of an equal volume of water at 4°C.   The determination of specific gravity is explained in detail in Sec. 2.1.   For preliminary identification purposes, the specific gravity can be roughly estimated by bouncing the mineral or rock in the hand.   It will "feel" heavy or light (e.g. lead and graphite, respectively) according to whether its specific gravity is high or low.   For the most common nonmetallic minerals in the earth crust, the average specific gravity may be assumed to be between 2.65 and 2.75 (see Table 1.2).

Luster.   Most minerals exhibit a certain characteristic appearance (luster) under reflected light.   The luster may be metallic, nonmetallic, or submetallic.   The nonmetallic lusters are described as *vitreous*—the appearance of glass; *greasy*—an oily appearance; *adamantine*—hard brilliance common in diamonds; *pearly*—iridescent appearance of a pearl; *silky*—similar to the sheen of silk; and *resinous*—appearance of resin.

Ability to Transmit Light (Diaphaneity).   A mineral is *transparent* if objects can be clearly seen through it, *translucent* if it transmits light but objects cannot be clearly seen through it, or *opaque* if no light can be transmitted through the mineral or through its thinnest edges.

1.3. Rock-forming Minerals.[2,3]   The properties of the *most important* rock-forming minerals are given in Table 1.2.   If minerals are classified according to their chemical composition, they may be subdivided into silicates, oxides, carbonates, and sulfates.   The most important minerals belonging to each group are listed in the second column of Table 1.2.

Water in minerals may be attracted (sorbed, Sec. 2.3) or chemically bound to the substance of the mineral (structural).   The *sorbed water* is just sitting at the surface of a mineral particle and in some rare cases penetrates into its interior.   It can be removed by evaporation, either in open air or in an oven with a temperature of about 105°C, as is done in the laboratory. *Structural water* forms a part of the structure of a mineral and can be removed only by breaking the structure, for example, by high heat (Sec. 1.5). Water shown in the last column of Table 1.2 is all structural. In some chemical formulas in that column, water is shown as H_2O, in some others as hydrogen and oxygen in different combinations. The chemical formulas in the last column of Table 1.2 contain only a few simple chemical radicals. Such a radical is silica, SiO_2, in a number of silicates and in quartz, which is pure crystalline silica. Another radical, alumina, Al_2O_3, is found in feldspars, kaolinite, and micas and forms compact, heavy corundum ($G = 4$, $H = 9$). Carbonates contain the CO_3 radical. Magnesium, Mg, or potassium, K, may be found in many silicates together with silicon, Si. Sodium, Na, is not very common in rock-forming minerals, but in the form of common salt, NaCl, it constitutes the mineral halite.

The presence of iron, Fe, increases the specific gravity of the mineral. The luster of such minerals is mostly metallic to submetallic. These minerals are important in mining but are rarely found in foundations.

As shown in the last column of Table 1.2, the silicates are subdivided into *hydrous* (containing hydrogen and oxygen in their chemical formulas) and *anhydrous* (no water). The most frequently found anhydrous silicates are members of the feldspar group. The feldspars usually are divided into "orthoclase-microcline," or potash, feldspars and "plagioclase," or soda-lime (or sodic), feldspars. Under the action of water carrying carbon dioxide, CO_2, the feldspars alter into clay minerals or muscovite (white mica). Classification of most igneous rocks requires the ability to distinguish between the presence or absence of orthoclase or plagioclase in a rock. Very often the difference between these two can be determined for certain only by use of a microscope.

The *micas* are easily recognized by their flexible translucent thin sheets or flakes, which are easily separated. In soils they appear in the form of small shining scales. Under weathering action, biotite (black mica) alters more rapidly than muscovite, or white mica. Rock containing much mica is of dubious value as a foundation or in the walls of a deep cut. The easy cleavage of rocks containing mica may result in rapid deterioration, especially if the rock is freshly exposed.

The hydrous silicates include serpentine, chlorite, talc, illite, and kaolinite (the last is discussed in detail in Sec. 1.4). The mineral *serpentine* makes the rock of the same name, which has a variety of appearances and shades, although generally of a green color. It may be hard and competent,* or it may be soft, greasy, and dangerous to construction.

* *Competent* rock may be defined as a rock capable of safely supporting a particular structure or the overlying rocks and soils or, on some occasions, both. The term is rather loose but is widely used.

TABLE 1.2. Mineral Properties

Chemical subdivision	Mineral name	Color	Hardness	Fracture and tenacity	Luster	Streak	Crystal form	Specific gravity	Chemical composition
Silicates	Chlorite	Green	2–2.5	Nonelastic, platy	Pearly	Pale green to white	Monoclinic	2.7	Hydrosilicates of Al with ferrous iron and Mg
	Feldspar	Colorless, white, reddish, yellowish, (when impure)	6	Uneven, subconchoidal. Brittle	Vitreous with pearly cleavage	White	Monoclinic or triclinic. Twinning common	2.54–2.76	$KAlSi_3O_8$ $NaAlSi_3O_8$ to $CaAl_2Si_2O_8$
	Hornblende—Can be distinguished from pyroxene only when it occurs as long, needlelike crystals and more difficult to obtain cleavage								
	Kaolinite	White. Yellow to brown tinges		Requires microscope or X-ray diffraction for identification.			Monoclinic	Requires microscope or X-ray diffraction for identification	$H_4Al_2Si_2O_9$
	Micas	White (muscovite), black (biotite). Impure colorations	2–3	Uneven. Flexible (elastic)	Vitreous to silky		Monoclinic, often as 6-sided flat tablets	2.76–3.2	$H_2K(MgFe)Al(SiO_4)_3$ (biotite) $H_2KAl_3(SiO_4)_3$ Muscovite
	Olivine		6.5–7		Glassy	Can be recognized only if found as phenocrysts. Common in dark rocks			
	Pyroxenes	White-green-black	5–7	Uneven. Brittle	Vitreous to pearly	White to gray-green	Monoclinic to orthorhombic (see text)	3.2–3.6	Mg, Fe, Ca-Mg, Ca-Fe, Na-Fe, Al-Fe, and Li-Al. Oxides and silicates
	Talc	White to greenish white	1	Soft, soapy feel. Foliated. Massive	Silvery to greasy	White to greenish white	Monoclinic	2.7–2.8	$H_2Mg_3(SiO_3)_4$
	Serpentine	Greenish, usually in varied shades	2–5	Often fibrous to massive. Platy	Greasy-massive Silky-fibrous		Monoclinic (crystals unknown)	2.2–2.7	$H_4Mg_3Si_2O_9$
	Zeolites		3½–5½		Vitreous	Colorless or white	Usually requires petrographic methods for identification	2.0–2.4	Na, Al, and Si oxides and sometimes Ca and H_2O

TABLE 1.2. Mineral Properties (Continued)

Chemical subdivision	Mineral name	Color	Hardness	Fracture and tenacity	Luster	Streak	Crystal form	Specific gravity	Chemical composition
Oxides	Corundum	Shades of gray. Blue	9	Brittle to tough	Adamantine, vitreous to dull and greasy		Hexagonal	4	Al_2O_3
	Hematite	Gray to black	5.5–6.5	Platy. Subconchoidal (no cleavage)	Metallic to dull	Red	Hexagonal (rhombohedral)	5.2	Fe_2O_3
	Ilmenite	Reddish to brownish black	5–6	Conchoidal. Brittle (no cleavage)	Submetallic	Black to brownish red	Hexagonal	4.5–5	$Fe\,TiO_3$
	Limonite	Brown to brownish yellow	5–5.5		Dull to submetallic	Yellowish brown	Amorphous. Concretionary or earthy mass	3.6–4.0	$Fe_2O_3 \cdot H_2O$
	Magnetite	Dark gray to black	5.5–6.5	Uneven. Brittle (poor cleavage)	Metallic		Isometric	5.2	Fe_3O_4
	Quartz	White, gray, black, rose	7	Conchoidal, uneven, splintery. Brittle to tough	Oily to glassy	White	Hexagonal	2.65	SiO_2
Carbonates	Calcite	Colorless or white but may be stained	3		Vitreous. If massive, dull	Colorless or white	Rhombohedral	2.71	$CaCO_3$
	Dolomite	White to multicolored	3.5–4		Vitreous to pearly to dull		Rhombohedral (Crystals are rare)	2.8–2.9	$CaMg(CO_3)_2$
Sulfates	Anhydrite	White, but may be tinted	3–3.5		Pearly, glassy Massive-dull		Orthorhombic. Usually granular to compact masses	2.95	$CaSO_4$
	Gypsum	Colorless to white. May be tinted various colors	1.5–2	Foliated. Conchoidal	Glassy, pearly white Fibrous-satiny Massive-dull		Monoclinic. Twins common	2.32	$CaSO_4 \cdot 2H_2O$

Note. Quartz, listed as an oxide in Table 1.2, is classified as a silicate by modern mineralogists. Cryptocrystalline forms of silica are described in the text.

Whenever encountered in deep cuts or tunnels, it is to be regarded with suspicion, as it can rapidly alter from a competent to a soft, incompetent material. The process of *serpentinization* involves transformation of other minerals into serpentine. *Chlorites,* common in schists and slates (see Sec. 1.11), also are green but not so hazardous in construction as serpentine. *Talc* (see Table 1.2) is a poor foundation material because of its low density and general instability.

In the group of silicates, the most important mineral is *quartz,* the crystalline form of silica, SiO_2. Large quartz crystals are easily identified megascopically, as they are hexagonal prisms with six-sided pyramids on both sides (Fig. 1.1a). In smaller grains, quartz is equally easy to identify by its conchoidal fracture, vitreous luster, great hardness ($H = 7$), and, generally, its lack of definite color. Quartz is one of the most durable minerals and does not alter under the action of common weathering agents, such as water, or changes in temperature. Thus quartz is found in many rocks and soils. The cryptocrystalline and amorphous forms of quartz (opal, chert, and others) occasionally are of concern to engineering. Opal, for instance, may cause alkali reactivity in concrete and, therefore, failure of the structure (Sec. 8.18).

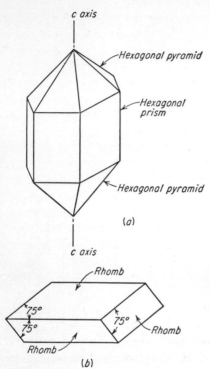

FIG. 1.1. Crystals of (a) quartz, (b) calcite.

The most important of the carbonates is calcite, $CaCO_3$. Its crystals are easily recognized by their softness ($H = 3$) and their six rhomb-shaped sides (Fig. 1.1b). Calcite is a *secondary* mineral, i.e., a mineral formed after the rock containing it has been formed. It is soluble in water containing carbon dioxide. Application of hydrochloric acid, HCl, to a rock containing even a small amount of calcite will cause a strong effervescence, and thus it is used as an identification method. Vinegar will give a weak effervescence with calcite. *Dolomite* possesses similar physical properties but differs from calcite in its greater hardness and less intense reaction to hydrochloric acid. Effervescence upon application of the acid occurs only in powdered dolomite material.

In the sulfate group, *gypsum* and *anhydrite* differ only by the content of structural water.   Because it contains such water, gypsum has a lower specific gravity and lower hardness than anhydrite.   Structural water is not very tightly held in gypsum; a small piece heated in a test tube will cause small drops of water to appear on the tube walls.   No water will be observed if a piece of anhydrite is similarly heated.

1.4. The Clay Minerals.[5-7]   Whenever clay is encountered in the foundation, the engineer and the geologist are placed on guard; past experience has shown that under some circumstances, "unpredictable" phenomena may take place.   Research on clay minerals, started about 1930, has given evidence, however, that some prediction of clay properties for the benefit of the engineer can be made.   A definite identification of clay minerals, which is required in major engineering works, involves complicated analytical methods briefly discussed in Sec. 1.5.

The clay minerals are essentially *hydrous aluminum silicates* or occasionally hydrous magnesium or iron silicates.   Clay minerals are crystalline with a very few exceptions.   In engineering practice, clays are described as consisting of *particles*, although actually the particles are minute *flakes*.   As in all crystalline substances, the atoms in these flakes are arranged in units, in this case *sheets*.   In clay minerals sheets are of two varieties: silica sheets and alumina sheets.   A *silica sheet* is made of tetrahedra, each tetrahedron being bounded by four triangular plane surfaces with oxygen atoms at the vertices and a silicon atom in the interior of the tetrahedron (Fig. 1.2a, b, c).   All oxygen atoms are equally spaced, and all are equally distant from the silicon atom.   The tetrahedra are combined into hexagonal units (Fig. 1.2d).   In each unit, oxygen atoms 1 are connected with oxygen atoms 2 and 3 and, by means of protruding atoms 4, are connected to six neighboring units of identical size and shape.   By repeating themselves indefinitely and still sharing oxygen atoms (4 in Fig. 1.2c), these units form a *lattice* of the mineral.   An *alumina sheet* consists of two-row units shown in Fig. 1.3a, b, and c, arranged in octahedrons with oxygen, O, atoms or hydroxyl, OH, groups at the vertices of alternate rows, respectively, and with an embedded aluminum atom at the middle (Fig. 1.3c).   The lattice structure of clay minerals is essentially the basis for their classification into three major groups, namely, the *kaolinites*, the *montmorillonites* (or montmorillinoids), and the *illites*, or "hydromicas" (so termed because of their structural similarity to micas).[5]

The Kaolinites.   The minerals of this group are formed of a single tetrahedral silica sheet and a single octahedral alumina sheet, a combination that repeats itself indefinitely.   Particles forming the kaolinite minerals may be regarded generally as aggregates of small, sometimes approximately hexagonal flakes.   Figure 1.5 shows the shapes of the kaolinite flakes when observed through an electron microscope (electron streams substitute for lenses, and the enlargement is many times greater than that obtained with an optical microscope). The kaolinites form very stable clays because their tight inexpandible structure resists the introduction of water into the lattices and its consequent destabilizing effect. Furthermore, when wet, the kaolinites are but moderately plastic and tend to have a larger coefficient of internal friction than other clay

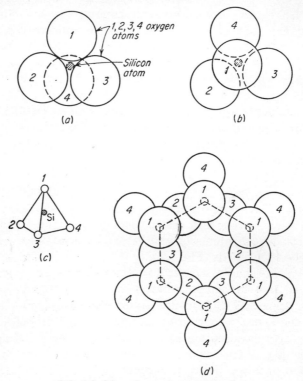

FIG. 1.2. Silica sheet of clay minerals.

minerals. Generally, unless they contain expandible impurities, the kaolinites themselves are not subject to expansion or heaving when saturated. (It should be noted that the clay minerals are seldom found in "pure" form; usually more than one type are present, and the percentage and size of each type present will determine the ultimate characteristics of the combined mineral.) *Halloysite*, one of the kaolinite minerals, occurs in round or flattened tubes (Fig. 1.4). When this mineral is wet, the tubelike structure apparently acts like a pile of roller bearings and the mass flows or creeps. As a material for embankments, the halloysites generally are regarded as unsuitable.

The chemical formula of the kaolinites is $Al_2O_3 \cdot 2SiO_2 \cdot 2H_2O$ or $H_4Al_2Si_2O_9$. The latter formula is more correct, since it shows that the structural water is bound to the lattices in the form of a hydroxyl, OH, and not as H_2O.

The Montmorillonites. According to prevailing views, montmorillonite is composed of conjoining identical units made of an alumina octahedral sheet between two silica tetrahedral sheets. The sheets are bound together rather loosely, and thus an unstable mineral results, especially in the presence of water. In fact, the attracted water molecules easily

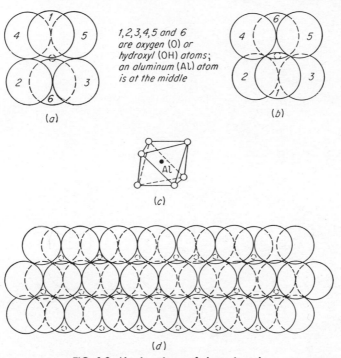

1,2,3,4,5 and 6 are oxygen (O) or hydroxyl (OH) atoms; an aluminum (AL) atom is at the middle

(a)

(b)

(c)

(d)

FIG. 1.3. Alumina sheet of clay minerals.

insert themselves between the sheets, causing swelling or expansion. In such cases, individual montmorillonite flakes are enclosed (wrapped) in water films; thus wet montmorillonites have a high plasticity and a low coefficient of internal friction. When a saturated montmorillonite is drying out, it is subject to high shrinkage and cracking.

The expansive properties of the montmorillonite clays are a matter of engineering concern. Heavy structures founded on such clays may be lifted and damaged, not to mention possible failures of pavements and building slabs placed directly on the ground. Slopes, both artificial and natural, made primarily of montmorillonite clays are subject to sliding and flowing in wet weather.

The well-known *bentonites* usually formed from volcanic ashes are a form of montmorillonite clays noted for their expansive properties. In engineering practice these properties can be put to beneficial use, however, particularly for preventing leakage from reservoirs and canals. In canals, the canal lining is made of a mixture of local soil and montmorillonite-type clay. When the canal is put into operation, water makes the clay swell and thus seal the canal banks and bottom.

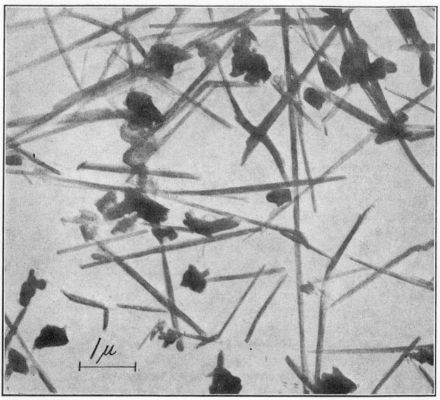

FIG. 1.4. Electron micrograph of halloysite from Webster, North Carolina. (*Photograph by Dr. T. F. Bates.*)

The chemical formula of the montmorillonites is $(OH)_4Si_8Al_4O_{20} \cdot nH_2O$. The term nH_2O refers to the *interlayer* water which may be present between the sheets of a natural montmorillonite in the form of one, two, three, or more layers of water, each layer being one molecule thick.

The Illites. A structural unit of an illite is similar to that of montmorillonite with some changes in chemical composition. Whereas the montmorillonites consist of exceedingly fine particles that appear like fog, even under the greatly enlarging electron microscope, the flakes composing the illites frequently form aggregates. The aggregate-like structure exposes less surface to attract water than the montmorillonites. Hence, in comparison with montmorillonites, the illites have a more limited hydration capacity. The expansive properties of illites also are less and their coefficients of internal friction higher than the montmorillonites.

FIG. 1.5. Electron micrograph of kaolinite from Banda, India. (*Photograph by Dr. T. F. Bates.*)

The generalized chemical formula for illites is $(OH)_4 K_y (Si_{8-y} \cdot Al_y)(Al_4 \cdot Fe_4 \cdot Mg_4 \cdot Mg_6)O_{20}$. In muscovites (mica) y equals 2, whereas in illites, y is less than 2 and generally equals 1.5.[5]

The range of specific gravity values in clay minerals is as follows: illites from 2.64 to 3.0, kaolinites from 2.60 to 2.68, and montmorillonites from 2.2 to 2.7.

1.5. Identification of Clay Minerals. Electron-microscope observations supply some data on the shape of clay particles. The atomic arrangement of the clay flakes may be determined by *X-ray diffraction methods* which provide means to measure the spacing of the atomic planes, i.e., sheet thickness.

The distances thus measured are expressed in angstroms. [An angstrom (symbol A) equals one ten-millionth part of one millimeter, or one ten-thousandth part of one micron.] The powdered specimen can be mounted on a small pedestal connected with a continuously rotating base. The X rays are permitted to hit the specimen and be diffracted upon a film. Each line of the pattern represents a reflection from dif-

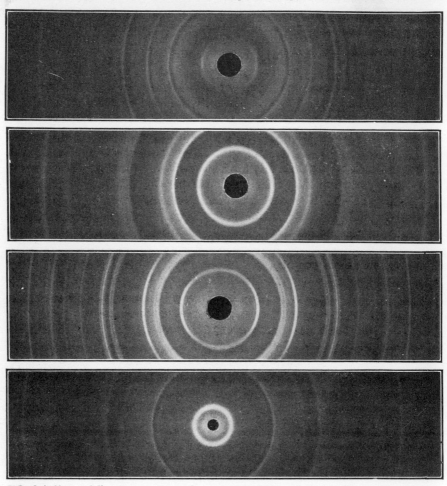

FIG. 1.6. X-ray diffraction patterns of some typical clay minerals. From top to bottom: illite, Yangtse River, China; halloysite, Lawrence County, Indiana; kaolinite, Langley, South Carolina; montmorillonite, Granby, Colorado. (*USBR.*)

ferent atomic planes within the mineral. The thickness of a sheet is 5.05 A and 4.93 A for octahedral and tetrahedral sheets, respectively. As an average, this makes over 50,000,000 sheets per inch of clay thickness.[7]

The clay minerals can be identified further by the temperatures required to extract structural water from their lattices (or for transformation of one clay mineral into another). The effect of high temperatures on clay minerals is studied by *differential thermal analysis*.   The amount of water released by a mineral when heated to a high temperature may be measured by using a special *thermal balance*.

  In differential thermal analysis a small sample of the soil material to be tested and calcined alumina (heat-treated and thus reduced to powdered aluminum oxide) are placed in cavities of a nickel block and heated to 1100°C at the rate of about 12°C/min. In the improved devices, up to six samples may be tested simultaneously.   Calcined alumina is thermally inert (or inactive; i.e., it does not change its chemical composition when heated) up to 1100°C.   A curve of temperature difference between the material

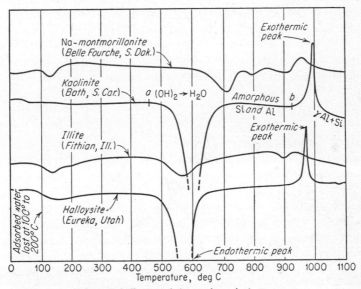

FIG. 1.7. Differential thermal analysis curves.

under test and calcined alumina is drawn on an automatic recording drum, and each kind of material yields a more or less characteristic curve.   Figure 1.7 gives examples of such curves for some major types of clay minerals.   In the case of kaolinite, the structural water starts to be released at a temperature of about 500°C, which corresponds to the breaking point of the endothermic curve (point *a*, in Fig. 1.7).   Where water tightly bound within the lattice in the form of hydroxyl, OH, anions is released, the bulk of the mineral is reduced to amorphous alumina, Al_2O_3, and silica, SiO_2. Further heating causes recrystallization, which starts at a temperature somewhat below 1000°C, corresponding to the starting point *b* of the exothermic curve (Fig. 1.7). In the process of recrystallization a γ isotope of alumina (γ Al_2O_3) is formed, accompanied by amorphous silica.   If the test is continued beyond completion of the exothermic curve, still another mineral is formed (mullite, Al_2SiO_5).

  1.6. Base Exchange.   If a mineral contains a large number of bases (for instance, sodium cations, Na+) and is acted upon by a liquid containing a high number of bases of some other kind (for instance, potassium cations, K+), the mineral and the liquid may interchange their cations, sometimes in a rapid reaction. This is the *base exchange* or, more accurately, *cation exchange*, which usually is more intense in the clay minerals than in other minerals. The commonly exchangeable cations in clay minerals are $Na+$, $K+$, $Ca++$, $Mg++$, $H+$, and $(NH_4)+$. The base-exchange property varies considerably for different clay minerals. The kaolinites are the least susceptible and the montmorillonites the most susceptible to cation exchange. Also, when in an extremely fine state, all inorganic minerals reveal a weak cation-exchange capacity. Some organic soils also have considerable cation-exchange capacity.

The base-exchange properties of soils are the basis of stabilization of soils by certain chemicals or by electroosmotic phenomena. Water softeners depend upon the base-exchange capacity, and acid soils containing $H+$ ions can be "sweetened" to improve their agricultural value by adding lime, which causes an exchange of $H+$ and $Ca++$ ions.

ROCKS

1.7. Rock Classifications. The term "rock" as used in engineering geology means a compact, semihard to hard mass of natural material composed of one or several minerals. Rocks were and are being continuously formed in various ways: by cooling of magma, i.e., the hot liquid (molten rock) which issues from considerable depths below the earth surface; by precipitation of inorganic materials from water; by deposition of shells of various organisms; by condensation of a gas containing mineral particles; by disintegration of other rocks and subsequent recombination of the resulting minerals to form new rock types; and by action of intense heat and/or pressure on preexisting rock types.

Geologists have classified the earth's rocks on the basis of origin into three major groups: igneous, sedimentary, and metamorphic. All rocks can, by petrographic study, be placed into their correct classification group. This grouping, however, gives little or no inkling of the engineering properties of a particular rock. For example, to the engineer, the term "granite" generally means a hard, reliable bearing material; however, it is known that the bearing properties of a granite are exceedingly variable from locale to locale and from outcrop to outcrop. Atmospheric agents (weathering) may transform a granite into a poorly coherent mass, selective weathering may alter its constituent feldspar crystals into clay minerals, or heat and pressure from faulting or other major geological disturbances may crush the constituent grains to form a cohesive but relatively soft material that still resembles the original granite. Thus, if a structure load is placed upon a granite so transformed, considerable settlement of the structure may result. In other words, although the aforementioned natural forces can completely alter the engineering properties of a rock, the alteration product may be called by the same geological name as the original rock. Engineers need a classification or classifications of rocks by their *engineering properties*, and such classifications either do not exist or are yet in their infancy.

Modification of existing rock classifications by use of descriptive adjectives appended to the rock name is the best method for describing a rock for engineering purposes. By this means a "word picture" of the rock condition is obtained, e.g., "completely decomposed, partly cohesive granite." Such adjectives should denote (1) the primary structure and mineral make-up, (2) the extent and nature of alterations, (3) structural features of strength or weakness, (4) weakening effects of weathering, (5) continuous processes that may cause further change, and (6) surface conditions.

In his analysis of rocks, the engineering geologist should also establish the formation* to which a rock belongs and, often, the sequence of processes that are responsible for having made the rock. It is important that the geological analysis be made from the point of view of the proposed service of the rock, i.e., whether it will be used as a bearing material, as a building stone, or to make concrete.

1.8. Texture, Structure, Fabric. The terminology defined in this section generally applies to igneous rocks, but some of these terms also may be used to describe sedimentary or metamorphic rocks. The *texture* of a rock is the arrangement of its grains or particles as seen on the freshly exposed surface. An igneous rock (Sec. 1.9) which contains numerous large crystals easily seen by the naked eye is coarse-textured or coarse-grained and is said to have a *phaneric texture* (Fig. 1.8). If grains cannot be seen without magnification, the igneous rock is fine-grained and has an *aphanitic texture*. If some large crystals are embedded into a rock of otherwise aphanitic texture, the igneous rock is a *porphyry* (Fig. 1.9), although this term also describes igneous rocks which have occasional very large crystals embedded in a phaneric ground mass.

In sedimentary rocks (Sec. 1.10) the counterparts of porphyries are *conglomerates* (Fig. 1.18) formed of coarse-grained pebbles embedded in a finer *matrix* of sand or clay. Sedimentary medium-grained rocks made of grains visible to the naked eye (e.g., sandstones) are counterparts of the igneous phanerites. The very fine-grained sedimentary rocks (siltstones, claystones) are equivalent in texture to the igneous aphanites.

Structure. The use of the term "structure" is reserved for more pronounced features of a rock than those described by the term texture. In igneous rocks the structure may signify a relative arrangement of different spatial features of the rock, both small (microscopic) and large (macroscopic). For example, holes in the body of an igneous rock may be char-

* A *formation* is a geological bed or combination of beds or successive strata sufficiently distinctive from others to be regarded as a unit (see Sec. 3.1).

FIG. 1.8. Phaneric texture (coarse-grained granite), ¾ ×.   (*Photograph by H. Howe.*)

FIG. 1.9. Felsite porphyry, ⅘ ×.   (*Photograph by H. Howe.*)

acteristic of its structure. A *vesicular* structure denotes the presence of small holes, or vesicles, throughout the igneous rock, such as are found in pumices and some basalts (Fig. 1.10). Holes larger than vesicles are *vugs*. The latter generally are filled with minerals other than those forming the given rock (Fig. 1.11). An important macroscopic structural feature is *jointing* of the rock. Joints are fissures that may be open or

FIG. 1.10. Vesicular structure (basalt), 1 ✕. *(Photograph by H. Howe.)*

closed and run in different directions. They usually represent a more or less regular system and may tend to break the rock into cubes or regular blocks (Fig. 1.12). For further discussion of joints see Sec. 2.5. *Fractures* also are macroscopic features. They may run in any direction and may intersect each other at any angle, if they intersect at all. A fracture usually is irregular as contrasted to the usual plane or even surface of a joint.

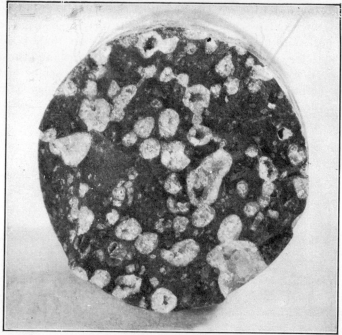

FIG. 1.11. Vugs in basalt filled with minerals, $1\frac{3}{4}$ ×.   (*Photograph by H. Howe.*)

FIG. 1.12. Joints in granite from Needle Mountains, Colorado.   (*Photograph by W. Cross, USGS.*)

Fabric. This is a controversial term which sometimes is considered as a generalization of the term texture. In this text, fabric denotes the spatial pattern of the rock particles which includes grain sizes and their ratios, grain shapes, grain orientation, microfracturing, packing and interlocking of particles, the character of the matrix, etc.[6] Further details on fabric definition and use may be found in Refs. 6, 9, and 11.

1.9. Igneous Rocks.[3,9-11] Igneous rocks have been formed on or at various depths below the earth surface. This fact influences their texture

TABLE 1.3. Igneous Rock Classification

Depth of formation—texture—structure	Light colored		Dark colored	
	+ Quartz	− Quartz	+ Quartz	− Quartz
Fine, medium, and coarse-grained (granitic texture) (formed at considerable depths* in deep-seated dikes, batholiths, stocks, and laccoliths.)	Granite +O,H Granodiorite +P,M Granite porph. Pegmatite Quartz-monzonite P,O,B,H	Monzonite (monzonite aplite, monzonite pegmatite) P,O,B,H Syenite	Quartz diorite P,B,H, ± X	Diorite P,B,H,± X Gabbro P,+X,+H Peridotite ± P,+X,+H
Fine to medium-grained (formed at moderate depths† in dikes and sills)	Granite porph.	Monzonite porph.		Diorite porph. Diorite aplite Dolerite (diabase) +X, P, some B, and augite
Very fine-grained to glassy, may be spongy, glassy, or vesicular (formed at surface flows, dikes, plugs, ash deposits) ‡	Compact	Felsites (glassy) Rhyolite +O,B,H	Trachyte	Obsidian (glassy)	Andesite +P (glassy felsite) Basalt (vesicular) +P,X
		Volcanic breccia			
	Spongy to loose	Pumice, volcanic ash, tuff			

Explanation of symbols: O = orthoclase feldspar P = plagioclase feldspar B = biotite
  M = muscovite H = hornblende X = pyroxene
  + means a predominance of the mineral
  ± means the mineral may be present
  − means the mineral is absent
* Also termed "plutonic, intrusive, or abyssal" rocks.
† Also termed "hypabyssal" rocks.
‡ Also termed "volcanic or extrusive" rocks.

and, in some cases, their structure. In Table 1.3, rocks are classified vertically according to origin and depth of formation and horizontally according to the color—light or dark—and presence or absence of quartz. The difference in color is controlled by the mineralogical composition of the rocks. Light-colored rocks generally are feldspathic, whereas dark rocks generally contain ferromagnesian minerals, particularly pyroxenes (symbol X in Table 1.3).

Igneous rocks formed at *great depths* occur in deep-seated dikes, batholiths, stocks, and laccoliths.   These terms refer to various types of magma intrusion into the earth crust and are illustrated in Fig. 1.13.   These rocks generally provide excellent foundation materials, but they should be closely examined if they are to bear the weight of heavy structures such as masonry dams.   In fact, alteration of constituent minerals may considerably reduce their strength.

Granite is primarily composed of feldspar, quartz, and mica.   The feldspar, usually orthoclase (O in Table 1.3), predominates.   In the *granodiorite*, plagioclase predominates.   For engineering purposes the difference between granite and granodiorite often is neglected, however (see also Sec. 1.12).   The granitic texture usually is from medium to

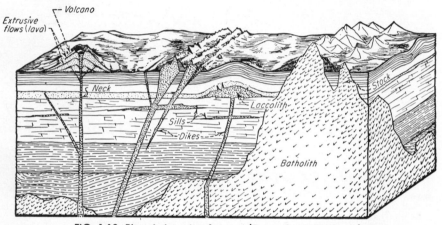

FIG. 1.13. Plutonic intrusive forms.   (Drawn by J. Vitaliano.)

coarse.   Coarse grains in the *granite porphyry* (Fig. 1.14) are feldspathic. The color of granite generally depends on the predominating feldspar color and may range from white to dark gray, with reds and pinks not uncommon.   The structure of granite is *massive*, although jointing in granite commonly breaks it into fairly large blocks.   If one set of joints is exceedingly prominent and closely spaced, the granite will have a sheeted appearance (Fig. 1.15).   *Pegmatite* is coarse-grained granite, often occurring in dikes or inclusions in the main mass.

Monzonite is a fine- to coarse-grained rock of light gray color, composed of feldspar and some biotite and hornblende.   The variety containing no quartz is monzonite aplite or monzonite pegmatite.   The term "pegmatite" if used without qualification generally means granite pegmatite.

Syenite is composed of the same minerals as granite but has little or no quartz.   Otherwise it is similar to granite, into which it often grades.

Other members of this group (Table 1.3) are dark heavy rocks, *diorite*, *gabbro*, and *peridotite*, the last two rocks often being found associated. As far as suitability for construction is concerned, peridotite should be regarded with caution as it commonly alters to serpentine (refer to Sec. 1.3).

In the group of igneous rocks formed at *moderate depths*, diabase (also termed dolerite) should be mentioned. This is a tough rock very suitable for use in roads and pavements, but it should be regarded with caution if

FIG. 1.14. Porphyritic granite, ⅘ ×. *(Photograph by H. Howe.)*

encountered in dam abutments, as its method of formation may have resulted in numerous cavities or fissures (which would cause leakage of the dam).

Volcanic rocks are formed at or near the surface. They may be hard and competent or may be interbedded with noncoherent volcanic materials such as tuff, ashes, or even sand. These loose materials can produce difficulties in construction, such as caving in tunnels, high seepage losses in dams, and unstable slopes. On the credit side is the possibility that the noncoherent materials can be used as pozzolans to partially replace cement in concrete (see Sec. 8.23).

In the felsite group, *rhyolite* and *trachyte* are very similar, differing only in quartz content. Their color ranges from light to dark gray. *Andesite*, a dark-colored rock, belongs to the same group. Very often where differentiation is not possible without a microscope, these rocks are merely classified as "felsites." They often have a glassy ground mass with embedded crystals (Fig. 1.16).

FIG. 1.15. Sheeted granite from North Platte River, Wyoming.

In building practice the term *basalt* or "traprock" is used to designate dark-colored, fine-grained rocks. Petrographically, basalt is a rock characterized by the predominance of plagioclase, the presence of considerable amounts of pyroxene and some olivine, and the absence of quartz. It is fine-grained, the grains being sometimes too fine to be seen under a hand lens. Its color varies from dark gray to black. A well-recognized feature (though not always present) is the columnar, practically hexagonal jointing (Fig. 1.17).   When weathered, basalt becomes rusty red, as may be observed in the abandoned parts of a basalt quarry.

(A)   (B)   (C)

FIG. 1.16. Various porphyries with aphanitic matrix, ⅓ ×.   (a) Rhyolite p., (b) latite p., (c) felsite p.   (*Photograph by H. Howe.*)

FIG. 1.17. Columnar jointing in basalt (near Klamath Falls, Oregon).   See also Fig. 2.9a.

Pumice is one of the glassy volcanic rocks but possesses stony or earthy texture with complete absence of crystals.   Its structure is highly vesicular or spongy (Fig. 1.18), and the resulting low specific gravity enables pumice lumps to float on water.   It is a dubious foundation material because of its low density and high porosity.

All noncoherent materials ejected from a volcano generally are described as *ashes*. (Occasional large rocks ejected from a volcano may be termed "bombs.") If individual fragments of ash are welded together, either by pressure or by some other agency, *tuff* is formed. This is a fine-grained, lightweight material, gray to yellow in color. It may be soft or very hard, depending upon the degree of "welding" which it has undergone. Larger size particles ejected from a volcano may be strongly

FIG. 1.18. Pumice, 1 ✕. *(Photograph by H. Howe.)*

welded together to form a *volcanic breccia* or *agglomerate*. Breccia generally consists of sharp angular fragments of all kinds of volcanic rocks and others, such as sandstones or granites, through which the volcanic products have been ejected.

Volcanic and *depositional breccia* (Fig. 1.19), the latter being classified as a sedimentary rock (Sec. 1.10), generally are good foundation materials. A *fault breccia* composed of fragments of the rocks which have been broken in faulting is rarely accepted as a foundation material unless it is well cemented and competent.

FIG. 1.19. Depositional breccia, 1⅕ ✕. (*Photograph by H. Howe.*)

1.10. Sedimentary Rocks.[3,12-15] When the products of the disintegration and decomposition (Sec. 3.3) of any rock type are transported, redeposited, and partly or fully consolidated or cemented into a new rock type, the resulting material is classified as a *sedimentary rock*. This classification also includes those rocks which result from chemical precipitation or deposition of organic remains in water. Deposits laid down by sedimentary action usually can be recognized by their layered (stratified) structure as contrasted to the usual massive structure of igneous rocks. Also, both animal (mostly invertebrate) and plant fossils commonly are found in sedimentary rocks, whereas they are nonexistent in igneous rocks. The general term "sediments" commonly is applied to deposits from water, wind, or glacial action.

The minerals, rock fragments, or organic remains which comprise a sedimentary rock are referred to as its "constituents." The rock is made of particles or grains formed of these constituents. The sizes and, occasionally, the distribution of these particles (or grains) are the basis for classifying (or subdividing) sedimentary rocks. As a rock mass is eroded by atmospheric or chemical agents, it is gradually destroyed, and the resulting fragments vary greatly in size. The more complete the abrasion process, the smaller will be the size of the resulting fragments. In fact, the size of the fragments (clay excepted) often offers a clue as to the distance which they have been transported (by wind, water, or glaciers); the smaller the size, the farther they have been transported. The various grain sizes found in sedimentary rocks are, in descending order of size, boulders, cobbles, gravel, sand, silt, and clay.

Boulders and Cobbles. The larger fragments resulting from the erosion process are termed "boulders" or "cobbles" depending upon their size. According to Wentworth's classification [12a] (accepted by many geologists and presented in Table 1.4), fragments from 64 (about $2\frac{1}{2}$ in.) to 256 mm

TABLE 1.4. Fragment and Grain Sizes and Corresponding Sedimentary Rock

Material	Unified classification grading*	Wentworth grading	Rock
Boulders.................	12 in.†	10 in.	Boulderstone
Cobbles.................	3–12 in.	64–256 mm	Cobblestone
Gravel..................	$\frac{1}{4}$–3 in.	4–64 mm	Conglomerate
Granules...............		2–4 mm	Conglomerate
Sand...................	0.074 mm–$\frac{1}{4}$ in.	$\frac{1}{16}$–2 mm	Sandstone
Silt....................	Less than 0.074 mm	$\frac{1}{256}$–$\frac{1}{16}$ mm	Siltstone
Clay...................	Less than 0.074 mm	Less than $\frac{1}{256}$ mm	Claystone

* Except boulders and cobbles (see Sec. 4.11).
† 1 in. = 2.54 cm or 25.4 mm.

(about 10 in.) are cobbles and all fragments larger than 10 in. are boulders. Two of the largest engineering organizations in the United States have agreed to classify all rounded rock fragments from 3 to 12 in. in diameter as *cobbles* and all such fragments over 12 in. in diameter as *boulders*.[12c]

In preparing specifications for excavation contracts, these agencies often limit the specified size of boulders as "a rock not exceeding 1 cu yd in volume." Fragments larger than 1 cu yd are then specified as "rock," and the contractor is paid the same price for removing these very large rocks as for excavating the same rock from its natural bed.

Gravel, Sand, Silt, Clay. These classifications are discussed in detail in Chap. 4.

Rock Cement. If any of the previously described rock fragments are bound firmly together, the bond may be the result of *cementing* action. The cementing of an incoherent rock mass may result from (1) the infiltration of water carrying various chemicals or (2) the disintegration of certain minerals in the mass to form new bonding minerals. The three most common cements found in sedimentary rocks are silica or siliceous cement, SiO_2; calcium carbonate or calcareous cement, $CaCO_3$; and clay or argillaceous cement. Siliceous cement is the most resistant to weathering and water action, and clay is the least durable bonding material. Calcareous cement generally makes a durable rock but may be leached by water containing carbon dioxide or acids.

Cements introduced by ground water or any other means into the sedimentary material after it has been consolidated into rock are termed *secondary* cements. In some cases the apparent cementing action in a sedimentary rock is due only to adhesion between argillaceous particles. This adhesion is caused by the weight of overlying materials, as no secondary cement has been introduced. If the rock is cemented by the introduction of iron-bearing minerals, the cement and occasionally the rock are termed ferruginous. (The effect of the type of cement on rock strength is discussed in Sec. 2.5.)

Slaking Characteristics. When freshly exposed to air, some rocks will crumble into flakes or granular particles, a process known as *slaking*. A similar effect is produced when certain rocks are immersed in water (particularly if the rock specimen has been dried out before its immersion). If numerous small particles crumble away from the rock while under water, the rock should be regarded with suspicion if it is to be used as a foundation material. To ascertain its water-slaking characteristics, the rock should be kept submerged for several days. Sedimentary rocks containing high percentages of clay minerals or clay cements (argillaceous rocks) are the most susceptible to either air or water slaking.

Classification. In Table 1.5, the most important sedimentary rocks are classified on the basis of grain size, texture, and structure. These rocks are divided into *clastic* (or "detrital") rocks composed of fragments of preexisting rocks and minerals; *fine crystalline* rocks, mostly with organic admixtures; and some *amorphous* and *biofragmental* rocks, composed of the fragments of tiny invertebrates. The term "rock" is used in Table 1.5 in a purely geological sense (see Introduction to Chap. 1). The siltstones and claystones in the table often are termed "shales" in engineering practice. Several rock classifications in the table are discussed as soils throughout the book.

From an engineering viewpoint, the most important sedimentary rocks are sandstones, limestones, and shales. The size of grains in *sandstones* varies (Table 1.4); accordingly their texture may vary from fine- to coarse-grained. The structure of a sandstone may be massive, horizontally bedded, or cross-bedded. In cross-bedded structure, bedding is inclined to the original upper and lower boundaries of the rock body. Generally, sandstones are assumed to be fairly competent bearing materials. They should be examined, however, for the presence of softer materials (such as siltstones) intercalated between the hard layers. Argillaceous sandstones may be subject to air or water slaking.

Quartzite. In a modern geological sense, quartzite is a sandstone containing more than 90 per cent quartz. The old-fashioned definition,

TABLE 1.5. Textural Classification of Sedimentary Rocks*

Fabric (texture)	Essential components	Identification characteristics	Rock name
Clastic (detrital)	Volcanic material	{ Frag. > 32 mm { Particles < 4 mm	Agglomerate. Breccia Tuff
	Gravel	Abraded pieces > 4 mm > 50 %, clay < 25 %	Conglomerate
	Rock, mineral frag.	Angular pieces > 4 mm > 50 %, clay < 25 %	Breccia
	Rock frag. and clay	High size range, usually unsorted; matrix of clay, sometimes sand, but usually in excess of frag.	Till or tillite (depends on compaction)
	Sand	Particles < 4 mm > 1/16 mm > 50 %, clay < 25 %	Sandstone, arkose, quartzite, graywacke
	Detrital calcite grains	Calcite > 50 %, clay < 25 %	Limestone
	Silt	Particles < 1/16 mm > 50 %, clay < 25 %, massive or stratified	Siltstone
	Clay minerals	(Clay > 25 %, massive to stratified { Mostly clays and sericite, incipient recrystallization { Montmorillonite clays 75 % (Kaolinite clays 75 %	Claystone Argillite Bentonite Kaolin
	Clay and calcite	Very fine-grained; carbonates 25–75 %	Marl, marlstone
Crystalline	Calcite	Carbonate > 50 %, of which calcite > 50 % { Coarse to mc., compact { Fine to mc., porous, firm, friable	Limestone Chalk
	Calcite and clay	Very fine-grained; calcite 25–75 %	Marl, marlstone
	Carbonates	Carbonates > 25 %, compact to earthy	Caliche
	Dolomite	Carbonate > 50 %, of which dolomite > 50 %, coarse to fine, compact	Dolomite
	Chalcedony	Chalcedony > 25 %, mc. to crc., conchoidal fracture, compact	Chalcedonic chert
Amorphous	Opal	Opal > 50 %, massive to banded, compact	Opal. Opaline chert
	Amorphous carbon	Fibrous or spongy or compact, carbonized plant remains > 50 %, black-brown	Coal
Biofragmental	Calcareous shells	Whole or fragmental shells 50 %	Coquina
	Diatom tests†	Diatom tests 50 %	Diatomite. Diatomaceous earth
	Foraminifera tests†	Foraminifera tests > 50 %	Foraminiferal limestone
	Algal structures†	Algal structures > 50 %	Algal limestone
	Coral structures†	Coral structures > 50 %	Coral limestone
	Partially or completely carbonized plant remains	Brown to black, spongy or compact, plant remains easily visible	Peat
		Black, massive to banded, compact, almost metallic appearance	Bituminous or anthracite coal
		Brown to black, fibrous to compact	Lignite

* After R. C. Mielenz.

† *Tests* are the protective covering of some invertebrates. *Foraminifera* are very tiny, many-celled shells which have numerous holes or pores. *Algae* are cellular aquatic plants. *Coral* is the solid secretion made of calcium carbonate from small invertebrates. *Diatoms* are microscopic plants that secrete siliceous materials.

Abbreviations used in table: frag. = fragments; mc. = microcrystalline; crc. = cryptocrystalline; > = greater than or over; < = less than or under.

which still is widely used, restricts the term to a sandstone that is firmly cemented with silica, the grains being almost entirely of silica. Upon cursory examination, the rock appears to have interlocking texture, but close study will reveal individual grains. Quartzites formed by the application of very high pressures and temperatures usually are classified as metamorphic rocks (Sec. 1.11), but most are regarded as normal sedimentary rocks.[14] Some quártzites may resemble limestone but are much harder ($H = 6$) and do not react to hydrochloric acid.

Arkose is a variety of sandstone containing as many feldspar as quartz particles. It may contain mica and, if well-cemented, may resemble a

FIG. 1.20. Conglomerate, ⅓ ×. *(Photograph by H. Howe.)*

granite (so much so, in fact, that microscope studies may be required for differentiation). The grains in arkose generally are angular to subangular but do not have the interlocking of a granite.

Graywacke is a sandstone, frequently dark colored or black, cemented with silica or clay. It may contain particles and flakes of slates and shales. The dark color and occasional extreme induration are its most distinctive characteristics and are the reason why graywacke often is mistaken for igneous rock. Perhaps one of the most important means of identification is the fact that graywackes commonly are found in association with slates and shales.[12b] Generally, graywackes are competent bearing rocks.

Conglomerate (Fig. 1.20) is a term generally applied to a rock composed predominantly of cemented, very coarse (¼ to 3 in.), rather rounded rock chunks; sometimes it has a good grading* down to very small sized parti-cles.   By definition,[12b] at least 10 per cent of the rounded fragments must be over 2 mm in diameter.   If these fragments are not rounded, the rock is classified as a "breccia."   If there is a scattering of large rounded pebbles or boulders embedded in an indurated clay, the resulting forma-tion is a *tillite* or *fanglomerate*.   If the large fragments have resulted from volcanic ejection, i.e., rocks which were thrown from a volcano, and are cemented with small fragments, the resultant mass is an *agglomerate* or *volcanic breccia*.   Rocks that contain a high percentage of cobbles and boulders may be classified as *cobblestones* and *boulderstones* (Table 1.4). Unless they are well cemented, any of these materials may weather severely in deep cuts.   It always is advisable to perform a slaking test if the conglomerate is to be considered for engineering use.

In the group of rocks made of clay minerals, *shales* sometimes are of the most concern to engineers.   A *shale* is a laminated sedimentary rock, often dark in color, and composed predominantly of clay-sized particles, although a small percentage of sand or silt sizes also may occur.   The degree of induration of a shale is extremely variable and may range from a rock soft enough to be scratched by a fingernail to a rock that can be scratched only by a knife blade.   The predominant characteristic of a shale is the high percentage of fine clay particles (2μ† or less in size), and therefore many rocks classified as shales are in reality *claystones* or *mud-stones*.   These rocks may possess all the same characteristics as a shale but will *not* have the fissility (easy cleavage) or laminations.   For this reason, considerable confusion exists in the classification of shales.   It is not desirable, however, to eliminate the use of the term "shale" because of its frequent occurrence in practice and in literature (though it is done in Table 1.5).   Clays and silts are changed into shales by the process of adhesion, compaction, and cementation.   When subjected to alternate wetting and drying, insufficiently indurated shales may revert to the original clayey, often sticky mass from which they had been formed. Natural shale deposits can vary from soft weathered layers that can be excavated by rippers (or rooters, equipment for removing roots and tree stumps) to indurated rocks requiring explosives for removal.   A compe-tent shale will give a clear ring if struck with a hammer.   Shale deposits often lie next to and intergrade with sandstones and siltstones.

Siltstones are composed primarily of silt-sized grains (0.062 to 0.0039 mm in size)[12b] and may be similar to a claystone in appearance.   Siltstones can be distinguished, however, by their gritty feel, especially if a small

---

* In the engineering sense, a well-graded material contains particles of coarse, medium, and fine sizes in approximately equal proportions.

† Symbol μ means "micron," or one thousandth of one millimeter.   Symbol μ is also used for Poisson's ratio (Sec. 2.13).

piece is ground between the teeth.   Claystones appear smooth, with little or no gritty feeling.

The main sedimentary rocks in the group of fine so-called "crystalline" sediments are limestone and dolomite.   The latter rock consists of the

FIG. 1.21. Solution cavities in limestone drill core, 1 ×.   (USBR.)

mineral of the same name (Table 1.1) and is a competent foundation material.   *Limestone* is primarily composed of calcium carbonate.   It may be fine and crystalline as classified in Table 1.5, but its texture is variable and even may be brecciated.   In some limestones, calcite grains are held together with clay cement.   The color varies, white, yellow, brown, and gray being most common.   The rock also may vary from impervious to water to loose-textured and porous (Fig. 1.21).   Generally limestone is a good foundation material unless it is cavernous. Considerable leakage may be expected if a cavernous limestone occurs under a dam or in a reservoir. Planes using jet engines would ruin runways built with limestone aggregate (Sec. 8.21) because the intense heat from the jets would decompose the limestone. *Chalk* is a weak variety of limestone and generally is not regarded as a competent bearing material for heavy structures. Marl, another calcareous "rock," is described in Sec. 4.12.

1.11. Metamorphic Rocks.[3,15-17] Rocks formed by the complete or incomplete recrystallization (change in crystal shape or in composition) of igneous or sedimentary rocks by high temperatures, high pressures, and/or high shearing stresses are *metamorphic rocks*. A platy or foliated

TABLE 1.6. Metamorphic Rock Classification

Structure and texture	Composition	Rock name
Massive:		
Banded, consisting of alternating lenses	Various tabular, prismatic, and granular minerals (frequently elongated)	Gneiss
Granular, consisting mostly of equidimensional grains	Calcite, dolomite, quartz, in small particles	Marble or quartzite
Foliated or platy	Various tabular and/or prismatic minerals (generally elongated)	Schist, serpentine (rock), slate, phyllite

structure in such rocks indicates that high shearing stresses have been the principal agency in their formation. Foliation is not always visible to the naked eye, but individual grains may exhibit strain lines under the microscope. Metamorphic rocks formed without intense shear action have a massive structure. In Table 1.6 the most common metamorphic rocks are subdivided into two basic classes according to their structure.

Gneiss. A metamorphic rock in which the constituents have a *banded* distribution is generally known as gneiss. The characteristic feature of a gneiss is its structure (Fig. 1.22); the mineral grains are elongated, or platy, and banding prevails. Sometimes gneisses grade into schists. Usually gneisses represent good engineering materials, except for those with a superabundance of mica flakes. These types cannot be used for building stones because of air-slaking and raveling or for concrete aggregate because of considerable cleavage and its weakening effect on concrete.

Schist is a finely foliated rock containing a high percentage of mica, which controls its structure. Sometimes crystals of other minerals are incorporated in the *planes of schistosity* separating the layers (Fig. 1.23*a*). Some schists may be composed almost entirely of silica and have an almost massive structure (Fig. 1.23*b*).   Depending upon the amount of pressure applied in the metamorphic process (which, in the metamorphic rocks, is analogous to the prestressing of engineering materials), schist may be a very competent material.   During excavation, however, blocks

FIG. 1.22. Gneiss, ¾ ×.   (*Photograph by H. Howe.*)

may tend to separate along the planes of schistosity.   If schist is acted upon by fast-running water, it may require some protection to prevent "quarrying" (plucking action) by the water.   In general the dip of the planes of schistosity in schists is different from the dip of the whole formation (Fig. 1.23*c*).   Both dips are of importance, particularly in the investigation of high dam abutments, since sliding may be caused by the thrust of the arch along either dip (Sec. 15.17).   This type of dip also is observed in other platy rocks such as slate.

Slate is a dark-colored, platy rock with extremely fine texture and easy cleavage. Because of this easy cleavage, slate is split into very thin sheets and used as roofing material. As foundation material, slate is excellent; however, in excavations, large slate blocks may suddenly become detached when undermined. *Phyllite*, although physically similar to slate, differs somewhat by a shiny luster imparted by mica flakes, by more pronounced brittleness, and by a tendency to air slake. Cases

FIG. 1.23. Schist, ⅓ ✕. (a) Garnetiferous schist, (b) quartzite schist, (c) muscovite-biotite schist. (*Photograph by H. Howe.*)

of swelling were observed in phyllite when the pressure from the overburden was relieved by tunnel excavation.[18]

Argillite is an intermediate metamorphic stage in the hardening of clays from shales to slates. The only common distinguishing feature is that the parting planes in argillite are always parallel to the bedding of the formation as a whole whereas in slates or phyllites, these planes may be at angles to the formation bedding.

Marble is the end product of the metamorphism of limestones and other sedimentary rocks composed of calcium or magnesium carbonate. It is very dense and exhibits a wide variety of colors, depending upon the impurities present. When the rock is broken, a highly brilliant (luster) surface is apparent because of the large size of the crystals normally found in marbles. Marble can be scratched with a knife, and hydrochloric acid will cause effervescence when dropped on a piece of marble. Dolomitic marble effervesces with hydrochloric acid only if powdered. In construction, marble is used for facing concrete or masonry exterior and interior walls and floors.

TABLE 1.7. Field Identification of Rocks (without hand lens)
(Specimens should be unweathered and not altered)

Light-colored

Grains or crystals visible to naked eye

Angular particles				Rounded particles			Erratic large grains	
Large	Fine to medium	Very fine	Foliated or banded	Large	Fine to medium	Very fine	Rounded	Angular
Pegmatite Granite (+Q, +F)* Granodiorite (+Q,+F)* Monzonite (−Q,+F)* Syenite (−Q,+F)* Marble (reacts with HCl) Arkose (usually bedded)	Tuff (contains glasslike fragments but has earthy appearance)	Felsite* (rhyolite +Q and trachyte −Q)	Schist (shiny) Gneiss (may have subangular particles)	Conglomerate (+10% of grains over 2mm dia.) Cobblestone (large number of grains over 64 mm) Sandstone (bedded) (if it reacts to HCl = calcareous sandstone; if it gets slick when wet = argillaceous sandstone)	Quartzite (not friable and very hard)	Siltstone	Depositional breccia	Volcanic breccia and agglomerate fault breccia (may have clay)

No grains or sparse crystals visible to naked eye

Breaks with rough surface					Breaks with smooth or platy surface (usually has bedding)				
Glassy luster	Dull luster	Shiny luster	Earthy appearance		Laminated		Not laminated		
			Spongy, light wt.	Porous, moderate wt.	Slick when wet	Not slick	Slick when wet	Not slick	
Quartzite	Felsites* (rhyolite, trachyte)	Schist (foliated)	Pumice Volcanic ash	Chalk (HCl reaction)	Shale	Shale Slate (dull) Phyllite (shiny)	Claystone Mudstone Serpentine (usually greasy and may be banded)	Reaction to HCl → Limestone Chalk (earthy)	No reaction to cold HCl → Dolomite

Dark-colored (dark gray or green to black)

Grains or crystals visible to naked eye

Angular particles		Rounded to subangular particles
Fine to medium	Very fine to glassy	Graywacke (fine to medium-grained) Dark sandstones
Peridotite (−Q,−B)* Gabbro (−Q,−B)* Diorite (−Q,+B)* Diabase (−Q,+B)* (sometimes called dolerite)	Andesite* Basalt (usually vesicular)*	

No grains or sparse crystals visible to naked eye

Glassy luster	Dull luster-laminated		Dull luster—not laminated
	Slick when wet	Not slick	
Obsidian	Shale	Shale (flexible) Slate (brittle) (dull) Phyllite (shiny)	Basalt* Serpentine (usually greasy and may be banded)

* Rocks so marked may contain occasional large crystals embedded in a very fine-grained matrix or occasional very large crystals in a medium-grained matrix—in either case the term *porphyry* is appended to the rock name, e.g., *syenite porphyry*.

(+Q) = contains numerous white or colorless quartz crystals   (+B) = contains numerous flakes of black mica (biotite)
(−Q) = contains little or no quartz   (−B) = contains little or no black mica
(+F) = contains numerous white to pink feldspar crystals
(−F) = contains little or no feldspar

1.12. Field Identification of Rocks.   In Table 1.7 is presented a simplified method of rock identification for the observer who does not have microscopes or hand lenses and has only a casual knowledge of geology. The table has been devised to represent those features that a person will first see when picking up a rock.   It is based primarily on texture and structure.   The user must realize, however, that many of the rock varieties shown may occasionally have color variations or textural relationships which will place the rock in the "Dark-colored" portion of the table rather than in the "Light-colored" portion or place it in a "Fine to medium" category instead of a "Very fine" category.   All rocks that are being classified should be fresh specimens; i.e., they should have had very little weathering or alteration.   It will be noted that some igneous rocks such as granite, granodiorite, monzonite, etc., cannot be separated on the basis of textural-structural classification.   Such a separation rarely is necessary as far as the engineering properties of the rock are concerned. For example, a granite and monzonite probably will have approximately the same strength characteristics when regarded as foundation material.

It is realized that many petrographers will disagree with the classification method presented and can give numerous examples where the table is not absolutely accurate. This table has been designed, however, to enable the average engineer to establish an approximate classification of a rock and, on the basis of past experience, estimate its engineering properties.   It is believed that the descriptions given for the various rock types will fit the most *common* occurrences of these rocks.

REFERENCES

1. Brush, G. J., and S. L. Penfield: "Manual of Determinative Mineralogy and Blowpipe Analysis," John Wiley & Sons, Inc., New York, 1926.
2. Dana, E. S., and F. Hurlburt: "Manual of Mineralogy," 16th ed., John Wiley & Sons, Inc., New York, 1952.   Also see E. S. Dana and W. E. Ford "Textbook of Mineralogy," 4th ed., John Wiley & Sons, Inc., New York, 1932.
3. Pirrson, L. V., and A. Knopf: "Rocks and Rock Minerals," 3d ed., John Wiley & Sons, Inc., New York, 1947.
4. Rock Color Charts, Committee of National Research Council, Geological Society of America, New York.
5. Grim, R. E.: "Clay Mineralogy," McGraw-Hill Book Company, Inc., New York, 1953.
6. Mielenz, R. C., and M. E. King: Physical-Chemical Properties and Engineering Performance of Clays, *Proc. 1st Natl. Conf. on Clays and Clay Tech.*, California Division of Mines and Geology, 1954.
7. Brindley, G. W. (ed.): "X-ray Identification and Crystal Structure of Clay Minerals," Taylor and Francis, Ltd., London, 1951.
8. Fairbairn, H. W.: Introduction to Petrofabric Analysis, mimeographed lecture notes, Department of Geology, Queen's University, Kingston, Canada.
9. Mielenz, R. C.: Petrography and Engineering Properties of Igneous Rocks, *USBR Eng. Monog.* Number 1, Denver, Colo., 1948.

10. Johannsen, A.: "A Descriptive Petrography of Igneous Rocks," 4 vols., University of Chicago Press, Chicago, 1932–1938.

11. Daly, R. A.: "Igneous Rocks and the Depths of the Earth," McGraw-Hill Book Company, Inc., New York, 1933.

12a. Wentworth, C. K.: A Scale of Grade and Class Terms for Clastic Sediments, *J. Geol.*, vol. 30, 1922.

12b. Quoted in Pettijohn, E. J.: "Sedimentary Rocks," Harper & Brothers, New York, 1940.

12c. Symposium on the Identification and Classification of Soils, *ASTM Spec. Tech. Publ.* 113, 1951.

13. Krumbein, W. C., and F. J. Pettijohn: "Manual of Sedimentary Petrography," Appleton-Century-Crofts, Inc., New York, 1938.

14. Krynine, P. D.: The Megascopic Study and Field Classification of Sedimentary Rocks, *J. Geol.*, vol. 56, no. 2, pp. 130–165, March, 1948; also *Penn. State Coll. Tech. Paper* 130, 1948.

15. Lovering, T. S.: Report of the Committee on Structural Petrology, Division of Geology and Geography, National Research Council, Washington, D. C., 1938.

16. Kemp, J. F.: "A Handbook of Rocks," 6th ed., D. Van Nostrand Company, Inc., 1940.

17. Harker, A.: "Metamorphism," Methuen & Co., Ltd., London, 1932.

18. Judd, W. R.: Foundation Problems of the Eklutna Project, *Proc. ASCE Sep.* 444, June, 1954.

CHAPTER 2

ENGINEERING PROPERTIES OF ROCKS

Rocks and stones loaded by a structure, or otherwise, undergo displacements and, if overloaded, may be damaged; e.g., they may crack and break. The possible effects of loads on rocks and stones depend on the physical properties of these materials and should be known to the designer of a structure. These properties are discussed in this chapter.

The terms "rock" and "stone" are sometimes synonymous. In reality, however, there is some difference in their meaning: the term rock means a geological formation in its crude form as it exists in the earth, i.e., "in place"; stone is more properly applied to individual blocks, masses, or fragments that have been broken from their original massive ledges for application, mostly in construction.[1]

WEIGHT, POROSITY, AND SORPTION

2.1. Specific Gravity. In dealing with any material, including rock, it is necessary to know its unit weight as expressed in pounds per cubic foot (pcf) or in tons per cubic meter if the metric system is used. The unit weight of a rock depends on the specific gravity (density) of its constituents, on its porosity, and on the amount of water in the pores.

To avoid confusion in the interpretation of the formulas of this chapter, it should be remembered that laboratories use metric measures. Weights are expressed in grams (g), and volumes in cubic centimeters (cc). One cubic centimeter of water at 4°C weighs 1 g. In current engineering computations it is always assumed that the volume of water in cubic centimeters and the weight of this volume in grams are numerically identical. For very precise computations, however, correction should be introduced for changes of water density that occur with temperature.

Specific gravity may be determined in the laboratory as follows: (1) the rock specimen is dried for 24 hr in an oven at 105°C, cooled, and weighed (weight W_0); (2) it is then completely immersed in water, e.g., for 48 hr, and weighed in saturated condition (weight W_w); and (3) still in a, soaked condition, it is weighed while suspended in water (weight W_s). Then the specific gravity

$$G = \frac{W_0}{W_w - W_s} \tag{2.1}$$

This value, called bulk or *apparent specific gravity*, is the one most frequently used in studying the physical properties of the rock.

To ensure complete saturation of a rock sample during the test, water is introduced into the sample under pressure. Suction (vacuum) up to 20 in. Hg, as read on a manometer, may be applied instead. In any case, it is desirable to determine the specific gravity as an average from three or more tests.

If an actual test is performed, it is advisable to consult the ASTM (American Society of Testing Materials) standard, Designation C97—47, or some other standard that may succeed it in laboratory usage.

The amount of water filling the pores of the sample in the test described (in grams or in cubic centimeters) is

$$A = W_w - W_0 \tag{2.2}$$

This amount of water should be deducted from the weight W_w in order to obtain the *true specific gravity* of the material of the rock:

$$G_{\text{true}} = \frac{W_0}{W_w - A - W_s} \tag{2.1a}$$

Evidently the true specific gravity G_{true} is larger than the bulk specific gravity G. In rocks of low porosity this difference is negligible. The true specific gravity also may be found by the pycnometer method using pulverized rock material as described in physics texts.

Numerical information on the specific gravity of rock constituents and on the rock porosity may be found in the tables of this chapter and in Refs. 2 and 3.

Rocks containing heavy metals possess high densities (4.5 and over). Of the rocks commonly encountered in civil engineering projects, the highest specific gravities are found in some igneous and some metamorphic rocks. Thus some gabbro varieties have shown a density of 3.08, whereas the density of a gneiss variety was 2.97. Granites often have a specific gravity of about 2.65. Sedimentary rocks usually have lower specific gravities. Tests on one sample of Ohio sandstone showed the average specific gravity of its constituents to be 2.06.[5]

2.2. Porosity. The *porosity* of a rock is the ratio of the volume of voids (pores) to the over-all volume of the rock specimen. To determine the porosity of a rock specimen, the volume of the water filling the pores $W_w - W_0$ (using the symbols of Sec. 2.1) is divided by the total volume of the sample V that is obtained by direct measurement. Obviously, a sample of regular shape is preferable for convenience in measurement.

The value of porosity n as expressed in per cent of the volume of the sample is

$$n = \frac{W_w - W_0}{V} \, 100 \qquad (2.3)$$

In precise computations the value of $W_w - W_0$ should be divided by the density of water at the temperature of the test. It goes without saying that the temperature of the water used in the test should be the same as the temperature in the room. It should be noted that symbol n [Eq. (2.3)] is also used to designate the porosity of the soils.

Some studies have shown that there is a definite relationship between the porosity and specific gravity of a rock and its mode of origin.[4] Thus, crystalline rocks such as granite formed under great pressures have low porosities and high specific gravities. The same is true of some metamorphic rocks. On the other hand, some (Table 2.1) sedimentary rocks, such as coquina or tuff (which is a sedimentary volcanic), have a high porosity and a low specific gravity. The various relationships are apparent in Table 2.1. When the table is used, Secs. 2.3 and 2.4, respectively, should be consulted for the terms "sorption" and "dry unit weight."

2.3. Sorption: Sorbed Water.

Water filling the pores of an immersed rock specimen may be attracted by the rock or remain unattracted (free). In a rock of low and medium porosity all water is probably attracted. In this book, all attracted water is termed sorbed (or sorpted) water rather than, as is often done, subdividing it into absorbed and adsorbed water.

When immersed in water, a rock sample does not sorb as much water as its porosity permits. During immersion, a part of the air in the specimen is "entrapped" by the water and cannot find an exit; thus water is prevented from filling a certain percentage of pores. Also, some clay within the pores may swell upon contact with water and thus act as a stopper against further water penetration. If a rock is immersed in water for a specified period of time and at a specified temperature, the ratio between the volume of the specimen and the volume of sorbed water is the *per cent sorption* by volume of that rock under the specified conditions.

Using the symbols of Sec. 2.1 the per cent sorption may be expressed also *by weight*. For this purpose the weight of sorbed water A is divided by the weight of the specimen thus:

$$\frac{W_w - W_0}{W_0} \, 100 = \frac{A}{W_0} \, 100 \qquad (2.4)$$

Under natural conditions water fills only a certain portion of the total volume of pores. The *degree of saturation* is the ratio of the volume of the pores filled with water to the total volume of pores; it is proportionately less than 100 per cent. For example, the degree of saturation of granite samples immersed in water for 1 year was 66 per cent for Eastern granites and 44 per cent for Middle Western granites.[2]

TABLE 2.1. Specific Gravity, Porosity, Sorption, and Unit Weight of
Various Rocks*

Rock type	No. of specimens	Apparent sp. gr.	Porosity, %	Sorption, %	Dry unit wt, pcf
Igneous:					
Andesite..................	4	2.22	10.77	4.86	174.3
		2.70	0.72	0.28	168.5
		2.79	0.10	0.05	138.8
Aplite..................	1	2.50†	4.11	1.67	156.1
Basalt.........	4	2.77	22.06	9.97	172.3
		2.75	1.10	0.13	171.6
		2.21	0.22	0.66	138.2
Dacite..................	1	2.46	3.50	1.44	153.6
Diabase..................	2	2.95	0.17	0.06	184.4
		2.82	1.00	0.38	176.0
Gabbro..................	3	3.00	0.00	0.00	187.3
		2.86	0.29	0.13	178.4
		2.72	0.62	0.25	169.6
Granite..................	17	2.67	3.98	1.55	166.6
		2.60	1.11	0.44	162.7
		2.53	0.44	0.20	158.2
Granite (fluorite).........	1	2.99	1.67	0.58	186.6
Granodiorite..............	1	2.70	0.50	0.19	168.5
Quartz Syenite...........	1	2.63	1.54	0.62	164.1
Rhyolite..................	1	2.49	4.13	1.67	155.2
Sedimentary:					
Breccia (syenite)...........	1	3.10	0.78	0.27	193.3
Breccia (limestone)........	1	2.28	18.73	8.26	142.1
Chert....................	1	2.48	4.10	1.69	155.1
Coquina..................	1	1.19	56.70	47.80	74.0
Coral....................	2	2.66	1.06	0.41	166.2
Limestone...............	7	2.54	4.36	1.73	158.9
		2.67	1.70	0.65	166.8
		2.72	0.27	0.12	169.7
Limestone (dolomitic)......	2	2.69	2.08	0.80	167.9
Limestone (oolitic).........	3	2.66	1.06	0.42	166.3
Sandstone...............	6	2.58	1.62	0.66	160.9
		2.35	9.25	4.12	147.2
		1.91	26.40	13.80	119.5
Sandstone (calcareous).....	1	2.31	11.85	5.14	144.5
Sandstone (clayey).........	1	2.48	6.10	2.48	155.1
Travertine (onyx)..........	2	2.63	1.07	0.42	164.7
Metamorphic:					
Gneiss...................	5	3.12	2.23	0.84	195.0
		2.66	0.78	0.30	166.1
		2.61	0.30	0.12	162.9
Marble..................	7	2.73	2.02	0.77	170.2
		2.61	0.62	0.23	163.1
		2.49	0.31	0.13	155.5
Marble (dolomitic).........	2	2.84	0.60	0.21	177.2
Quartzite................	3	2.64	0.46	0.17	164.7
Slate....................	3	2.77	0.00	0.00	173.0
		2.74	1.06	0.40	171.5

* From J. H. Griffith, Physical Properties of Typical American Rocks, *Iowa Eng. Expt. Sta. Bull.* 131, March, 1937.

† Values in italics are the results of tests on only one rock specimen.

Where three values are given for one rock type, the middle value is the average for that particular physical property. However, the highest and lowest values shown were not used in computing the average value.

2.4. Unit Weight of Rocks. The unit weight of a porous substance may be referred to as *dry, saturated* (weighed in the open air with all pores filled with water), *partly saturated,* and *submerged* (immersed and weighed in water). If the porosity of a rock is slight, the difference between the dry and the saturated weight generally is negligible. In the case of high porosity, weight of water in the pores cannot be disregarded (e.g., see the values for coquina in Table 2.1).

If a rock is under water, it apparently loses weight because of buoyancy (62.4 pcf or 1 metric ton/cu m, of the volume actually occupied by the rock material, but not by the pores). For example, if the porosity of a rock is 20 per cent, the loss in weight due to buoyancy would be $0.80 \times 62.4 =$ about 50 pcf. Thus, if the dry weight of such a rock is 140 pcf, its submerged weight would be only $140 - 50$, or 90 pcf. As can be seen, buoyancy becomes a very important factor in analyzing the bearing properties of a foundation material that is likely to become saturated, such as under a dam or hydroelectric power plant. The unit weight can be calculated by either of the following two mutually checking formulas in which the symbols of Sec. 2.1 are used:

$$\text{Unit weight} = \frac{1.728 W_0}{\text{vol. cu in.}} \tag{2.5}$$

or
$$\text{Unit weight} = 62.4 G \tag{2.6}$$

This value of unit weight is also sometimes called "dry density" or "specific weight." In practice the term "density" is loosely used to designate the unit weight of the rock. Referring again to Table 2.1, it may be stated that a rock with a high unit weight generally has a low porosity, low sorption, and a high specific gravity.

STRENGTH OF ROCKS

2.5. Stresses in Rock: Compressive Strength. Generally three kinds of stresses are considered in studying the resistivity of rock: *compressive stresses,* which tend to decrease the volume of the material; *shear stresses,* which tend to move one part of a specimen with respect to the other or make it flow; and *tensile stresses,* which tend to produce cracks and fissures in the material. Accordingly the rock may have a compressive strength and a shearing strength. The tensile strength of both rocks and soils is negligible. Consequently, those structures or parts of structures that have to undergo tension are not made of rock but of other acceptable materials, e.g., reinforced concrete or steel. Besides the three kinds of stresses mentioned, rocks in natural conditions are sometimes subject to *torsion,* or twisting.

Stresses are measures in pounds per square foot or pounds per square inch.   The *compressive strength* of a material, such as rock, is the stress required to break a loaded sample that is unconfined at the sides.   The simplest case of failure in compression is shown in Fig. 2.1, in which the load tending to break the sample is designated as P.   If the cross section of the sample is 2 by 2 in., or 4 sq in., and the compressive strength of the rock is 10,000 psi, a load of $10,000 \times 4 = 40,000$ lb will break the sample.   Generally if the applied load (Fig. 2.1) is P lb and the cross section of the sample is A sq in., the compressive stress p (in pounds per square inch) in the sample is

$$p = \frac{P}{A} \qquad (2.7)$$

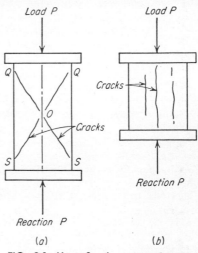

FIG. 2.1. Unconfined compression test. (a) Shear failure, (b) tension failure.

As shown in Table 2.2, igneous rocks, some quartzites, and some sandstones generally have the highest compressive strengths. Basalt, if unweathered, may reach an unconfined compressive strength of 60,000 psi.   The compressive strength of porphyritic igneous rocks depends on their porosity: The more compact the porphyry, the higher its compressive strength.

TABLE 2.2. Compressive Strengths of Rocks (General Values)*

Compressive strength, psi	Type of rock
Over 40,000	Some basalts, diabase, some quartzites
25,000–40,000	Fine-grained granites, diorite, basalt, compact, well-cemented sandstones and limestones, quartzite
10,000–25,000	Average sandstones and limestones, medium- and coarse-grained granites, gneiss
5,000–10,000	Porous sandstones and limestones, shales
Under 5,000	Tuff, chalk, very porous sandstone, siltstone

* From A. von Moos and F. de Quervain, "Technische Gesteinkunde," Birkhaeuser Verlag, Basel, Switzerland, 1948.

Probably there is some analogy between the factors influencing the strength of rocks and those influencing the strength of metals.   It has been observed, for example, that a metal with a comparatively rough surface will have less strength than a metal with a practically smooth surface.   Thus, the compressive strength of rocks will be influenced by their texture, particularly by coarseness of the grains; e.g., fine-grained sandstones are stronger than the coarse-grained.   Igneous and metamorphic rocks which under the microscope show a strong interlocking between the crystals will be stronger than those which show a poor interlocking. This factor is known as the *crystallinity* of the rock. However, in sedimentary rocks, the strength of the interstitial bonding material—the "cement"—may have as much influence on the compressive strength

TABLE 2.3. Compressive Strength of Rocks (Detailed Values)*

Type of rock	No. of tests†	Compressive strength, psi (averages)	Standard deviation D % range R‡
Andesite (USBR).................	3 and 3	18,710–19,150	3,360– 3,510 (R)
Basalt (USBR)....................	3 and 3	24,450–31,850	9,879–16,780 (R)
Gneiss (USBR)...................	1 and 1	9,310–15,140	0 0 (R)
Granite..........................	12	33,200	3.2 (D)
Granite, slightly altered (USBR)...	3 and 3	8,250– 9,400	3,820– 4,820 (R)
Limestone.......................	12	10,900	5.3 (D)
Limestone, reef breccia (USBR)....	2 and 4	860– 4,960	190– 4,080 (R)
Marble..........................	12	30,800	9.0 (D)
Sandstone (two types).............	12 and 12	10,400 and 6,100	4.6 and 14.5 (D)
Sandstone (USBR)................	9 and 8	8,810–12,200	7,470– 5,210 (R)
Schist, biotite (USBR)............	1 and 1	7,750–12,010	0 0 (R)
Schist, biotite-chlorite (USBR).....	1 and 1	5,290–17,000	0 0 (R)
Schist, biotite-sillimanite (USBR)..	1 and 1	1,160– 4,930	0 0 (R)
Schist, biotite-sillimanite-quartz (USBR)......................	1 and 2	1,250– 4,520	0– 2,800 (R)
Slate...........................	12	30,400	5.2 (D)
Tuff (USBR).....................	3	530	190 (R)

* From S. L. Windes, Physical Properties of Mine Rocks, Part I, *USBM Rept. Invest.* 4459, 1949, and Physical Properties of Some Typical Foundation Rocks, *USBR Concrete Lab. Rept.* SP 39, 1954.

† The first figure refers to the number of specimens tested to arrive at the first average value given under "Compressive strength"; the second figure refers to the number of specimens tested to arrive at the second average value given under "Compressive strength."

‡ The first figure given refers to the total range between minimum and maximum compressive-strength values found in testing the first group of specimens; the second figure refers to the total range found in testing the second group of specimens.

as the texture. This is particularly true for sandstones, conglomerates, and breccias. If the cementing material is entirely or partially clay, the compressive strength will be rather low. Seams filled with clay are frequent in limestones, and thus limestones with low compressive strengths are not unusual. The highest compressive strength is obtained when the cementing medium of a sedimentary rock is quartz (secondary silica), in which case the rock is called a "silicified" rock, or a combination of silica and recrystallized fine hydromicas (illite, chlorite, etc.). The

presence of fissures, often microscopic, and seams (i.e., more or less wide, fissurelike inclusions of some foreign material in the rock) is detrimental to the compressive strength, especially if the direction of these fissures coincides with the failure planes (*QS* in Fig. 2.1*a*).

The compressive strength of a rock depends on the direction of the acting compressive stress with relation to bedding; i.e., the highest compressive strength is obtained when the compressive stress is normal to the bedding.

TABLE 2.4. Microseism Tests of the Bureau of Mines*

Kind of rock	Grain size	Total compressive strength at detecting, %		Total compressive strength, kg/sq cm	Type of failure
		First microseism	Second microseism		
Granite....... {	Coarse	26–36	80–96	45,600	Violent shatter
	Coarse	13	17	41,700	Violent shatter
Basalt........	Aphanitic	16	25	25,000	Crush
Dolomite.....	Medium to coarse	8	16	34,900	Violent shatter
Breccia.......	Large fragments				
Limestone....	Fine to medium	10–22	16–22	30,700	Shatter
Sandstone....	Fine to medium	33	37	27,900	Violent shatter
Schist........	Medium to coarse	25	33	12,200	Crush

* From L. Obert and W. J. DuVall, The Microseismic Method of Predicting Rock Failure in Underground Mining, *USBM Rept. Invest.* 3797 and 3803, 1945.

Table 2.3 shows the results of several compression tests. The abbreviation after the rock name indicates the tests by the Bureau of Reclamation,[6] whereas other tests were made by the Bureau of Mines.[3] In the USBR tests the vertical load was applied at random directions to the bedding of the sedimentary rocks, whereas in the USBM tests the load was applied perpendicularly to the bedding and the compressive strengths were generally higher.

Saturation decreases the compressive strength of a rock. In one study[2] the decrease in compressive strength because of a wet condition was 12 per cent for granite; in another[3] from 4 to 8 per cent for granite and marble and from 10 to 20 per cent in layered sandstones, limestones, and slates. The higher the porosity—and thus the greater the chance for saturation—the lower will be the strength of the rock when saturated. There is a definite relationship between the per cent sorption and compressive strength:[4] *As the per cent sorption increases, the compressive strength decreases.*

2.6. Some Phenomena Accompanying Rock Compression. The USBM has published the results of special field and laboratory compression tests on various rocks. Even before the first fissure in the compressed sample can be observed, there is a very small movement of the material within the sample along eventual surfaces of failure. This motion is analogous to a tiny earthquake within the sample, and it can be detected, i.e., "heard," by using a special geophone. During the test, the value of the load acting on the sample gradually increases until failure occurs. Columns 3 and 4 of Table 2.4 list those fractions of the total compressive strength (in per cent thereof) at which the first microseism (i.e., the tiny earthquake) and the first fissure are detected.

One purpose of these tests was to determine if special geophones could be used to predict rock falls or roof collapses in mines. It has been shown by the tests that pillar spacing can be gradually increased until the geophones detect microseisms of sufficient strength to indicate approaching roof failure. Further research may prove that this method can be used to determine the necessity for using tunnel supports in transportation or water tunnels.

2.7. Tensile Strength of Rocks. The tensile strength of granite is about 1,000 psi,[3] only a fraction of its compressive strength. Other rocks have still lower tensile strength: marble 700 to 900 psi, limestone about 500 psi, and sandstone 100 to 200 psi.

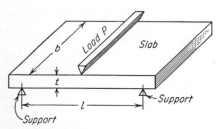

FIG. 2.2. Modulus of rupture $3Pl/2bt^2$.

If a stone slab is placed on two practically immovable supports and subjected to the action of a load P (Fig. 2.2), the slab deflects and there are tension at the bottom of the slab and compression at its top. If the load P is gradually increased, the slab fails in tension. It becomes evident from this example why engineers prefer to use reinforced concrete slabs instead of stone slabs if the slab has to be supported as shown in Fig. 2.2. Tensile stresses may develop in a stone slab because of not only load action but also settlement of the structure, earthquakes, or temperature effects.

Figure 2.2 also shows data for the computation of the "modulus of rupture" of the slab.

2.8. Mechanics of Shear in Rocks. Figure 2.3 represents a prism consisting of two cubes cut from a rock mass. The upper cube is pressed to the lower by a compressive stress p and is pushed horizontally by a tangential (shearing) stress t. If the coefficient of friction of rock on rock is f, the upper cube with respect to the lower would be on the verge of sliding if

$$t = pf \tag{2.8}$$

The motion of one rock cube with respect to the other as in Fig. 2.3 is a relatively simple phenomenon. Shear across the rock material itself is a much more complicated phenomenon, however. In this case the coefficient f covers not only friction between shear surfaces but also the breaking of the bonds between particles and other disturbances. Though it is often called "coefficient of internal friction," it is more proper to call it coefficient of shearing resistance. This coefficient is visualized as the tangent of an *angle of shearing resistance* ϕ:

$$f = \tan \phi \qquad (2.9)$$

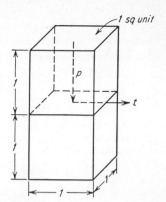

FIG. 2.3. Normal and tangential stresses.

The shearing resistance of a rock (and of many other materials such as clay) is too complicated to be expressed by the simple formula (2.8) which states that the shearing resistance of a rock is directly proportional to the pressure p on the shearing surface. For better results, empirical formula (2.10) often is used:

$$t = p \tan \phi + c \qquad (2.10)$$

In this formula, the term c is known as *cohesion*. The values t, p, and c are expressed in terms of stress, i.e., in pounds per square inch, pounds per square foot, or kilograms per square centimeter. Equation (2.10) was proposed by the French mathematician and engineer Coulomb in 1773. This formula is further discussed in Sec. 2.9.

The effect of a shearing stress on a rock material may be twofold: Either one part of the material slides along a plane surface, separating this part from the rest of the rock mass, or the rock flows plastically without forming perceptible surfaces of separation. Plastic flow is characteristic of rocks possessing considerable cohesion and a small (or zero) angle of internal friction, such as partially cooled lava. Sandstones with plastic cementing material such as clay also are liable to plastic flow. A case of plastic rock flow is shown in Fig. 2.4, which represents the cross section of a deep quarry in a plastic rock such as basalt. The bottom of the quarry may rise because of (1) the *elastic rebound*, or upward motion, of the material following its unloading (see

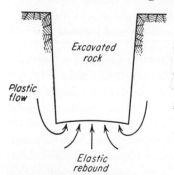

FIG. 2.4. Rise of bottom of a deep basalt quarry.

Sec. 2.11) and (2) the *plastic flow* from the sides of the quarry toward its bottom caused by the weight of the adjacent rock.

2.9. Compression and Shear Tests on Rock. The compressive (or tensile) and shear stresses in rock and any other material act simultaneously, and if a failure in a compressed rock occurs, this is generally a result of their common action. Compression tests duplicate the behavior of a material under this combined compression and shear stress action. These are the unconfined compression test and the triaxial compression tests; the former is a particular case of the latter. The triaxial compression test also is termed triaxial shear test or cylinder test. The difference between the unconfined compression test and the triaxial test may be

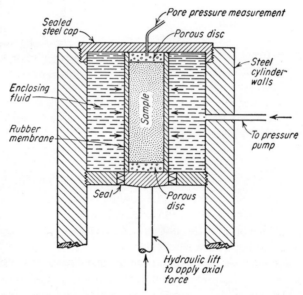

FIG. 2.5. Diagrammatic sketch of triaxial shear test on a rock sample.

made clear by comparing Figs. 2.1 and 2.5. In both cases a rock sample is acted upon by a vertical stress which increases gradually until a failure occurs. In the triaxial test the specimen is enclosed in a cylinder containing fluid (generally water, glycerol, kerosene, or oil) under pressure. Thus the specimen is acted upon not only by a vertical stress but also by a horizontal pressure which provides *lateral support* to the specimen. The unconfined compression test is made in the open air, and the lateral support to the specimen is zero. In both cases not only compressive strength but also shear strength can be determined as explained hereafter. The *compressive strength* is the unit axial load at failure either in the unconfined compression test or in the triaxial test with zero lateral pressure.

Unconfined Compression Test.   Specimens used in this test are small prisms 2 or 3 in. on a side or cylinders about 2 in. in diameter and approximately 2 in. high. The failure is caused primarily by shearing stresses acting along failure surfaces *QS* (Fig. 2.1).   These surfaces make an approximate angle of $45° - (\phi/2)$ with the vertical (ϕ being the angle of shearing resistance).   If the failure surfaces *QS* are more or less regular (which is not usual), the value of the angle of shearing resistance may be estimated by measuring the angle $45° - (\phi/2)$ on the specimen.   If the angle of shearing resistance is very small, the specimen may bulge because of plastic flow of the material.   Occasionally the specimen fissures or cracks vertically along its center line or surface (Fig. 2.1*b*); there also may be several vertical cracks of this nature. Such vertical cracking or fissuring is a failure *in tension* caused by horizontal tensile stresses.   Similar tension failures may be observed in stone slab pedestals under overloaded stone (e.g., marble) columns in older buildings.   In performing an actual unconfined compression test, the ASTM standard, Designation C170—50 (or some other standard that may replace it) should be consulted.

Triaxial Compression Test.   The triaxial device used by the USBR[6] and shown in Fig. 2.5 consists of a heavy steel cylinder filled with kerosene.   The fluid is placed under variable pressures that duplicate the actual confining pressures in natural rock masses.   These confining pressures are known as the surcharge or overburden pressures.   The highest lateral pressures used in the device described are of the order of about one-fourth to one-fifth of the compression strength.   The rock specimen in a rubber jacket is placed at the axis of the cylinder and pressed by an upward-acting force against the heavy cover of the cylinder.   The rock samples used are from $1\frac{5}{8}$ to 6 in. in diameter, and their length is about twice this diameter.   The failure lines are similar to those in the unconfined compression test (Fig. 2.1); the angle *QOQ* is acute in brittle rock and obtuse in plastic rock.   In the latter case the sample tends to bulge.   For rocks with a moisture content of 75 per cent or more (by dry weight), pore pressure also has to be measured.

Mohr's Diagram.   When Eq. (2.10) is used, it should be clearly understood that the values of ϕ and *c* are *not* constant characteristics of a rock (or a soil material) but depend on the character of the shear phenomenon, particularly on the rate of application of the shearing force.   This is true both for the shear failures in the field and for the laboratory shear tests.

If in a triaxial test on rock (or soil) the lateral pressure and the axial load are applied at a moderately slow rate, a reasonably correct value of the angle of shearing resistance may be obtained.   Figure 2.6 shows the results of the triaxial tests described in Ref. 6 plotted graphically.   Distances *OA* and *OA'* are unit lateral pressures (in pounds per square inch or pounds per square foot) for two consecutive triaxial tests on a material. Distances *AB* and *A'B'* are critical unit axial loads (in pounds per square inch or pounds per square foot) required to break the sample in the same tests.   The circles using *AB* and *A'B'* as diameters are Mohr's circles. Figure 2.6 shows the common practice of tracing semicircles rather than circles.   The common tangent to both circles intersects the vertical axis at some point *D*; the distance *OD* is the unit cohesion *c* (in pounds per square inch or pounds per square foot) as in Eq. (2.10).   This tangent is termed *Mohr's envelope*, and according to the evidence available, it is a curve rather than a straight line.   In practice several tests are made on the same material; if the Mohr envelope is visualized as a straight line, its final position is determined as the average of several available straight lines, e.g., by the method of least squares[6] or by some simpler method.

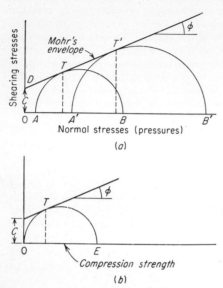

(a)

(b)

FIG. 2.6. Mohr's diagram (a) for two triaxial tests, (b) for a triaxial test without restraint or for an unconfined compression test.

If perpendiculars on the horizontal axis are dropped (dotted vertical lines in Fig. 2.6) from the points of tangency T and T' of the Mohr circles with the Mohr envelope, these perpendiculars show the values of the *shearing stresses* in the corresponding tests.   The shearing stresses are measured on the same scale on which the pressures are plotted in the horizontal axis.

In the triaxial tests of several specimens of slightly altered granite from the Grand Coulee pumping plant back wall,[6] the average compressive strength was 9,400 psi with a range of 4,820 psi, the range being the difference between the maximum and the minimum experimental value.   The unit cohesion c was 1,420 psi, and the order of magnitude of the angle of shearing resistance ϕ was about 55°. When the lateral pressure was 1,000 psi, the critical axial stress required to break the sample was about 17,000 psi, but it increased to about 22,000 psi when the lateral pressure was 2,000 psi.

Conclusions from the Compression Tests.   The compressive strength of a rock increases as the lateral restraint increases.   In the field this restraint is due to the lateral pressure that the confined rocks exert on one another. The main cause of the lateral pressure in the field is the weight of the overburden covering a given rock.   The deeper the rock, the larger the lateral pressure, provided, of course, that the rock is thoroughly confined and cannot escape or flow laterally.   In a laboratory triaxial test the lateral pressure is provided by the fluid in the cylinder placed under pressure. It follows that the compressive strength of rock located at a depth is greater than at the ground surface but drops when the rock is excavated and the restraint is removed.   It also follows that for the design of deep foundations, all values of compressive strength given in this chapter are on the safe side, since these values have been determined from the results of unconfined compression tests (i.e., without restraint).

2.10. Safety Factor. To be on the safe side, stones placed in a structure or rock supporting a structure are loaded to only a fraction of the ultimate crushing load or shearing stress as determined in the laboratory or in the field. In other words, when such loadings are designed, a *safety factor* is used as generally is done in other fields of engineering. In the case of stones, the safety factor generally is from 6 to 10; in the case of foundation rock, it usually is much higher. For example, an average value of the compressive strength for granites is about 10,000 psi, or about 720 tons/sq ft, whereas the New York Building Code establishes the "bearing value" of granite as only 25 to 40 tons/sq ft. Usually only major structures can economically justify the performance of special laboratory compression tests on rock and stone samples. Generally, for medium- and small-size buildings, the foundation rock is only visually inspected and its bearing value established by consulting the local building code. This value usually is expressed in tons or kips per square foot or pounds per square foot. "Bearing value," "bearing power," "bearing capacity," and other similar terms are synonymously used to express the unit load that can be placed "safely," i.e., without detrimental deformations to the structure, on the surface of a rock (or soil). In both rock and soil foundations the value of the bearing power depends not only on the type of the underlying material but also on the type of the structure. More information along these lines may be found in Chaps. 13 and 14.

ELASTICITY OF ROCKS

2.11. Modulus of Elasticity E for Rocks. Referring to Fig. 2.7, a certain load P' smaller than the crushing load decreases the height L of the square prismatic sample vertically by a value of ΔL and increases its width horizontally from B to $B + \Delta B$. If, after removal of the load, the sample tends to recover its original shape and size, the rock is said to possess elastic properties. Rarely, however, does the sample recover its original shape and size after one loading and unloading; a part of the deformation generally remains. This is *plastic* or irreversible deformation. After one or more consecutive loadings and unloadings, the sample may become *elastic;* i.e., after each loading and unloading it recovers exactly the same shape and size as before that particular loading. If, in addition, at each loading the deformation ΔL is proportional to the load P' that caused it, the material is said to obey Hooke's law of proportionality of stress to strain ("perfect" elasticity):

$$E = \frac{P'/B \times B}{\Delta L/L} = \frac{\text{unit stress}}{\text{unit strain}} \qquad (2.11)$$

Symbol E is the *modulus of elasticity* for the given rock in compression and is also known as Young's modulus. It is expressed in terms of stress, i.e., pounds per square inch, pounds per square foot, kips per square inch,* kilograms per square centimeter, or metric tons per square inch. Some numerical values of E are given in Table 2.5.

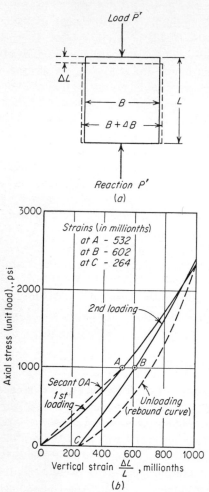

FIG. 2.7. (a) Deformations of a loaded prismatic rock sample. (b) Stress-strain diagram for a granite.

Strictly speaking, Eq. (2.11) applies to an *isotropic rock*† in which the elastic properties are the same in all directions. In an isotropic material (rock in this case) the modulus of elasticity is constant; i.e., it is the same in all directions. The elastic properties of actual rocks are usually variable in different directions, and the value of the modulus of elasticity depends on the direction of measurement. For example, the smallest value of the modulus of elasticity generally occurs when E is measured *perpendicular* (normal) *to the bedding of a rock*. This means that the greatest deflections in a loaded rock occur when the load is placed perpendicular to bedding. In a general case the elastic properties of a foundation rock are adversely affected by various geological conditions such as fault zones, fractures, small igneous intrusions in sedimentary rocks, and lenticularity of bedding, which permits the introduction of occasional soft zones into generally hard rocks. Most of these conditions are described in the subsequent sections of this chapter.

Variances in the interlocking of the constituent grains account in part for the considerable range in modulus of elasticity and compressive strength found in *granitic rocks*. When the grains are poorly interlocked, even if the granites are composed of firm, hard crystals of quartz and

---

* One kip (kilopound) equals 1,000 lb.

† Nonisotropic materials are usually termed *anisotropic*.

feldspar, the rock may be weak and relatively inelastic. On the other hand, if their grains are well interlocked, some granites and especially gabbros and diabases will show high strength and elasticity values. Data compiled from many tests[7] indicate an average value of E of 12,400 ksi for 20 gabbros and diabases and an average of 6,400 ksi for 23 granites. The average compressive strengths in the same tests were 25,600 psi for 41 gabbros and diabases and 21,050 psi for 154 granites. Generally, the higher the compressive strength of a rock, the higher its E value, though there are exceptions.

TABLE 2.5. Modulus of Elasticity for Rocks

Kind of rock	Modulus of elasticity, ksi		
	Refs. 2 and 7	Ref. 3	Ref. 6
Granite..................	4,500–8,300	4,410–6,410	800–5,200
Quartzite................	5,700–8,300		
Sandstone...............	2,300	870–1,620	1,900–3,200
Limestone...............	....	4,840–5,390	1,000–5,500
Marble..................	7,600–11,000	7,150–10,400	
Slate....................	....	12,110–13,660	

Laboratory and field loading tests in Portugal gave the following values of the modulus of elasticity (in ksi): Altered granite 284–711; schist 569–2,845; altered schist 284–569.[31]

The position of the water table has an influence on the *moisture content* of a rock. According to the same data[3] the value of E in both granite and marble increases when the moisture content of the specimen *slightly* increases. In the case of limestone, however, the situation is reversed. Larger moisture contents may decrease the value of the modulus of elasticity of a rock. Thus, "dry" E values, i.e., obtained in the laboratory, may be misleading. This fact should be seriously considered in the design of dams (particularly arch dams, Sec 15.12) and foundations of other heavy structures that may be affected by water.

Observations have also shown that the E value of a rock apparently increases after the rock has been loaded for a long period of time. It also has been observed that many rocks when placed under high loads in the laboratory and then unloaded after several years will return almost completely to their original shape and size.

The degree of integrity of the test sample influences the test results. If incipient fractures are present (and thus minute pores or fissures are numerous), the E values may be considerably lower than if the test had been performed on the rock in place. The surcharge on the rock in place tends to keep all such incipient fractures tightly sealed and thus is of little influence on the value of the modulus of elasticity, whereas in the laboratory, the fissures are open and thus influence the E value.

Field Determination of the E values for Rock. An excavation is made, and a wall of it is loaded by using a jack working against a very large concrete block poured at the opposite wall of the excavation. The pressure exerted by the jack (total pressure P) is applied to the rock through a heavily reinforced rubber disk (diameter d). The deflection w at the center of the disk is measured. Designating by E the modulus of elasticity and by μ (greek letter mu, Sec. 2.13) its Poisson's ratio (Sec. 2.12), the following formula of the theory of elasticity may be used:

$$w = \frac{4P}{\pi d} \frac{1 - \mu^2}{E} \tag{2.12}$$

If the average value of μ is $\frac{1}{6}$, neglecting it completely would involve an error of about 3 per cent in the E value. Then

$$E = \frac{4P}{\pi} \frac{1}{wd} = \frac{1.27P}{wd} \tag{2.13}$$

If P is expressed in pounds and both w and d in inches, the E value computed using Eq. (2.13) will be expressed in pounds per square inch. When these methods are used, it should be remembered that the excavation destroys the stress condition existing in the rock in its natural state. Therefore the modulus of elasticity actually measured will be somewhat different from that in the natural rock. Furthermore, in this test, the E value measured applies only in the experimental direction of the jack action, though it may give a fair idea of the *order of magnitude* of the average modulus of elasticity in the given rock. The E and μ values are predominantly used for computing settlement of structures built on a given rock. Equations similar to (2.12) are then applied, varying according to the load distribution and configuration of the loaded area.

The following method furnishes an *average* E value of the rock in all directions and, as in the preceding method, may indicate the order of magnitude of the modulus of elasticity for a given rock.

A portion of an existing water tunnel with concrete lining is closed on both sides by concrete head walls and subjected to water pressure. The test is done under the assumption that the moduli of elasticity of the rock and the concrete are practically identical. In many cases, however, the elastic properties of these materials far from satisfy this assumption.

Designate by p (in pounds per square inch) the water pressure in the tunnel, by D the diameter of the tunnel, and by ΔD its increase due to pressure; then

$$E = p(1 + \mu) \frac{D}{\Delta D} \tag{2.14}$$

or, neglecting μ,

$$E = \frac{pD}{\Delta D} \qquad (2.15)$$

A more correct formula is (2.16). Designating by t the thickness of the concrete lining and by E' its modulus of elasticity (from 2×10^6 to 6×10^6 psi) and using the largest value of E' to be on the conservative side,[21]

$$E = \frac{pD}{\Delta D} - \frac{2tE'}{D} \qquad (2.16)$$

A study of actual strain in two French arch dams and their abutments[22] has shown that though the strains in the rock were in the same direction as those in the adjacent concrete, at a short distance from the rock-concrete interface, the rock was strained from 5 to 13 times more than the concrete. At certain points there were strains in rock while the adjacent concrete showed none. The rock was granulite* in one case and fissured mica schist in the other. The schistosity of the latter rock was almost vertical.

Other methods of measuring the modulus of elasticity for rock in special excavations (galleries), both lined and unlined, may be found in Refs. 21 and 23. Among other things, the application of the loads to the rock using diametric invar rods with acoustic cells[21] and the use of special pressure cushions[23,24] are described. (Invar is an alloy of iron with 36 per cent of nickel; it is practically insensitive to temperature changes.)

Somewhat apart from the methods mentioned is the seismic method described in Ref. 25. The values of the moduli of elasticity for a great number of rocks and the corresponding velocities of propagation of longitudinal (compressional) waves have been assembled and correlated in Ref. 26. If E is the value of the modulus of elasticity in pounds per square inch and V the longitudinal wave velocity in feet per second, the following approximate relation holds:

$$E = 10^3(V - 8,000) \qquad (2.17)$$

For a more accurate graph, see Ref. 25. In this method, an elastic disturbance is set up in the rock by detonating a small charge of explosive or by striking the rock a sharp blow with a sledge hammer. The velocity of the wave thus produced in the rock may be measured as explained in Sec. 6.32. It is obvious that the E value measured by this method refers to the horizontally loaded rocks, though in fairly homogeneous rocks the results may indicate the order of magnitude of the average E value.

2.12. Modulus of Compression. *Stress-Strain Diagram.* The *stress-strain diagram* in Fig. 2.7b represents the results of an unconfined compression test performed to determine the modulus of elasticity E of a granite. The sample was loaded, unloaded, and loaded again. In each loading, the load was gradually increased. Point A corresponds to the stage when the increasing unit load was 1,000 psi. The corresponding deformation ΔL (Fig. 2.7a) divided by the height L of the sample at that time is the *strain* at that stage (532 millionths). The modulus of elas-

* Granulite is a metamorphic rock composed of even-sized interlocking granular minerals such as feldspars or garnets.

ticity E_c corresponding to that stage is the slope of the secant OA, i.e., the straight line joining points O and A:

$$E_c = \frac{1,000}{0.000532} = 1,880,000 \text{ psi} = 1.88 \times 10^6 \text{ psi}$$

In reality what is obtained in the first loading of a sample is the *modulus of compression* (or modulus of deformation) of the material, which includes both elastic and plastic deformations. To obtain the true value of the modulus of elasticity the sample should be unloaded and loaded again and this cycle repeated until the stress-strain diagram becomes a straight line. In the present case the second loading gives a fair approximation to a straight line, and

$$E = \frac{1,000}{0.000602 - 0.000264} = 2.97 \times 10^6 \text{ psi}$$

At the first loading, the stress-strain diagram may be either a convex curve as in Fig. 2.7b for granite or a concave one as was found for *some* basalts and schists.[6] In the former case, the largest experimental (secant) E_c value occurred; in the latter case the opposite took place. Accordingly, if the average secant E_c value for a given loading range (e.g., from 0 to 1,000 psi) is considered, this average value may be larger or smaller than the final experimental secant value of E_c (for 1,000 psi in this case). In the same way, straightening out the curved stress-strain diagram by alternate loadings and unloadings may lead to secant E values which are larger or smaller than the secant E_c values for convex and concave stress-strain diagrams, respectively.

In practice, often no distinction is made between the modulus of compression and the modulus of elasticity, and the former modulus is called modulus of elasticity at the first loading or the first loading cycle. If a heavy structure, e.g., a large masonry dam with a large initial dead load and a rather small live (moving) load, is considered, the modulus of compression should be used. If the live load is relatively large, as in the case of an old high bridge pier and high tides when a large portion of the weight of the pier is periodically relieved by the tide, the use of a true E value should be recommended.

Field Determination of Rock Compressibility. To determine the possible settlement of a structure to be built on rock, field loading tests may be made by loading rigid plates and observing their settlement. Such tests, however, should be correlated with the results of laboratory tests on moduli of elasticity and compression. The width or the diameter of the load plate used in such tests has considerable influence on the results (Fig. 2.8), as the *zone of considerable load influence* will penetrate only a certain depth beneath the plate. The larger the area of the plate, the greater will be the depth of penetration of this zone. If a small-diameter plate is used, considerable load effects may be imposed only a short depth below the plate, and deeper, soft materials will be only slightly compressed. As the foundation of the actual structure will cover a much larger area than the loading plate, the building load may compress this soft material considerably. Obviously there also are pressures on the rock outside the zone of considerable load influence shown in Fig. 2.8, but these pressures are relatively small. Other field load tests are described in Sec. 13.5.

During the construction of Baldwin Hills Dam (Los Angeles, California) the changes in the diameter of a special tunnel constructed under the dam were measured, and from them the gradual settlement of the dam estimated.

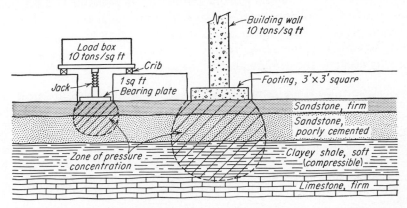

FIG. 2.8. Relative zones of load influence under bearing plate and under building footings.

2.13. Poisson's Ratio. Poisson's ratio (symbol μ or mu) is the ratio of the lateral strain $\Delta B/B$ to the longitudinal strain $\Delta L/L$ (Fig. 2.7a). As in the case of the modulus of elasticity, the experimental values of this ratio also depend on the character of the stress-strain diagram at first loading and on the manner of performing the tests. Though the modulus of elasticity and Poisson's ratio often are called *elastic constants*, their experimental values are variable and approximate. Values available for Poisson's ratio fluctuate between 0.15 and 0.24 for granite, between 0.16 and 0.23 for limestone, between 0.08 and 0.20 for schist,[6] and between 0.25 and 0.38 for marble.[8] The following values have been found in isolated tests: gneiss, 0.11; monzonite, 0.17; sandstone, 0.17; and tuff, 0.11.[6]

The Bureau of Mines used the sonic method for determining both the modulus of elasticity and Poisson's ratio.[3,5] These values, together with rather high values of Poisson's ratio in the direction parallel to the bedding of the rock (0.27 for marble, 0.31 for granite), resulted in exceedingly small and even negative values in other directions. The sonic method for measuring elastic properties of concrete and rock is given in ASTM Standard C215—51T and in Ref. 10.

2.14. Residual Stresses in Rock. Additional stresses created in rock by excessive loading or some disturbance, either natural or man-made such as blasting, may remain in the rock for a rather long time after the disturbance is over or the loading is removed. These are *residual stresses*. Their nature and possible change with time are not clearly understood; their presence, however, can be detected. Generally a stress cannot produce a deformation if the material on which it acts is so confined that the movement of rock particles is prevented. Removal of some rock materials close to the rock face "relieves" the rock. The rock material then starts to move slowly toward the relieved spot, and strains can be measured. The order of magnitude of the residual stresses may be estimated by multiplying the measured *strain* values by the modulus of elasticity E. Knowledge of residual stresses is especially necessary in tunnel construction for predicting and preventing detrimental pressures that may occur as the confinement of the rock is relieved by construction.

For measuring the strains, strainmeters, usually of the SR-4 type, are used. An SR-4 gauge consists of a tiny rectangular grid of resistance wires; these tiny grids can be placed in groups of four to form a "rosette." In these types of gauges three gauges are oriented along the sides of a $1\frac{1}{4}$-in. equilateral triangle to permit the measurement of strains in three directions. The rosette is cemented to the smoothed rock; the rock is relieved as explained; then the strain is measured by reading the galvanometer connected to the SR-4 gauges. In reality, the gauges measure the resistance of the wires to the passage of the electric current and are calibrated against strain.

In an experimental tunnel under Boulder Dam built during the construction of that dam, large rosettes were used. The rock was relieved by drilling with a jackhammer a series of holes 30 in. deep on a 4 ft diameter around the gauge points and then chipping out between the holes.[27] At Prospect Mountain in Colorado, rosettes of four SR-4 grids as described were used.[28] The rock was relieved by core drilling around the rosette to a depth of 5 in. In both cases the rosettes were cemented to the tunnel walls.

A combination of a Freyssinet jack and SR-4 gauges also is used to measure residual rock stresses.[29] A *Freyssinet jack* consists of two soft-steel plates welded together to form a sealed pocket. The plates are usually rectangular but may also be circular. A flat, rectangular hole is cut in the rock face of the tunnel; the jack is inserted in it and firmly grouted. A pair (or more) of strainmeters are placed across the hole at its periphery, and the distance between them carefully measured. Residual stresses in the rock make the strainmeters move, and the distance between them decreases. It may be reestablished, however, by forcing water under pressure into the jack; thus a heavy pressure is exerted on the rock parallel to its face. The *unit pressure* exerted on the rock in order to return the strainmeters to their original position may be assumed as approximately equal to the residual compressive stresses in the rock. The residual stresses also may be estimated from the readings of the strainmeters. The normal travel of the Freyssinet jack is 2 cm; the highest advisable pressure in the jack is 150 kg/sq cm.

Instead of the distance being measured between the gauges, careful measurements can be made on a length of piano wire stretched across a hole at the rock surface between two invar rods set into the rock.

ROCK DEFORMATIONS IN NATURE

2.15. Fractures in Rock. Any break in a mass of rock can be defined as a *fracture*, regardless of its size. Since strains are caused by stresses, it is obvious that fracturing is caused by stresses which are greater than the strength of the rock that resists them. When a series of fractures are more or less continuous and seem to form well-defined patterns bearing a relation to each other or to other elements of the rock mass, the fractures then may be described as *joints*. If the rock mass on each side of a fracture indicates that there has been displacement (horizontal or vertical or both) along the fracture plane, the plane is classified as a *fault*. Such displacements may be barely noticeable or may be several thousand feet or many miles in extent. When several faults occur in close proximity and usually parallel to each other, the resulting zone of broken rock commonly is called a *shear* or *fault zone*. The often-used terms "cracks" and "fissures" usually designate minor fractures.

Tensile stresses are primarily responsible for major fracturing involving faults and joints. These stresses generally are the result of a decrease in volume (shrinkage) due to (1) a drop in temperature, (2) a loss of moisture in the rock, or (3) both. In certain soft rocks (and in some soils), shrinkage also may be produced by recrystallization of gels. Usually, tension is combined with compression, a characteristic example being the generally hexagonal columnar structure of basalt (Fig. 2.9a). Mud cracks (Fig. 2.9b) are caused by shrinkage due to evaporation. Because of the nonuniform drying of a suspension (such as muddy water), the crack pattern in this case is more irregular than in basalt.

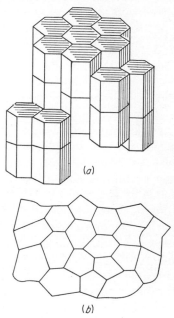

FIG. 2.9. Shrinkage and cracking: (a) basalt columns, (b) mud cracking.

Jointing. Joints in rocks may appear to be somewhat random in direction. However, a close study will disclose that they have a definite relationship either with the bedding and flow lines in the rock (apparent bedding showing flow of liquid magma) or with faults in the rock mass. (The flow lines are indicated by the orientation of the individual crystals in an igneous rock; i.e. the long axes of nearly all crystals are parallel or almost so.)   In sedimentary rocks, there generally are two systems of mutually perpendicular joints, both being perpendicular to the bedding plane.   In igneous rocks, there generally are three regular sets of joints

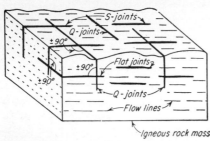
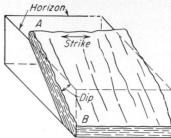

FIG. 2.10. Types of joints.   (After Hills,[11] and Cloos.[13])

FIG. 2.11. Strike and dip.

(Fig. 2.10).   One set lies approximately horizontal and parallel to the flow lines and is termed *flat-lying joints*.   Another set is about perpendicular to the flow lines and is termed *Q joints* or *cross joints*.   The third set, *S joints* or *longitudinal joints*,[11] dips steeply and strikes parallel to the flow lines if the latter are projected to a plane surface such as a map.

Joints also may be grouped into strike joints and dip joints.   Figure 2.11 illustrates the terms "strike" and "dip" when the rock surface is assumed to be an oblique plane.   *Strike* is the direction of contour lines or lines of equal elevations on the surface of the rock mass, whereas *dip* is the maximum slope of its surface.   (In Fig. 2.11, the dip is the angle made by the line AB with the horizon.)   In measurements of dip, it is important to measure the "true" dip, i.e., the angle located in a plane perpendicular to the strike; otherwise, a misleading "apparent" dip is recorded.   Strike is given as N (°) W or E or S (°) W or E.   For example, N30°E means 30 degrees east from the north direction.   Dip is indicated as (°) N, (°) NE, etc.   For example, 35°S means a dip in the southerly direction.   Smaller dips such as 2 or 3° are often expressed in feet per mile rather than in degrees.

A *sheeted zone* (Fig. 1.15), or sheeting, is composed of several closely spaced joints that divide the rock into flat slabs only a few inches in width.   By strict definition, sheeting slabs should lie approximately parallel to the local topography; in the geotechnical use, this definition is not used.

Rift and Grain.   Igneous rocks may split readily along two directions termed rift and grain (or run) (Fig. 2.12).   The plane along which the rock most easily splits, the *rift*, may be almost vertical or almost horizontal

(Fig. 2.12a and b). The rift generally is almost perpendicular to the grain and usually coincides with the direction of the longitudinal joints.[11] The grain usually is parallel to the foliation planes in the rock. In quarrying, the plane of difficult parting is termed the *hardway plane*. The three planes mentioned are almost mutually perpendicular. Their existence is often attributed to microscopic fissures caused by cooling or to the parallelism between flat particles of mica and feldspar ("orientation"). In sedimentary rocks, rift generally coincides with the bedding plane caused by shearing stresses similar to those occurring in the contact plane between two timber beams (Fig. 2.12c). Definitions of rift, grain, and hardway may vary locally.

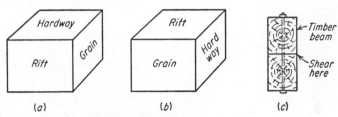

FIG. 2.12. Rift and grain in rocks.

2.16. Folding in Rocks. Both vertical and horizontal forces acting at the earth surface produce distortions of rock masses termed *folds*. Some folds are formed when rocks are *bent* (bending folds) by vertical stresses originated by magmatic forces. These forces deform and stretch the rock mass with a lengthening of the earth surface. If horizontal stresses act at the ends of a rock layer, they may produce *buckling* of the strata with a foreshortening of the original rock surface (buckling folds). Horizontal forces uniformly distributed along the length of a layer produce the same effect. If a rock under stress acts as a viscous fluid, *flow folds* are produced. If the rock layer contains numerous, closely spaced, vertical sheeting planes, either type of folding stress on the layer may result in a *slip* or *shear fold*. The shear type is not regarded as a true fold, as although an

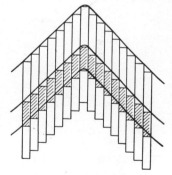

FIG. 2.13. Shear fold (schematic sketch).

apparent curvature develops in the rock layer, careful examination will disclose it to be the result of vertical slip movements between closely spaced shear zones (Fig. 2.13).

The simplest kind of a bending fold is the monocline (Fig. 2.14b). If a flexible beam is firmly embedded at points M and N (Fig. 2.14a) and support M sinks, a point of inflection (counterflexion) A is formed and the beam deflects as a monoclinal fold. Such folds often are found in regions of low topographic relief that are underlain by nearly horizontal beds; the analogy of Fig. 2.14a suggests that they were formed as a result of the sinking of a part of the basin. The term "monocline" also is used to describe a layer of rock which has a practically uniform dip in one direction.

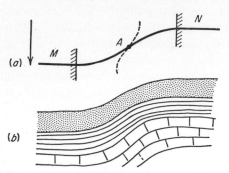

FIG. 2.14. (a) Settlement of a support of an embedded beam. (b) Monoclinical fold.

Figure 2.15c represents an upfold, or *anticline*, and a downfold, or *syncline*. They may be complementary (as shown) or may occur individually. The buckling of overloaded columns may be considered as an engineering analogy of individual anticlines and synclines. An overloaded column may assume either of the shapes shown in Fig. 2.15a and b; and theoretically, a buckled column may have even more than the

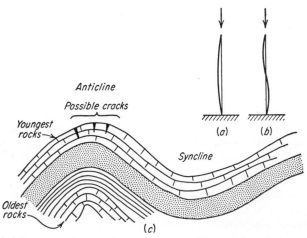

FIG. 2.15. (a), (b) Buckling of overloaded columns. (c) Folding: anticline and syncline.

two waves shown, just as is often the case of anticlines and synclines. This analogy shows that both individual anticlines and synclines have been produced by a horizontal force acting at the ends of a layer or several superimposed layers. This is a true buckling fold. In a combined "anticline-syncline" system, the origin can be traced to the action of horizontal shearing stresses that respond (as friction does) to the action of some major tensile force at the upper strata of the earth crust, such as an over-

all shrinkage drag.[30] In fact, according to the theories of mechanics, a
change in the shape of a mass—which is very pronounced in this case—is
due to the action of shearing
stresses and not to compression,
as such. It should be added that
the action of the horizontal shears
may be visualized as that of a large
horizontal couple, *which has to be
balanced* by a vertical couple. The
upward-acting force of the vertical
couple causes a system of vertical

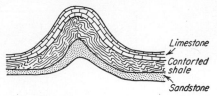

FIG. 2.16. Distortion in soft rock due to
folding.

stresses and an upward warping (anticline), whereas the downward-acting
force of the couple is responsible for creating the syncline.

There is tension at the top of an anticline and at the bottom of a
syncline. As there is less resistance to the upward movement of an anticline than to the downwarp of a syncline, there is a greater possibility for
cracks to develop at the top of an anticline (Fig. 2.15c).

The cross sections of folds are not always formed by smooth curves as
in Fig. 2.15; the beds may be locally thinned, twisted, and contorted,
especially in folds in weak sedimentary materials such as shale (Fig. 2.16).
Where soft rocks are enclosed by hard rocks, the soft rocks are distorted
and folded, often in contorted forms having little or no relation to the
main warping of the mass.

It should be noted that soft rocks which deform under stress are termed "incompetent" whereas rocks that buckle and fracture are termed "competent." These
terms should not be confused, however, with similar terms describing the bearing
capacity of foundation rocks.

The terms anticline and syncline describe convex and concave flexures, respectively,
regardless of the stratigraphic sequence of the beds. According to a strictly technical definition, however, in an anticline the oldest beds are near the core of the fold
and in a syncline the youngest beds are at the core.

Terminology. The geological terms that describe the parts of a fold are as follows
(Fig. 2.17): The flanks, or sides, are the *limbs;* the *axial plane* (or axial surface) is the
surface dividing the fold in two approximately equal parts. If this surface is a vertical plane, the fold is *symmetrical;* if the surface is tilted, the fold is *asymmetrical* or
inclined (Fig. 2.18a). If one limb of the fold is turned past the vertical, the fold is
said to be *overturned* (Fig. 2.18b). The intersection of the axial surface with crest of
an anticline or the trough of a syncline is the *axis* of the fold (Fig. 2.17). The slope
of the axis, i.e., the angle made by it with the horizontal, is called the *pitch* or *plunge*
of the fold (Fig. 2.17). The amount of vertical deformation caused by the folding is
termed the *closure* of the anticline (Fig. 2.17). If an anticlinal-type fold has approximately equal dips on all sides away from the center (or highest point), the fold is
termed a *dome* and is said to have "quaquaversal" dip; conversely, if a synclinal-
type fold has equal dips in all directions down toward the center (or lowest point),
it is termed a *basin*. (The terms "geosyncline" and "geoanticline" refer to very
large depressions or very large elevations, respectively, in the earth's crust and may

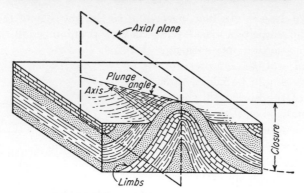

FIG. 2.17. Symmetrical plunging anticline.

(a)   (b)

FIG. 2.18. Asymmetrical anticlines.

be many miles in areal extent. The use of these terms does not necessarily imply, however, that the rocks within them have been folded.)

2.17. Faults. Figure 2.19 shows a rock deformation known as a *fault*. In the simplest case, rock may be fractured practically along a plane; hence the commonly used term *fault plane*, although in nature no fault surface is a plane. The rock masses adjacent to the fault plane are fault walls. The rock mass located above the fault plane is the *hanging wall*, and the rock mass below the fault plane is the *foot wall*. In a *normal fault* (Fig. 2.19a) the dip of the fault plane *usually* is over 45° and the hanging wall appears to have moved down. In a *reverse fault* the hanging wall appears to have risen and the dip of the fault plane *usually* is less than 45° (Fig. 2.19b).

In both cases the fracturing along a fault surface has the characteristics of a *shear failure*. Besides the shear movement proper, the fault walls may move away from each other, and in many instances instead of a single fracture, there is a damaged *fault zone*, frequently 10 to 20 ft wide and sometimes much larger. The walls of a fault also may move along each other, sometimes for a considerable distance. In some spectacular cases, brooks have been offset to considerable distances, although originally they ran perpendicular to the fault zone. In a reverse fault the ground surface is actually shortened, whereas it is lengthened by a normal fault. Generally, it is impossible to ascertain which of the fault walls have moved with respect to the other. Faults may develop gradually, in a slow manner, or may occur suddenly. Sudden faulting is generally accompanied by an earthquake (Chap. 18).

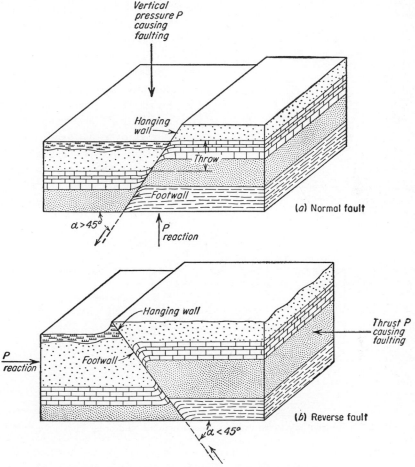

FIG. 2.19. (a) Normal fault. (b) Reverse fault.

In a *cylindrical fault* the rock mass rotates in such a way as to produce a reverse fault near the ground surface whereas at a depth the fault is characteristically normal (Fig. 2.20). The various terms used in describing and measuring a fault are shown in Fig. 2.21. An *overthrust fault* is one in which the hanging wall has actually thrust over the foot wall. Other fault terminology may be found in Ref. 11.

Causes of Faults. Since compressive (or tensile) stresses and the shear stresses act simultaneously, a fault, which is basically a shear failure, in reality may be caused by pressures or tensions. Thus, a normal fault may be produced by a vertical pressure (Fig. 2.19*a*), and the reverse fault may be due to a horizontal thrust (Fig. 2.19*b*). Among other possibilities, a vertical pressure may be due to the removal of the vertical support at a depth, e.g., if magma flows away from under a portion of the earth crust. Soft rocks and noncohesive granular materials are very sensitive to removals of such vertical support. "Gravity faults" occurring in such cases are akin to landslides (Chap. 18). Tension in deeper strata may act practically the same as the removal of the vertical support. Tensile forces of all kinds in the upper strata are favorable to fault occurrence because of the low tensile strength of rocks.

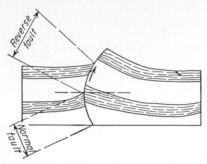

FIG. 2.20. Cylindrical fault. (*After Hills.*[11])

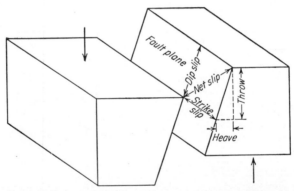

FIG. 2.21. Terminology to be used in describing and measuring fault movement.

Mechanics of Faults. The fault planes are analogous to the shear surfaces (or failure surfaces) in an unconfined compression or triaxial compression test (Sec. 2.9 and Fig. 2.1). The angle made by these surfaces (1) is acute in the case of brittle rocks and (2) tends to be obtuse in plastic or ductile rocks before the final failure in the form of a barrellike bulging occurs. If a complete analogy between the laboratory results and the actual faulting exists, the following conclusions can be drawn from Fig. 2.22*a*, *b*, and *c*. The first two parts duplicate a normal fault; the third corresponds to a reverse fault. The angle made by the shearing surfaces with the horizontal is large (45° and over) in the case of Fig. 2.22*a*, which correspond to a steep dip and steep fault planes in a normal fault. The same reasoning would lead to a conclusion that in the case of brittle rocks, the dip of reverse faults would be less than 45°. In the plastic and ductile rocks the opposite would be true, however; namely, in normal faults the dip would be less than 45°, whereas the dip of reverse faults would be over 45°. However, a field investigation may find that the dip in a normal fault in brittle rock is less than 45° and over 45° in a reverse fault. This probably would mean that the fault occurred when the rock was plastic or ductile and became brittle afterward. Such may be the case of some igneous rocks in which minor faults occur while the igneous body is in the final stage of cooling and thus still under the action of the magmatic forces. The depth of the igneous layer below the surface during the faulting is also of importance, as many rocks which are brittle under ordinary conditions

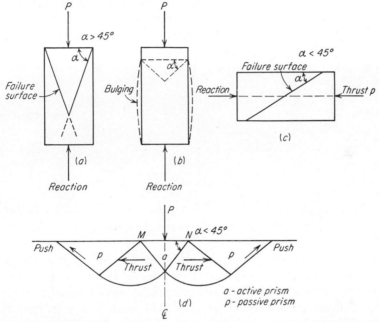

FIG. 2.22. (a), (b), (c) Shear failures. (d) Prandtl's shear pattern (P is uniformly distributed load).

of pressure and temperature become plastic if buried at great depths. A sedimentary rock may have been faulted when there was comparatively little cementing material between the grains, but the secondary (i.e., subsequent) cementation by silicates, carbonates, or iron oxides made it brittle.

Figure 2.22d shows a shear pattern proposed by the German scientist Prandtl.[15] Under the action of the vertical load P uniformly distributed along the surface MN, the active prism a pushes out the passive prism p. Essentially the Prandtl pattern used in this case gives the same information as the compression tests. The pattern also shows how a vertical load P may cause a horizontal thrust within a mass.

2.18. Field Location of Faults. The first step in locating a fault in the field is a surface examination. If it is suspected that a fault may be hidden by a thin covering of soil, some "hit-and-miss" digging should be done. Generally, faults are most easily recognized in stratified rocks (when outcrops occur) by the offsetting of strata.   If a fault is perpendicular to the strike of the bedding or is oblique in relation to that strike (oblique faults), such offsets are quite apparent, but in the *strike faults*, which are parallel to the bedding plane, there are no apparent offsets. In the latter case, the presence of a fault may be established only from occasionally found (1) slickensides (polished and often striated or scratched surfaces that occur in rock or clay as a result of friction, Fig. 2.23); (2) gouge, or pulverized rock that looks and feels like clay; (3) brecciation (Sec. 1.9); and (4) crushing of the rock.   Depending on local conditions, the surface studies may be supplemented by subsurface work in the form of trenches, bore holes, shafts, and tunnels or by the use of geophysical methods (Chap. 6).

FIG. 2.23. Slickensides in clay (disregard tire prints in foreground).   (*Courtesy of Dr. J. L. Sherard.*)

In igneous rocks, faults may be detected by abrupt offsets in small stringers of pegmatite, calcite, or other mineral bands in the rock.   In massive igneous rocks without such mineralization or injections, it may be very difficult to determine whether a fracture is due to faulting or jointing.   From the engineering point of view, however, such determination may not be necessary, *unless brecciation or gouge seams occur*. Airphotos (Chap. 7) may aid in the detection of faulting in all rock types, particularly when there is a heavy vegetal cover.   In metamorphic rocks, fault detection may be very difficult without the aid of an expert stratigrapher and petrographer if the above-mentioned obvious fault signs are not present.   One reason for detecting faults in igneous and metamorphic rocks is to determine if incompetent (unsound) rocks have been displaced by faulting and lie under the proposed foundation.

Another aid in locating faults is to study available geologic publications on the area in question.   Very often, the general trend, frequency, and possibility of fault occurrence are given.   Breccia, gouge, or slickensides on pieces of core from a bore hole indicate that the bore hole has passed through a fault, but the strike and often the dip of the fault cannot be determined because the rotation of the drill has twisted and disturbed the core.   Faults encompassing very large areas often are recognized from the measurement of numerous strikes and dips in the region and plotting them on the map.   Abrupt changes in direction of the strikes in a certain zone indicate some disturbance of the beds.

Folds can be recognized by obvious flexures in outcrops or by measuring dips and strikes and plotting them on a map.   A large anticline or a large syncline, termed "anticlinorium" and "synclinorium," respectively,* usually can be detected only by such mapping or by an examination of airphotos.   For further information on the recognition and study of folds and faults, see Refs. 16 to 19.

Determination of the Direction of Fault Movement.   Although the displacement in the beds on both sides of a fault can often be seen, it may be difficult to determine in which direction the blocks have moved.   The fact that bedding is continuous over a long distance may be misleading, as a mass several thousand feet long may have been downdropped by faulting.   The determination of direction of movement is of importance in engineering work as can be demonstrated by the following examples: (1) A sequence of sedimentary rocks may contain a highly pervious sandstone, and faulting may have caused this pervious layer to appear only at one abutment of a damsite.   It then becomes necessary to know if this pervious layer is present at a shallow depth under the other abutment or was moved upward and subsequently removed by erosion.   If the latter were the case, remedial measures to prevent seepage of water from the reservoir through this layer would be unnecessary.   (2) Faulting may have introduced very soft rocks below a large building site whereas a surface study on nearby outcrops may have indicated that generally the rocks under the building site were competent.

One method of determining direction of movement is to slide the hand gently along the surface of a fault parallel to the direction of the slickensides.   If the surface feels rough, the block of rock under the hand has moved in the same direction as the sliding motion of the hand.   Conversely, if the surface feels smooth, then the rock face has moved in the direction opposite to the sliding motion of the hand.   This method is not foolproof, as weathering may remove or smooth the slickensided surface or may have caused it to be rough in all directions.   However, visual inspection in such cases usually will disclose whether or not there are sufficient slickensides present to be used for this determination.   Another more certain method can be used if feather joints occur along the edges of the fault.   Feather joints, in reality, are tension joints that

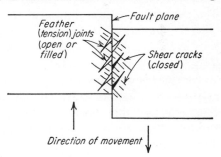

FIG. 2.24. Use of feather joints to determine direction of fault movement. (*After Cloos.*[14])

develop in the zone of deformation between two masses that have moved with relation to each other.   The acute angle between the feather joint and the fault plane generally points in the direction that the block (in which the joint occurs) has moved.[14]   This is illustrated in Fig. 2.24.   The feather joints generally are open or filled with minerals

* The plurals of these words are "anticlinoria" and "synclinoria," respectively.

or weathered materials, whereas the "shear" joints that also occur adjacent to the fault plane (Fig. 2.24) generally are tightly closed fractures.

2.19. Significance of Faults and Folds in Engineering.   Fractures in rock masses accompanied by differential displacements on both sides of the fracture often cut a site irrespective of the dip and strike of the rock. Search for faults is not always effective, and faults are discovered sometimes during construction or even afterward, resulting in a considerable increase in cost of the structure.   Faults may be deeply buried, and if the excavation floor is intersected by small faults containing gouge and brecciated rock, in many cases it is advisable from both technical and economic viewpoints to abandon the site.   In other cases, if a fault is disclosed when the bottom of the excavation has almost reached the design elevation, the site may be made usable by removing a large portion of faulted rock, obviously with a considerable increase in the cost of both earth and concrete work.

A point of great concern to engineers is the determination of whether a fault (in his building site) is "active" or "inactive," or "passive." *Active* faults are those in which movements have occurred during the recorded history of man and along which further movements can be expected at any time (such as the San Andreas and some other faults in California).   *Inactive*, or *passive*, faults are ruptures that have no recorded history of movement and thus are assumed to be and *probably* will remain in a static condition.   In some localities, e.g., in Nevada, both types are found.   Unfortunately, from an engineering (and geological) viewpoint, it is impossible to state definitely if an apparently inactive fault will remain in that condition.   The fault may reopen, either because of a new strain accumulation in the locality or from the effect of earthquake vibrations.   It is reassuring to a certain extent to know that in the United States there are no known instances of failures of dams which were built across active faults.   This does not mean, however, that such an accident cannot occur.

From the basic products of faulting, *gouge* is probably of the most concern in foundation problems.   This is usually an impervious material and may hinder or stop the movement of ground water from one side of the fault to the other and thus create disastrous hydrostatic heads, e.g., if encountered in a tunnel (Chap. 9).   Also, it may reduce the coefficient of sliding friction along the fault plane; thus any heavy load (such as a building) placed upon beds overlying a gouge seam may start translating laterally, ultimately causing failure.   The presence of soft breccia may cause sudden "squeezes" in a tunnel that intersects a fault (Chap. 9).

Of the various types of folds, the synclines are perhaps the most significant in engineering because of their capacity to convey and accumulate fluids.   Serious water problems may arise in the construction and maintenance of tunnels intersecting synclines containing water-bearing strata (Sec. 9.8). If such a syncline is discovered before the design period, the elevation of the planned tunnel may be changed in order to place it in drier strata. In deep highway or railroad cuts analogous water problems arise that may create continuous worries to the maintenance engineers. In foundations proper, folds are not so critical as faults. Occasionally, folds may influence the selection of a damsite. For example, when the reservoir is located over a monocline containing pervious strata, there may be excessive seepage if the monocline dips downstream. If the monocline were to dip upstream, the reservoir probably would have little seepage, providing, however, that the monocline contains some impervious layers such as shale. The significance of monoclines in excavation work is discussed in Chap. 13.

REFERENCES

1. Bowles, O.: "The Stone Industries," McGraw-Hill Book Company, Inc., New York, 1939.
2. Kessler, D. W., H. Insley, and W. H. Sligh: Physical, Mineralogical, and Durability Studies on the Building and Monumental Granites of the United States, *J. Research Natl. Bur. Standards, Research Paper* 1320, vol. 25, 1940.
3. Obert, L., S. L. Windes, and W. I. Duvall: Standardized Tests for Determining the Physical Properties of Mine Rock, *USBM Rept. Invest.* 3891, 1946.
4. Griffith, J. H.: Physical Properties of Typical American Rocks, *Iowa State Coll. Eng. Expt. Sta. Bull.* 131, March, 1937.
5. Windes, S. L.: Physical Properties of Mine Rocks, Part 1, *USBM Rept. Invest.* 4459, 1949.
6. Physical Properties of Some Typical Foundation Rocks, *USBR Concrete Lab. Rept.* SP 39, 1954.
7. Birch, F.: Handbook of Physical Constants, *GSA Spec. Paper* 36, 1942.
8. Adams, F. D., and E. G. Coker: An Investigation into the Elastic Constants of Rock with Reference to Cubic Compressibility, *Carnegie Inst. Wash. Publ.*, 1906.
9. Obert, L., and W. J. Duvall: The Microseismic Method of Predicting Rock Failure in Underground Mining, *USBM Rept. Invest.* 3797 and 3803, 1945.
10. Wegel, R. L., and H. Walther: Internal Dissipation in Solids for Small Cyclical Strains, *Physics*, vol. 6, 1935.
11. Hills, E. S.: "Outlines of Structural Geology," John Wiley & Sons, Inc., New York, 1953.
12. Swanson, C. O.: Notes on Stress, Strain, and Joints, *J. Geol.*, vol. 35, pp. 193–223, 1927.
13. Cloos, H.: Tektonik und Magma: Bd. 1', *Abhandl. Preuss. geol. Landesanstalt*, vol. 89, 1922.
14. Cloos, E.: Feather Joints as Indicators of the Direction of Movements on Faults, Thrusts, Joints, and Magmatic Contacts, *Proc. Natl. Acad. Sci. U.S.*, vol. 18, pp. 387–395, 1932.
15. Prandtl, L.: Ueber die Härte plastischer Körper, *Nachr. Ges. Wiss. Göttingen*, Math. physik. *Kl.*, 1920.
16. Louderback, G. D.: Faults and Engineering Geology, "Berkey Volume," 1950.

17. Willis, B., and R. Willis: "Geologic Structures," 3d ed., McGraw-Hill Book Company, Inc., New York, 1934.
18. Lahee, F. H.: "Field Geology," 5th ed., McGraw-Hill Book Company, Inc., New York, 1952.
19. Report of the Committee on the Nomenclature of Faults, *Bull. GSA*, vol. 24, pp. 163–186, 1913.
20. von Moos, A., and F. de Quervain: "Technische Gesteinkunde," Birkhaeuser Verlag, Basel, Switzerland, 1948.
21. Bernard, P.: Mésure des modules élastiques et application au calcul des galéries en charge, *Proc. 3d Conf. on Soil Mech. and Foundation Eng.*, vol II, Zurich, 1953.
22. Bellier, J., *et al.*: Compressibilité du rocher sous les appuis des barrages, *Proc. 3d Conf. on Soil Mech. and Foundation Eng.*, vol. I, Zurich, 1953.
23. Nonveiller, E.: The Determination of the Deformation of Loaded Rock in Tunnels, *Proc. Yugoslav Soc. Soil Mech. Meeting*, Bled, 1951.
24. Hörlinmann, S.: Kolbenlose Druckpressen (Druckkissen) im Bauwesen, *Bautechnik Arch.*, vol. 1, Berlin, 1949.
25. Brown, P. D., and J. Robertshaw: The In-Situ Measurement of Young's Modulus for Rock by a Dynamic Method, *Géotechnique*, vol. 3, 1953.
26. Reich, H.: Geologische Unterlagen der angewandten Geophysik, "Handbuch der Experimentalphysik," vol. 25, part 3, 1930.
27. Lieurance, R. S.: Stresses in Foundation of Boulder (Hoover) Dam, *USBR Tech. Memo.* 346, 1933.
28. Olsen, O. J.: "Residual Stresses in Rock as Determined from Strain-relief Measurement in Tunnel Walls," University of Colorado Thesis, 1949.
29. Tincelin, M. E.: Measurement of Earth Pressure in the Iron-ore Mines of Eastern France, *USGS Trans.* 35, 1953 (from *Ann. inst. tech. bâtiment et trav. publ.*, série X, no. 58, pp. 972–990, 1952); also P. Habib and R. Marchand, Measurement of Earth Pressures by means of the Flat Jack Test, *USGS Trans.* 34, 1953 (from *ibid.*, pp. 966–971).
30. Mead, W. J.: Notes on the Mechanics of Geologic Structures, *J. Geol.*, vol. 28, 1920; also F. H. Lahee, "Field Geology," 5th ed., p. 179, McGraw-Hill Book Company, Inc., New York, 1952.
31. Rocha, M., J. L. Serafim, and A. F. da Silveira: Arch Dam Design and Observation of Arch Dams in Portugal, *Proc. ASCE Sep.* 997, vol. 82, June, 1956.

CHAPTER 3

FORMATION AND ENGINEERING USE OF SOILS

It is rare that the civil engineer has to be concerned with depths of 250 or 300 ft from the ground surface; his everyday practice is mostly concerned with a 50-ft depth or less. Most small- and medium-sized engineering structures are built on soils; excavations in the rock underlying soils generally are required only for founding major structures such as large bridges or concrete dams. Airport runways and highways are built predominantly on soils, although deep cuts in rock have been required on modern freeways and turnpikes. The engineer should learn what kind of materials are overlying deep rock and in what shapes (landforms) they are distributed over the surface of the globe. He then is in a position to know how to handle these materials efficiently and economically. In this chapter, various landforms made of soils are discussed.

DESTRUCTIVE PROCESSES IN ROCK

3.1. Processes Acting on the Earth Surface: Landforms.[1–4] A *landform* may be defined as a portion of the earth surface differing by its shape and other structural features from the neighboring portions. Mountains, valleys, plains, and even swamps are landforms.

The various processes that continually are acting on the earth's surface are gradation, diastrophism, and vulcanism. *Gradation* is the building up or tearing down of existing landforms (including mountains). Erosion, for example, is a particular case of gradation by the action of water, air, or ice. *Diastrophism* is a term used to designate the process wherein solid, and usually large, portions of the earth move with respect to one another. The term *vulcanism* refers to the action of molten rock, both on the earth's surface and within the earth. With the exception of vulcanism and sometimes erosion, these processes may take hundreds and even millions of years to change the face of the earth significantly. The sudden eruption of a volcano, however, with the ensuing flow of molten rock (lava) or deposition of volcanic ash, can abruptly change land almost overnight. For example, the city of Pompeii in Italy was buried by the ashes from Mt. Vesuvius in A.D. 79, and a farmer's land was destroyed by a rising volcano in Parícutin, Mexico, in 1943.

Whereas a landform is just a sculptural feature at the earth surface, a *formation* is a certain sequence of strata or *beds* formed by materials with similar characteristics and often named for a given locality (e.g., Trenton

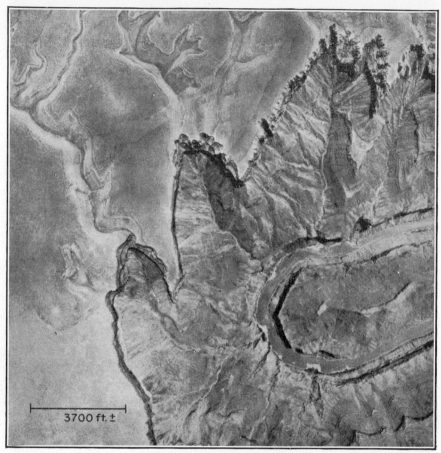

FIG. 3.1. Deep erosion of flat sediments, Colorado River, Arizona. (*USFS airphoto.*)

formation in New Jersey or Orinda formation in California). A formation is not necessarily concentrated in one geographical area but may be totally remote from the locality that gave it its name. The manner in which the materials are placed in the formation is defined by certain geological terms. *Stratification*, or *bedding*, means that the materials are placed in layers bounded by planes approximately parallel to each other. The thickness of beds is generally of a considerable range; an **average** thickness is between 1 and 12 in. The beds may be subdivided by more or less parallel planes into thin *laminae.* (The singular of the word is lamina.) The order of magnitude of the thickness of a lamina is ¼ in.

The term *conformable contact* between formations (conformity) is used when the overlying bed is in its normal geological position with respect to the underlying bed, i.e., the stratification of the overlying bed is parallel to the stratification of the underlying bed, and no appreciable geologic time interval has passed between the deposition of the overlying bed on the underlying bed. (Obviously, a rather short geologic time may be very long from the layman's point of view.) The term *disconformity* is occasionally employed to describe a contact where the overlying beds lie parallel to the underlying beds but a long period of geologic time is known to have passed before the deposition of the overlying beds. If a long period of geologic time has passed before the overlying beds were deposited, the contact is said to be *unconformable.* An unconformity often may be represented either by a deeply weathered or severely eroded surface on the underlying beds at their contact with the overlying beds or by the fact the strata in the underlying beds slope at a considerably different angle from the strata above the contact.

3.2. Products of Rock Destruction: Residual and Transported Soils.
The products of rock destruction are spread all over the globe. A concentration of such products in a rather short distance can be observed in mountainous regions of the so-called "young geology." This phrase indicates that the destructive processes are still proceeding at a considerable rate, such as in the San Rafael Mountains in southern California. If the observer descends a trail from a mountain peak in such a region, he notices first that the trail is very steep near the top, it gradually flattens out, and at a great distance from the peak, sometimes many miles away, the *gradient* (slope) of the trail becomes quite mild. Accordingly, next to the top, large *boulders* are lying; somewhat down the slope, the boulders become mixed with smaller *cobbles* (Sec. 1.10). Below this elevation, cobbles only or mixed with *gravel* are observed. Descending further, gravel mixed with *sand* is seen, and then sand deposits alone are encountered. In this particular case, the products of rock destruction have been brought downhill by violent storm water, which, however, decreases its velocity and its carrying capacity as the gradient decreases. For this reason, at a certain distance from the peak, the sand grains fall to the bottom of the stream carrying them; finer particles such as *silt* and *clay* are carried farther before being deposited in a basin such as a pond, lake, or sea remote from the original mountain.

Exposed rock surfaces without soil cover are *outcrops.* The products of rock destruction spread, as previously explained in this section, are *debris.* As noted in Chap. 1, the products of rock destruction of gravel size and smaller are termed *soils* in engineering usage, whereas cobbles and boulders are considered to be fragments of rock.   The soils may be residual or transported.   Residual soils are those which remain in place directly over the *parent material* from which they are derived.   Transported soils, on the other hand, are those which have been moved from the original bedrock and redeposited at another location.   Transporting agencies may be ice (glacial soils), wind (aeolian soils), water (alluvial or fluvial soils), and finally the force of gravity (colluvial soils such as talus). *Talus* is spalled material fallen from rock outcrops and accumulated at the toe of a slope.   Soils formed from volcanic eruptions, such as ash, also may be considered as transported.

3.3. Weathering.[2]   Where the composition or structure of rocks near or at the earth's surface is altered by physical and chemical agents resulting from atmospheric processes, the rock is said to be *weathered*.   The atmospheric agents so involved are basically *air* and *water*.   The term weathering sometimes is used synonymously with the term *alteration*. Engineering geologists, however, commonly refer to an altered rock as one wherein the chemical composition or structure has been changed by the action of a geologic process originated at considerable depth.   Such a process is, for example, burning or melting of the rock by the extrusion of molten lava or the passing over the rock of acid-containing waters from considerable depths.

The processes of weathering are subdivided into those which cause disintegration and those which cause decomposition.   *Disintegration* refers to weathering of rock by *physical* agents, such as (1) periodical temperature changes, (2) freezing and thawing, and (3) the physical effects of plants and animals on rocks.   *Decomposition* refers to the changes in rocks produced by *chemical* agents, such as (1) oxidation, (2) hydration, (3) carbonation, and (4) the chemical effects of vegetation.

Disintegration.   Thermal destruction of rocks is accomplished by alternate heating and cooling due to the daily or seasonal temperature changes. Such destruction is especially pronounced in deserts, where days are hot and nights are cold.   The rock fails then primarily *by fatigue*, because of the continuous reversal of compression and tensile stresses produced in it by the temperature changes.   If the pores of a rock are filled with water, the water expands at freezing and the rock fails *in tension*, with further widening of the fissures and joints.   In this connection, the so-called hard rocks are often separated into more or less regular blocks along existing joints.   Blocks thus formed may move slightly under the weight of a heavy structure such as a concrete dam.   To ensure a stable foundation for the structure, such blocks often are removed ("scaled").

The disintegrating process is activated by any exposure of the rock surface (primarily in quarries and road cuts or by removal of the soil cover by erosion) and to a minor degree by the burrowing activities of small animals and earthworms. Vegetation is but a minor factor in rock disintegration, though tree roots may actually split even a large rock block.

Decomposition. The various processes that contribute to decomposition of a rock are chemical in nature. *Oxidation* means that oxygen ions are added to the minerals composing the rock; rocks containing iron, for example, are greatly affected by oxidation. Conversely, *reduction* occurs when oxygen ions are removed from the minerals in the rock. The decoloration of certain rocks (a "pseudo-weathered surface") often is attributed to the oxidation or possibly reduction of some of the chemical constituents of the rock. *Hydration* is usually the chemical addition of water to the minerals. Water thus added is structural (Sec. 1.3) and should be distinguished from the type of water that promotes disintegration processes. An important hydration effect is the decomposition of the mineral feldspar in granite to form clay of the kaolinite type. *Carbonation* is the solution of the rock material by water containing a considerable amount of carbon dioxide; it may be detrimental, particularly in the limestone areas. All surface water contains a small percentage of carbon dioxide.

Organic acids, developed when vegetation decays, tend to increase the solution power of the natural water. Some types of vegetable matter, such as lichens, tend to extract certain chemical elements from the rocks. On the other hand, vegetation tends to retain moisture in the rocks and thus protects the rock, decreasing the rate of the rock weathering. Decomposition may proceed to great depths; in Brazil, shales are known to be decayed to a depth of 400 ft.

The way in which rocks are destroyed by weathering processes largely depends on the climatic conditions. Rocks are weathered predominantly (1) by decomposition in warm, humid climates; (2) by disintegration in warm, dry climates; (3) by a combination of both disintegration and decomposition in temperate climates; and finally (4) by expansion of freezing water causing disintegration in the so-called cool, dry climates such as arctic and antarctic.

Wind is a very important atmospheric weathering agent. Wind not only transports weathered rock materials from one place on the earth's surface to another place but also tends to erode existing rocks. The removal of surface materials by the wind is called *deflation*, and erosion by the wind is called *corrasion* and *abrasion*. As a result of the latter process, but of minor concern in engineering, weird forms such as natural bridges are developed in desert regions. Of importance from the engineering point of view is the deposition of loess, sand dunes, and volcanic ash by the transportation effects of the wind (Secs. 3.15 and 3.17).

VALLEYS AND RIVERS

3.4. Valley Formation. Because of topographic irregularities, water running on the surface first creates small rivulets which gradually increase in size. During rainy seasons these rivulets are transformed into larger

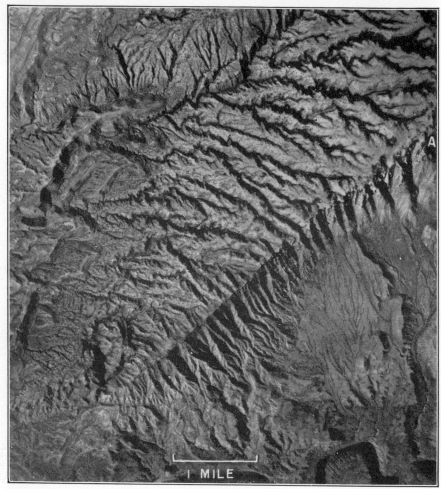

FIG. 3.2. Drainage divide (A), Wyoming. Note development of dendritic drainage pattern, left side of photograph. (*USBR airphoto.*)

streams and torrents; the latter often carry large amounts of water moving with great speed and thus erode deep gorges and canyons. Eventually, the irregular ground depressions thus formed become *valleys* with gentle slopes. From the engineering point of view, a valley is characterized by its longitudinal profile and its cross sections. The longitudinal profile of a valley is the profile along its *thalweg*. A thalweg may be defined as the path of a theoretical water drop that flows, without changing in volume, along the bottom of the valley from the point where the valley originates to its lowest point. The average slope of the thalweg is the average *gradient* of the valley. The cross sections of the valley for engineering purposes are taken from bank to bank perpendicular (normal) to the thalweg.

Because of continuous action by the running surface water, erosion contributes to the growth of the valley during its whole life span (Fig. 3.1). The lengthening of the valley is accomplished mostly by head erosion, or gradual destruction of rocks and soil masses in the upstream direction. This retreating action is accompanied by the destruction of the steep slopes of the original gorges and the downstream transportation of the eroded material. This backward movement of the valley stops, however, when a divide (Fig. 3.2), i.e., a conspicuous downbreak of the thalweg, is reached and water starts to flow in two opposite directions. Other obstacles such as rocks that are not easily eroded may also stop the progressive upstream (backward) growth of the valley.

While cutting backward, the stream flowing in the valley also wears its bed downward, at least locally. In the course of time, the floor of the valley will be cut down to the ground-water level (water table). In such a case, the reserve of water in the stream would depend not only on the runoff in the drainage area but also on the support from the ground water. Thus, a *permanent* stream is formed that flows in both dry and wet seasons. *Intermittent* streams periodically dry out but have a permanent water table below the stream bed. *Ephemeral* streams contain water only after rainfall or snow melting and have no connection with the water table. It should be noted that if the valley of a small stream is filled with earth material and its dry surface is used for engineering purposes, for instance for a housing project, the water flow under the fill may persist for many years. In such a locality the flow of water may be diverted to a new, artificially built channel. If this locality is subject to torrential floods by cloudbursts, even far upstream, too much water may be introduced in the new stream channel and, thus, the water again may take its original course.

Generally, the valley is much wider than the stream flowing in it, except during flood periods. The width of a valley varies according to the erodability of the material on the valley slopes. The depth limit of the valley, or *base level*, is controlled by the water level in the body of water (ocean, sea, lake) into which the stream flows. Only the lower end of the valley attains base level; the rest of it is higher, except, perhaps, for local depressions.

3.5. Cycle of Valley Erosion. A river flowing along the valley erodes the material in its bed and, because of the surface runoff, indirectly contributes to the erosion of the walls of the valley. The materials thus derived are transported in the form of the so-called "sediment" and finally deposited. Rivers and the valleys along which they flow may be *youthful, mature,* and *old.* At these three stages in the life of a river or valley, its longitudinal profile, cross section, and plan undergo gradual

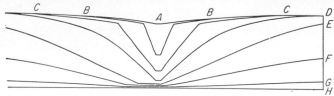

FIG. 3.3. Gradual changes in the transverse section of a valley: AA = original profile, BB = youth, CC = late youth, DD = early maturity, EE = middle maturity, FF = late maturity, GG = old age, HH = base level. (*From F. H. Lahee, "Field Geology," 4th ed., McGraw-Hill Book Company, Inc., New York, 1941.*)

FIG. 3.4. Meandering stream in a youthful to mature valley, Colorado. Note oxbow development at right center of photograph. (*Photograph by P. N. Davies.*)

changes. At the *youthful* stage of a valley, its longitudinal profile is irregular and contains rapids, falls, and even lakes because of local obstructions, such as landslides, and its cross section tends to be V-shaped (Fig. 3.3). The plan of a youthful valley or river is somewhat angular or zigzag. As erosion progresses, the river reaches *maturity.* The irregularities gradually disappear, and the plan acquires the shape of a smooth sinusoidal curve (Fig. 3.4). The longitudinal profile also becomes reduced in gradient, decreasing gradually toward the mouth of the river

(Fig. 3.5).   The valley at its mature stage is wide; its slopes are flatter than in its youth and often covered with talus.

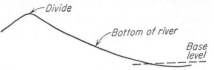

Periodical floods (freshets) contribute to the gradual widening of the valley until at its *old age* it becomes a wide peneplain.   (*Pene* means "almost" in Greek; hence peneplain means "almost a plain.")

FIG. 3.5. Longitudinal profile of the thalweg of a mature valley (vertical scale exaggerated).

Between the floods, the old river meanders, changes it plan, but stays within a certain "meander belt" at the central part of the peneplain.   In

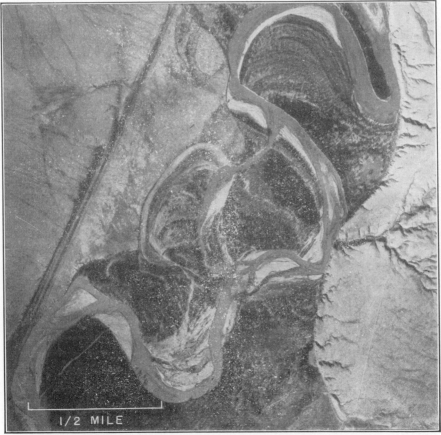

FIG. 3.6. Airphoto illustrating stream meanders, Montana.   (*USBR airphoto.*)

shifting from one location to another, a meandering river may leave behind *oxbow lakes* or abandoned oxbow-shaped depressions (Fig. 3.6).   A classic example of a meandering river is the Mississippi River.[13]

3.6. Regional Erosion.   The description of the cycle of river erosion presented in Fig. 3.3 refers to one river or valley only.   Actually, erosion covers an entire region in which rivers are the principal agents of erosion. The stream pattern is then made up of master streams and their tributaries.   In the case of an individual river, the region under consideration also passes through the stages of youth, maturity, and old age.   Individual streams may be at different stages of life, however.   For instance, a tributary may be young, whereas the corresponding master stream may have already reached maturity.   Figure 3.7*a* represents schematically the cross section of a region in which the erosion process is at its youthful stage.   Broad   divides,   as   shown   by   letter   *D*   in   Fig.   3.7*a*,   separate

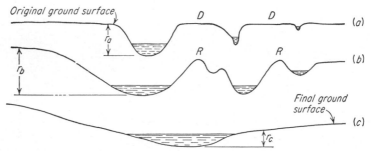

FIG. 3.7. Region erosion: (a) youth, (b) maturity, (c) old age (letter *r* indicates "relief").

V-shaped valleys.   As erosion progresses, divides become narrower and sharper and are gradually reduced to ridges between streams (*R* in Fig. 3.7*b*).   The *relief* r_a, or the average difference of elevations in the region, gradually increases (from r_a to r_b, Fig. 3.7*b*).   Stream patterns become both digital, i.e., like open fingers of the hand, and *dendritic* (Fig. 3.2), or treelike.*   In the next stage, which is represented in Fig. 3.7*c*, relief decreases again (from r_b to r_c), and the entire region is brought to a common slope.   This is the maturity stage.   At old age, the whole region tends to become a peneplain with a limited number of master streams. During geological time, development of such peneplains may occur time and time again in a region.   Each time, a new peneplain cycle is started when the old one is uplifted by diastrophism.

Because of different flow conditions, such as differences in gradient or in bed material, neighboring streams do not always erode their respective valleys equally.   As a result, the stronger stream pushes the divide *D* or ridge *R* (Fig. 3.7) toward its weaker neighbor.   In this way, ridges are eroded vertically and moved laterally by the streams.   The divide between the two streams may disappear completely, thus causing water in both neighboring streams to come in contact; then the head of the weaker stream (or one of its tributaries) becomes diverted into the channel

* For other stream and erosion patterns see Sec. 7.17.

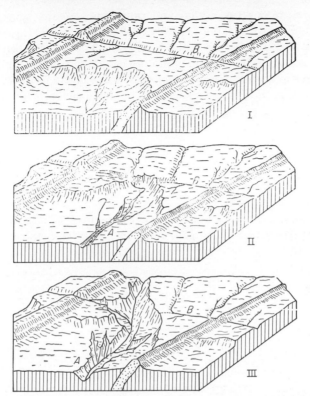

FIG. 3.8. Stream piracy: (I) The tributaries at A are advancing by headward erosion toward stream B. (II) Stream B is captured and diverted to the pirate stream A. (III) The A valley is extended and deepened. (*From W. H. Emmons et al., "Geology: Principles and Processes," 3d ed., McGraw-Hill Book Company, Inc., New York, 1949.*)

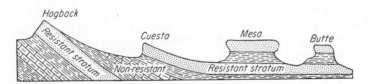

FIG. 3.9. Landforms produced by erosion. (*From W. H. Emmons et al., "Geology: Principles and Processes," 3d ed., McGraw-Hill Book Company, Inc., New York, 1949.*)

of the stronger stream. This phenomenon is known as *river piracy* (see also Fig. 3.8). The details of erosion and of the intermediate sculptured forms through which a region passes before it reaches the final peneplain stage differ considerably with differing environments (Fig. 3.9), climate being particularly critical. There also may be interruptions of the regular erosion process caused, for instance, by diastrophism. Regional uplift may greatly accelerate erosion, but downwarping (down-dropping) of the regions may stop erosion and cause the streams to aggrade, i.e., build up their beds by rapid deposition of the bed load.

3.7. Drowned and Rejuvenated Valleys. The earth's crust does not remain at a constant elevation but occasionally rises or sinks locally. Such changes in the level in the proximity of the ocean or a sea may result in the formation of *drowned* valleys, submerged beneath the sea level.

Examples of drowned valleys are the Hudson River Valley from its mouth to as far upstream as Troy, New York, and San Francisco Bay, California, which represents a large drowned-valley system.

Local rising of the thalweg may substantially increase the gradient of the valley and the velocity of the flow in the stream, with a subsequent intensification of erosion. Such an occurrence, termed the *rejuvenation* of a river, may excavate a new channel in the floor of a valley.

3.8. River Terraces. The geological terms *terraces* and *benches* have practically the same meaning. Both are more or less horizontal surfaces

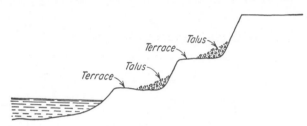

FIG. 3.10. Terraces and talus.

between slopes as in Fig. 3.10 (in which talus is also shown). The term "bench," however, is used mostly to designate such forms in solid rock. Also, terraces are long in comparison with their width, whereas benches are relatively short.

There are various ways in which terraces, particularly river terraces, can be formed. A landslide may produce a terrace. A large river with an alluvial plain at its lower end may incise a new channel in that old deposit. By meandering, it may broaden its flood plain until terraces are the only remnants of the old higher plain (Fig. 3.11). Many river terraces were formed at the end of the glacial epoch (glacial terraces, Sec. 3.13).

The engineering significance of river terraces is considerable, particularly in locating highways and railroads. If two cities or regions that have to be connected by a railroad have considerable difference in elevations, it is often convenient to follow the valleys of the streams or their tributaries flowing between those cities or regions. If the walls of a valley are steep, the construction of a route along that valley may be difficult and expensive. The presence of terraces, however, avoids (or at least decreases) excavation in the walls of the valley and thus reduces the construction costs. In addition, river terraces generally are good sources of sand and gravel for the construction (Sec. 8.9).

FIG. 3.11. River valley in late youth, New Mexico. Note terraces to right center of photograph. *(Photograph by Dr. E. N. Harshman.)*

3.9. Flood Plains and Deltas. Stream deposits are generally termed alluvial deposits or *alluvium.* The deposition is primarily due to the decrease in velocity of the water carrying the sediment. This occurs, for example, when a stream reaches a *flood plain,* a wide flat part of the valley subject to floods when the stream carries an excess of water, e.g., after rains or snow melting, especially when combined with rain. The velocity of water flowing along a flood plain is at a maximum at the channel and at a minimum at the borders of the plain. This reduction in velocity causes sand and silt to deposit at the borders of the plain with formation of ridges or *natural levees* several feet high and occasionally reaching 15 ft or more in height. The flood plain of the Mississippi River, including the natural levees and the low swampy lands, covers about 25,000 sq miles.

Reduction in velocity and, hence, deposition of sediment occur also when the running water of the stream reaches the standing water of the ocean or other basin into which it flows. This causes the sediment load in the stream to drop to the bottom. As a result, a large deposit of alluvial material is formed with its top generally close to the water surface. If the water of the basin contains salts (such as sodium chloride or calcium chloride, which are electrolytes), the finer clay particles flocculate and precipitate, thus increasing the volume of the deposit. As this type of deposition in plan (i.e., observed from above) is a crude triangle (similar to the greek letter delta Δ, with the apex pointing upstream), the deposit

FIG. 3.12. Delta deposits, Strawberry Valley, Utah. (*Courtesy of Dr. R. C. Mielenz.*)

is called a *delta* or *deltaic deposit*.   New channels are continually formed in the region of the delta, which gradually shifts from one place to another. The delta deposits shown in Fig. 3.12*a* and *b* have been formed in fresh water without benefit of electrolytic action and contain mostly sand and silt.

The Mississippi delta in the Gulf of Mexico is one of the largest in the world. Apparently, in former ages, this delta occupied positions both east and west (mostly west) of the present-day channel that passes by the city of New Orleans.   Some of the largest deltas in the world other than the Mississippi delta are the deltas of the Nile (Egypt), Po (Italy), Ganges (India), and Rhone (France and Switzerland).

GLACIAL SOILS AND GLACIAL DEPOSITS

3.10. Glacial Soils: Drift.[5]   There were several periods of glaciation in the Northern Hemisphere when growing ice masses (ice sheets) moved

FIG. 3.13. Glacial drift.

FIG. 3.14. Unsorted till left by glacier, near Fort Niagara, New York. (*Photograph by G. K. Gilbert, USGS.*)

south from the Arctic Ocean and melted away after a certain more or less long time.   The latest glaciation is believed to have occurred during the Pleistocene period and ended roughly 5,000 to 15,000 years ago.   Similar glacial invasions occurred in all the northern parts of the globe and probably in the Southern Hemisphere.   Areas invaded by the ice sheets are *glacial areas*, and those portions of the glacial areas which were actually covered with ice are *glaciated areas*.   The ice left traces of its existence and movement in the form of the present-day glacial and partly aeolian (loessial) soils.   It also changed the topography of the invaded area by creating new landforms, such as moraines, drumlins, eskers, and kames, as discussed hereafter.

FIG. 3.15. Glacial drift (lateral moraine), Roaring Fork River, Colorado.   Note irregular sorting and angularity of boulders.

The term "glacier" is used to define a stream or moving accumulation of ice.*   The mantle of earth and rock material (clay, sand, gravel, boulders) that at the present time covers the area occupied by the glaciers is called *glacial drift* or simply *drift* (Fig. 3.13).   Drift comprises all material deposited by the glaciers either directly by the ice or indirectly by the melt water issuing from the glacier.   *Glacier till* (Fig. 3.14) comprises that part of the material which was deposited only by and underneath the ice. "Stratified drift" is some (but not all) material deposited by melt water. Till and drift in general are heterogeneous and unsorted, with little or no stratification (Fig. 3.15).

3.11. Ice Sheets.   The original ice sheets were several thousand feet thick.   The present-day Swiss Alps continental glaciers may be con-

_____

* There also are "rock" glaciers created by the slow downhill creep of large masses of talus due to the action of gravity and possibly some lubrication by underlying water streams.

sidered to a certain extent as diminutive analogues of the ice sheets of the glacial epoch.[19,20]   The ice structure was crystalline in the whole sheet, and in the upper zones of the sheet there were voids, fissures (crevasses), and tunnels filled with free water, sometimes in the form of streams.   There was also water under the sheet.   This was provided by the streams and brooks crossed by the ice; a small amount of water was also produced by the melting of ice in contact with warmer soil.   Water under the ice sheet served as a lubricant, facilitating its translation, and in those places where drift was trapped in rock depressions (Fig. 3.16), it was moistened and compacted, much in the same way as soil compaction is done at the

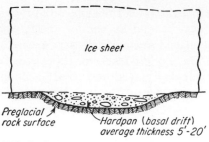

FIG. 3.16. Formation of glacial hardpan.

FIG. 3.17. Granite glacial erratic 89 ft in circumference. $1\frac{1}{2}$ m SW of Hardwick, Mass. (Photo by W. C. Alden, courtesy USGS.)

present time (Chap. 16).   Thus *hardpan* or *basal drift* was formed, a foundation material often as strong as rock.

Close to the southern terminus of the ice movement (e.g., in Connecticut), highway practice, prior to the widespread use of the embankment compaction, indicated an average shrinkage of glacial material extracted from the borrow pit and placed into embankment at about 12 per cent by volume.   Thus, it can be concluded that the drift at that region was simply dropped from the melting ice, which agrees with the commonly accepted hypothesis.

The thickness of drift varies considerably.   For example, in southern Wisconsin, it has been estimated at 45 ft; in Illinois, at 115 ft; in Iowa, at 150 to 200 ft; in central Ohio, at 95 ft (although a maximum thickness of 763 ft has been noted).   Because of the hardness and resistivity to abrasion of the rocks in New England, the thickness of the drift cover there is relatively insignificant, being only 15 to 20 ft.

The earth mass in front of the moving ice sheet was "harrowed" by high shearing stresses.   Because of the plasticity of the lower portion of the ice sheet, boulders, gravel, and soil materials from the harrowed zones were incorporated into that portion.   The boulders were also able to sink through the plastic zone or be wrapped with plastic ice and pushed forward (Fig. 3.17).   In passing over a rock outcrop, hard rock fragments in the ice either abraded or scratched it.   In the latter case, the lines produced at the surface of the rock—*striae* or *striations*—indicate the direction of the ice movement (Fig. 3.18).

3.12. Unstratified Glacial Deposits.   Landforms left by both the ice and melt water usually are unsorted and unstratified.   These are moraines, drumlins, eskers, and kames.

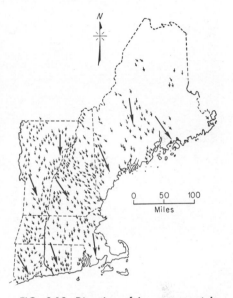

FIG. 3.18. Direction of ice movement in New England as indicated by striated bedrock. (*From W. H. Emmons et al., "Geology: Principles and Processes,"* 3d ed., McGraw-Hill Book Company, Inc., New York, 1949.)

Terminal and Ground Moraines. It is believed that the front end of the ice sheet started to melt when it reached a sufficiently warm region. The arriving drift was pushed through it and gradually piled up to form an *end* or *terminal moraine* (Fig. 3.19).   This is characterized by a generally discontinuous crest or ridge roughly perpendicular to the direction of the ice motion.   The height of a moraine crest rarely exceeds 100 to 250 ft.[5]

Ground or Recessional Moraines.   These were formed by the drift dropping when the front of the ice sheet gradually receded because of the ice melting.   Sometimes this drift is piled up in irregular oblong masses, with axes parallel to the original, southernmost ice front.   These masses are rarely continuous, however, and often, because of the occasional depressions between the recessional morainal masses, a *knob-and-kettle* topography is developed. This topography is characterized by many irregularly placed knobs of drift interspaced with many depressions usually filled with small lakes. The drift materials in knob masses are

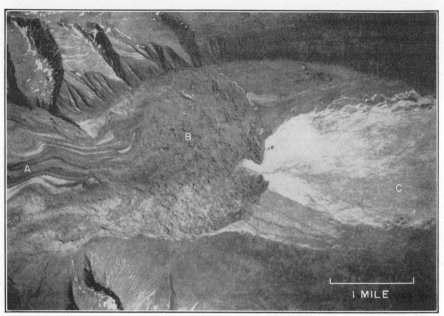

FIG. 3.19. Glacier (A), terminal moraine being deposited (B), and braided outwash stream (C), Alaska. (*USAF airphoto.*)

exceedingly heterogeneous and range from clay to boulders, without good gradation, however. Drift also was dropped laterally along the edges of an ice sheet, and thus *lateral moraines*, similar in constitution and appearance to the terminal moraine, were formed. The axes of the lateral moraine are roughly parallel to the direction of the ice motion.

Drumlins are elongated hills of glacial drift with ellipse-like contour lines, in the majority of cases with a steep northern slope and gentle southern slope (Fig. 3.20). The major axis of the ellipse coincides with the general direction of the moving ice. The sizes of drumlins vary; e.g., they may be 1 mile long, 1,200 to 1,800 ft wide, and 60 to 80 ft high.[5] The material composing a drumlin is generally rich in clay, although some drumlins contain considerable amounts of sand. In some areas, drumlins consist of bedrock covered with till (rock drumlins). Groups of several drumlins are called "drumlin fields." Such fields are found in west-central New York, east-central Wisconsin, and southern New England, particularly in Massachusetts. Drumlins probably have been formed

FIG. 3.20. Drumlin, south of Newark, New York. (*Photograph by G. K. Gilbert, USGS.*)

FIG. 3.21. Esker. Dodge County, Wisconsin. (*From W. H. Emmons et al., "Geology: Principles and Processes," 3d ed., McGraw-Hill Book Company, Inc., New York, 1949.*)

from irregular accumulations of excess drift under the moving ice and overridden by the latter.

Ice Contact Deposits. This term refers to isolated masses of drift accumulation in depressions in immediate contact with thin, wasting ice. *Eskers* (Figs. 3.13 and 3.21) and *kames* (Fig. 3.22) were formed by the bed loads of the streams that flowed in the glacier. These deposits are usually characterized by an extreme range and abrupt change in the grain size and by irregular layering.

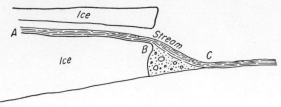

FIG. 3.22. Formation of a kame. Part *AB* may produce an esker, and part *BC* produces a kame.

3.13. Stratified Glacial Deposits. Glacial deposits of this type usually occurred in the lakes formed in close association with glaciers (glacial lacustrine deposits). Glacial lakes were formed when melt water from the glacier was blocked by the ice and filled depressions to form lakes or when streams were blocked in their natural flow, for instance, by accumulated outwash. The usual characteristic of the glacial lake deposits is the extreme fineness of the materials, which usually are clay, silt, or rock flour (finely pulverized rock resulting from the abrasion of rock by the moving ice). Huge glacial clay deposits possess practically horizontal surfaces, and clays themselves are often *varved*, i.e., consist of intermittent, very uniform laminae of silt and clay, the silt being light-colored (generally buff to pink) and the clay darker colored (generally brownish). The thickness of a varve can vary from minute fractions of an inch to over $\frac{1}{2}$ in. During the glacial epoch varved clays apparently were laid down by melt water that reached the lake during and after the melting season, i.e., in the summer. The coarser particles (silt) settled down quickly, leaving clay in suspension until the following winter. According to this explanation, each pair of varves corresponds to one year of deposition. Varves do not form in salt water, since its electrolytic action makes the clay particles coagulate and settle down together with the silt.

The horizontal permeability of a varve in the silt layer is greater than in the clay layer. Since the vertical permeability of the varved clay is low because of the resistance of the clay layers to the passage of water, the *over-all* horizontal permeability of such materials is higher than the vertical. The shearing resistance of a varved clay is higher normally to the varves than along the varves. Owing to the difference in texture and porosity, the natural moisture content by dry weight (Sec. 4.2) in two adjacent horizontal layers of a varved clay may be very different, e.g., 35.5 per cent in silt and 57.9 per cent in clay (Connecticut River, Hartford, Connecticut).

Glacial Terraces. Sediment carried by the streams in glacial zones settled out in some places because of local decrease in velocity; as the stream narrowed, these deposits formed *terraces* (Fig. 3.23). The shelf or bench of a glacial terrace may be several hundred feet wide, and the slope of the terrace toward the middle of the stream may be (1) steep, as in the case of ordinary river terraces made mostly of granular materials, or (2) rounded, as in predominantly clayey river terraces.

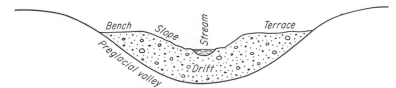

FIG. 3.23. Glacial river terrace.

FIG. 3.24. Glacier and outwash stream, Alaska.

Lateral moraines differ from glacial terraces by their hump-backed shape and complete lack of stratification. Because of their gradual deposition in water, the material in the terraces shows more stratifications than in the lateral moraines, though in both cases the material is heterogeneous and poorly graded.

Fluvioglacial Deposits. These are deposits made by the streams carrying the outwash from the glaciers (Fig. 3.24). Often it is difficult to distinguish between a fluvioglacial deposit and a simple alluvial deposit. However, the proximity of glaciated zones and highly heterogeneous material indicate the fluvioglacial character of the deposit.

3.14. Engineering Problems in Glacial Zones. Glacial deposits contain numerous beds of pervious sand and gravel and may also contain considerable quantities of impervious clay or clayey silt. Sand and gravel beds are often quite conspicuous but also may constitute covered lenslike deposits, either isolated or connected with long stringers of pervious material. Inclusions or pockets of impervious and often highly compressible soft clay may be interlaced with pervious and competent sand or gravel strata. Perhaps one of the most expressive descriptions of glacial materials was given by an eminent geologist who was asked by a group of engineers as to what materials constituted the foundation for a structure to be placed in a certain area covered with glacial drift. He said that the foundation materials in that case could be compared with the contents of a garbage can. He meant, of course, that any type of soil or rock material may be present in a glacial deposit, in almost any form, and in any proportion.

For many structures the high permeability of the underlying materials is an asset rather than a problem. Such are, for instance, highway and runway subgrades that require expensive artificial drainage if the local foundation material is impervious and, conversely, are perfectly safe if built of and in pervious materials. Structures that have to hold water such as dams and their reservoirs, though built on apparently impervious material, may be endangered by the presence of hidden pervious beds. In a general case it is difficult to locate all such beds economically by conventional exploration methods as described in Chap. 6. Therefore, if a large hydraulic structure is to be built, for instance, on a moraine, it should be assumed that the material of the latter is pervious and that suitable design precautions will be required. If, under the circumstances, a prohibitively expensive design is obtained, another location for the structure has to be sought.

Design of *foundations* for light structures in glacial zones generally does not present difficulties. Heavy structures built in glacial zones may be endangered by the possibility of *differential* (i.e., nonuniform) *settlement*. The structure may be built on sand and gravel strata quite competent to bear the loads produced by the structure, but the differential settlement may be caused by the undetected soft clay pockets. The extent of exploration required to answer the queries of the designer as to the possible settlement of a heavy structure built in a glacial zone is difficult to determine. Some suggestions along these lines may be found in Chap. 13.

AEOLIAN SOILS AND AEOLIAN DEPOSITS

3.15. Loess.[2,6,7] Aeolian, or wind-blown, soils may be subdivided into two groups, namely, (1) loessial soils, such as primary and secondary *loess,* and (2) sands forming *sand dunes* (Sec. 3.17).   Loess may be either of glacial or of desert origin.   It is believed that most of the loess in the United States is associated with the glacial age.   Summer winds of high velocity blowing from the melting ice sheet over dry outwash collected

FIG. 3.25. Typical loessial terrain, Nebraska.

FIG. 3.26. Characteristic loess landform, usually called "cat-steps," Kansas.

dust and carried it farther south to be deposited many miles from the ice sheet.   *Primary loess* is that wind-blown material which still is in the same location where it was originally deposited by the wind and has undergone little, if any, chemical decomposition (Figs. 3.25 and 3.26).   *Secondary loess* is that loessial-type soil that after the original deposition either has been transported over short distances by water (or otherwise) or has undergone intense chemical decomposition without practically changing the location, e.g., loess along the Mississippi Valley (Figs. 3.27 and 3.28).

Constituents of Loess.   From the purely scientific point of view loess is defined[8] as "a quartzose, somewhat feldspathic, clastic sediment composed of a uniformly sorted mixture of silt, fine sand, and clay particles arranged in an open, cohesive fabric." True, or primary, loess contains

FIG. 3.27. Airphoto of loess terrain, Nebraska.   *(USDA-PMA airphoto.)*

a high percentage of silt-size materials and is poorly graded.   The secondary loess is composed mostly of clay-size particles; i.e., it is finer than the primary.

Some Properties of Loess.   In nature, loess deposits are of variable thickness, and layers 200 ft thick have been found.   It is very porous.

According to data of the U.S. Army Engineers,[9] a Nebraska loess had a dry unit weight of 85 pcf, which at a density of grains of 2.64 amounts to 49 per cent porosity.

The material is slightly or moderately plastic.   Its permeability in the vertical direction is greater than its horizontal permeability because of the presence of long vertical tubes in the loess fabric (probably casts of plant roots). In most soils, however, the horizontal permeability is greater than the vertical because of stratification.

If a loaded loess deposit is wetted, it rapidly consolidates and the structure constructed on it settles. This property of loess is sometimes termed hydroconsolidation. There are two explanations of this property. Most of the loess in the United States has clay hulls or films around silt grains. Added water lubricates the clay and makes the silt grains slide

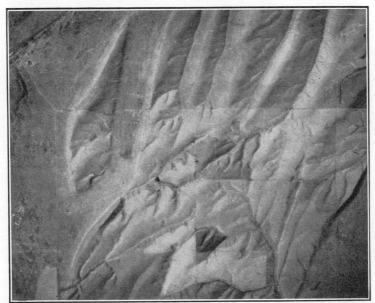

FIG. 3.28. Aeolian palouse soil overlying lava, Washington. (*USDA airphoto.*)

with respect to one another, which causes the structure built on the loess deposit to settle. According to the other explanation,[6] the severe settlements exhibited by loess in other parts of the world can be attributed both to clay lubrication and also to the removal by water of weak calcium carbonate cement from the loess. Studies in the United States have shown, however, that calcium carbonate exists in loess, generally in the form of disseminated grains and *not as cement*, at least not in appreciable quantities.

It should be noted also that hydroconsolidation has been observed not only in loess but also in dry alluvial soils. It is quite possible that in such cases the natural soil mass is close to the state of limit equilibrium; i.e., the shearing stresses acting on it are just balanced by the shearing strength of the soil mass. Addition of water decreases the shearing strength of the mass and causes a general collapse of the weakened fabric.

3.16. Engineering Problems in Loess Areas.   Because of its hydro-consolidation properties, loess may be a dangerous foundation material if brought in contact with water.   In the case of dams and specifically their reservoirs, wetting of the underlying material by the water from the reservoir means considerable settlement of the dam.   Failures of smaller structures on loess that became saturated are also known.   In a spectacular case of an overnight settling and cracking of a house, the accident was caused by the discharge of water from a hose forgotten on the lawn.

FIG. 3.29. Canal in loess, Nebraska.   Note steep side slopes.

Another difficulty with loess is its ready ability to "pipe" under the water action.   If water starts to leak from an excavation or a canal, it forms a path inside the loess mass which gradually progresses and widens, often irregularly, until a failure occurs.   Similar accidents also may take place in the case of steel pipes placed in loess.   Water finds its way around the pipe, and cavities as large as 9 ft in diameter have been known to develop.   Presumably, an accident of this sort can be prevented by a careful placing of the pipe and the backfill around it.

The loess settlement problem does not appear too serious if a concrete structure is built on a foundation that is not in contact with excessive amounts of water.   Such are, for instance, the footings of the towers of power-transmission lines or similar installations.   Such towers, however, should not be placed in local depressions which would permit water to accumulate around the tower.

In central Iowa, radio relay towers for transmission of microwaves have been constructed on loess using concrete footings 60 by 60 ft in plan.   Over a period of 3 years the total or differential settlement was under ½ in. (data of American Telegraph and Telephone Co.).

Remolding of several upper feet of loess at the surface and careful recompaction of the remolded material (using the methods described in Chap. 16) may create a reliable platform for building footings.   Embankments made of compacted loess proved to be entirely satisfactory as was shown by the example of Arkabutla Dam in the lower Mississippi Valley,[10] built by the U.S. Army Engineers.   If *properly compacted*, the loess material acquires a considerable shearing strength and resistance to erosion.

FIG. 3.30. Vertical slopes in loess, Iowa.

In order to control settlement of earth dams on loess, grouting procedures have been devised.   Such procedures consist in pumping a slurry of loess and bentonite or loess only into the bore holes made in the foundation (see Sec. 16.7 for more details).

Pile driving in loess is not easy, either in the case of timber piles or in that of H piles (Sec. 13.6).   Experimental work[11] has shown that displacement piles (Sec. 13.6) can most easily be driven in prewetted loess. Pile loading tests have proved also that piles driven in the prewetted loess show greater bearing power than piles jetted down or driven in prebored holes.

Excavation in loess usually is not difficult because of the capacity of the loess material to stand on almost vertical slopes (Figs. 3.29 and 3.30). Steep loess banks, however, are subject to the formation of gullies and accumulation of the spalled material at the toe of the bank.   As a gully advances toward the top of the slope, the slope may slide.

Several state highway departments have experimented with various types of loess slopes.   It was found in Iowa, for example,[12] that slopes in deep loess cuts could be excavated and inexpensively maintained if the bank was cut in a series of steps and risers (Fig. 3.30).   The steps were drained toward the toe of the next higher riser, and water was made to flow longitudinally by giving a convenient longitudinal gradient to the steps.   In other experimental cases, the loess cuts with 2:1 (2 horizontal to 1 vertical, see also Fig. 16.1 for designations) and 3:1 slopes were seeded in order to stabilize the slopes with vegetation.   The results were fairly satisfactory.   Canal excavations in loess with steep slopes (often 1:4) are feasible.   The spalled material is removed by the water flow, and though water undermining the toe of the slopes of a canal may cause them to slide, such accidents are not too common.

3.17. Sand Dunes.   Loose sand swept up from the plains generally is transported close to the earth's surface, and if it is stopped in its motion, it is deposited in the form of *dunes*. Different shapes of dunes are shown in Figs. 3.31 to 3.33.   The dune type labeled *A* in Fig. 3.31 is termed "Barkhan," a Russian term of oriental origin.   Notice that for all types of dunes shown in Fig. 3.31, the prevailing wind direction is from left to the right of the figure.

FIG. 3.31. Sand dunes, in plan.   Distance at the right lower corner (labeled "scale") equals 500 ft for A, B, and D and 4,000 ft for C and E.   (From F. H. Lahee, "Field Geology," 4th ed., McGraw-Hill Book Company, Inc., New York, 1941.)

During the first period of the formation of a dune, its height increases, but the vertical growth stops when the amount of sand brought in by the wind balances that taken away by the same wind.   At this new stage of the existence of a dune, sand particles are blown up and dropped over the crest, after which they slide down.   Thus, the dune moves ("migrates").   Sand dunes come in groups, sometimes 30 to 40 per mile.   (There also are loess dunes, generally widely spaced, e.g., 2 or 3 per mile, and exceptionally, clay dunes or ridges are encountered.)

The height of sand dunes commonly is between a few feet and 200 or 300 ft.   The windward slope is gentle (Fig. 3.34), making from 5 to 12° with the horizontal.   The leeward slope is steep, approaching an angle of

34°. Dune sand is a very uniform material, generally consisting of small quartz fragments with some mica. Particles smaller than 0.15 mm (sieve No. 100) fluctuate from 2 to 15 per cent, being higher in exceptional cases (for sieve sizes see Sec. 4.11).

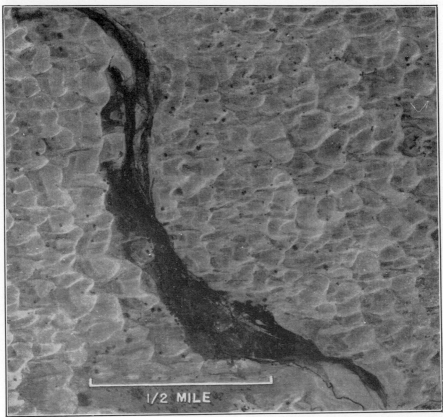

FIG. 3.32. Airphoto of sand dunes, Washington.   (USDA-PMA airphoto.)

Engineering Problems in Sand-dune Areas. Stabilizing movable sand is a major problem in construction and maintenance of highways and railroads crossing dune zones in deserts. This can be done, for example, by seeding the dune with such grass varieties as may thrive on sand when partly covered by the wind-blown material. Planting heather also is a method of stabilizing a dune. Recently, more substantial methods have been used, such as planting young trees (mostly conifers, such as pine) or treating the sand with crude road oil. Highways through sand-dune terrain require a careful maintenance service, since pavements, if any, are alternately buried and uncovered by the moving sand and unsurfaced roads generally are passable for automobiles only after rainstorms when

1/2 MILE

FIG. 3.33. Airphoto of sand dunes modified by grass growth, Nebraska. (*USDA-PMA airphoto.*)

the sand acquires some stability.   Highway and railroad cuts in moving-sand areas are traps for the sand and should be avoided.

The lower portions of *transmission-line towers* built in a sand-dune

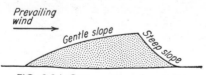

FIG. 3.34. Cross section of a dune.

locality may be alternately buried and uncovered, with a corresponding variation in the stress condition of the tower.   The dune on which the tower is placed may even move away, leaving the foundation of the tower completely exposed and hence in an unstable condition.   Seeding or application of road oil is also suitable for use in these cases.   For the foundations of a very expensive structure, however, piles should be driven to a depth unaffected by any shifting of the dunes.

The surface runoff in sand-dune areas is very low because of the considerable sorbing capacity of the sand.   For this reason the stream flow in such areas depends predominantly on the ground water, and occasional heavy rains do not cause any appreciable fluctuation of the water level in those stretches of rivers located in the dune terrain.   The water level in certain portions of the Loup Rivers in central Nebraska, for example, varies less than 1 ft from season to season.   Because of the water-sorbing capacity of the dune sand, the building and maintenance of *hydraulic structures* in sand-dune terrain are exceedingly expensive, and reservoirs in sand dunes can be built only if high water losses can be tolerated. *Irrigation canals* in sand-dune terrain should be lined.

ALLUVIAL AND WATER-FORMED SOILS

3.18. Alluvial Soils.   Eroded soil carried by water and deposited is alluvial (water-laid) soil, or *alluvium*.   As explained in Sec. 3.2, immediately adjacent to the steep portion of the valley and the headwaters of a

FIG. 3.35. Torrential stream deposition, California.   Note coarse materials and lack of small sizes.   *(Photograph by E. B. Waggoner.)*

stream, boulders and coarser gravel might be expected and there will be a minimum of fine sizes (Fig. 3.35).   At a distance of several miles from the place of original erosion, small sizes ("fines") predominate as shown in Fig. 3.36.

Alluvial deposits are in many respects similar to glacial.   They are more stratified, however, and thus their properties can be determined from a fewer number of drill holes than under equal conditions in the glacial zones. The alluvial deposits are not less heterogeneous than the glacial. It is not unusual, for example, to find a bed of alluvial clay several feet long, although it may be fairly narrow and only a few feet thick. Rather uniform sand and gravel beds of considerable dimensions may be found, and although there may be lenslike inclusions of sand in gravel beds and vice versa, these deposits as a whole are fairly continuous.

FIG. 3.36. Alluvium.   Note cross-bedding, stratification, and occasional folds.

Besides forming terraces and benches in the valley itself, deposition of alluvium also may occur on *plains*, thus forming relatively flat deposits. Large plains are not necessarily continuous but may be interrupted by isolated hills and occasional valleys as in the case of the Coastal Plain of the United States.

Near the river boundary of this plain, there are elevations of as much as 700 ft east of the Mississippi River and more than 1,000 ft in Texas. The outer portion of this plain is characterized by low, rather level terraces with a maximum elevation of 200 ft and, exceptionally, 300 ft as in southern Texas; in some places, these terraces penetrate deep into the inland.

The sediment carried by a flow moving across a plain during a freshet (high water) may be spread if the gradient of the stream decreases gradually. In this case a *flood plain* is formed. If, however, the gradient decreases abruptly, a larger part of the sediment carried by the stream drops in one place and forms an *alluvial fan,* a broad cone with the apex at the point where the gradient breaks (Fig. 3.37).

FIG. 3.37. Alluvial fan (topographic aspects according to position of observer): (a) vertical airphoto, (b) oblique airphoto, Alaska.   (*USBR.*)

An example of an alluvial fan is the site of the campus of the University of California at Los Angeles. This fan was formed by debris (outwash) from the adjacent Santa Monica Mountains.

The alluvial soil materials of the United States can be subdivided as follows: (1) soils of the Coastal Plain bordering the Atlantic Ocean (south of glacial soils) and the Gulf of Mexico; (2) mountain outwash, primarily the Great Plains outwash mantle coming from the Rocky Mountains; (3) valley fill; and finally, (4) recent alluvium deposited by the streams of the present geologic epoch.

Particular cases of recent alluvium are *organic silt* and *mud*. These are fine outwash from the littoral hills and mountain ridges, deposited along the ocean and bay shores and in the rivers flowing into them, especially in the lower reaches of these rivers. Along the East Coast of the United States, considerable deposits of organic silt are found in the Hudson River, in the Potomac River, in the Thames River, and in others. The greater part of the organic silt consists of angular fragments of quartz and feldspar, slivers and flakes of mica, abundant sericite (fine mica), and clayey matter; numerous microorganisms are also present. In the natural state, organic silt is dark and smells unpleasant; after drying, it becomes light gray and loses its characteristic odor.

Organic silt next to the Lincoln Tunnel crossing of the Hudson River in New York City contains about one-third (by weight) of clay sizes; the remainder is practically all silt sizes. The average plastic limit is about 25.

A counterpart of this material on the West Coast of the United States is the San Francisco "Bay mud" that in reality may be identified as silty clay. It is a gray-bluish in color and often is identified as "blue clay." Strictly speaking, the term "mud" should be referred to a slimy or pasty mixture of soil and organic material at the bottom of a river or lake. This sharply distinguishes true mud from the Bay mud, as the latter deposits reach 70 ft or more in thickness and often practically reach the water level in the Bay. A crust a couple of feet thick develops at the exposed mud surface, together with some vegetation growth.

3.19. Engineering Significance of Alluvial Deposits. Because of the similarity of alluvial and glacial deposits, the major engineering problems in both types of deposits are practically the same.

Alluvial deposits generally provide an excellent source for coarse construction materials, such as concrete aggregate, or pervious material for highway subgrade or highway embankments. Alluvial borrow pits generally contain a rather limited proportion of clay and silt sizes and therefore rarely make good sources of impervious material for embankments that have to hold water, such as earth dams. Blending of materials may correct this deficiency, however.

Deposits of *organic silt* in large Eastern United States rivers generally are covered with water, often 15 to 20 ft deep. Such deposits present problems in bridge pier construction.   The material has a very limited bearing power, and it is necessary to pass through it to a reliable underlying stratum (Sec. 14.10).   Also see Sec. 13.18 for the construction of fills on mud.

3.20. Openwork Gravel.[14]   The term *openwork gravel* is used to describe a gravel structure that seems to be most generally found in the United States in regions that have either been glaciated or are affected by glacial outwash deposition.   In European countries, occurrences have been reported that apparently are purely alluvial in origin (see Ref. 14, discussion).   For this reason, openwork gravel is presented in this section on alluvial soils.   The material in this particular type of gravel deposit generally is poorly graded, and fines are completely lacking.   Thus the interstices, or voids, are usually open (unfilled with fine material).   Clay, that apparently develops after the gravel has been formed, often works its way into these openings, however, but no sand is found in the interstices.   Openwork gravel apparently occurs as large lenticular bodies often interconnected with stringers of the same material and may be associated with common gravel and other kinds of soil.   Openwork gravel has a very high permeability; thus, excessive seepage losses through these gravels could be expected if they passed underneath a dam or through abutments of a dam.   Several methods have been used in attempts to seal such deposits when encountered in dam reservoirs.   One of these has been to inject grout or a silt or clay into openwork gravel.   Another has been to attempt to seal the deposits by placing a layer of silt over their surface.   On the credit side for openwork gravel is that it often furnishes very prolific supplies of water when penetrated by water wells.

Bad slides in highway cuts have occurred in openwork gravel and associated materials in several Eastern states.   The bearing capacity of an openwork gravel deposit should be determined in each individual case by a loading test or tests (see Ref. 14, discussion).

3.21. Swamps, Muskeg, and Peat Bogs.   Swamps or marshes are areas of wet, soggy ground, saturated or almost saturated, but generally not entirely covered with water.   They are filled with decaying vegetable matter, with different varieties of characteristic swamp grasses, often of river green color, growing at the surface.   Some swamps have been formed by lakes and sluggish streams.   In such cases, water may be displaced by the sinking vegetable matter, with the formation of gradually disappearing ponds at the surface, or deep water may be covered by a gradually thickening mat of vegetation.   Swamps often are developed at sea or lake coasts.   In the former case, salty *tidal flats* are often intermittent with swampy land.

It is difficult to establish the difference between the terms "swamp" and "muskeg."   The latter term probably is derived from the Indian word *maskeg*, which means grassy bog.   It has been proposed that muskeg be defined as "organic terrain."[15]   Large organic terrains, or muskegs, are found in Canada, Siberia, and Alaska and in the United States in Washington, Oregon, North Dakota, Michigan, New York, and New England.   Smaller areas also occur in other states.   Organic terrain in Europe occurs mostly in the British Isles, Western Russia, and Eastern Germany.   In the arctic regions, muskeg is intermixed with "tundra" (Sec. 10.8).

Organic terrain, or muskeg, may be classified according to either coverage pattern or topographic features.[15,16]   From the point of view of coverage, muskegs may be woody or nonwoody.   The growth habit (i.e., vegetation) in the woody muskegs varies from shrub and cranberry plants to bush, dwarf trees, and regular trees of different stature, up to 15 ft or more high.   Nonwoody muskegs may be covered with grasses 2 to 5 ft high, with smaller grasses and sedges or with continuous mats of lichens and mosses.

Topographic classification of organic terrain, or muskeg, establishes classification features from mound, ridge, and rock gravel plain to peaks, plateaus, and different kinds of swampy ponds.   Swamps may be formed at high places as well as in the lowlands.

The soil material in organic terrain, or muskeg, generally is black muck in various stages of decomposition, cohesive when dry, i.e., remaining in chunks or cakes.   The zone of contact between the muskeg material and the underlying regular soil mass (the organic-inorganic interface) may be thin and sharply marked, or the upper organic material may work into the lower soil sometimes to a depth of nearly a foot.   The muskeg material generally smells unpleasant because of the presence of methane (swamp gas) in its interstices.

Peat[18,19] is a mass of plant remnants in which the process of *humification* (or humus formation) is under way or already completed.   In the former case peat is *fibrous* and dark brown in color.   In the early stages of decomposition, stems and roots are still distinguishable.   At the late stages of the humification process, peat is a black, slimy or soft, cheesy substance, practically without recognizable structure.   According to the character of vegetation and place of formation, there may be *bog peat* formed from mosses (usually of the sphagnum type), *meadow peat* (grasses), *sea peat* (sea weeds), etc.   If peat is flooded by water containing soil material in suspension, vegetable matter will be either intimately mixed or interstratified with the soil.   These peat-soil mixtures are sometimes erroneously termed peat.   When a peat-soil mixture approaches uniformity and individual vegetation ingredients are no longer distinguishable, the material is classified as *organic clay*.

The natural moisture content of peat material is several hundred per cent by dry weight, and bog peat, in particular cases, may reach 1,000 per cent.   When peat is air-dried for fuel, its moisture content is reduced to 35 to 40 per cent and a 60 per cent shrinkage takes place.

The specific gravity of the peat solid matter ranges from 1.2 to 1.7.   The dry density of pure peat may be as low as from 4.0 to 7.5 pcf (Ireland); United States practice has found average values of about 12 to 15 pcf dry density for peat-soil admixtures (northern California).   Peat porosity is very high, and the material is very compressible.   The elastic rebound after one loading and unloading is negligible, and the material remains plastically compressed.   Shear tests (using vane, Sec. 4.19) have shown the shear strength of peat of different localities to be similar, ranging from 1 to 3 and, exceptionally, 4 psi if the material contains stiffer constituents such as rhizomes and roots of reeds.   The results of unconfined compression tests on northern California peat varieties basically confirm these European data.

To the authors' knowledge, peat deposits are from 1 to 80 ft in thickness, apparently according to the geologic age of the locality.   Swampy landscape in the proximity of a seashore should suggest the presence of peat deposits in the area.

3.22. Engineering Problems in Swamps, Muskegs, and Peat Deposits. The bearing values of the materials that may be classified as organic terrain (Sec. 3.21) are very low, and only very light structures such as secondary roads can be built on them.   Such were the early-day "corduroy" roads built of logs and that in recent years were used to cross swamps in wartime.   Scotland has developed a good technique of building *secondary roads* on peat, particularly on the so-called "hilly peat" with peat deposits between and over rock outcrops.   Drainage prior to construction is highly recommended in order to decrease the moisture content and increase the shearing resistance of the material.   Embankments are preferable to cuts, and deep lateral ditches should be avoided, as they will jeopardize the lateral support of the roadway.[17]   In building *important highways* across organic terrains, particularly peat areas, the practice of some American engineers is *to remove the peat and replace it* with adequate fill material, e.g., sand, gravel, sandy clay, etc.[10]   The peat may be totally or partly excavated; in the latter case the remaining peat is displaced, usually by applying surcharge.   "Jetting" of embankments may help, i.e., saturation with water under 20 to 30 psi pressure to consolidate the fill and squeeze out some water from the foundation of the embankment (because of the increase of the weight of the embankment).   It should be noted that jetting of embankments may be used not only on organic terrain but in any locality.   Blasting of the peat material is used if excavation with conventional earthwork equipment proves inefficient.

Heavy structures in peat areas should be built on piles.   If differential settlement and accidental slides during construction can be tolerated, as, for instance, in the barricades surrounding an explosive storage magazine, such barricades may be simply made in the form of high earth embankments.

3.23. Coral Reefs.[21]   These are found around many tropical and sub-tropical islands in the Pacific Ocean and Caribbean Sea.   *Atolls* are discontinuous ring-shaped reefs, enclosing more or less circular lagoons. Figure 3.38 shows a cross section of a *coral reef* formed during a subsidence of the shore line.   The reef consists of a central cemented mass, mostly with cavities, flanked by a talus slope to seaward, by a lagoon filled with sediments to landward, and by growing corals and nullipores on the top and upper seaward slope.   Corals are coral polyps (anemones) or animals, whereas nullipore algae are plants.   Their secretions consist mostly of calcium carbonate, and the central part of the reef is primarily composed of cemented skeletal remains that by gradual solution and recrystallization become limestone.   Materials of engineering value obtained from coral reefs are (1) *coral*, composed of skeletons with some sand and clay,

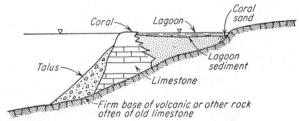

FIG. 3.38. Schematic cross section of a coral reef.   (*After Duke.*)

which is used for road surfacing and medium- and low-grade concrete; (2) *limestone* and limestone sands; and (3) *cascajo*, or weathered limestone, generally in fragments, with admixtures of clay or loam.   Some varieties of cascajo possess sufficient supporting power to be used in foundation material.

RESIDUAL SOILS.   NONSOIL AREAS

3.24. Residual Soils.   Large areas of residual soils are located primarily on both sides (i.e., approximately east and west) of the principal mountain chains in the United States, namely, the Appalachians, the Rockies, the Sierra Nevada, and the Columbia River Plateau in Washington, Oregon, and southern Idaho.

Residual soils on igneous and metamorphic rocks generally are sheets of a thickness insufficient to be important to the engineer.   If, however, the exposure of parent rocks of these soil types is sufficiently level, the soils formed on them may serve for the location of such engineering installations as airports.

Residual soils develop easily on *sedimentary rocks*, particularly on limestone, because of their comparatively weak resistance to solution and weathering.   Signs of destruction in limestone are *sinkholes* and *caves* (Fig. 3.39).   The former are caused by vertical erosion, and water accumulated in them may often be drained by perforating the limestone.   A series of sinkholes located along a curved line may indicate a subterranean stream (e.g., the Lost River in Indiana).   Caves in limestone also are a product of solution and erosion (Carlsbad Caverns in New Mexico, Mammoth Caves in Kentucky).   Subsurface investigations in limestone and overlying soils should be carried out in considerable detail.   Two buildings spaced about 200 ft apart and based on limestone may be perfectly

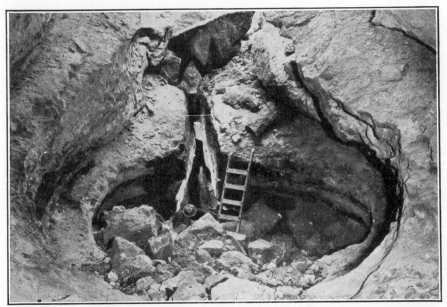

FIG. 3.39. Solution channels in limestone, Little Colorado River, Arizona.   (*USBR photograph.*)

safe, whereas the rock between them may be filled with caverns.   Generally speaking, the soils formed on limestone are well drained, though they may be highly plastic.   The contact between residual soil and the underlying limestone usually is highly irregular.

The cementing material of the rock has considerable influence on the properties of residual soils formed on sandstone.   If a sandstone with strong cement is penetrated by a highway cut, the rock will eventually accumulate a thin veneer of soils.   On the other hand in an excavation in a humid climate, a sandstone cemented with calcite or clay probably will become almost entirely clay over a period of years.   Plasticity of soil derived from sandstone generally is low.   Soils derived from fine-grained sandstones or sandstones interbedded with shales are moderately plastic. Completely nonplastic soils, such as cohesionless sands, also may be formed on sandstones.

Residual soils on shales are formed when the surface layers of the shale revert back to the clay or silt from which it was originally made. *Adobe* is a particular residual soil which develops on sedimentary rocks (often on shale), on gravelly and clayey deposits, and on recent alluvium. This

FIG. 3.40. Badland development in a fault scarp in sedimentary beds, California. (*Photograph by E. B. Waggoner.*)

term as used in the West and Southwest United States designates surficial strata, mostly of black, brown, or gray-brown soil material, very fine grained, and highly plastic. On drying, adobe assumes a coarse, black appearance, with wide fissures. The term was borrowed from the Mexicans, who used this material, with or without admixtures, for brickmaking and primitive buildings. The term has been used by the U.S. Department of Agriculture as an adjective (adobe soil, adobe clay).

Adobe varieties of high plasticity possess detrimental swelling properties. Rain water may form pools at the exposed surface of the adobe clay and make it swell, thus sealing it. The moisture content of the adobe-clay deposit at a certain depth then becomes constant, provided the material does not dry out and is not affected by the fluctuations of the ground-water table.

A common occurrence is a layer of black adobe, say 3 or 4 ft thick, underlain by a layer of gray adobe of less thickness.   There are, however, thicker adobe deposits. Black adobe is generally finer than the gray one; the percentage of fines is high in both layers (80 to 90 per cent of fines passing sieve No. 200) and generally decreases in the downward direction.   As an example, the liquid limit of adobe may be 60, with a plastic limit of about 25.   The average natural density of adobe clay is rather high (95 to 100 pcf).   The unconfined compressive strength is also generally high (5,000 to 11,000 psf).   The numerical values given are just examples and may vary in particular cases.

3.25. Nonsoil Areas.   These are areas where the soil is thin or otherwise has no engineering significance, such as rough topography, canyons, scablands, and badlands.   *Scablands* are level or slightly rolling rock outcrops that have practically no soil cover.   *Badlands* topography (Fig. 3.40) is produced by excessive erosion of poorly consolidated materials in a semiarid climate, such as are found in North and South Dakota. Throughout the Eastern United States, nonsoil areas are insignificant as compared with the West.   According to the authors' knowledge, in the East, scablands are sparse and there are no badlands.

REFERENCES

1. Miller, W. J.: "An Introduction to Historical Geology—with Special Reference to North America," 6th ed., D. Van Nostrand Company, Inc., New York, 1952.
2. Longwell, C. R., A. Knopf, and R. F. Flint: "A Textbook of Geology," Part I, Physical Geology, John Wiley & Sons, Inc., New York, 1932.
3. Lahee, F. H.: "Field Geology," 5th ed., McGraw-Hill Book Company, Inc., New York, 1952.
4. Emmons, W. H., *et al.*: "Geology: Principles and Processes," 4th ed., McGraw-Hill Book Company, Inc., New York, 1955.
5. Flint, R. F.: "Glacial Geology and the Pleistocene Epoch," John Wiley & Sons, Inc., New York, 1947.
6. Scheidig, A.: "Der Loess und seine geotechnische Eigenschaften," T. Steinkopf, Leipzig, 1934.
7. Denisov, N. J.: "On the Nature of Subsidences in Loess-like Soils" (in Russian), Soviet Science Publishing House, Moscow, 1946.
8. King, M. H., and W. R. Judd: Loess, Its Petrography, Physical Behavior, and Relationship to Engineering Structures, Paper presented at the annual meeting of the Geological Society of America, Boston, Mass., October, 1952.
9. Turnbull, W. J.: Utility of Loess as a Construction Material, *Proc. 2d Conf. on Soil Mech. and Foundation Eng.*, vol. V, Rotterdam, paper IV, p. 3, 1948.
10. Turnbull, W. J., *et al.*: Sedimentary Geology of the Alluvial Valley of the Lower Mississippi River and Its Influence on Foundation Problems, "Applied Sedimentation," John Wiley & Sons, Inc., New York, 1950.
11. Report of Loess Research Studies for the Ashton Pile Testing Program, *USBR Earth Lab. Rept.* EM-278, 1951.
12. Gwynne, G. S.: Terraced Highway Side Slopes in Loess, Southwestern Iowa, *Bull. GSA*, vol. 61, 1950.

13. Fisk, H. N.: Geological Investigation of the Alluvial Valley of the Lower Mississippi River, *Mississippi River Commun.*, Vicksburg, Miss., 1944; also Mississippi River Valley Geology Relation to River Regions, *ASCE Trans.*, vol. 117, 1952.
14. Cary, A. S.: Origin and Significance of Openwork Gravel, and discussion by J. Feld *et al.*, *ASCE Trans.*, vol. 116, 1951.
15. Radforth, N. W.: Suggested Classification of Muskeg for the Engineer, *Eng. J. (Can.)*, vol. 35, November, 1952.
16. Radforth, N. W.: The Use of Plant Material in the Recognition of Northern Organic Terrain Characteristics, *Natl. Research Council Can., Assoc. Comm. on Soil and Snow Mech. Tech. Memo.* 28, Ottawa, 1954.
17. Dryburgh, F. B., and E. R. McKillop: Construction and Maintenance of Roads over Peat, *Natl. Research Council Can., Assoc. Comm. on Soil and Snow Mech. Tech. Memo.* 29, Ottawa, 1954.
18. Hanrahan, E. T.: The Mechanical Properties of Peat with Special Reference to Road Construction, *Trans. Inst. Civil Engrs.* (Ireland), vol. 78, 1952.
19. Glossop, R., and G. C. Wilson: Soil Stability Problems in Road Engineering, *Proc. Inst. Civil Engrs.*, part II, June, 1953.
20. "Field Manual of Soil Engineering," rev. ed., Michigan State Highway Department, Lansing, Mich., 1946.
21. Duke, C. M.: Engineering Properties of Coral Materials, *Proc. ASTM*, vol. 49, 1949.

CHAPTER 4

ELEMENTS OF SOIL MECHANICS

This chapter outlines those properties of soils important in civil engineering. Special emphasis is placed on those soil characteristics that are presented incompletely, if at all, in available textbooks. The portion of the chapter dealing with stresses and strains is directed to (1) geologists and other earth scientists who want to acquire the "working knowledge" of soil mechanics essential in their work with the engineers and to do so with a minimum of mathematics and (2) those engineers who are not familiar with soil mechanics but wish to get acquainted with it in a general way.

SOIL PROFILE. SOIL MOISTURE

4.1. Soil Profile.[1-3] The various soils formed at the earth surface are continuously changing at various rates and sometimes form new soil varieties because of (1) *intrinsic factors*, such as the properties inherited by the soil from its *parent* materials, and (2) *environmental factors* characteristic.of the locality in which the soil has been placed in the course of its geologic history. Environmental factors involve *geology* of the locality, including rising and sinking of the terrain or loading and unloading of it by alluvial processes and sedimentation; *hydrology*, particularly local runoff conditions; *flora and fauna*, or remnants of vegetal and animal life, respectively; and *climate*, particularly its basic components—temperature and humidity. For example, weathering of granite in a humid northern climate produces podzols or ashlike soils, whereas in humid tropic regions, red laterites are formed. The influence of the climate may vary at a given locality according to its *topography*. For example, water may accumulate in depressions of the earth surface and hence control the degree of soil saturation, or exposure of a slope to the sun rays may activate or change the trend of formation of new soils.

To the engineer, the *soil profile* is a cross section of an actual soil deposit from the earth surface down to the depth which he considers important for his purposes. If this cross section reaches into the underlying rock, the term "geological section" sometimes is used (compare Sec. 7.6).   If represented graphically, an engineering soil profile shows the sequence and thicknesses of soil strata and describes the kinds of soil materials, often in terms of engineering soil classification systems (Sec. 4.11).   According to the *pedological* viewpoint, the combination of the new soils gradually formed on the original soils together with the latter constitutes the soil profile that finally becomes mature.

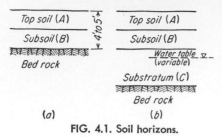

FIG. 4.1. Soil horizons.

From the ground surface down, different soil strata, or *horizons*, may be distinguished (Fig. 4.1).   The surface horizon is *topsoil*, or A horizon, often with subdivisions A_0, A_1,

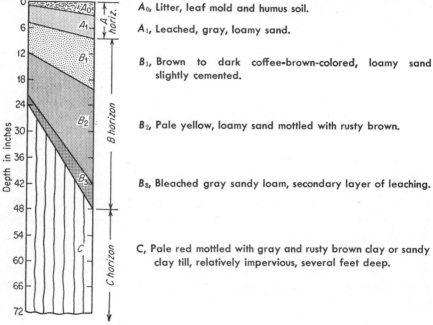

A_0, Litter, leaf mold and humus soil.

A_1, Leached, gray, loamy sand.

B_1, Brown to dark coffee-brown-colored, loamy sand slightly cemented.

B_2, Pale yellow, loamy sand mottled with rusty brown.

B_3, Bleached gray sandy loam, secondary layer of leaching.

C, Pale red mottled with gray and rusty brown clay or sandy clay till, relatively impervious, several feet deep.

FIG. 4.2. Soil profile of a glacial soil of the podzol group, Michigan.   Oblique lines show the thickness ranges.[3]

A_2, A_3, etc.   It is underlain by the subsoil of B horizon, also with subdivisions B_1, B_2, etc.   The zone under the B horizon is the *substratum* (C horizon).   In the case of *residual soils* (Sec. 3.24) the substratum is absent.   Fine material leached from A horizon down into B horizon reinforces the latter to form *hardpan* (or clay pan).   This should not be confused with glacial hardpan (Sec. 3.11).

The very top of the A horizon is *humus* (designation A_0) formed by the mixture of soil with decomposed organic matter. Although valuable in agriculture and gardening, humus generally is wasted in engineering except in irrigation projects or earth (dirt) roads. *Loam* is a transitional material from humus to the underlying soil. This term gradually is being eliminated from engineering usage. Figure 4.2 illustrates pedological soil nomenclature.

4.2. Soil Moisture.[4-7] Soil properties are essentially affected by *sorbed water* present in the soil. A particular kind of sorbed water is

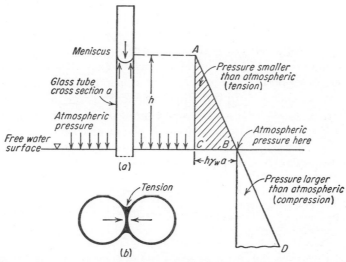

FIG. 4.3. Capillary movement and capillary forces: (a) tension in water caused by a concave meniscus, (b) contact pressure caused by capillary water ring.

capillary water (or capillary moisture) that generally moves from wet soil to dry soils when both materials are in contact. Capillary movement in natural earth starts from a free-water surface ("water table") and generally moves vertically upward but may move in any direction. In clays, water may slowly rise to considerable heights, e.g., 50 to 60 ft, though generally less. In sands the capillary water moves only several inches, but very quickly. Capillary water may move also through a completely saturated soil mass (in which all pores are filled with water) if the mass has an exposed face from which water can evaporate. Such is the case of swampy lands where capillary water moves continually through saturated soils to replace water lost by evaporation at the surface.

Figure 4.3a depicts the capillary rise in a glass tube of a small diameter. The top of the moving water forms a cuplike *meniscus*. Its shape is exactly that of a half sphere if water and the walls of the tube are clean; otherwise the meniscus is flatter. If the capillary movement stops at a certain height h (Fig. 4.3a), it is obvious that water has been pulled up by a force equal to its weight. This pulling force is represented by the triangle ABC. It is at a maximum at the free-water surface (line BC) and gradually decreases as it is spent in producing the work of lifting the water. Atmospheric pressure compresses water at the free-water level, and below this level, water is compressed also by its own weight. This additional compression (hydrostatic pressure) increases directly with the weight of water (line BD). When water in the tube is lifted up, it is placed under tension as in the case of any substance being pulled apart. Though the atmospheric pressure in the water within the tube still exists, it is *decreased* by tension. It has been said that thus there is a *pressure deficiency* in the lifted water. The cause of the force pulling the capillary water may be explained in an elementary way by the surface tension of the water and the pulling force of the meniscus. The surface-tension hypothesis presupposes that a very thin upper layer (a membrane) of any liquid is extended in all directions similar to a very tight glove on a hand. For water at 68°F, this force is about 0.073 g/cm along any line mentally traced on the water surface. If the meniscus is a half sphere in shape, the surface tension acts upward at the periphery of the meniscus and pulls water up.

The foregoing is only a "working hypothesis," since the true cause of capillary movement cannot be considered as established. The hypothesis of *capillary potential*[5] is based on the analogy of the capillary flow to that of heat or electricity. The capillary potential is measured by the height h of the moved water column (Fig. 4.3a). Symbol pF stands for its logarithm.[7] The theory of *free energy*[6] states that matter possesses free (or available) energy and that two or more "coexisting" systems can be in equilibrium if their unit-free-energy values are equal; otherwise a movement starts from the system possessing higher value of unit free energy toward the system with a lower value. For example, water possesses more unit free energy than a moist soil mass, hence the capillary movement from the free-water surface into the moist soil in contact with that surface.

In an actual earth mass there are no straight canals but capillary water moves through an intricate net of pores. The analogy of Fig. 4.3a gives a fair idea of the capillary movement, however.

When capillary moisture invades a dry, powdered soil, the particles are held together by an over-all compressive force. This is the *capillary pressure*. The finer the pores, the stronger is the pulling force of the meniscus and the greater the capillary pressure. Putting it another way, the thinner the capillary films on the soil particles, the stronger the capillary pressure. When a clay lump dries out, the capillary films thin out and the clay *hardens*.

Figure 4.3b illustrates the phenomenon of *contact pressure* between two sand grains connected by a capillary ring. The concave meniscus (as in

Fig. 4.3a) is associated with tension in the water between the grains, whereas the thin moisture film around the grains causes compression in them. If loosely deposited, moist sand (with water rings, Fig. 4.3b) occupies a larger volume than that occupied by the same grains in dry condition. This is known as *bulking of sand*, a phenomenon completely different from swelling of clay as discussed in Secs. 4.6 and 4.10.

Moisture-content Determination. A field sample is dried in the air or, to accelerate the procedure, in a laboratory oven at about 100°C until its weight becomes constant (often 4 hr is sufficient). The sample is weighed before and after drying, and the weight of the moisture thus lost is expressed in per cent of the dry weight, accurate to 0.1 per cent, e.g., 26.1 per cent. Another method is to add alcohol to a sample to dissolve the water of the sample. The inflammable fluid thus formed is then burned out. In either of these two methods, the structural (or chemically bound) moisture of the soil minerals is *not* removed.

In field soil investigations, the soils are identified as *dry, moist, wet,* and *saturated.* In practice this is done by visual inspection and sometimes by feel. These characteristics have no definite mathematical correlation with numerical values of the moisture content of different soils; for example, a sand often is saturated at a moisture content of 16 per cent or so, whereas a clay at this moisture content may be only damp.

The *permeability* of a soil is its capacity to permit free (not sorbed) water to pass through it. It is characterized by the coefficient of permeability of the soil discussed in Chap. 5. The greater the value of the coefficient of permeability, all other conditions being equal, the more permeable (or pervious) is the soil. Generally sands are more pervious than silts, and silts are more pervious than clays.

If all different soil varieties were placed in a long row according to their properties, sands would be at one end and clays at the other, so different are their properties.

SOIL PARTICLES

4.3. Size of Soil Particles, Gradation.* It is customary to express the size of soil particles in metric measures, generally in millimeters (1 in. = 25.4 mm). The size of very fine particles is in microns (designation μ, Greek letter mu); 1 μ = 0.001 mm.

The terms "particle" and "grain" are often used interchangeably, though there is a tendency to reserve the term grain for sands; clays and similar materials often are referred to as "fine-grained" soils. However,

---

* "Gradation" is used by engineers to refer to the relative size of soil particles and should not be confused with the same term used by geologists to describe the geological process of gradation (Sec. 3.1). The term "texture" of a soil, which is practically equivalent to gradation, is rarely used.

a sharp distinction should be made between the terms (1) sand, clay, and silt sizes and (2) the actual soils of these names. The latter may be and generally are mixtures of these sizes and may or may not be plastic (Sec. 4.9). Numerical values of different sizes are given in Sec. 4.11.

Colloidal matter in soil (sizes smaller than 2 μ) is included in clay content. There are two kinds of colloids: (1) *gels*, or jellylike colloids, and (2) *sols*, or colloids similar to a liquid. Some gels upon shock or vibrations become sols, but after a period of rest, sometimes a few hours or more, the former state of gel is restored. Such are the so-called *thixotropic* clays (Sec 4.6).

Gradation and Mechanical Analysis. In *well-graded* soil, the finer particles tend to fit between the coarse ones, thus reducing the amount of voids to a minimum. Soil in which practically all particles are of the same size is usually termed *uniform*, though by definition it is *poorly graded*. Also, if in the range of sizes from a maximum to a minimum certain sizes are lacking or overabundant, the soil is termed poorly graded.

The objective of the *mechanical analysis* is separation of soil particles into *fractions*, each fraction containing grains or particles of approximately the same size. The soil material first is passed through a set of sieves; usually the U.S. Standard sieves are used. The soil passed through a sieve is designated by the number of that sieve preceded by the word "minus." Thus "minus 40" material has passed through sieve No. 40. · The number of a sieve indicates the number of meshes per inch of sieve bottom. Sieves may have square or round openings; thus when a sieve analysis is made, the make of the sieve and the shape of the openings should be recorded. Also, it is preferable to wash the material through a sieve rather than passing it through in a dry condition. The percentage of the weight of the whole sample tested as *passed* through each sieve of the set is computed. There is always a small loss of material in sieving. Therefore, it is advisable to weigh the sample before sieving and then compare this weight with the total weight of the material passed through or retained on the sieves.

Wet Analysis. Hydrometers are used to separate particles finer than 0.074 mm (sizes minus 200). The action of these devices is based on Stokes' sedimentation formula. In current engineering practice the hydrometer method is rarely used.

Gradation Curves. The results of a mechanical analysis may be represented graphically in the form of a "gradation curve" or "size-distribution curve." The abscissas of this curve are diameters of particles plotted on a logarithmic scale, and the corresponding ordinates show the percentage of particles "finer than" a given diameter contained in the given soil materials. The *effective size* of the given soil is the maximum diameter of the smallest 10 per cent (d_{10}); the *uniformity coefficient* is the ratio obtained by dividing the maximum size of the smallest 60 per cent (d_{60}) by the effective size d_{10}. The uniformity coefficient is greater in clays than in sands.

Figure 4.4 represents gradation curves of five soils. Gradation curves of poorly graded soils include oblique portions approaching either vertical or horizontal lines. Note the soil names at the bottom of Fig. 4.4 and the size of sieve openings at the top.

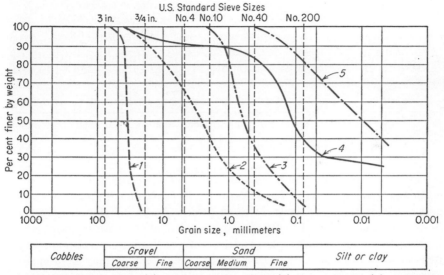

FIG. 4.4. Gradation curves: (1) uniform coarse gravel, (2) gravelly sand, (3) medium to fine sand, (4) sandy clay, (5) clayey silt.[27]

4.4. Shape of Soil Particles.

Sands. Sand grains may be angular, subangular, rounded, or subrounded according to the degree of wear caused by rolling and abrasion. Generally, angular grains mean that the sand has been exposed to wear for only a short time if at all, e.g., residual sands that have remained at the place of their formation. Volcanic sands contain angular fragments of rock glass; ice-worn sand grains may have flat faces abraded by ice; marine sands often are angular. Rounded grains are typical for river and beach sands (Fig. 4.5). A river sand, however, may contain coarser rounded grains and smaller angular grains because the beach sands, being in suspension most of the time, escape abrasion. Aeolian (wind-blown) sands possess fine, rounded grains.

Clays. The clay minerals are made of microscopically and submicroscopically thin sheets (Sec. 1.4); a considerable percentage of clay particles are usually flat or *scalelike* tiny fragments of those sheets. These flat particles are mixed with others of irregular shape and with colloidal matter. The latter often is seen only as dense fog, even under modern electron microscopes (Secs. 1.4 and 1.5).

FIG. 4.5. Two California sands: coarse beach sand and fine river sand (note scale). (*Courtesy of Raymond Lundgren.*)

SOIL STRUCTURE

4.5. Structure Types. Cohesion.[8-11] Dry, clean, granular soils such as gravels or sands, at best, have *unstable* structure. If perfectly confined they may be made very dense by compaction. Sand grains may hold together by contact pressure (Sec. 4.2) or "apparent cohesion" but become unstable again upon drying out. Sand grains may be *cemented* and thus approach conglomerates (Fig. 1.18).

The *structure of clay* may be defined as the pattern established by the arrangement and mutual relationships of the constituent particles, amorphous masses, and discontinuities (voids, fissures, cracks). This definition is very similar to the definition of the term soil "fabric" as used by some investigators.[8,9] Relatively large structural features constitute the *macrostructure* of clay.* A microstructure is characterized by a uniform monotonous pattern of very small particles held together by *attractive forces* which probably are electrostatic and known in chemistry as *van der Waals' forces.*[10] These forces are operative in all kinds of matter and rapidly increase as the distance between the particles decreases. True cohesion holding solid matter (e.g., rocks) together may be explained by the action of these forces.

The structure of the clay often reflects its origin. A residual clay may preserve the microstructure of the parent rock; occasional addition of organic matter, which is amorphous, contributes to the monotony of the

* "Macro" means large; "micro" means minute.

structure and the dark shade of the material.   Inorganic clays formed by sedimentation often show a *flocculent structure*.   A stream flowing into a basin such as a lake drops coarser particles to the bottom because of a decrease of velocity, and if smaller colloids approach one another too closely, they are mutually attracted and form *flocs*.   If the basin into which the stream flows contains salt water, the floc formation is more intensive because sodium chloride is an electrolyte and favors flocculation. Spectacular flocculation and precipitation may be observed if some common salt is added to a clay suspension.   This electrolytic action causes

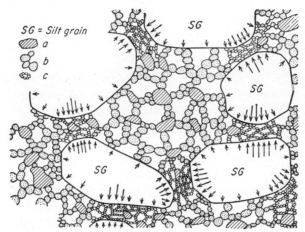

FIG. 4.6. Hypothesis of bonds between coarser particles in a clay.   (*After A. Casagrande.*)

the marine clays (those formed in the sea) to be *more porous* than freshwater sedimentary clays.   This is because the salt water causes the relatively remote colloidal particles as well as the close ones to flocculate and form large flocs and larger voids.   Figure 4.6 is a simplified visualization of the flocculent clay structure.

Stratified, or laminated, structure may be formed when clay particles, which are mostly flat, fall through quiet water periodically as in the case of varved clays (Sec. 3.13).

4.6. Alterations of Clay Structure.   Clay structure generally is somewhat unstable, as it is subject to changes and alterations, especially in those portions of the deposit close to the *exposed* surface.

Hardening, Shrinkage, and Swelling.   When clay is drying out, it is subjected to compression (Sec. 4.2); it decreases in volume, or *shrinks*. As the particles come closer to each other, van der Waals' forces increase automatically and true cohesion increases.   Fissures may close up in such cases.   If hard clay is *submerged* in water, there is no surface tension at the clay surface and hence no capillary compression.   This results in expansion (swelling) of the clay. Clay is at least partially an elastic material, and there is an elastic rebound upon removal of the compressive forces (Secs. 2.8 and 2.11). Another explanation for this swelling is the restoration of the *sorbing capacity* of the clay that during the drying-out process was overcome by evaporative forces. When the latter vanish, the clay pores again are filled with sorbed water. An analogous phenomenon occurs when a submerged clay deposit is loaded. It then *consolidates* (Sec. 4.22); i.e., water is squeezed out of the pores with a resultant over-all decrease in volume of the deposit. When the load is removed, the sorbing capacity of the clay is restored and swelling takes place.

Alternate cracking and swelling, because of cyclical wetting and drying of the material, occur in adobe-type clays (Sec. 3.24) in arid and semiarid regions (California, Texas, and others).

Fissuring of Clay. Not all clay fissuring is caused by drying. For example, London clay is fissured in nature. A sample of this clay remolded at liquid limit (Sec. 4.9) and stored under water for one year cracked along a pattern similar to the natural pattern.[12] Stiff clays may be subdivided by hair cracks (fissures or, by analogy with rock, joints) into fragments. Fissuring often is observed in flood-plain clays consisting of layers which were exposed independently of one another.

Desiccation. If the surface of a natural soft-clay deposit or a hydraulic fill is exposed and flooded only for rare, short periods, the surface hardens and a *crust* is formed. The thickness of the crust may be from less than 1 to 4 or 5 ft (Boston, Chicago, and other locations). The bearing power of the crust generally is greater than that of the underlying material; sometimes the crust will safely support one-story buildings, although the underlying mass is still soft.

In arid and semiarid regions with long dry periods, water is drawn by evaporation from considerable depths (such as 15 ft), and clay is fissured and cracked to those depths. Rain water makes clay swell to only a rather insignificant depth, so deep fissuring caused by drying remains. The bearing power of the clay material damaged by cracking is less than that of the same material in a sound state.

Thixotropy and Syneresis. The gels of a *thixotropic* clay affected by vibrations turn liquid, i.e., become sols, without addition of water. Thus, the soil loses its shearing strength (Sec. 4.16), since liquids possess little or no shearing strength. After a period of rest, however, sols again become gels and the clay rehardens. A thixotropic clay may become fluid during an earthquake or because of nearby pile driving. *Syneresis* is a spontaneous separation of an initially homogeneous colloidal system into both a coherent gel and a liquid. The phenomenon basically is the drawing together of particles under the action of increased van der Waals' forces.[8]

Remolding of Clay. Destruction or damage of the clay structure produced practically instantaneously, by either natural or man-made forces, is called *remolding of clay.* Unconfined clay remolding in the laboratory may be done simply by rubbing or kneading small clay portions between the fingers. This results in a general decrease in cohesion and volume. *Sensitivity* (or sensitivity value) of clay to unconfined remolding is measured by the ratio of the unconfined compressive strength of clay before and *immediately* after remolding. The clay sample remolded in the laboratory is repacked to its original density and tested; if this test is postponed, the van der Waals' forces are restored, at least partially. Such a restoration process sometimes is termed "natural hardening" of clay.[13]

The sensitivity value varies from about 1.0 for heavily overconsolidated clays to over 100 for the so-called extrasensitive or "quick" clays. Clays of the lowest sensitivity are rare, but sensitivities of 2 to 4 are very common among normally loaded clays,* and sensitivities of 4 to 8 are quite frequently encountered.[12] The extrasensitive clays may not only alter their structure but lose it completely and become fluid (Sec. 17.14) without addition of water.[14] This phenomenon is different, however, from thixotropy.

A *confined* clay remolding in the laboratory may result in an increase of the unconfined compressive strength. In this case the clay particles are rearranged in a tightly packed structure, fissures closed, the volumes of voids decreased, and the number of contact points between particles increased.

The two principal types of clay remolding under field conditions are landslides and pile driving. Whereas the laboratory remolding is associated with the orientation of the particles and, hence, a decrease in volume, the yardage of the clay moved by the landslide and spread on the slope (Sec. 17.3) is generally larger than in the original state. The shearing strength of an earth mass involved in a landslide generally is as badly damaged as in an unconfined laboratory remolding.

A pile driven into clay of fair sensitivity, such as 3 or 4, may produce complete remolding within a narrow annular space around the pile and a less disturbed zone with an outside radius of up to three diameters of the pile. Cases have been reported, however, where there was no complete remolding of the material around the pile. In clays of low sensitivity, pile driving may even cause an increase in cohesion (and generally in shearing strength) of clay, especially at the lower part of long piles where the clay is confined by the weight of the overlying material and hence cannot easily move or expand.

A special case of remolding occurs along the shearing or separation sur-

* For the term "normally loaded clays" see Sec. 4.18.

face when a portion of the mass tends to separate from the other portion to which it is tightly pressed. The shearing surfaces (slickensides) are quite smooth and look almost polished in such cases (Fig. 2.23).

Slaking is the complete disintegration and loss of the soil structure when dry material is immersed or flooded. The outer portions of a soil mass subject to slaking then become saturated and prevent the air inside the mass from leaving. If, in addition, the soil is of the expandable type, the destruction proceeds rapidly. Kaolinite, halloysite, and many mica clays defy penetration of water and remain intact when submerged, nor does sodium bentonite disintegrate in water.[8] Shales and clays in excavations permitted to dry before concrete is placed may slake when wetted by concrete. A slaking test always should be done if the soil in the structure or its foundation is likely to be attacked by water. A dry soil lump placed on metallic mesh is submerged and, if not immediately destroyed, is left for one or more days under water. (Compare Sec. 1.10, Slaking Characteristics of Rocks.)

POROSITY AND DENSITY

4.7. Porosity, Voids Ratio, and Degree of Saturation. (See Sec. 2.2.) Porosity of soil is analogous to porosity of rocks and is designated by the same symbol n. Figure 4.7 is a conventional representation of one cubic unit of soil, e.g., 1 cu ft. In 1 cu ft of soil there are n cu ft of voids and hence $(1 - n)$ cu ft of solid matter (particles, grains). The ratio of the volume of pores to the volume of solid matter is the *voids ratio* e of the soil:

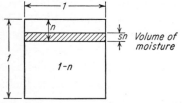

Volume of moisture

$$e = \frac{n}{1 - n} \qquad (4.1)$$

or

$$n = \frac{e}{1 + e} \qquad (4.2)$$

FIG. 4.7. Porosity, voids ratio, and degree of saturation.

In a pile of coarse, dry sand, porosity is close to 33 per cent ($n = 0.33$). According to Eq. (4.1), the voids ratio of such material is about 0.5. The pores in clays are very small but very numerous; this explains the high porosity of clays (50 or 60 per cent or more). Small diameters of clay pores are also responsible for the low permeability of clays. Swamp soils such as peat and muck may have a porosity of 90 per cent or more and consequently a very high e value.

Figure 4.8 shows schematically the loosest and the closest packing of spheroidal grains. In the loosest packing, each sphere touches six others; in the closest packing, each sphere has twelve points of contact with others. The pore space in either case is bounded by complicated curved surfaces.

Part of the voids of a soil mass is occupied by moisture (crosshatched in Fig. 4.7). As in rocks (Sec. 2.3) the ratio of the volume of moisture to the total volume of pores is the *degree of saturation* S. If $S = 1$ (100 per cent), the soil is *saturated* (or fully saturated).

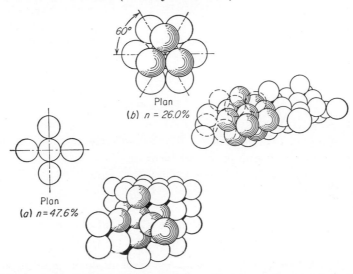

FIG. 4.8. Packing of spherical particles: (a) loosest, (b) closest. *(Drawing by J. Vitaliano.)*

4.8. Density and Unit Weight of Soil. The *specific gravity* G of soil particles depends on the specific gravity of the parent rocks (Sec. 2.1). For engineering computations, the G value is often assumed at 2.65 or 2.70. Organic soils have smaller G values; soils containing heavy minerals have high G values but are practically never encountered in civil engineering. The unit weight of water (symbol γ_w) is taken as 62.4 pcf or 1 g/cc. If the specific gravity of grains has to be determined, methods given in physics textbooks (e.g., pycnometer method) are used.

If the material is perfectly dry, i.e., its moisture content is zero, the weight of one cubic unit, e.g., 1 cu ft, is its dry unit weight; in practice the term "dry density" is used (symbol D). If a soil sample has a volume V_s cu in. under field conditions and in the laboratory weighs W_d after drying, the dry density as expressed in pounds per cubic foot is

$$D = 3.81 \frac{W_d}{V_s} \tag{4.3}$$

Example: A soil sample taken in a brass container 2 in. in diameter and 4 in. long has a volume of 12.57 cu in.; it weighs 352 g after drying out; its dry density is

$$D = 3.81 \frac{352}{12.57} = 107 \text{ pcf}$$

If the specific gravity of grains is assumed at $G = 2.65$, 1 cu ft of material (without pores) weighs $2.65 \times 62.4 = 166$ lb.   Then its porosity and voids ratio are

$$n = 1 - \frac{D}{166}$$

$$e = \frac{n}{1 - n} = \frac{166}{D} - 1 \tag{4.4}$$

Example: If the dry density is 107 pcf, its porosity and voids ratio under field conditions, regardless of whether or not the material is saturated, are

$$n = 1 - {}^{107}\!/_{166} = 0.355$$

and
$$e = \frac{0.355}{1 - 0.355} = 0.550$$

Still assuming $G = 2.65$ and a saturated soil (e.g., sample taken under water), the moisture content w_0 would be

$$w_0 = \frac{62.4}{D} - 0.381 \tag{4.5}$$

If the actual field moisture content is w, the ratio w/w_0 is the degree of saturation.

Example: Dry density is 107 pcf; moisture content at saturation would be

$$w_0 = (62.4 \div 107) - 0.381 = 0.206,$$

or 20.6 per cent.   If the actual field moisture content is only 11.2 per cent, the degree of saturation would be $S = 11.2 \div 20.6 = 0.54 = 54$ per cent.

If the specific gravity of grains G differs from 2.65, the following formulas should be used:

$$n = 1 - \frac{D}{\gamma_w G} \tag{4.4a}$$

$$e = \frac{\gamma_w G}{D} - 1$$

$$w_0 = \frac{\gamma_w}{D} - \frac{1}{G} \tag{4.5a}$$

It cannot be overemphasized that Eqs. (4.5) and (4.5a) refer to *saturated* soils only.   In this case

$$e = w_0 G \tag{4.6}$$

or approximately (if $G = 2.65$)

$$e = 2.65 w_0 \tag{4.6a}$$

Saturated and Submerged Unit Weight. If all pores of a material are filled with water, the *saturated* unit weight γ_{sat} equals dry density D plus the weight of the water in the pores $w_0 D$, where w_0 is the moisture content at saturation:

$$\gamma_{sat} = D + w_0 D = D - \frac{D}{G} + \gamma_w \qquad (4.7)$$

If saturated earth material is placed under water (submerged), *buoyancy* equal to the unit weight of water, $\gamma_w = 62.4$ pcf, acts on each cubic foot

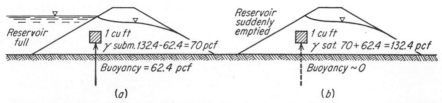

FIG. 4.9. Difference between (a) submerged or buoyed and (b) saturated unit weights of a soil material.

of the material. To obtain the submerged unit weight from the saturated unit weight γ_{sat}, the unit weight of water γ_w should be subtracted:

$$\gamma_{subm} = D - \frac{D}{G} = D\left(1 - \frac{1}{G}\right) \qquad (4.8)$$

For approximate computations, placing $G = 2.65$, the submerged unit weight is

$$\gamma_{subm} = 0.624D \qquad (4.8a)$$

Figure 4.9 shows the differences between the saturated and submerged weights. In an earth dam having a full reservoir, buoyancy equal to $\gamma_w = 62.4$ pcf acts on every cubic foot of the material under water. If the reservoir suddenly is emptied ("sudden drawdown"), buoyancy is discontinued and the apparent weight of the soil suddenly is increased by 62.4 pcf, after which the water starts to drain out of the soil.

Actual Density. The dry density of a sand with a grain specific gravity of 2.65 and a 33 per cent porosity is about 110 pcf. This is a *medium-dense* sand. Sands with less dry density would be loose or very loose, and those with greater dry density would be *dense* or very dense.

Since clays are usually more porous than sands, their dry density is usually less than that of sand. For example, the dry density of an average clay with a particle specific gravity of 2.70 and a 45 per cent porosity would be about 95 pcf. For the purpose of visual inspection, clays can be described as *soft*, *firm*, *stiff*, and *hard*, depending upon their apparent dry density and apparent resistivity to compression.

The methods of determining the dry density of a fill or a natural deposit in place are standardized (ASTM Designation D420—45). Briefly, the procedure is to smooth the ground surface and take a sample, e.g., with a spade. The extracted material is promptly weighed, its moisture content determined, and dry weight computed. The volume of the hole is measured by pouring soil or sand from a calibrated flask (Fig. 16.15). Dry density D then is obtained by dividing the dry weight of the extracted material by the volume of the hole.

Relative Density of a Sand. To determine the relative density of a *sand*, its average dry density D (e.g., determined in the field as described above) and its maximum and minimum dry densities (D_{max} and D_{min}) should be known. The procedures for determining the latter values are not standardized but vary in each laboratory (e.g., see "Earth Manual" of the USBR, 1951 ed., pp. 174–177).

$$\text{Relative density, } \% = \frac{D_{max}}{D} \frac{D - D_{min}}{D_{max} - D_{min}} 100 \qquad (4.9)$$

The relative density varies from zero for $D = D_{min}$ to 100 per cent for $D = D_{max}$.

PLASTICITY AND SWELLING

4.9. Plasticity and Atterberg Limits. Plasticity as discussed in this section is the property of a soil mass to be deformed continuously and permanently *without* rupture during the application of a stress that exceeds, even slightly, the shearing strength of the soil (Sec. 4.16). A *plastic* body deformed by external forces keeps that shape after these forces have been removed. Conversely, *elastic* bodies regain their original shape and size after the removal of acting forces. Such recovery is *elastic rebound.* Actual soil bodies generally are only partly elastic; when they are relieved from a compressive stress, they tend to return to their original shape and size but do not reach them. The total deformation produced by a compressive stress in this case consists of an elastic (or reversible) deformation and a *plastic* (or irreversible) deformation. Obviously the term "elastic rebound" refers to the elastic part of the deformation only. A loose sand mass, if compressed, undergoes an apparent "plastic" deformation that in reality is a decrease in the total volume of the mass because of the mutual accommodation of the particles and decrease in pore volume. In this case, the *individual* sand grains are elastically compressed, and there may be a small elastic rebound after the removal of the compression force. For computations in soil mechanics, it is assumed that all soil particles are *incompressible.*

Atterberg Limits. A clay suspension flows as a liquid and has practically no shearing strength. The latter is gradually built up as the suspension dries and the clay mass passes through the following stages:

TABLE 4.1

Stage of the mass	Limits between stages and symbols used
Liquid	
	Liquid limit (LL or w_l)
Plastic	
	Plastic limit (PL or w_p)
Semisolid	
	Shrinkage limit (SL or w_s)
Solid	

The above limits are called Atterberg limits or limits of consistency. These are *moisture contents* of the mass as it passes from stage to stage. When clay is in the plastic stage, it may flow plastically if overloaded. *Plastic flow* may be defined as a slow movement of an overloaded plastic substance without change in volume. Only those soils with scalelike particles, e.g., clays, become plastic upon addition of water or some other fluid *with bipolar molecules.* (There is no plasticity of clay with carbon tetrachloride, CCl_4, for

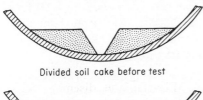

Divided soil cake before test

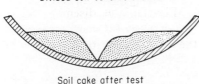

Soil cake after test

FIG. 4.10. Liquid limit determination.

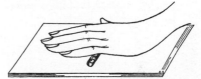

FIG. 4.11. Plastic limit determination.

example.) Sand and silts are not plastic unless mixed with some plastic substance (e.g., organic silt).

The consistency limits tests are made only on minus 40 material. The liquid limit (LL) is the moisture content at which two sections of a standard clay cake just barely touch each other when jarred in a standard manner (Fig. 4.10). At the *plastic limit* (PL) a clay just begins to crumble when rolled out into threads of about 1/8 in. diameter (Fig. 4.11). The difference LL-PL is the *plasticity index* (PI or I_p). This important soil characteristic indicates the *range* of moisture contents within which the soil has plastic properties. PI, as it is simply termed in current practice, is of importance in specifications for the use of soils as engineering materials and in soil classifications (Sec. 4.11). The *shrinkage-limit* concept finds considerable application in testing water-sensitive, expandable clays

(Sec. 4.10).   It is a stage when a saturated clay exposed to drying starts to lose its water content; then air gradually invades the pores, and the soil changes its color from dark to light.   It is obvious that the shrinkage limit is the moisture content at saturation and can be computed by Eq. (4.5) or (4.5a).

If metric measures are used, $\gamma_w = 1$.   A small soil sample is dried out, and its weight in grams and volume in cubic centimeters determined, the latter by immersing the dry sample into mercury that fills a container flush with its top.   The volume of the sample equals the weight of the displaced mercury in grams divided by 13.6, the specific gravity of mercury.

About 1948, the stiffening limit concept was introduced in Norway[15] primarily for the extrasensitive Norwegian clays (see Sec. 17.14).   At water contents greater than the conventional liquid limit, such clays are apparently stable but can be liquefied suddenly by factors which are not well understood.   The LL of such Norwegian clays is about 40 per cent, whereas their stiffening limit is from 65 per cent up.   There are such clays in the United States also.   The varved clay and silt (Nespelem formation) near Coulee City, Washington, has LL = 60 and a stiffening limit of 120.   Claystone near Osage, Wyoming, has LL = 501 and a stiffening limit of 1,560.

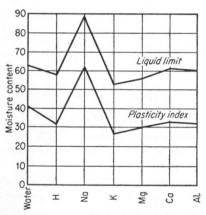

FIG. 4.12. Consistency limits as influenced by variations in ionization.[28]

The stiffening limit as determined in the laboratory is defined as the minimum water content at which a thoroughly stirred clay suspension still flows under its own weight in a standard test tube 11 mm diameter after exactly 1 min of rest.[8]

Ionization of Clays.[5,12,17,18]   Consistency-limit values determined by ionized solutions are different from those determined with pure faucet water.   (Distilled water is very rarely used.)   Figure 4.12 shows the influence of the various ions on the consistency-limit values.   The influence of the sodium ion contained in common table-salt solution is conspicuous.   It is necessary to note that salt water considerably *decreases* soil permeability.[16]

4.10. Clay Swelling.[19-25]   There are three basic causes of soil swelling. One is a combination of elastic rebound and sometimes moisture content recovery in a compressed soil mass after the compressive force has been removed.   This cause is explained in Sec. 4.6; it is conspicuous in a consolidation test (Sec. 4.22) and is common for any material, particularly a saturated clay. The two other causes depend on the property of some clays to attract water intensively and keep it with a resultant over-all increase in volume. These latter-type clays are (1) frost-sensitive clays and (2) water-sensitive clays. The frost-sensitive clays that swell at low temperatures are discussed in Chap. 10; water-sensitive expandable clays are discussed hereafter.

The problem of foundation soils that swell upon wetting is present throughout the Western United States but is practically unknown in Eastern United States. There are expandable soils in Burma,[19] South Africa,[20] Israel,[21] and other localities. In arid and semiarid regions, particularly in California and Texas, the soils subject to swelling are predominantly (but not exclusively) adobe and bentonite. Adobe is described in Sec. 3.24. Construction practice has shown that only the upper strata of adobe are subject to swelling (e.g., about 3 ft from the ground surface in the San Francisco Bay area). Apparently, at a certain depth, adobe is at its expansion limit and further access of water does not increase its volume. Adobe thus is "sealed," and occasionally pools are formed at its surface during the rainy season. *Bentonite* is described in Sec. 1.4. It is found in the Western United States, e.g., in Montana, South Dakota, and Wyoming. In industry bentonite is known under the trade name of Volclay.

Expandability (or expansivity) of clay is dependent primarily upon the amount of montmorillonite and some types of illite present. The basic cause of swelling is the attraction and sorption of water ("osmotic imbibition") by the expanding clay lattices. A contributing factor is the relief of the capillary pressure because of the thickening of the capillary films. As is known, the thinner the attracted capillary film, the greater its compressive effect on the particle (Sec. 4.2). During the wetting of the soil the capillary films are automatically enlarged. Thus, the acting compression stress is relieved, which permits further opening of the expandable lattices.

Swelling in clays is associated with sorption of moisture either from a liquid that comes in contact with the clay or from the ambient humid air. Water may contact the top of the clay deposit (rain and melting snow) or be lifted up by suction forces from the lower strata. Surface water outside a building may penetrate into its interior under continuous shallow footings. Chunks of stiff but expandable clay placed in the body of an embankment may swell in contact with wet ground and fissure the embankment. Generally, according to the rate of water affluent, the swelling will be gradual and last a long time until a limit is attained. Laboratory tests have proved that the water intake is greatest at the start of the swelling process and proceeds at a decreasing rate. As the attracted water is accumulated, *expansion pressure* builds up in it. In reality this is a particular case of *pore pressure* where water is forced into insufficient room.

Identification of Expansive Clays. Large engineering organizations identify expandable minerals by petrographic tests such as microscope analysis, X-ray diffraction, and differential thermal analysis. In ordinary engineering practice such methods are usually economically infeasible; instead the simple tests discussed in succeeding paragraphs are used.

Free-swell Test.[22] Ten cubic centimeters of clay passing No. 40 sieve is slowly added to 100 cc of water in a graduate. The volume of settled and expanded material is read after 24 hr in terms of the graduation of the cylinder. By this procedure, bentonite may swell 1500 per cent, kaolinite about 80 per cent. and illite from 30 to 80 per cent. Some investigators consider that soils having free-swell values as low as 100 may exhibit considerable volume change when wetted under light loading and, therefore, should be viewed with caution. Certain Texas clays with free-swell values in the range of only 50 per cent caused difficulties due to expansion.[22a] The free-swell test alone cannot sufficiently depict the expandability of a clay; therefore, the following various tests should be performed in more or less important cases.

Colloidal Content and Consistency-limit Values as Swelling Indicators. High colloidal content indicates only a *possibility* of the presence of expandable colloids. The *probability* of such presence of expandable colloids is high, however, in regions where montmorillonite-type clays are known to be found and where frequent structure and slab heavings are observed. Also, in such regions, combinations of a low shrinkage limit and a high plasticity index indicate probable swelling. This is because swelling can start only if all clay voids are filled with water; hence, a low-shrinkage-limit value indicates that swelling may start at a low moisture content. A high value of the plasticity index indicates that the clay is capable of sorbing large amounts of water without bursting or cracking at high moisture contents.

Load-expansion Tests. These tests generally are performed in *consolidometers* or similar devices used in the study of compressibility and consolidation of soils (Sec. 4.22 and Fig. 4.31). Basically, a consolidometer is a ring in which the clay sample is placed between two porous stones. The lower stone is immovable, and the upward movements of the upper stone are recorded. In the methods used by the USBR[22] two identical samples are cut from natural or remolded material and placed in consolidometers. After the samples have been dried, at least to the shrinkage limit, the volume of one of them is determined by immersion in mercury (Sec. 4.9). A unit load equal to the anticipated unit soil pressure to be exerted by the structure on its foundation is applied to the top surface of the other sample. The latter sample is wetted from below, and as it swells, its thickness is measured. Thus the change in volume of the given material from natural (or remolded) condition to the air-dried and saturated condition, respectively, is found. An ingenious Swedish device for this test is described in Ref. 24.

Expansion Pressure. The swelling clay of a natural deposit exerts a certain expansion pressure on the overlying materials and structures. If the weight of the earth or structure is greater than the expansion pressure, considerable upward displacement (rise) should not be expected. A maximum field expansion pressure of 147 psi (about 21,000 psf) was recorded in the highly expansion clays of the Central Valley of California.[22]

Consolidometer tests may be used to determine the expansion pressure by preventing the vertical expansion of laterally confined specimens. In this case, potential expansion even greater than described above has been found (about 22,000 psf in the alluvium near Yuma, Arizona; about 30,000 psf in Eagle Ford shale near Dallas, Texas; both are montmorillonites[8]). If a sample of moist expansive clay is precompressed by a very heavy load and then wetted, expansive pressures of an extraordinary magnitude may occur.[8] Conversely, if a naturally precompressed clay is dried and pulverized, its laboratory expansion pressure may be too low. When clay material is tested for an embankment, it should be packed into the apparatus at the moisture content and density to be used in the prototype.

FIG. 4.13. Expansion pressure device. (*California Division of Highways.*)

Figure 4.13 is an expansion pressure apparatus originated by the California Division of Highways for testing expandability of compacted embankment material.[23] The material is conventionally compacted by the kneading action of a special compactor that duplicates field compaction equipment. Then a perforated plunger with a filter paper and a central vertical stem is placed on top of the sample; the plunger is placed under water for a period of not less than 24 hr. Expansion pressure makes the stem press against a device similar to a proving ring. An expansion pressure of 0.5 psi, as registered on the dial, produces a vertical motion of the stem of only 0.001 in. Therefore, the expansion pressure is measured at a virtually constant volume of the sample.

A somewhat similar device is used in India.[25] When a sample of "black cotton soil" similar to Russian "chernozem" (black earth) was tested, an expansion pressure over 6,000 psf was gradually developed in 70 days (Fig. 4.14).

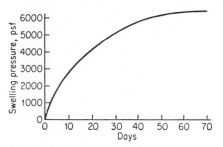

FIG. 4.14. Typical curve of increase of swelling pressure with time.[25]

One technical difficulty in testing expansive soils in containers is the tendency of the sample to swell unevenly, more at the center, less at the periphery because of the wall friction. An attempt to remedy this difficulty has been made in the Swedish apparatus mentioned in Ref. 23. Generally, properly conducted tests of expansive clays in containers are time-consuming, especially with clays of low permeability.

Shearing Strength of the Expanded Soil. The shearing strength of clay decreases as the moisture content is increased above a certain limit. This is, of course, true of swelling soil and rock materials. Though the shearing strength of a swelling clay sometimes may be reduced to practically zero, the decrease in dry density may be relatively small. The same is true of any remolded and recompacted material.

The clay from the Welton-Mohawk canal system, Arizona,[22] had a dry density of 103 pcf and was 93 per cent saturated in its natural state. Its unconfined compressive strength was about 22,400 psf. When saturated (to 99 per cent) under no load, the clay expanded to an average dry density of about 86 pcf and lost practically all its shearing strength.

SOIL CLASSIFICATION SYSTEMS

The two most used classifications in the United States are (1) Unified Soil Classification System (USC) and (2) Public Roads Classification (PR). For other systems see Ref. 26.

4.11. Unified Soil Classification System.[27] This system, originally known under the initials AC (Airfield Classification), was designed by Prof. A. Casagrande for the Army Engineers. It was modified to its present USC form and adopted by the U.S. Armed Forces and the USBR in 1952. Rock fragments above 3 in. in size* that may form part of a soil mass are not considered in this system. The basic soil types considered in the USC are given in Table 4.2.

TABLE 4.2. Unified Soil Classification System: Basic Soil Types

Component	Passing sieve	Retained on sieve	Liquid limit	Plastic limit
Gravel............	3 in.	No. 4 (4.76 mm)		
Coarse..........	3 in.	¾ in.		
Fine.............	¾ in.	No. 4		
Sand..............	No. 4	No. 200 (0.074 mm)		
Coarse..........	No. 4	No. 10 (2 mm)		
Medium..........	No. 10	No. 40 (0.42 mm)		
Fine.............	No. 40	No. 200		
Fines:				
Silt..............	No. 200	....	28 or less	6 or less
Clay.............	No. 200	....	....	Above 6

* The smallest dimension of the particle defines its over-all size. Thus a particle that measures roughly 3 by 5 by 4 in. is classified as a 3-in. rock.

Soil Groups. Natural soils are mixtures of gravel, sand, and fines in variable proportions. All natural soils are subdivided into coarse-grained, fine-grained, and highly organic soils. Coarse-grained soils are those having 50 per cent or less fines; fine-grained soils have more than 50 per cent fines. All percentages are by weight. Highly organic soils are identifiable by their black or dark color and sometimes by the odor. The coarse- and fine-grained soils are subdivided as shown in Table 4.3.

TABLE 4.3. Unified Soil Classification System Subdivisions

Main soil type	General symbol	Subdivisions of main soil type			
		Well-graded, little or no fines	Poorly graded, little or no fines	Coarse, with fines nonplastic or of low plasticity	Coarse with plastic fines
Coarse-grained soils:					
Gravel and gravelly.........	G	GW	GP	GM	GC
Sand and sandy............	S	SW	SP	SM	SC
Fine-grained soils:		Subdivided on the basis of LL			
		50 or less		Over 50	
Inorganic silts, very fine sands, silty or clayey fine sands, micaceous and diatomaceous soils, elastic silts.	M	ML		MH	
Inorganic clays............	C	CL		CH	
Organic clays and silts.......	O	OL		OH	
Peat, humus, swamp soil......	Pt				

Identification of Soil Groups in the Field.[27] The field investigations are mostly visual and manual. If the material is coarse-grained, it is spread on a flat surface and examined for gradation, grain size and shape, and, if possible, mineral composition. It is difficult without experience to differentiate in the field between well-graded and poorly graded soils, and in ambiguous cases, gradation curves should be constructed (Fig. 4.4). Pebbles and sand grains of sound rock are easily identified; weathered rock is discolored and crushes easily between fingers. To disclose objectionable shale, the slaking test should be done (Sec. 4.6). Percentage of fines in sand or gravel can be estimated by thoroughly shaking a soil and water mixture in a graduated cylinder and observing settlement. The sand sizes settle quickly to the bottom, and their volume can be estimated from the divisions on the graduate and compared with the original volume of the sample.

The principal procedures for field investigation of fine-grained soils are (1) the dilatancy test (reaction of the soil to shaking) and (2) estimation of the plasticity characteristics. In the *dilatancy* test, a soft cake about ½ cu in. in size is made of the minus 40 portion of the soil. The cake is placed in the open palm of one hand and shaken horizontally, striking vigorously against the other hand several times. A nonplastic (silty) cake becomes "livery" in this test and shows moisture beads at its surface. When squeezed, the cake becomes dry from the outside and brittle; repeated shaking again gives it the wet appearance. *Plasticity* may be investigated by rolling moist material between the palms. The material is plastic if it can be rolled into fine threads about ⅛ in. in diameter (compare PL test, Sec. 4.9).

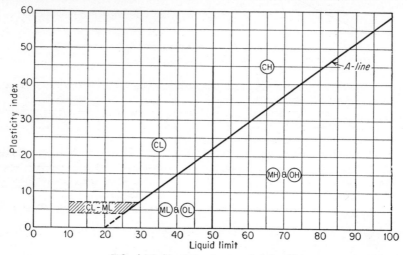

FIG. 4.15. Plasticity chart and A line.[27]

Laboratory Investigations and the A Line. The main laboratory tests required in the USC are the sieve analysis and the plasticity limits determination. Figure 4.15 represents the *plasticity chart* and the so-called A line. The latter is a plot of the PI against LL defined by the relationship PI = 0.73(LL − 20). Note the deviation of this rule in the shaded areas. Clays are *above* the A line, and silts are below it. Clays located close to the A line are silty clays; silts close to the A line are clayey silts. The vertical line through LL = 50 separates silts and clays with low LL from those with high LL. The LL and PI of a given soil found in the laboratory determine the position of that soil on the chart. If the position of a soil is within the shaded area, a double classification such as ML-CL should be used.

4.12. Public Roads Soil Classification System.[28]

In a recent version of this classification system, all soils except peat (which is excluded) are subdivided into two large classes: (1) granular materials containing up to 35 per cent fines and (2) silt-clay materials containing over 35 per cent fines. Granular materials are considered excellent-to-good highway subgrade (Sec. 8.27), whereas the silt-clay materials are fair-to-poor subgrade. All soils are subdivided into seven groups. Groups A-1 and A-3 contain granular materials, whereas groups A-4, A-5, A-6, and A-7 correspond to silt-clay materials. Group A-2 contains borderline materials and is subdivided into four subgroups. Materials belonging to the groups other than A-2 are designated by the number of the group and a "group index" computed in each case from the results of the laboratory tests. As in the USC, the laboratory tests required for soil-identification purposes are the sieve analysis and the determination of the consistency limits (except shrinkage limit). The hydrometer analysis is not used in either system.

The soil groups of the PR system are as follows:

A-1. Well-graded mixture of stone fragments or gravel, sand, and a nonplastic or feebly plastic binder (fines). Subgroup A-1-a includes predominantly stone fragments or gravel; subgroup A-1-b consists predominantly of coarse sand.

A-3. Fine beach sand, desert blow sand without fines or with a very small amount of nonplastic silt.

A-2. Wide variety of granular materials which in reality are borderline materials between A-1 and A-3, and the silty clays.

A-4. Nonplastic or moderately plastic silty soil, 75 per cent or more of minus 200.

A-5. Similar to group A-4 except that it is usually diatomaceous or micaceous and may be highly elastic as indicated by a high LL.

A-6. Plastic clay, 75 per cent or more of minus 200. Usually a high volume change between wet and dry states. The group index (see next paragraph) ranges from 1 to 16 with increasing values indicating the combined effect of increasing PIs and decreasing percentages of coarse materials.

A-7. Similar to A-6 except that it has a high LL (as in A-5). Range of group indices is 1 to 20.

A-2-4 and A-2-5. Various granular materials with 35 per cent or less of minus 200 and with minus 40 having the characteristics of A-4 and A-5 groups (i.e., gravel and coarse sand with silt content or PI in excess of limitations of group A-1 and fine sand with nonplastic silt content in excess of limitations of A-3).

A-2-6 and A-2-7. Materials similar to A-2-5 and A-2-4 except that the fine portions contain plastic clay with the characteristics of the A-6 or A-7 groups.

The group index equals $0.2a + 0.005ac + 0.01bd$, wherein a is the per cent of minus 200 less 35, expressed as a positive whole number not exceeding 40; b is the per cent of the minus 200 less 15, expressed as a positive whole number not exceeding 40; c is the numerical liquid limit value less 40 (i.e., $LL - 40$), expressed as a whole positive number not exceeding 20; and d is the numerical plasticity index less 10 (i.e., $PI - 10$), expressed as a whole positive number not exceeding 20.

Example: The minus 200 fraction is 75, LL is 40 per cent, and PI is 17 per cent. The group index is calculated as follows: $a = 40$ (i.e., $75 - 35$); $b = 40$ (although $75 - 15 = 60$); $c = 0$, since LL equals 40; and $d = 7$ (i.e., $17 - 10$). Group index = 11 when rounded to the nearest whole number.

After the laboratory tests are made and the group index computed, Table 4.4 is used from left to right, working by elimination until the proper classification is found.

Thus the soil of the preceding example belongs to the A-6 group and should be designated as A-6-(11), where (11) is its group index as previously determined.

Addendum. The following are a few soil terms not described elsewhere in the book: *Caliche* designates formations of clays, sands, and gravel cemented by calcium carbonate deposited by the evaporation of ground waters during changes in their level; *marl* is a stiff, calcareous clay of marine origin; *gumbo* is a fine-grained clayey material, impervious but sticky when saturated and waxy in appearance and touch; *diatomaceous earths* are deposits of fine, generally white, siliceous powder formed by diatoms, or microscopic unicellular marine or fresh-water algae.

TABLE 4.4. Soil Identification in the PR System

Group classification	Granular materials			Silt-clay materials			
	A-1	A-3	A-2	A-4	A-5	A-6	A-7
Sieve analysis:							
Passing sieve No. 40..	50 max	51 min					
Passing sieve No. 200.	25 max	10 min	35 max	36 min	36 min	36 min	36 min
Liquid limit (LL).......	......	......	......	40 max	41 min	40 max	41 min
Plasticity index (PI)....	6 max	......	......	10 max	10 max	11 min	11 min
Group index.............	......	......	4 max	8 max	12 max	16 max	20 max

STRESSES IN EARTH MASSES

4.13. Forces and Stresses Considered in Soil Mechanics. In soil mechanics considerations, the earth surface often is assumed to be hori-

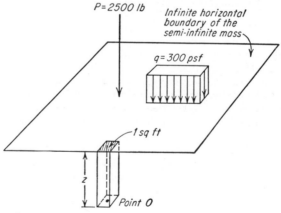

FIG. 4.16. Semi-infinite mass. *P* = concentrated load, *q* = uniformly distributed load.

zontal and infinite in all directions.   A mass thus bounded at the top only is a *semi-infinite mass*.   For simplicity in computations, structures generally are replaced either by *concentrated loads* (*P* in Fig. 4.16) or by *distributed loads q*, generally uniformly distributed at the earth surface (Fig. 4.16).   The weight of a structure is a vertical *force*; there also may be horizontal or oblique forces such as the earth pressure exerted on a retaining wall by its backfill.

Forces acting on an earth mass produce *stresses* in it.   It is very important to distinguish stresses from forces.   As soon as a structure is built
at or close to the assumed horizontal surface of the earth mass or ground
surface, stresses occur in the
whole semi-infinite mass.   At each
point of the earth mass, a small
cube may be imagined with horizontal and vertical sides (Fig.
4.17).   The weight of the structure
affects each small cube and causes
vertical pressures on its horizontal
side (p_v in Fig. 4.17); symbol p
stands for pressure, and subscript v
means vertical.   Each small cube

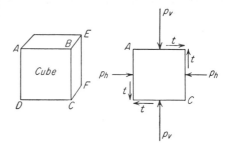

FIG. 4.17. Stresses at a point.

being loaded vertically pushes the neighboring cubes horizontally, hence
horizontal pressure p_h acting on the vertical sides of the cube.   The deeper
under the ground surface and the farther the cube is from the structure,
the smaller are both the vertical pressure p_v and horizontal pressure p_h.
It is said that the pressures "dissipate" with the distance from the point
where the main force acts and finally become negligible.   Now, imagine a
tilted cube; in this case all pressures on its sides would be oblique.   Thus
at every point of a loaded earth mass, there is an infinite number of *pressures in all directions* and of variable values.   The pressures within an
earth mass also are called normal stresses, the term "normal" in this case
meaning "perpendicular."

Tension also is a normal stress, but all soils except compact clays and cemented
sands have poor tensile resistance.   Fissuring of clays is caused by tensile stresses.
Since the tensile resistance of clays increases with depth because of the compressive
influence of the overburden, fissures are produced only at the upper strata of a clay
deposit.

Besides normal stresses, there are *tangential* or shearing stresses, or
simply "shears," at each point of a loaded mass (symbol t, meaning
"tangential").   Figure 4.18a shows in a simplified manner how shearing
stresses within an earth mass originate owing to surface loading of the
mass.   It may be assumed that a concentrated load P produces stresses
in all directions starting from its point of application.   One of these
stresses acting along the direction OA strikes a horizontal plane (MN in
Fig. 4.18a) and is broken into a vertical pressure p_v and a horizontal
shearing stress t along plane MN.   (Imagine that the small cube at point
A is pressed down and simultaneously pushed away from point O using
OA as a long stick.)   Besides a horizontal plane MN, a vertical plane also
may be imagined at point A.   Stress OA will strike it too and will be

broken into a horizontal pressure p_h and vertical shearing stress t (not shown in Fig. 4.18a).

Figure 4.18b shows that shearing stress is akin to friction. In fact, if a block B is pulled by a horizontal force T along a plane MN, there is friction opposing the motion, this friction being equal to force T. If,

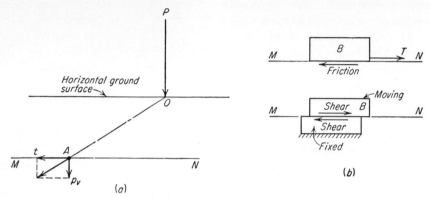

FIG. 4.18. Concept of shear in earth masses.

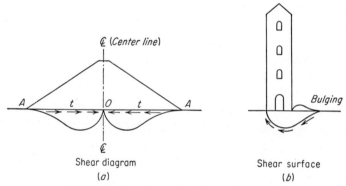

FIG. 4.19. Shearing stresses in earth masses.

however, the upper portion of body B tends to break the bond connecting it to the lower (fixed) portion and move, there is (1) a "driving" shearing stress above the plane of sliding that pushes the upper portion forward and (2) a shearing stress that is numerically equal but acts in an opposite direction and resists that action.

There are no shearing stresses at *unloaded surfaces* and at *axes of symmetry* of structures and imposed loads. An embankment (Fig. 4.19a) tends to spread or flatten out; this tendency is opposed by the shearing stresses acting against the impending motion, i.e., from the edges A toward the "center line" of a symmetrical embankment through point O. There are no shears either at the edges A or at the center line (point O).

As the diagram below the base of the embankment shows, the shearing stresses reach a maximum somewhere between points O and A on both sides of the center line.

In Fig. 4.17 it should be noted that the numerical values of the shears t are all equal. This is according to the general rule of mechanics that the shearing stresses acting at a point along two mutually perpendicular planes *are equal*. For example, at each point of the embankment base (Fig. 4.19*a*), besides a horizontal shear, there also is a vertical shear of equal magnitude. As these vertical shears are small in comparison with the weight of the embankment they are neglected in computations.

All stresses, both normal and tangential, are often measured in tons per square foot or, as done in this book, in pounds per square foot (psf). In many branches of engineering the designation "psi" (pounds per square inch) is popular. In metric measures the stresses are measured in metric tons per square meter or in grams per square centimeter.

4.14. Stressed Condition in an Unloaded Earth Mass. Before a semi-infinite mass is loaded, there are vertical and horizontal *pressures* on it, but *no* vertical or horizontal *shears*. For example, if a cubic foot of earth weighs 130 pcf, the vertical pressure at a depth of 20 ft would be $130 \times 20 = 2{,}600$ psf, or 1.3 tons/sq ft. At every point of an earth mass there also is a horizontal pressure which is a fraction of the vertical pressure. This fraction, called *coefficient of earth pressure at rest* (symbol K), may be assumed roughly at 0.4 in sands and more in clays. Thus in the preceding example the horizontal pressure at a depth of 20 ft (acting in all horizontal directions) would be $2{,}600 \times 0.4 = 1{,}040$ psf, approximately. If a natural earth mass is in equilibrium, all its parts are in equilibrium. Pressures and shears produce *displacements* of particles within the earth mass.

Generally structures are built on earth masses in equilibrium. In this case, in computing stresses, those caused by the structure only are considered.

4.15. Stressed Condition in a Loaded Earth Mass. A compressed soil layer decreases in thickness; the value of this decrease is a *deformation*, though the term "decrease in thickness" is more correct. *Settlement* is a vertical downward movement of a structure built on a compressible soil deposit (a horizontal movement of a structure would be a *lateral displacement*). Except for very rigid and sometimes very small ones, structures rarely settle uniformly across the whole area occupied by the structure. Very often, because of nonuniform loading and the heterogeneous character of the supporting soils, various portions of the structure settle by different amounts; i.e., *differential settlement* takes place. If, for example, the settlement values in two adjacent portions of a building are $1\frac{1}{2}$ and $\frac{1}{2}$ in., the value of the differential settlement would be 1 in. Large but uniform settlement of a structure generally is less detrimental than a differential settlement of the same or even of a smaller magnitude.

Normal stresses, particularly compression, change the thickness of a stratum but affect its shape very little, if at all. Conversely, shearing stresses are responsible for all cases of change in shape of an earth mass. For instance, squeezing of a clay in a semifluid state (plastic flow) from underneath an overloaded structure is caused by shearing stresses. Shearing stresses in many cases tend to separate a portion of the earth mass (a "wedge") from the rest of the mass. The surface of separation is termed *shearing surface* (or sliding or failure surface). A high building (e.g., a tower, Fig. 4.19b) tends to tip, and a potential shear surface is developed under it. As in Fig. 4.18b, there are two systems of shears along the potential shearing surface. The "driving" shearing stresses as caused by the weight of the structure tend to move the wedge out. The earth mass at every stage of the shearing process mobilizes "resisting" stresses that act against the direction of the pending motion of the wedge (as shown by the arrows in the figure). These resisting stresses are exactly equal to the driving stresses but cannot be mobilized beyond a certain limit controlled by the *shearing strength* (or shearing resistance) of the material. When the driving shearing stresses balance the shearing strength, the mass is said to be in a state of *limit equilibrium*. An additional even small increase of the driving shearing stresses (or drop in resistance) will cause a *shear failure*. In the case of Fig. 4.19b there will an outer movement of the wedge and bulging of the displaced earth.

It is generally considered that an earth mass "fails" if a considerable part of it (a "wedge," as in Fig. 4.19b) is cut off by shears or a considerable amount of the material supporting a structure is squeezed out. A structure may fail, i.e., become unable to serve its purpose, either because of (1) the *shear failure* of the supporting mass or (2) considerable differential *settlement*. Of course, there also is a great number of structural failures that cannot be discussed here. Accordingly, soils should be tested for (1) their resistance to shear and (2) their compressibility.

SHEAR AND SHEAR TESTS OF SOILS

4.16. Shear in Cohesionless Soils. As with rock, a cohesionless clean sand or gravel mass is in the state of limit equilibrium if (also see Sec. 2.8)

$$t = p \tan \phi \qquad (4.10)$$

The value of t in Eq. (4.10) is the maximum shear stress that the given material is capable of sustaining. In other words, the equation expresses the shearing strength s of a given material ($s = t$). The symbol ϕ in this case stands for the angle of internal friction of the material. This value varies from sand to sand or gravel to gravel and generally depends on the density of the material and the size and shape of its grains or particles. Larger values of ϕ are associated with (1) angular rather than with rounded materials because of the interlocking of the grains in the former type, (2) well-graded rather than with poorly graded materials, and (3) dense rather than loose materials. Engineers often use a ϕ value of 30° for sand. In this case tan $\phi = 0.577$. It may be assumed that in natural sand deposits, the angle of internal friction varies between 30 and 40° and sometimes more in dense deposits. Under water, ϕ is 1 or 2° less, though in engineering computations it sometimes is assumed to be 25°. The smaller the value of ϕ assumed in the design, the more on the safe side are the computed results, but the probability of overdesign and thus unnecessary expense is greater.

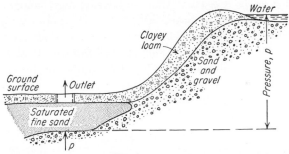

FIG. 4.20. A case of quicksand action. (*From D. P. Krynine, "Soil Mechanics," 2d ed., McGraw-Hill Book Company, Inc., New York, 1947.*)

If dry, clean sand or gravel is poured carefully and without impact on the earth surface, it will form a pile. The slope the pile makes with the horizontal ground surface ("natural slope") is called the *angle of repose*. The angle of repose is slightly smaller than the angle of internal friction ϕ. The use of the angle of repose instead of the angle of friction is on the safe side, but slightly less economical.

It is evident that at the top of a dry, clean sand mass there is no shearing strength, since there is no pressure on the grains and the sand is loose. The shearing strength of such a mass increases with depth because of the increase of pressure. For the same reason, any uniformly distributed load at the surface of a sand mass automatically increases the shearing strength of the sand.

Pore Pressure in Fine Sands. In projects located at the foot of hills or similar situations, the pore water may be under pressure (Fig. 4.20). If an excavation cuts through an impervious soil layer to a saturated fine sand under pressure, a dense, fine sand suspension will flow out. This is a *quicksand condition*. If water under pressure is enclosed in a material that will not flow out with the water, water is ejected from the excavation as artesian water (Sec. 5.22). In laymen's terms, "quicksand" designates dry or moist, loose, fine sand deposits in which a pedestrian may sink. The term quicksand should not be used in this sense by engineers or geologists.

The value of the pore pressure generally is called "neutral stress" (symbol u). Since the pore pressure acts in all directions, it decreases the pressure between sand grains. In this case, Eq. (4.10) should be rewritten as

$$s = (p - u) \tan \phi \qquad (4.11)$$

If $u = p$, the sand mass loses its shearing strength completely and flows. This is *liquefaction* of a fine sand mass or of a mass made of a similar material.[29] Piles of chemical wastes often consist of saturated fine powders with a dry surface crust. Rain water seeping through fine fissures increases the pressure in the pore water; when the pore pressures balance the normal pressures, the waste piles flow. *Saturated silts* fail in the same way if subjected to vibrations. This kind of failure in cohesionless fine-grained soils such as silts and fine sands should not be confused with thixotropic phenomena in clays (Sec. 4.6).

4.17. Shear in Cohesive Soils. In Sec. 2.8, the Coulomb formula (2.10) may be simply expressed as

$$s = c + p \tan \phi \qquad (4.12)$$

where c is unit cohesion and p is unit pressure between particles. Plotting normal pressure p horizontally and shearing resistance s vertically, the plot in Fig. 4.21 may be obtained. If $c = 0$, Eq. (4.12) becomes Eq. (4.10) and the plot passes through the origin. In both cases in Fig. 4.21, the shearing strength of the clay should increase with depth. At the present state of our knowledge, Eq. (4.12) may be tentatively used for partially saturated clay. Pure shear in nature practically never occurs, as it generally is combined with compression, particularly in laboratory tests (Sec. 4.18).

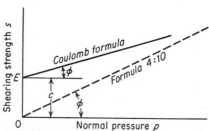

FIG. 4.21. Graphs of shearing strength.

4.18. Laboratory Shear Tests.* The following clay terminology is introduced. A *precompressed*† clay is believed to have been compressed

---

* Before reading Sec. 4.18, it is advisable to reread Sec. 2.9 in which compression and shear tests on rocks are described. These latter tests are very similar to those used in soil testing.

† The term "preconsolidation" often is used in the same sense.

by a heavy overburden during its geologic history, and possibly, this over-burden has subsequently been removed by erosion or other forces. "Pre-compression" by desiccation (Sec. 4.6), by moving ice sheets, or by chemi-cal action also is a possibility.   Clays that have never been subjected to a pressure larger than the weight of the existing overburden are *normally loaded* (Terzaghi's terminology).   The tests are as follows:

1. *Triaxial Compression Test.* This test is performed on soils in transparent, generally lucite, cylinders.   The soil samples are from 1.4 in. and up in diameter and are enclosed in a rubber membrane.   At least two tests with variable lateral pressure (i.e., pressure in the liquid filling the cylinder) should be performed.   If the soil sample is only partially

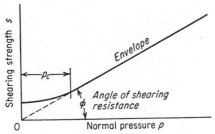

FIG. 4.22. Results of a slow triaxial test (p_c = preconsolidation pressure).

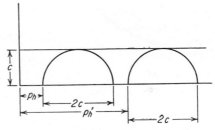

FIG. 4.23. Results of a quick triaxial test (Q).   In this case $p_v = p_v' = q_u = 2_c$.

saturated and Eq. (4.12) holds, the Mohr diagram is constructed as in Fig. 2.6.   If, however, the sample is saturated, the test is done in one of the following ways (A. Casagrande's nomenclature):

Slow Test (S): After full consolidation of the sample in the test cylinder under a certain lateral pressure, the axial (vertical) load on the sample is applied to destruc-tion, but so slowly that water from the sample drains freely enough to prevent detri-mental pore pressures.   This test imitates the behavior of a soil mass on which a structure, e.g., a dam, is *slowly* constructed.   If the clay has not been compressed in nature, the Mohr envelope will pass through the origin O of the experimental dia-gram (Fig. 4.22).   If the clay is precompressed, the shearing strength of the clay being tested practically does not increase and the deformations under load are very small, until the load (i.e., unit normal pressure on the material) reaches a certain value p_c, called *preconsolidation unit load* (Fig. 4.22).

Quick Test (Q): In this test no consolidation of the sample or drainage of water from the pores is permitted.   Whereas the S test is a drained test, the Q test is an *undrained* one.   The axial load is applied quickly, and the Mohr envelope is a horizontal line (Fig. 4.23), which means that the angle of shearing resistance $\phi = 0$.   The critical unit load p_v in this case equals the compressive strength q_u obtained in the unconfined compression test explained hereinafter.

The so-called "$\phi = 0$" method of design proposed by Skempton[30] and used in England is based on the Q triaxial test, i.e., on the assumption that the angle of shearing resistance of a soil is zero, though in reality it may be quite different from zero. This corresponds to the situation when there is no drainage under the structure.

2. *Unconfined Compression Test.* This test is used for cohesive materials either with some friction or completely frictionless ($\phi = 0$). The unit load that destroys the sample is the *unconfined compressive strength* of the material (symbol q_u). Generally, for clays and materials behaving like clays, the shearing strength is about one-half of the unconfined compressive strength ($\frac{1}{2}q_u$). Mixtures of sand and clay reveal *too small* a value of the compressive strength, which, however, is not indicative of unfitness of the material. The destruction pattern of the sample in this test is the same as with rocks (Fig. 2.1). Plastic samples bulge in a barrel-like fashion. In this case the unit load that causes a predetermined decrease in the height of the sample (usually 15 or 20 per cent) is assumed to be the unconfined compressive strength.

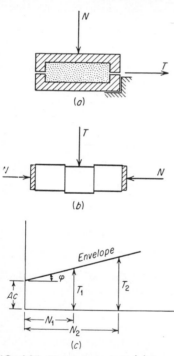

FIG. 4.24. Direct shear test: (a) box test, (b) double-shear ring test, (c) construction of the envelope.

3. *Direct Shear Tests.* In these tests the normal load P tends to keep the sample in place, whereas the directly applied tangential force T tends to separate one portion of the sample from the other. In the device of Fig. 4.24a, this is done by placing the sample in two superimposed boxes, one of which is movable. In the double shear device of Fig. 4.24b, a cylindrical sample is confined in the three rings which constitute an integral part of a field soil sampler and the middle ring is pulled out. Figure 4.24c shows a graphical procedure for determining the ϕ and c values. Because of the irregular stress distribution and change of the effective cross section of the sample during the test, only a *rough* idea is obtained of the shearing strength of the material.

4. *Penetration Tests. CBR Tests.* Penetration of a steel rod of a small diameter into a soil mass is resisted by the shearing stresses at its periphery if a small point resistance is neglected. Hence, such a penetration is a shear test. In penetration tests on natural soil deposits, the shearing strength in a vertical direction is measured; because of strati-

fication, this strength may be greater than that in a horizontal direction. Field penetration tests are described in Sec. 6.11.

A particular case of the laboratory penetration test is the California Bearing Ratio test known as the CBR test.[31] It is widely used in the design of flexible airport and highway pavements and sometimes in the specifications for the use of earth materials for various construction purposes. The features of the purely empirical test are standardized,[31a] and experience has shown that even minor deviations from standard procedures will cause wide variations in the test results. Therefore, those wishing to perform this test should closely follow the standard.

Basically the test consists of compacting about 8 lb of material at optimum moisture content (Sec. 16.11) and expected field density in a cylindrical mold 6 in. in diameter and 7 in. high. The sample is soaked for 4 days, its expansion measured, and penetration resistance tested by a 3-sq-in. piston of a testing machine. The pounds per square inch unit loads required to drive the piston 0.1 and 0.2 in. into the soil mass are divided by 1,000 and 1,500 psi, respectively. The ratio, expressed in per cent, for the 0.1-in. penetration is generally critical, i.e., greater than the other. Exceptionally, the ratio for the 0.2-in. penetration may be critical.

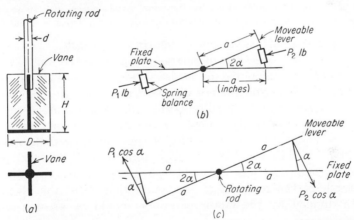

FIG. 4.25. Vane: (a) bottom of the device, elevation and plan; (b) schematic plan of the top; (c) data for the computation of the rotating moment.

4.19. Vane Shear Test.[32-35] The vane method of measuring the shear strength of the soils *in the field* was developed in Sweden[32,34] about 1948 and studied subsequently in England.[33,35] In its simplest form, the device consists of a four-winged vane fixed to a vertical rod (Fig. 4.25a). The vane may be sunk within a casing, as in sampling, or may be driven directly into the ground if soil conditions permit. The rod of the vane rotates by turning a lever at its top (Fig. 4.25b) at a rate of about 0.1°/sec.[33] The angle of rotation 2α is read on the horizontal graduated circle of the vane. The applied force is measured by one or two spring balances fixed to the lever and the fixed plate (readings P_1 and P_2, Fig. 4.25b). When the shearing strength of the earth material is overcome, the readings of the balances start to drop. To compute the shearing strength of the material, the corresponding critical readings of 2α, P_1, and P_2 are introduced into Eq. (4.14). To use this formula, a constant C that depends on the dimensions of the vane is computed once for all uses [Eq. (4.13)].

Approximate dimensions of the vane (Fig. 4.25a) are as follows: diameter $D = 2$ or 3 in., height $H = 4\frac{1}{2}$ in., thickness of the blade $= \frac{3}{32}$ in., diameter d of the rod $= \frac{1}{2}$ in., and that of the fixed plate $2a = 20$ in. (Fig. 4.25b). For all dimensions in inches, the value of C is

$$C = \frac{a}{\pi D^2 (H + \frac{1}{3}D)} \tag{4.13}$$

and the shearing strength s of the material is

$$s = C(P_1 + P_2) \cos \alpha \tag{4.14}$$

If P_1 and P_2 are expressed in pounds, Eq. (4.14) gives the shearing strength s of the given material in pounds per square inch. To express in pounds per square foot, multiply the result by 144.

It appears that the results given by the vane closely agree with those from the unconfined compression test on samples obtained down to a depth of about 45 ft.[32,33]

4.20. Shearing Stresses and Shearing Strength. The shearing stresses at all portions of a foundation or a proposed earth structure (e.g., a dam) are computed or at least estimated. This is a purely engineering problem which is not discussed here, though in some simple cases the shears may be readily estimated. For example, in a long flexible building, uniformly loaded, the shears at the long edges of the building are equal to about one-third of the uniformly distributed unit load. Thus, if that building is loaded uniformly with 3,000 psf, the shears at its long edges will be roughly 1,000 psf. In any case, the shearing strength of the material of the foundation or of the given earth structure divided by the factor of safety should be greater than the shearing stress.

PRESSURES IN EARTH MASSES

4.21. Vertical Pressure at a Point. When the horizontal surface of a semi-infinite mass is loaded, there are pressures and shears at all points of the mass. These stresses may be estimated, using the formulas of the theory of elasticity developed for idealized elastic bodies. Though soils, especially sand masses, are not elastic bodies, the elastic formulas may give a fair idea of the stress patterns if applied at a certain depth below the structure. According to an arbitrary practice, this depth, for structures of limited dimensions, should not be less than three times the greatest dimension of the structure in plan.

Boussinesq Formula. This formula is used to compute (or, more accurately, to estimate) the vertical pressure p_v at a point O within a semi-infinite elastic mass, caused by a concentrated load P applied at point A of the surface (Fig. 4.26). Imagine that point O is located on a horizontal plane which is z units (e.g., feet) deep. If the line of action of the load P is extended within the earth, it will strike that horizontal plane at some point A'.

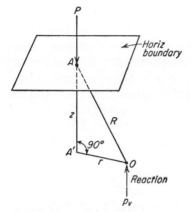

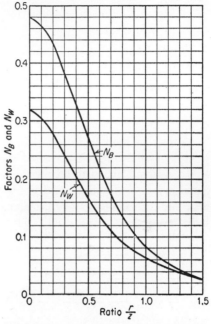

FIG. 4.26. Vertical pressure p_v at a point within an earth mass.

FIG. 4.27. Boussinesq's (N_B) and Wester-gaard's (N_W) factors for determining the vertical pressure p_v. (*Adapted from Taylor's "Fundamentals of Soil Mechanics," by permission.*)

Use symbol r for the distance from point A' to the given point O. The vertical pressure is

$$p_v = N_B \frac{P}{z^2} \tag{4.15}$$

wherein N_B is a coefficient found from the graph in Fig. 4.27. To use this graph the ratio r/z must first be computed.

Example: Assume a load of 4,500 tons applied as a point concentration at the surface. Determine the vertical pressure p_v at a point 40 ft below the load and 15 ft away horizontally, so that $r = 15$ ft, $z = 40$ ft, and $r/z = 0.38$. From graph, $N_B = 0.34$, and by Eq. (4.15)

$$p_v = 0.34 \frac{4,500}{40^2} = 0.96 \text{ ton/sq ft}$$

Westergaard Formula. This formula differs from the Boussinesq formula in the value of the coefficient N_w:

$$p_v = N_w \frac{P}{z^2} \tag{4.16}$$

The coefficient N_w is determined from the same graph in Fig. 4.27. Be careful not to confuse the coefficients N_B and N_w.

Example: If the Westergaard formula is used in the preceding example, the vertical pressure would be

$$p_v = 0.215 \frac{4,500}{40^2} = 0.60 \text{ ton/sq ft}$$

As may be seen from Fig. 4.27, the Westergaard formula gives smaller values of the vertical pressure than the Boussinesq formula. Westergaard derived his formula on the assumption that the mass is prevented from being strained laterally by the presence of exceedingly thin but very rigid horizontal sheets; this assumption approaches the condition of sedimentary rocks and soils. The original Westergaard formula contains Poisson's ratio, and Eq. (4.16) is a particular case when this ratio is zero.

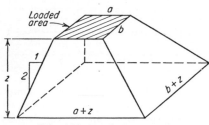

FIG. 4.28. The 2:1 method.

Two-to-one Method. This purely practical method is convenient to estimate roughly the vertical pressure p_v within an earth mass. If the dimensions of the loaded area are a and b, it is assumed that at a depth z, the load is spread over an area $(a + z)(b + z)$. The stresses are spread in the form of a pyramid (Fig. 4.28) with the sides sloping 1 horizontal to 2 vertical. If there are two or more loaded areas and their pyramids overlap, the load at the overlapping areas is not doubled.

Example: A raft 20 ft by 60 ft is loaded with 6,000 psf. The total load is

$$6,000 \times 20 \times 60 = 7,200,000 \text{ lb}$$

At a depth of 25 ft it is spread over an area of $(20 + 25)(60 + 25) = 3,825$ sq ft. Hence the *average* (but not maximum) vertical pressure at that level is

$$p_v = \frac{7,200,000}{3,825} = 1,880 \text{ psf}$$

Usually 50 per cent is added to the average pressure to obtain the maximum vertical pressure. Accordingly, in this case, the maximum vertical pressure under the center of the raft would be

$$p_v = 1,880 \times 1.5 = 2,820 \text{ psf}$$

Newmark's Chart. To compute the vertical pressure p_v caused within the earth mass by a flexible uniformly loaded area of any shape placed at or close to the ground surface, Newmark's chart may be used. The point of the surface under which the pressure in question is to be found may be under or outside the loaded area. The chart is simple and may be drawn in half an hour by a person familiar with drawing. The chart consists of nine concentric circles (Fig. 4.29). Their radii measured in inches are shown in that figure. The circles are subdivided into 20 equal sectors. The depth z where the vertical pressure p_v is to be determined

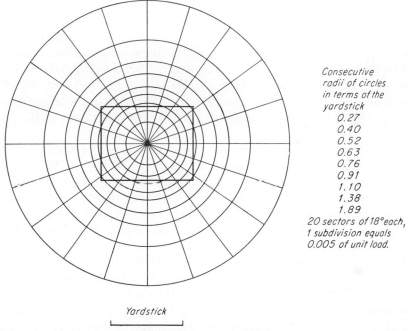

Consecutive radii of circles in terms of the yardstick
0.27
0.40
0.52
0.63
0.76
0.91
1.10
1.38
1.89
20 sectors of 18° each, 1 subdivision equals 0.005 of unit load.

Yardstick

FIG. 4.29. Newmark's chart.

is represented by 1 in. Draw the plan of the loaded area on transparent paper to that scale (e.g., if the depth is 40 ft, the scale will be 1 in. = 40 ft). Place the plan of the loaded area on the chart in such a way that the point of the surface under which the pressure has to be determined coincides exactly with the center of the chart. Count the number of subdivisions, both large and small, covered by the plan of the loaded area. If a subdivision is only partially covered, estimate what part of it is covered. Each subdivision equals 0.005 of the unit load at the area. Hence the vertical pressure in question will be found by multiplying 0.005 of the unit load at the area by the number of subdivisions counted. The Newmark chart is based on the Boussinesq formula.

Example: A flexible area 40 by 50 ft placed at the ground surface is uniformly loaded with 4,000 psf.   Determine the vertical pressure p_v at the center of the area at a depth of 40 ft.

Solution: The scale to which the plan of the given area should be prepared is 1 in. = 40 ft.   Since the area is symmetrical with respect to both center lines, it is necessary to count subdivisions in a quarter of the area only and multiply by 4.   The count is 20.3 subdivisions for a quarter of the area or 20.3 × 4 for 81.2 subdivisions for the whole area.   The vertical pressure in question is

$$p_v = 0.005 \times 4,000 \times 81.2 = 1,624 \text{ psf}$$

A quick check by the 2:1 method gives 1,725 psf for p_v.

Contact Pressures.   In actual design practice it generally is assumed that the pressure distribution under a uniformly loaded structure is also uniform.   Apparently this assumption is fair, since no failure caused by using this assumption has been reported.   Theoretically, soil pressures are uniform only under perfectly flexible structures, and even so, *settlements* of such structures are not uniform.   For example, a large built-up area in a compressible clay region would tend to settle more at its center than at its edges.

4.22. Theory of Consolidation.   Terzaghi's theory of consolidation is applicable to *saturated* clays or similar materials with flexible particles. The theory states that if a mass made of such material is suddenly loaded and the load remains acting on it for a very long time, an instantaneous *initial compression* first occurs as in any other loaded body.   Then the pore moisture will be gradually squeezed out with a consequent decrease in porosity and over-all volume.   This is the *primary compression* to which Terzaghi's theory can be applied.   The final stage of consolidation is the *secondary compression*, which is thought to be some plastic rearrangement of flexible particles; it is rather small in nonorganic clays but large in organic clays.   The consolidating clay deposit usually is covered with an overburden of various soil types; for computations, the compression deformations of the overburden are neglected, as usually they are relatively small in comparison with the decrease in thickness of the consolidating clay deposit.   The structure supported by the overburden gradually settles at a decreasing rate from year to year.   Theoretically the consolidation process lasts indefinitely; practically, it is finite, though it may last decades or even centuries.

The term "consolidated sediment" as used in geology means a hardened, rocklike material, e.g., shale that has been formed from sedimentary clay by gradual compression or other means.   To the engineer, however, consolidated clay is a clay in which the consolidation process, as he defines it, has come to an end.   If an additional load acts on a clay thus consolidated, the consolidation process will start again.

Figure 4.30 shows a typical soft-clay layer overlain by overburden. If the clay is "normally loaded" (Sec. 4.26), the unit weight of the over-burden is the preconsolidation unit load p_c. It should be computed from the ground surface to the middle plane of the soft layer. The weight of the structure "dissipates" with the depth, and the unit vertical pressure p_v exerted by the structure at the middle plane of the soft layer should be computed (Sec. 4.20). This is assumed to be the load that causes the soft clay to consolidate.

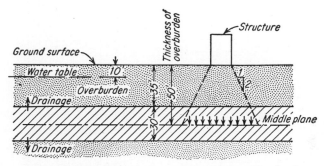

FIG. 4.30. Consolidation of a soft clay layer under load.

The voids ratio e_0 of the saturated clay before construction is computed by Eq. (4.6). The specific gravity G of particles to be used in this formula is determined or assumed (2.65 or 2.70). The final voids ratio e_f, that after consolidation is completed, is determined from the consolidation test described hereafter. Express the total thickness of the clay deposit H in inches; then its decrease in thickness after consolidation is completed, i.e., the expected settlement of the structure, in inches will be

$$\Delta H = H \frac{e_0 - e_f}{1 + e_0} \qquad (4.17)$$

Consolidation Test. A simple consolidation device, or *consolidometer*, is shown in Fig. 4.31. It consists of a ring 3 in. or more in diameter in which a clay sample, $\frac{1}{4}$ to 1 in. (or more), is placed between two porous stones. The porous stones permit the moisture to be squeezed out of the loaded sample, and measures should be taken to keep the sample saturated, e.g., by continuous maintenance of the proper water level in the standpipe. The downward movement of the piston is measured with a

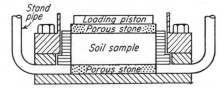

FIG. 4.31. Consolidometer.

dial gauge (precision of 0.0001 in.). The unit pressure acting on the piston is applied gradually in predetermined increments (e.g., 250, 500 psf, etc.). The unloading is done by decrements. Each load increment or decrement should stay on until full consolidation or full rebound for that loading is achieved and in any case not less than 24 hr.

Equation (4.17) can be rewritten as

$$e_f = e_0 - \frac{\Delta H}{H}(1 + e_0) \tag{4.18}$$

If, in Eq. (4.18), H is the thickness of the sample and ΔH is its decrease during the

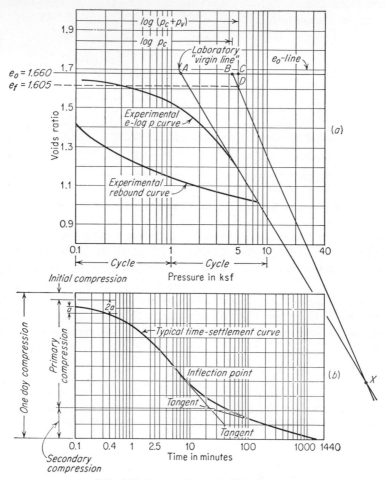

FIG. 4.32. Results of a consolidation test.

application of a load increment, the voids ratio e_f after the consolidation process is completed can be computed.

Unit pressures p at the end of application of each load increment are plotted on a logarithmic scale against voids ratios e_f. In this way the *laboratory e–p* (or *e* log *p*) curve is obtained (Fig. 4.32a). On the horizontal line passing through the division e_0, the loads p_c and $p_c + p_v$ on the logarithmic scale are plotted (points B and C, respectively). The tangent to the laboratory e–p curve (laboratory virgin line) should pass through point B, but because of the sample disturbance, it does not. Instead, point B is joined to a point X located on the laboratory virgin line by a certain procedure (p. 63 of Ref. 37; Ref. 36). Line BX is then the *field compression line* which in many cases may be replaced by a line parallel to the laboratory virgin line AX traced through point B. A vertical CD through point C determines the value of the final voids ratio e_f to be used in Eq. (4.17). In Fig. 4.32a, the loads are expressed in kips (thousands of pounds); the numerals at the end of each "cycle" of the paper mean 1,000, 10,000 lb, etc.

Figure 4.32b shows the time-settlement curve referred to the one-day (1,440-min) records of consolidation under one load increment. The three stages of consolidation are separated by conventional procedures (e.g., p. 71, Ref. 37).

To compare the times t_1 and t_2 of settlement of two layers h_1 and h_2 thick made exactly of the same soil, Eq. (4.19) may be used:

$$\frac{t_1}{t_2} = \frac{h_1{}^2}{h_2{}^2} \tag{4.19}$$

If one of the two layers referred to has a two-way drainage (i.e., both upward and downward) only one-half of its thickness should be used in Eq. (4.19).

The theory of consolidation is involved and controversial. Persons without thorough experience are advised not to use it in practical problems.

4.23. Active Pressure and Passive Resistance of Earth Materials. Figure 4.33a represents a rigid concrete retaining wall with a horizontal cohesionless backfill. The latter exerts a pressure on the wall, which in the case shown is horizontal and increases from top to bottom following a straight-line distribution (triangle ABC to the right of the figure). This is the *active pressure* (or thrust). To estimate its value it is necessary to know the *coefficient of active pressure K_a*, which depends on the value of ϕ of the material and is estimated by the Rankine formula:

$$K_a = \frac{1 - \sin \phi}{1 + \sin \phi} \tag{4.20}$$

For instance, for $\phi = 30°$, which is a very common value, $K_a = \frac{1}{3}$. The value of the thrust is

$$E_a = \frac{1}{2} K_a \gamma H^2 \tag{4.21}$$

and its resultant is applied at the "third point," i.e., at one-third of the height of the wall as shown in Fig. 4.33a. In practice the active pressure often is determined by the method of "equivalent fluid" wherein the earth pressure is replaced by the hydrostatic pressure of a fluid weighing 30 (or 35) psf.

If the backfill is not horizontal or the wall is flexible or the backfill is cohesive, the preceding rules for determination of active pressure do not hold.

Passive Resistance. Figure 4.33b represents a sheet piling wall formed by driving either heavy timber boards or steel piles of special shapes close together. The movement of such walls is opposed by the passive resistance of the earth into which the sheet piling is driven. To compute the value of passive resistance of a cohesionless material, Eq. (4.21) may be used by replacing the coefficient of active pressure K_a by its reciprocal K_p, the *coefficient of passive pressure*. Thus, if $K_a = \frac{1}{3}$, $K_p = 3$. The value of H also should be replaced by the depth of the earth toward which the wall is driven (letters h in Fig. 4.33). When the passive

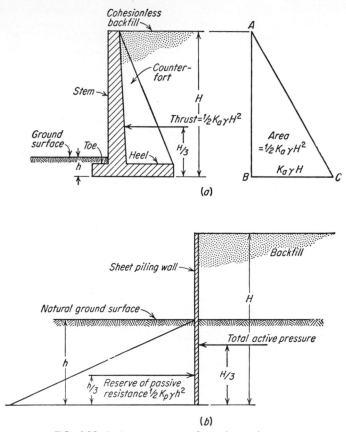

FIG. 4.33. Active pressure and passive resistance.

resistance thus computed is smaller than the active pressure, the wall may fail if there are no other means (e.g., efficient anchors) to prevent the failure. If the earth material surrounding the wall and into which the wall is pushed is cohesive, the passive resistance should be taken as at least equal to the unconfined compressive strength of the material.

4.24. Redistribution of Pressures and Arching. The active pressure on a wall (Fig. 4.33a) is triangular if the wall is rigid. Incomplete or irregular displacement or bending of a flexible wall causes a pressure

redistribution.   Figure 4.34 shows various possible patterns of pressure distribution with a flexible retaining wall (irregular dotted lines).   To take care of these possibilities, the triangular active pressure diagram AED is converted into a trapezoid $ABCD$ which is an envelope of possible pressure patterns found experimentally (by Terzaghi).   The area of tri-
angle AED and trapezoid $ABCD$
should be equal, since both express
the same total lateral pressure on
the wall.

Both vertical and horizontal pressures may be *relieved* at a certain
zone of the mass by the action of

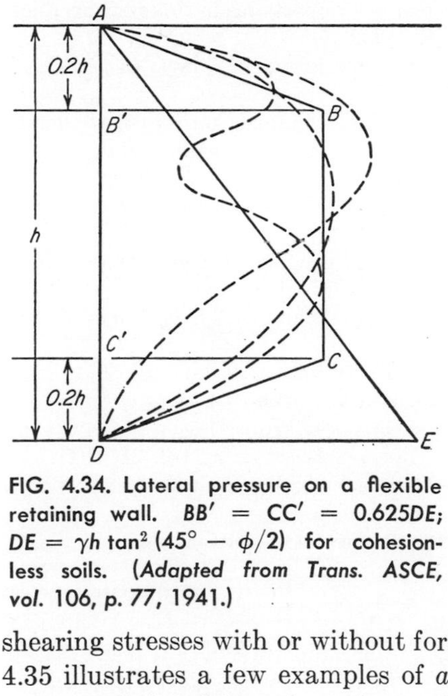

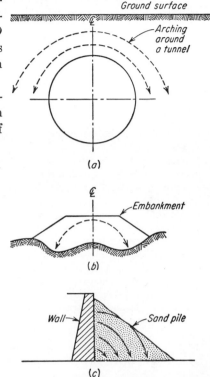

FIG. 4.34. Lateral pressure on a flexible retaining wall. $BB' = CC' = 0.625DE;$ $DE = \gamma h \tan^2 (45° - \phi/2)$ for cohesion-less soils. (*Adapted from Trans. ASCE, vol. 106, p. 77, 1941.*)

FIG. 4.35. Arching.

shearing stresses with or without formation of actual soil arches.   Figure 4.35 illustrates a few examples of *arching:* (*a*) a circular tunnel without concrete lining driven in rock or stiff clay, (*b*) arching within a highway embankment with redistribution of the pressure on the ground and a relief rather than common concentration of pressures at the "center line," and (*c*) arching within a pile of sand dumped against a vertical wall.

REFERENCES

1. Jenny, H.: "Factors of Soil Formation," McGraw-Hill Book Company, Inc., New York, 1941.

2. Joffe, J. S.: "Pedology," Rutgers University Press, New Brunswick, N.J., 1936.
3. "Field Manual of Soil Engineering," Michigan State Highway Department, Lansing, Mich., 1946.
4. Keen, B.: "The Physical Properties of Soil," Longmans, Green & Co., Inc., New York, 1931.
5. Baver, L. D.: "Soil Physics," 2d ed., John Wiley & Sons, Inc., New York, 1948.
6. Edlefsen, N. E., and A. Anderson: Thermodynamics of Soil Moisture, *Hilgardia*, vol. 15, 1943.
7. Schofield, R. K.: The pF of the Water in Soil, *Trans. 3d Intern. Congr. Soil Sci.*, vol. 2, Oxford, England, 1935.
8. Mielenz, R. C., and M. E. King: Physical-Chemical Properties and Engineering Performance of Clays, "Clays and Clay Technology," *Calif. Div. Mines Bull.* 169, 1955.
9. Kubiena, W. L.: "Micropedology," Collegiate Press, Inc., of Iowa State College, Ames, Iowa, 1938.
10. Lambe, T. W.: The Structure of Inorganic Clay, *Proc. ASCE Sep.* 315, October, 1953.
11. Casagrande, A.: The Structure of Clay and Its Importance in Foundation Engineering, *J. Boston Soc. Civil Engs.*, vol. 19, 1932.
12. Skempton, A. W., and R. D. Northey: The Sensitivity of Clays, *Géotechnique*, vol. 3, no. 1, March, 1952.
13. Moretto, O.: Effect of Natural Hardening on the Unconfined Compressive Strength of Remolded Clays, *Proc. 2d Conf. on Soil Mech. and Foundation Eng.*, vol. I, Rotterdam, paper II, p. 1, 1948.
14. Rosenquist, J. T.: Considerations on the Sensitivity of Norwegian Clays, *Géotechnique*, vol. 3, no. 5, March, 1953.
15. Ackermann, E.: Thixotropie und Fliesseigenschaften feinkoerniger Boeden, *Geol. Rundschau*, vol. 36, 1948 (*cf.* Ref. 8).
16. Lee, C. H.: Sealing the Lagoon Lining at Treasure Island with Salt, *Trans. ASCE*, vol. 106, 1941.
17. Holmes, A., *et al.*: Factors Involved in Stabilizing Soils with Asphaltic Materials, *Highway Research Board Proc.*, vol. 23, 1943.
18. Winterkorn, H. F.: A Study of Changes in Physical Properties of Putnam Clay Induced by Ionic Substitution, *Highway Research Board Proc.*, vol. 21, 1941.
19. Wooltorton, F. L. D.: Movements in the Desiccate Alkaline Soils of Burma, *Trans. ASCE*, vol. 116, 1951.
20. Jennings, J. E.: Foundations for Buildings in the Orange Free State Gold Fields, *J. S. Africa Inst. Engrs.*, vol. 49, 1950.
21. Reiner, M.: Stripfootings in Shrinkable Clay Soils, "In the Field of Building," Hebrew Institution of Technology, Haifa, Bull.
22. Holtz, W. G., and H. J. Gibbs: Engineering Properties of Expansive Clays, *Proc. ASCE Sep.* 516, October, 1954. Discussion by (a) R. F. Dawson, (b) W. T. Altmeyer, (c) W. T. Altmeyer, (d) E. S. Barber, and (e) L. A. du Bose.
23. Hveem, F. N.: Suggested Method of Test for Expansive Pressures of Remolded Soils, "Procedure for Testing Soils," *ASTM Comm. on Soils for Eng. Purposes*, Philadelphia, Pa., 1950.
24. Rengmark, F., and R. Ericson: Apparatus for Investigation of Swelling, Compression, and Elastic Properties of Soils, *Proc. 3d Intern. Conf. on Soil Mech. and Foundation Eng.*, vol. I, Zurich, paper 22, session 2, 1953.
25. Palit, R. M.: Determination of Swelling Pressure of Black Cotton Soil—A Method, *Proc. 3d Intern. Conf. Soil Mech. and Foundation Eng.*, vol I, Zurich, paper 19, session 2, 1953.

26. Casagrande, A.: Classification and Identification of Soils, *Trans. ASCE*, vol. 113, 1948.
27. The Unified Soil Classification System, *U.S. Corps Engrs. Waterways Expt. Sta. Tech. Memo.* 3–357 (3 vols.), Vicksburg, Miss., 1953.
28. Classification of Highway Subgrade Materials, *Highway Research Board Comm. Rept.*, vol. 25, 1945.
29. Casagrande, A.: Characteristics of Cohesionless Soils Affecting the Stability of Slopes and Earth Fills, *J. Boston Soc. Civil Engrs.*, vol. 23, 1936.
30. Skempton, A. W.: The $\phi = 0$ Analysis of Stability and Its Theoretical Basis, *Proc. 2d Conf. on Soil Mech. and Foundation Eng.* vol. I, Rotterdam, 1948.
31a. "Engineering Manual for Military Construction," part XII, chap. 2, Corps of Engineers, U.S. Army, July, 1951, rev. March, 1953.
31b. Development of CBR Flexible Pavement Design Method for Airfields—A Symposium, *Trans. ASCE*, vol. 115, 1950.
32. Carlson, L.: Determination in-situ of the Shear Strength of Undisturbed Clay by Means of a Rotating Auger, *Proc. 2d Conf. on Soil Mech. and Foundation Eng.*, vol. I, Rotterdam, 1948.
33. Skempton, A. W.: Vane Tests in the Alluvial Plain of the River Forth, *Géotechnique*, vol. 1, December, 1948.
34. Hansen, J. B.: Vane Tests in a Norwegian Quick-clay, *Géotechnique*, vol. 2, June, 1950.
35. Evans, J.: The Measurement of the Surface Bearing Capacity of Soils, *Géotechnique*, vol. 2, December, 1948.
36. Schmertmann, J. H.: Estimating the True Consolidation Behavior of Clay from Laboratory Test Results, *Proc. ASCE Sep.* 311, October, 1953.

For the study of basic principles of soil mechanics the following additional texts are recommended.

37. Terzaghi, K., and R. B. Peck: "Soil Mechanics in Engineering Practice," John Wiley & Sons, Inc., New York, 1948. The geological factor in soil mechanics is better discussed in this book than in other similar texts.
38. Peck, R. B., W. E. Hanson, and T. H. Thornburn: "Foundation Engineering," John Wiley & Sons, Inc., New York, 1953.
39. Sowers, G. B., and G. F. Sowers: "Introductory Soil Mechanics and Foundations," The Macmillan Company, New York, 1951.

CHAPTER 5

SUBSURFACE WATER

Water on, above, and below the ground surface may be in the liquid state, in the gaseous state (water vapor), or in the solid state (ice and snow). Water below the ground surface is *subsurface water*. Rocks and soils possessing the capacity to permit water (or oil or gas) to pass through them are *pervious* or permeable to a given fluid. Coarse, clean sands are pervious to practically all fluids.

GENERAL CONCEPTS[1-4]

5.1. Sources of Subsurface Water. Subsurface water originates chiefly from the *infiltration of meteoric water*, such as rain water or melted snow or ice, and from *seepage* from streams, lakes, ponds, channels, reservoirs, and other water bodies. Subsurface waters are not chemically pure H_2O but are solutions or suspensions of various substances, which may range from very dilute to highly concentrated. Also, subsurface water may contain air and other gases either dissolved in water or in the form of small, generally microscopic bubbles ("entrapped air").

Two more sources of subsurface water are *juvenile water*, which ascends from magma or lava, and *connate water*, that water which was buried at the same time as its igneous or sedimentary host rock.

5.2. Varieties of Subsurface Water. The simplest case of a flat country will be considered, with a thick, more or less uniform soil deposit covering rock. At varying depths under the ground surface there is a *zone of saturation* in which water fills all the pores in the soils and the openings in the underlying rock. The water in the zone of saturation is commonly termed *ground water*, its upper surface being the *water table*. As explained hereafter, under more complicated topographic and geologic conditions there may be more than one zone of saturation and more than one water table at a given locality.

Ground water is *free* or *gravitational;* i.e., it moves by obeying the law of gravity in contrast to the *attracted* (sorbed) water above the water table where attractive forces may be stronger than gravitation. In the gravitational flow, isolated soil particles are still covered with very thin films of attracted moisture, and in the fine-grained soils, even sandy, the films covering the particles are so close to one another that only a very narrow passage to the gravitational flow is left. In some clays and very fine sands, mostly silty or clayey in character, the gravitational flow may be practically stopped.

Water between the ground surface and the water table is called *vadose* by the geologists. A certain portion of the vadose water is attracted to the soil or rock particles; it moves from weaker centers of attraction to stronger ones by sliding along the surfaces of the particles and thus forming attracted films of variable thickness. A portion of the vadose water is gravitational, however. For example, the excess of rain or other meteoric water which the soils are not able to hold suspended moves slowly downward to the water table.

Another type of vadose water, gravitational in character, may occur during the rainy seasons in hilly localities where the hills are formed by impervious materials such as shale and covered with weathered, somewhat pervious soils. During rainy seasons the weathered material becomes fully saturated and water in it moves downhill, together with the surface flow. In this manner an *aquifer*, i.e., a water-bearing stratum, is formed close to the ground surface. This aquifer sometimes is several feet thick and persists for a long time.

The terms "water" and "moisture" may be interchangeably used to describe all kinds of subsurface water, except gravitational water, to which the term "moisture" is not applied.

The soils located above the water table possess a suction capacity (or free energy, Sec. 4.2) which produces the so-called *capillary movement* of water, or rise from the water table. Moisture thus attracted is mostly in liquid state, though a part of it is gaseous (water vapor). As explained in Sec. 4.2, in dry, coarse-grained soils (such as coarse sands) capillary moisture moves only a few inches above the ground-water table but does so rapidly. However, it may reach a height of 30 or 40 ft or more in fine-grained soils such as clays. In the latter case, months or even years may be required for water to reach the maximum capillary rise possible in a given material. Cases are known where it took capillary water some 2 or 3 years after construction to reach the top of a high embankment built on swampy ground.

In the *capillary fringe*, or zone formed by the rising capillary moisture, the degree of saturation (Sec. 4.7) decreases from the water table up. Only very close to the water table are all soil pores filled with water. Saturation of the soil by capillary water may take place, however, if the water table is shallow, i.e., close to the ground surface, and the maximum height of the capillary rise is large. For example, if the water table is 5 ft deep, but the maximum capillary rise in the given soil is 20 ft, there will be a continuous strong pull of liquid from the water table because of the evaporation at the ground surface and the soil will be fully (or nearly) saturated. This is one of the ways in which *swamps* are formed.

Soil particles above the capillary fringe attract vapor from the air of the pores and thus become covered with the films of *hygroscopic moisture*. Generally, moisture attracted from the air is of limited engineering significance only, but in some cases, it may be detrimental, e.g., (1) in frost and permafrost zones, when the water vapor freezes directly into ice and contributes to the formation of ice inclusions in the soil (Sec. 10.2), or (2) in the not entirely understood case of accumulation of water under pavements in otherwise perfectly dry regions (Sec. 14.19). Moisture on the surface of rock particles sometimes is termed *pellicular;* in reality, this is a variety of hygroscopic moisture.

According to the degree of saturation and moisture content, soil materials may be dry, moist, wet, saturated, and oversaturated. All these terms are relative only, except the term "saturated," which means that all pores are filled with water. A clay which is being gradually compressed still keeps all its pores filled with water though its moisture content may become very low and the material may feel and look only slightly moist or even dry. In the same way, there are no definite boundaries between gravitational, capillary, and hygroscopic water, and no further classification of subsurface water is attempted in this book.

Figure 5.1 shows the possible occurrence of the different kinds of subsurface water. The upper layer 1 represents the topsoil, where meteoric water is reached by the plant roots. Below this layer, rain or other meteoric water travels through layer 2 where it adheres to the surface of soil or rock particles if space is available. The moving water then reaches layer 3—the zone (fringe) of capillary water. Water moving from above may be trapped in a pocket formed by a less pervious layer in any zone above the water table, thus forming a "perched" water table (Sec. 5.3). Finally, layer 4 corresponds to the zone of saturation below the water table. This general picture, however, may be modified by infiltration of meteoric water, especially in poorly pervious soils after several days of continuous, violent rainstorms. In the latter case as the infiltrating meteoric water descends, it gradually fills all the pores of the soil deposit on its way and vacates the pores behind. Thus, a slowly sinking layer of gravitational water is formed. The rainy period may be followed by periods of limited rainfall or even drought, as in arid and semiarid regions of the Western United States, and the saturated layer in question may be encountered several feet below the ground surface in the otherwise relatively dry soil.

Notice that a portion of gravitational water below the water table is free to move and another portion may be confined between two impervious (confining) strata. The confined, or *artesian*, water has an intake zone (vertical dotted line in Fig. 5.1) but no exit and, hence, stands under pressure. For more information on artesian water, see Sec. 5.22.

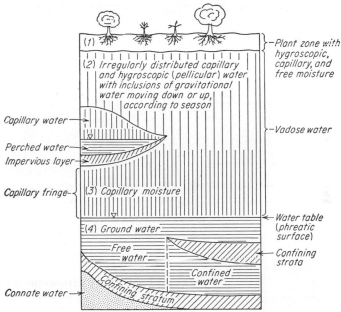

FIG. 5.1. Occurrence and distribution of subsurface water. Distribution is often erratic, and terminology used sometimes loose.

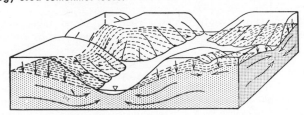

FIG. 5.2. Block showing single water table (general case in flat regions). (*From Longwell, Knopf, and Flint, "Physical Geology," 3d ed., John Wiley & Sons, Inc., New York.*)

5.3. Water Table. The water table is not horizontal either longitudinally, i.e., in the direction of flow, or transversely and, when located within soil, often conforms in a subdued manner to the configuration of the overlying ground surface. It is in constant movement, except at some isolated points. The relative position of the water table located within the earth mantle is shown in Fig. 5.2. This is the general case of a single water table. In addition to Fig. 5.1, in which the perched water table is shown, Fig. 5.3 explains how perched water may be drained by boring through a local impervious barrier (e.g., a clay deposit). Two or more water tables may be formed owing to the presence of continuous impervious barriers in a large sand deposit.

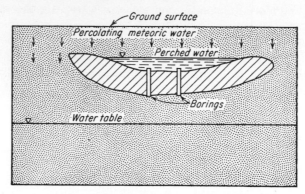

FIG. 5.3. Draining perched water by borings.

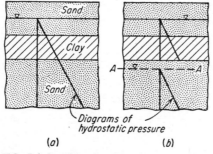

FIG. 5.4. Possible plurality of water tables. (After De Beer.)

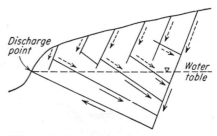

FIG. 5.5. Water table in fractured rock (seepage indicated by dotted arrows, flow by solid arrows). (From C. F. Tolman, "Ground Water," McGraw-Hill Book Company, Inc., New York, 1937.)

De Beer, a well-known Belgian engineer, found such hidden water tables from field pore pressure measurements.[5] A broken hydrostatic pressure diagram as in Fig. 5.4b (instead of a continuous one as in Fig. 5.4a) may indicate the presence of a hidden water table at level AA.

Figure 5.5 illustrates the formation of a water table in a rock outcrop by seepage or infiltration above the water table (dotted arrows) and actual flow upwards along fractures (solid arrows). Because of irregularity of fractures and discontinuity of permeable channels, there may be several apparent water tables in rock in close proximity. The water table in a superficially fractured rock zone is usually far from being parallel to the configuration of the outcrop.

5.4. Ground-water Mounds and Depressions. Figure 5.6 illustrates seepage from a reservoir with formation of a *mound* above the water table. In terms of earth surface topography, this would be described as a hill. If water seeps from a canal or a stream, the water table rises in the form of a *ridge*. Figure 5.7 shows the formation of a nonsymmetrical flat ridge below an irrigation canal. In contrast to mounds and ridges, there may be *depressions* in the water table; these can be caused by local pumping or other discharge from the aquifer. Such depressions reverse the gradient (slope) of the water table for a variable distance, depending upon aquifer characteristics and volume of discharge, and cause a local back flow of the ground water.

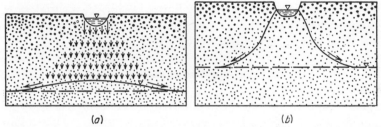

(a)   (b)

FIG. 5.6. Ground-water mounds. (From C. F. Tolman, "Ground Water," McGraw-Hill Book Company, Inc., New York, 1937.)

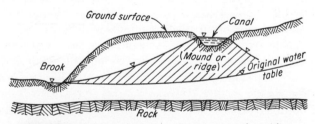

FIG. 5.7. Nonsymmetrical ground-water mound or ridge.

5.5. Fluctuations of the Water Table.[6] The elevation of the water table at a given locality depends on various factors such as fluctuations in precipitation and in the discharge of rivers. A yearly graph of the water-table fluctuations generally shows a maximum and a minimum, for instance, a maximum (or highest water table) in the fall and a minimum (or lowest water table) in the spring, though this order may be reversed. Besides these *seasonal fluctuations*, which may reach several feet during a given year, there are minor accidental fluctuations. Small or even microscopic bubbles of entrapped air may expand or contract according to the changes of barometric pressure, and this may influence the water level in wells at a given locality. Records of the water-table position in wells should be considered with great care. Apparent fluctuations of the water table may be recorded that were caused in reality by heavy loads (such as a railroad train) next to the observation well.

Marine tidal effect usually has little or no influence on the position of the water table beyond a distance of a few thousand feet from tidal waters. Earth tides have caused recorded diurnal fluctuations of the water table.

5.6. Ground-water Basins and Streams. Ground-water basins or reservoirs are areas with definite geological and, especially, hydrological boundaries. A ground-water basin may be several hundred miles long and wide (e.g., Dakota sandstone), or it may be relatively long and narrow, (e.g., a buried stream channel with ground-water flow). The river or brook flow in sand and gravel beds is accompanied by *underflow*, which is the lower part of the river or brook flow. As a rule, the underflow continues even when the surface channel dries out and sometimes even if the channel is backfilled or buried as later explained.

In a surface valley, water flows according to the direction of its thalweg (see Sec. 3.4). In this case, the subsurface flow also follows the general direction of the thalweg.

In flat regions the ground water moves in the form of a wide prism if the aquifer has regular hydrological characteristics. This prism dips slightly in the direction of the flow, and on limited areas in such localities, e.g., a few thousand feet square, the water table may be considered to be a plane. In hilly and mountainous regions the ground-water body is irregular, uneven, and often discontinuous with gaps between isolated currents. In such regions the presence of the ground water often may be guessed from the local topography, since the majority of local depressions, even those on the slopes, may be underlain by hidden ground-water channels (Fig. 5.8).

A more detailed explanation of the term "aquifer," already used, is that aquifers are rocks and soils containing considerable amounts of water which they will readily "yield" to wells. A sand yielding sufficient water for domestic use would be an aquifer in that sense but might not yield sufficient water to be considered an aquifer for irrigation. Sand and gravel deposits may yield considerable amounts of ground water, and so may some sandstones. An *aquiclude*, another term used by the geologists, simply means an impervious layer that does not yield water and usually acts as an upper or lower boundary to an aquifer.

5.7. Ground-water Temperature. The temperature of ground water within a few hundred feet beneath the surface is fairly constant, being 2 or 3°F above the mean air temperature at the given locality. It may be assumed for engineering purposes that the average ground-water temperatures at a reasonable depth vary from 40°F in the Dakotas and the region of the Great Lakes to 70°F in southern Florida or southern Texas.

At greater depths the temperature of the ground water follows the geo-thermal gradient of the enclosing earth and rocks; i.e., it increases because of the increasing proximity of hot lava (Sec. 9.10).

5.8. Fresh and Salt Ground Water. In the neighborhood of the ocean or a sea, salt water "encroaches" on fresh water and contaminates it, and

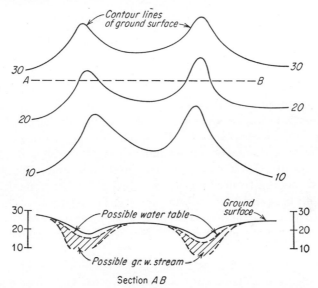

FIG. 5.8. Possible ground-water flow under surface depressions in hilly regions.

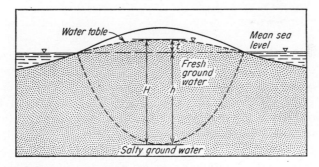

FIG. 5.9. Fresh ground water floating on salt ground water. (*From C. F. Tolman, "Ground Water," McGraw-Hill Book Company, Inc., New York, 1937.*)

vice versa, fresh water may override and displace underlying salt water. In particular cases (Long Island, New York; Hawaiian Islands) a body of fresh water floating on salt water may be formed. This state of equilibrium is known as the Ghyben-Hertzberg balance. It can be easily disturbed by overactive pumping of a well, which causes the salt water to contaminate the fresh water.

On an island, fresh ground water is often found at a certain elevation above the sea level (t in Fig. 5.9). The column of fresh water H ft high has to be balanced by salt water h ft high. Conditions of equilibrium require that the ratio H/h should equal the ratio of the unit weights of salt and fresh water (about 64 and 62.4 pcf, respectively, at 40 to 60°F). If t is known, H can be easily computed.

GROUND-WATER INVESTIGATIONS

5.9. Scope of Ground-water Investigations in Civil Engineering. Ground water is a *favorable factor* if it is used for human or stock consumption, for irrigation, or for industrial purposes such as air conditioning. In such cases, the engineer has to collect water from the earth and take measures for maintaining the ground-water supply in adequate condition.

In this book, however, more attention is paid to the cases where ground water is an *unfavorable* factor to the stability of engineering structures or tracts of land, particularly slopes, and seepage through and under dams. More specifically, such cases involve landslides and sliding of dams, heaving or flooding from rising water tables that threatens deep excavations, humid basements of buildings, and other hazards. The measures required to prevent (or at least to decrease) the possible damage that might be caused by ground water are discussed in this and other chapters in the book.

The engineer has first to establish the location of the ground water and second, to estimate the quantity of ground water likely to be encountered. For this purpose, a *ground-water survey* is made. Geological aid is of importance at all stages of ground-water investigation.

5.10. Ground-water Survey. If elevations of the water table at sufficient points and at a given time are known, *contour-line water-table maps* may be prepared. Though such maps are not always prepared from every set of measurements, determination of the ground-water level can be an everyday task of the engineer. Generally, elevations of the water table are determined in wells or bore holes (Sec. 6.25 and Fig. 6.33). Both high and low ground-water stages should be determined in any ground-water investigation, and because of the water-table fluctuations, the date of each ground-water determination always should be recorded. USBR requires that all bore holes drilled prior to construction be kept in good repair and be accessible for further observations during and after construction until the water table has reached a definite state of equilibrium. Geophysical methods also may be used for ground-water investigations.

5.11. Direction of Ground-water Flow. If a contour map is prepared using the elevations of the water table, the direction of the ground-water flow will be normal to the contour lines (i.e., perpendicular to the tangents of these lines). It should be remembered that the surface water also tends to flow normal to the contour lines of the ground surface. The velocity of ground water, other things being equal, is greatest where the contour lines are the closest together. An abrupt steepening of the ground-water flow is termed a *ground-water cascade.*

If the water table at a certain locality is considered as a plane, the direction of the water flow will coincide with that of the maximum slope of that plane. To determine the direction of the ground-water flow in this case, three borings A, B, and C (Fig. 5.10) are made and plotted on the map. Take the lowest elevation of the ground-water levels in the three borings (for instance, at C) as a datum or zero elevation plane and designate the elevations at A and B with h_a and h_b, respectively. Plot $AA' = h_a$ and

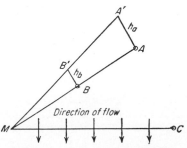

FIG. 5.10. Approximate determination of the ground-water flow direction.

$BB' = h_b$, and locate point M, the intersection of lines AB and $A'B'$. The elevation of point M is zero, and the approximate direction of the ground-water flow is roughly perpendicular to line MC, this line being a contour line of the plane ABC. This method is not desirable for determining the direction of the subsurface flow under valleys and in hilly and mountainous regions.

5.12. Hydraulic Gradient. The term "gradient" usually designates the slope of a line or a plane. In general, a *free water table* is curved as shown in Fig. 5.11a. Consider point A as being a distance l horizontally from point B. The vertical distance between points A and B is the *hydraulic head* lost (or spent) to produce flow between vertical planes passing through A and B. The *average* value of the hydraulic gradient between points A and B is

$$i = \frac{h}{l} \tag{5.1}$$

In reality the hydraulic gradient between points A and B (Fig. 5.11a) changes from point to point. The letter i in Eq. (5.1) stands for the word "inclination." Sometimes the hydraulic gradient is designated by the letter s, meaning "slope." Since h and l are lengths, i is an abstract number expressed either in fractions of a unit or in per cent. The gradient of the water table in flat country rarely exceeds 1 per cent, or 53 ft/mile. In hilly countries, larger gradients may be found. For short distances, when ground water descends a slope to join a water basin, such as a lake, gradients of 10 per cent, i.e., 100 ft in 1,000 ft, and more, are not uncommon.

Figure 5.11b gives an idea of the hydraulic gradient in the case of a con-

fined (or artesian) flow.   Water from a tank with a *constant water level*
percolates through a sand sample placed in a horizontal tube provided
with vertical glass tubes ("piezometric" tubes).   Should the stopcock *C*
be closed, the water level *AB* in the piezometric tubes should be horizontal.   This is the "static level."   In this case, water in the horizontal tube

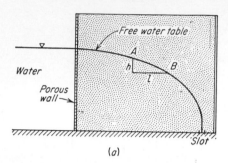

(a)

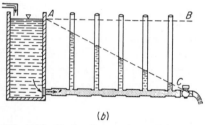

(b)

FIG. 5.11. Hydraulic gradient: (a) unconfined aquifer (free water table), (b) confined aquifer.

stands under the *hydraulic head*
equal to the elevation of level *AB*
over the center line of the horizontal tube.   As soon as stopcock *C*
is open, flow starts, and as water in
the horizontal tube proceeds toward the stopcock, the hydraulic
head is gradually lost and the water
level in the piezometric tubes
drops to form the *piezometric surface AC*.   Whereas in the case of
an unconfined aquifer the hydraulic
gradient is the slope of the free
water table, in a confined aquifer
it is the slope of the piezometric
surface.   The hydraulic gradient
in a confined aquifer is not necessarily constant along the flow and
may change from point to point.

**5.13. Velocity and Discharge
of Ground-water Flow.**   *Coefficient
of Permeability.*   Velocity *v*, when
laminar or nonturbulent flow prevails, is proportional to the hydraulic
gradient *i* and the coefficient of permeability *k*:

$$v = ki \qquad (5.2)$$

which is Darcy's formula.   The *k* of a given material at a certain standard
temperature is assumed to be constant for convenience in computations.
It depends on (1) the geometry of pores, i.e., their size and shape; (2)
mineral composition of the soil particles or grains; (3) occasionally, the
ionization of soil moisture (Sec. 4.2); (4) the properties of organic and
other admixtures, if any; and (5) the properties of the percolating fluid,
particularly its viscosity.   The latter, in turn, depends upon the temperature and, in exceptional cases, on atmospheric pressure.   For further
details on the correction for temperature, see Sec. 5.14.

For a more or less accurate estimation of the velocity of flow, the coefficient of permeability *should be determined experimentally*, preferably in the
field.   Laboratory methods using *permeameters* may also give some infor-

mation on the value of the coefficient of permeability (also see Sec. 5.17); still less accurate information is furnished by tables and formulas based on the size of particles only.[16,17] The coefficient of permeability is expressed in centimeters per second in the laboratory tests and in foot per day or per year for practical engineering computations. If expressed in metric measures, the coefficient of permeability is represented as a product of a number by a negative power of 10, usually the fourth negative power. Since 1 cm/sec equals 2,822 ft/day, the value of 1×10^{-4} cm/sec (or 0.0001 cm/sec) equals about 0.28 ft/day. Some very rough numerical values of the coefficient of permeability are given in the following table:

TABLE 5.1

Kind of soil material	Coefficient of permeability K	
	10^{-4} cm/sec	Ft/day
Uniform loose sand:		
Sieve No. 30–40............	3,000	847
Sieve No. 60–70......	300	35
Sieve No. 80–100..........	100	28
Fine sands..................	100–20	28–6
Silts.......................	2,000– 1	564–0.3
Clays......................	10– 0	3–0

The velocities determined by the Darcy formula are expressed in the same units as the coefficient of permeability used in that formula. For instance, if the value of k for a fine sand is 20 ft/day and the hydraulic gradient i is 0.01, water flows through that fine sand with a velocity $v = 0.20$ ft/day. Furthermore, if the velocity of flow is v ft/day, the discharge or the volume of the liquid passing through an area of 1 sq ft is v cu ft/day. Thus a velocity of 0.20 ft/day corresponds to a discharge of 0.20 cu ft (or $0.20 \times 7.48 = 1.5$ gal) per day per square foot of the area of discharge. The point is to *estimate* correctly the area of discharge A (which is not always easy) and to multiply it by the velocity v. The product Av is the discharge through the given area A under the given circumstances. In hydrogeological computations a somewhat different concept of the coefficient of permeability is used (Sec. 5.17).

5.14. Pumping Tests.[2e,7,8] Pumping tests are made on different projects mostly for determining the over-all coefficient of permeability of the soil materials in the area of influence of the test or for estimating the amount of water that may be collected from a given aquifer and the time required for replenishment, or *recharge*, of the aquifer after the pumping is discontinued.

Pumping from an Unconfined Aquifer. Figure 5.12 refers to a pumping test performed in an aquifer with a free water table, the saturated thickness of the aquifer being D. A well with screen or perforated casing is sunk to reach the impervious boundary underlying the aquifer. When water is pumped out from the well, the water table, originally horizontal, drops in the well through a distance termed the *drawdown*. Around the well the water table assumes the shape of a *depression curve* in every radial vertical plane to form a body of revolution about the vertical axis of the well. When the discharge is established, i.e., when the volume of pumped

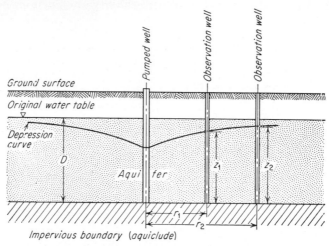

FIG. 5.12. Pumping from a free ground-water aquifer.

water in a unit of time, such as an hour, is more or less constant, the regular measurement starts. The test usually lasts 24 hr or more. Let Q represent the volume of water pumped out in 24 hr as measured in cubic feet. Several radial lines are traced through the center of the well, and at each line two borings (observation holes) are made, one at a distance r_1 and the other r_2 from the axis of the well (Fig. 5.12). The distance r_1 is often taken as equal to the thickness D of the aquifer or some round number, e.g., 50 ft. The ratio r_2/r_1 usually is an integer, 2 or 3, the same on all radial lines, whereas the distance r_1 varies. From the position of the water level in the borings, the vertical distances z_1 and z_2 are found. The differences $z_1{}^2 - z_2{}^2$ at different radial lines are approximately equal. Their average value is computed and placed in Eq. (5.3). Thus, the coefficient of permeability k, which in this case is expressed in feet per day, can be computed:

$$k = \frac{Q}{\pi(z_1{}^2 - z_2{}^2)} \left(2.3 \log^* \frac{r_2}{r_1} \right) \qquad (5.3)$$

* The expression "log" is equivalent to logarithm to the base 10; the expression "ln" is equivalent to logarithm to the base $e = 2.718$.

If there is sufficient space on the site, the ratio r_2/r_1 is made equal to 2.72. In this case the expression 2.3 log r_2/r_1 equals unity. Thus the value of the coefficient of permeability is

$$k = \frac{Q}{\pi(z_1^2 - z_2^2)} \tag{5.4}$$

The derivation of these formulas may be found in various books on soil mechanics, and it should be emphasized again that Eqs. (5.3) and (5.4) refer to pumping from *free ground water*. To be comparable, the field results of pumping tests performed on different soils or at different

TABLE 5.2. Reduction of the Coefficients of Permeability to Standard Temperatures

Water temperature at the test, °F	Coefficient of reduction	
	To 60°F	To 68°F (or 20°C)
40	1.37	1.54
45	1.26	1.41
50	1.16	1.30
55	1.08	1.22
60	1.00	1.12
65	0.93	1.04
70	0.87	0.97
75	0.82	0.92

localities should be reduced to a standard temperature of 60°F (or 68°F = 20°C for laboratory permeability tests). This is the correction for viscosity of the water, since the viscosity changes as the temperature of the water changes. If the only purpose of the pumping tests is to find a possible yield of an aquifer or aquifers in a given locality, correction for viscosity may be neglected.

To reduce the experimental value of the coefficient of permeability to a standard temperature, multiply it by the coefficients shown in Table 5.2. Interpolate for intermediate values.

Limitations of the Pumping Test. (1) In deriving Eqs. (5.3) and (5.4), the water table is assumed to be horizontal. These formulas also may be used with negligible error if there is a gradient in the flow direction provided there is no appreciable variation in thickness of the aquifer D. In the latter case, a radial line in the direction of flow is traced, and observation wells are placed *on both sides* of the experimental well. (2) If there is no underlying boundary, i.e., if the experimental well does not fully penetrate the aquifer, there is a considerable drawdown and the depression surface assumes a semispherical shape, at least in the proximity of the experimental well. It is therefore advisable to set the observation holes at a distance of $2D$ or farther from the observation well.   (3) It is considered by experience that if drawdowns are limited to 10 per cent of the thickness D of the aquifer and the slope of the depression curve at the periphery of the experimental well is less than 30°, the error in the determination of the coefficient of permeability is not large.

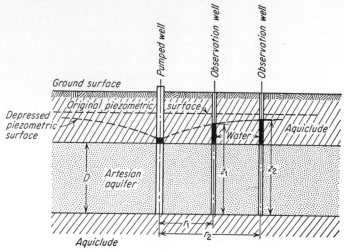

FIG. 5.13. Pumping from an artesian aquifer.

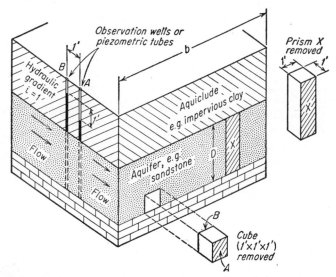

FIG. 5.14. Coefficient of permeability as used by hydrogeologists.

Pumping from a Confined (Artesian) Aquifer (Figs. 5.13, 5.14).   The aquifer is bounded at both top and bottom by impervious formations (aquicludes) and is saturated with water under pressure.   If piezometric tubes are introduced into the aquifer (Fig. 5.14, left), the water level in these tubes will form a piezometric surface completely in or above the upper impervious formation, or aquiclude. Before pumping starts, the piezometric surface is horizontal, but as the pumping proceeds, it acquires the shape of a depression curve entirely located in or above the upper aquiclude as shown by the dotted lines in Fig. 5.13. At all times the water table is at the top of the aquifer provided the water is still under pressure, i.e., if the piezometric surface is located above the top of the aquifer.

The value of the coefficient of permeability in this case is [2d]

$$k = \frac{Q}{2D(z_2 - z_1)} \tag{5.5}$$

if the ratio $r_2/r_1 = 2.72$. Otherwise the numerator of the fraction in Eq. (5.5) should be multiplied by log r_2/r_1. Correction for temperature should be used as explained previously. Note that pumping from a confined (artesian) aquifer also is discussed in Sec. 5.22.

5.15. Coefficient of Permeability from Pumping-in Tests. The coefficient of permeability under the water table may be determined in the field by drilling a hole in the ground, filling it with water, and measuring the rate of water flow from the hole into the ground. Also, water may be pumped from the hole and backfilling observed (Fig. 5.15b). Various approaches to the determination of k by this and similar methods exist. Perhaps the simplest one is given in Ref. 9. The difference of levels h inside and outside a pipe filled with water is recorded, and its decrement Δh in a small interval of time Δt, such as 1 min, is determined (Fig. 5.15a). Then if the diameter of the pipe

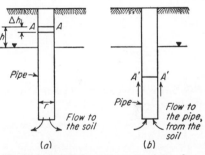

FIG. 5.15. Coefficient of permeability from water-level fluctuations in borings.

is r and the elevation of the water level in the pipe at the time of measuring is h, the value of the coefficient of permeability is

$$k = \frac{r}{2h} \frac{\Delta h}{\Delta t} \tag{5.6}$$

If h and r are expressed in feet and t in days (24 hr), the value of k in the formula is in feet per day.

This method of determining the coefficients of permeability, known as an "open-end" test, is subject to errors unless Δh and Δt are very small, and the ratio $\Delta h/\Delta t$ approaches the derivative dh/dt. Observations in this and similar tests start only when the flow has been established, i.e., when the soil mass around the hole is saturated, and this is difficult to accomplish. Particularly deficient are premature readings in highly pervious materials because of a very rapid drawdown during the first seconds after the hole has been filled.

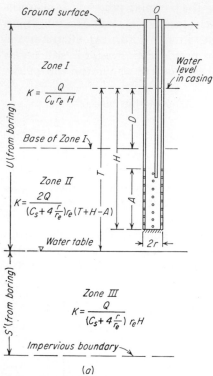

FIG. 5.16a. Permeability Test B of the USBR.
(Note: In Zone III, $H = T$.)

The USBR has devised several pumping-in tests using drill holes (preferably 6 in. in diameter). The hole may be cased or uncased and may or may not reach the water table. Hereafter, a test known as "Test B" is described.[13]

A 5-ft length of perforated casing, 3 to 6 in. in diameter, is sunk by drilling, jetting, or driving. A 6-in. coarse gravel cushion is poured into the casing, and the observation pipe (Fig. 5.16a) set on it. Water sufficient to maintain the casing water level above the perforations (in fact, if the water level is below the top of the perforations, there is no water in pipe O and the measurements cannot be performed) is pumped into the casing, and its level in pipe O is observed. The difference in water levels between measurements should not be over 0.2 ft in three or more consecutive observations spaced 5 min apart. As soon as the position of the water level in the casing is thus established, the value Q of steady flow into the casing (in cubic feet per second) and the values of T, H, and A/H become known.

Formulas for computing the coefficient of permeability k are given in Fig. 5.16a. To use these formulas, the coefficients C_u and C_s are taken from graphs (Fig. 5.16b), the former being the nearest A/H value. The symbol r in the formulas stands for the outside radius of the casing, and r_e is the ratio of the total area of the holes to the total area of the perforated section of casing, the latter being $2r\pi A$, wherein A is the length of the perforated area. The position of the base of zone 1 (Fig. 5.16a) is found from the diagram of Fig. 5.16c. All dimensions, including the radius of the casing, should be expressed in feet. As soon as the test in the upper 5 ft of the natural soil deposit is completed, the casing is driven farther and the test repeated.

5.16. Slichter's Field Method of Determining k.[15] Water in a boring or well, termed a "salt well," is charged with an electrolyte such as common table salt, NaCl, or, preferably, ammonium chloride, NH_4Cl. In the general downstream direction of the ground-water flow, a group of wells is located approximately along an arc of a circle, with the salt well at the center of the circle. The arrival of the electrolyte at one of these wells is announced by a short-circuit signal. From the time records and direct measurements in the field may be determined (1) the distance l between the salt well and the given well, (2) their difference of levels h, and (3) time of travel t of the elec-

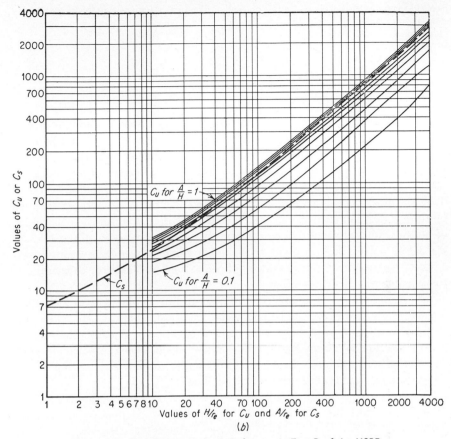

FIG. 5.16b. Coefficients C_u and C_s for use in Test B of the USBR.

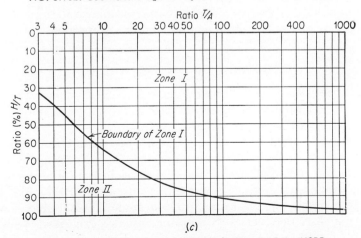

FIG. 5.16c. Nonsaturated soil portion in Test B of the USBR.

trolyte from the salt well to the given well. The symbol n is used to designate the porosity of the soil material between the two wells. The coefficient of permeability is determined as follows:[10]

$$k = \frac{l^2 n}{ht} \tag{5.7}$$

This method is satisfactory in very uniform materials but can give very erroneous results in heterogeneous or interstratified coarse- and fine-grained materials.

5.17. General Comments on the Permeability Tests. The laboratory permeability tests performed using *permeameters* (Sec. 5.13) are described in the texts on soil mechanics and elsewhere. These tests may reach a high degree of accuracy as far as a tested sample is concerned, but because of the extreme difficulty in taking truly representative samples in the field, the agreement between field and laboratory tests is generally unsatisfactory. Though the existing field methods of determining the coefficient of permeability are crude in character, they define the over-all permeability of the materials in a given area better than *isolated* laboratory tests. This is the reason why this text emphasizes field methods for determining the coefficient of permeability.

All permeability tests, both in the field and in the laboratory, are performed when the experimental flow has been established, i.e., when the material is saturated and the rate of flow constant. Also all theories of subsurface flow as given in the texts on soil mechanics and elsewhere are based on the so-called "steady-state" conditions and are not valid for nonsaturated flow unless otherwise specified.

As already stated (Sec. 5.13) there are formulas to determine the coefficient of permeability from one or a few isolated properties of the soil material, such as the effective grain-size diameter (Sec. 4.3). These formulas may furnish fair results in individual cases but, as a rule, should be considered with reservation, since the permeability does not depend on one or two but on a whole complex of soil properties.

COLLECTION OF GROUND WATER

Ground water may be collected from springs and wells. Technical details of the construction of collection systems are discussed in textbooks on water supply and thus are not given here.

5.18. Springs. Geologists use an elaborate system of classification for springs; however, as it is not in common use in engineering, it is not presented here. Usually, ground water emerges in the form of a spring where the ground-water table intersects the ground surface (Fig. 5.17, right). If a fissure intersects an artesian aquifer, water may emerge upward (Fig. 5.17, left, letter f). Springs usually issue from a hillside or in a valley. They often are formed where porous strata, such as sand, gravel, or sandstone, rest upon impervious beds. *Hot* (thermal) *springs* and geysers may be associated with vulcanism or result from percolation of meteoric waters to considerable depth. Geysers are springs in which hot or warm water, presumably under pressure, is ejected above the earth surface. A well-known geyser is the "Old Faithful" geyser in Yellowstone National Park. In *boiling springs*, the vigorous ejection of gases agitates the water.

5.19. Wells. A well is any artificial, vertical, or practically vertical excavation in the earth surface constructed for the purpose of capturing or releasing fluids. Wells may be dug, driven, jetted, bored, or drilled as explained in Chap. 6. Other holes similarly constructed are designated as test holes, observation holes, bore holes, etc., depending upon the use for which they are intended.

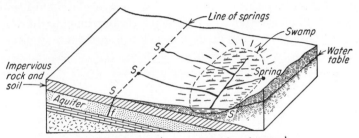

FIG. 5.17. Springs (S = spring, f = fracture).

Dug wells, generally wide and shallow, are excavated either by hand tools or by mechanical equipment. Such wells may be expensive in construction and also in maintenance, since frequent cleaning is required. They are advantageous in materials of low permeability, and their large diameter permits them to act as both storage reservoirs and wells. Recharge is slow, and for intermittent pumping, the rest periods should be long enough to permit recharge of the storage. *Driven wells* are constructed by driving a casing that at its lower end has a drive or "sand" point. The main difficulty with this type is that the pipe or casing becomes clogged with sand. Some wells are driven by the *water-jet* method: A current of water is forced through the drill pipe under high pressure, and the resulting jet at the lower end of the pipe loosens the surrounding material, forcing it to the top of the hole. *Bored wells* are made by the use of either hand or power augers. *Drilled wells* are the type of well most frequently used at the present time. Rotary or churn methods are used in drilling and described in Chap. 6.

Well Casing. A well may be cased wholly or in part with iron or steel pipe; brick, concrete, or tile casing; or wooden cribbing. Cased wells may be (1) open-end wells, where the water enters only through the open lower end, or (2) screened wells, which are provided with a screen or perforated casing at the lower end.   The screen prevents the infiltration of sand during pumping and increases the intake area to the well.   Either type of well may have a sand-gravel filter surrounding the intake area to prevent excessive inflow of fine sand and resultant clogging of the well. Deep rock wells may not require casing.

5.20. Performance of the Wells.   The yield of a well is increased very little by increasing its *diameter*.   Both theory and practice show that doubling the diameter increases the discharge at the same drawdown from only 15 to 25 per cent for wells less than 18 in. in diameter.   For larger diameters the 15 per cent increase usually holds.

For a given depth of the well, the increase in yield per each additional foot of *drawdown* becomes less and less as the pumping progresses.   At 50 per cent drawdown (i.e., when the water table in the well has dropped half the depth of the water in the well) the increase in yield begins to be quite small.   In an artesian well the yield is proportional to the drawdown.

At a given drawdown the yield of a well is approximately directly proportional to the length of the *intake area*.   The yield increases very rapidly as the *uniformity* and *average grain size* of the surrounding earth material increase.

Each well has a certain zone of influence, shaped as a circle, say 2,000 ft in diameter.   If the zone of influence of a well is overlapped by that of some neighboring well or wells, or if there are impermeable boundaries arresting or decreasing the flow, the yield decreases, and conversely, the yield increases if there is a recharge source within the zone of influence. For further information see Ref. 4.

5.21. Two Concepts of Coefficient of Permeability.   The letter k used in Secs. 5.13 through 5.17 is the coefficient of permeability expressed in terms of *velocity*.   In geohydrological computations, the coefficient of permeability is expressed in terms of *discharge* and designated by capital letter K.   It may be defined[2e] as the rate of flow of water in gallons per day through a cross-sectional area of 1 sq ft under a unit (100 per cent) hydraulic gradient at a temperature of 60°F.   This concept is clarified by Fig. 5.14.   Assume that the exposed side of the aquifer shown in the figure is sealed and then a cube of unit dimensions (1 by 1 by 1 ft) is removed.   The amount of water at 60°F that flows in 24 hr through the rear vertical wall of the opening thus formed is K, provided that during the flow there is 1 ft of difference in the levels of the two observation wells spaced 1 ft apart (unit hydraulic gradient).   Actually, the observation wells A and B should be placed at the opposite vertical sides A and B of the cube before the latter is removed.*

---

* Another concept of the coefficient of permeability is used by physicists.   It is closely allied to the engineering concept in that the coefficient is expressed in terms of velocity.

5.22. Artesian Water.[18,19] As already stated, artesian water is ground water confined between two impervious beds or aquicludes (Fig. 5.1). If the upper confining bed is broached by a well, artesian flow starts. In general, requisite conditions for artesian flow are (1) tapping an aquifer such as sandstone or sand containing water under pressure and (2) the presence around the well of an agency that offers more resistance to upward flow than the well. Such an agency may be impervious rock or impervious clay. Artesian aquifers may be deep or shallow. For example, in some parts of Colorado, they are at depths in excess of 2,500 ft, whereas along the Atlantic Coast, shallow artesian wells of only 100 ft in depth are found. At each point A of the confined ground water (Fig. 5.18), it stands under hydrostatic pressure equal to its unit weight (62.4 pcf) times the hydraulic head h, i.e., the vertical distance of the

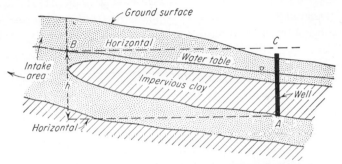

FiG. 5.18. Schematic representation of artesian flow.

point in question from level BC traced through point B at the boundary of the *intake area* (Fig. 5.18). When the upper confining bed is pierced by a well, confined water moves up toward level BC, a piezometric surface in this case, but does not quite reach it because of friction and leakage.

Because of the compressibility and elasticity of the aquifer and the overlying strata, the simplified theory of artesian flow as given above requires amplification. Consider a sandstone formation that was first covered with a thick deposit of impervious clay and then filled with water under pressure. Water pressure in this aquifer tends to lift the confining clay and to decrease slightly the vertical pressure acting on the skeleton or fabric of the sandstone. In this connection, the aquifer may even expand a minute amount. Water gradually fills in all the pores of the clay deposit. The water, under the action of the pressure and attraction to clay particles, is compressed and decreases in volume. Relief in pressure caused by the start of the artesian flow results in the clay gradually settling upon the aquifer and compressing it. Water is gradually released by both the aquifer and the clay deposit. In fact, the clay, now unsupported, consolidates under its own weight and slowly squeezes out water contained in it. At the same time, because of the relief in pressure, water previously compressed in the aquifer expands. The phenomena described produce a flow "from storage," sometimes without a noticeable drop of the water table at the intake area.

As artesian flow continues, pressure in the confined water gradually decreases; consequently, the discharge of the artesian well decreases. Hence, the initial discharge of an artesian well may not be indicative of its real capacity. Because of change in volume of rock and especially of soil material, artesian flow may be accompanied by depressions (settlements) of the earth surface.[35,36]

Artesian Aquifers. Assume that water flowing along an artesian aquifer is only *transmitted* from section to section of the aquifer and no water is *released* from storage. Consider prism X (Fig. 5.14); the amount of water transmitted from the rear wall to the front wall (crosshatched) of the prism under the action of a unit hydraulic gradient ($i = 1$) is

$$T = KD \tag{5.8}$$

T is the *coefficient of transmissibility*, which shows the ability of the aquifer to transmit water. If the width of the aquifer is b ft and the *actual* hydraulic gradient is i_0, the amount of water transmitted from section to section of the aquifer is

$$Q = Tbi_0 \tag{5.9}$$

If the coefficient of permeability K is expressed in gallons per day, the coefficient of transmissibility T is also expressed in gallons per day.[20-23]

The coefficients K and T are determined from pumping tests under the assumption that the amount of water Q pumped during the day is constant. The corresponding group of formulas are known as *equilibrium formulas* (Thiem formula[24,25] and others). The other approach to the analysis of an aquifer was proposed by Theis, his formula being known as the *nonequilibrium formula*,[20,25] in which the time factor t and the *coefficient of storage S* are introduced. For nonconfined aquifers the coefficient S is equivalent to the *specific yield* of the materials dewatered by pumping. In turn, the specific yield is the quantity of water yielded *by gravity only* from saturated water-bearing material and is expressed as a fraction of the volume of the material drained. In the case of an artesian aquifer, S is the ratio of the volume *of water* obtained from a prism of water-bearing material with a base 1 ft square and a height equal to the thickness of the aquifer D (prism X in Fig. 5.14), per unit drop of hydraulic head, to the volume of that prism. The actual drop in hydraulic head acting on prism X is the drop of the piezometric surface just above that prism as caused by the relief in pressure within the aquifer. The value of the coefficient S, for instance, equal to 0.00032 means that 320 cu ft of water would be released under an area of the aquifer 1,000 ft square (i.e., 1,000 by 1,000 ft) as the average head declines 1 ft. Theoretically the volume of water released is proportional to the change in head. Both coefficients T and S should be considered as empirical characteristics of a given artesian aquifer which can be determined from pumping tests.

Artesian Wells. An artesian well is one in which the water level rises to a higher elevation than that at which it was first encountered, this rise being caused by the pore water pressure within the tapped aquifer. A

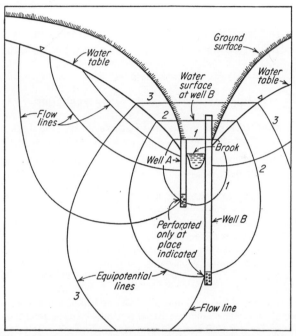

FIG. 5.19. Flowing or artesian wells in an unconfined aquifer.

flowing well is one in which water flows over the top of the casing owing to artesian pressure head. Flowing wells may occur also in unconfined (free water) aquifers, i.e., when no confining aquiclude is present. An example is given in Fig. 5.19 which depicts a continuous water table to a considerable depth and a narrow valley with a brook near which two wells A and B are sunk. As in any flow, the movement of water in this case may be visualized in the form of *flow lines* which start at the water table and run toward the thalweg of the valley or wells placed there. The flow lines are normal (perpendicular) to the *equipotential lines*, or lines of equal pressure. The equipotentials start normal to the water table and close up also at the water table at the opposite side of the valley. If at any point of equipotential a piezometric tube is inserted, water will rise in that tube to the level determined by the intersection of the equipotential with the water table.   In the case of Fig. 5.19, wells with perforations at the bottom only act as piezometric tubes inserted to the depths where perforations are.   If well A only is sunk, water will rise in it just to the top as indicated by level 1.   Water in well B would tend to reach level 2.   Should a considerable length of the well casings be perforated, the water level in the wells would be controlled by the water level in the brook, e.g., be lower than actually observed.   Figure 5.19 is patterned according to some actual observations in South Dakota* and elsewhere.

In general, flow lines and equipotential lines form a pattern known as a *flow net*.   Methods of constructing flow nets may be found in most books on soil mechanics.

PORE MOISTURE AND PORE PRESSURE

5.23. Determining Field Moisture Content by Nuclear Radiations.
The problem of measuring the moisture content at a point of the earth mass at different arbitrary times and without disturbing the earth mass has not been definitely solved.

The principle of neutron scattering may be used for approximate soil moisture content determination as follows.[26-28]   A nuclear reaction between radium D, or polonium, on one side and beryllium on the other results in freeing fast-traveling neutrons from the nuclei of the beryllium atoms in one installation.[28]   Such a neutron source is placed in a probe or thin-walled casing (Fig. 5.20a) about 1 in. in diameter and 7 in. long.   In its turn the probe is placed in an aluminum or stainless steel tube with a steel watertight point.   A detector such as a Geiger counter is also placed in the tube and wrapped in a thin silver foil.

The fast neutrons leave the tube and are scattered in all directions within the surrounding earth mass.   They collide with the atoms of the substances forming the surrounding earth mass.   Collisions with hydrogen atoms result in a decrease in kinetic energy, and hence velocity, of the fast neutrons, which thus become slow neutrons.   Slow neutrons returning to the vicinity of the probe react with silver atoms of the foil, thus causing an emission of γ rays which are detected by the Geiger counter and counted by the scaler (Fig. 5.20b).   A change in the count of the slow neutrons reflects primarily a change in water content, since hydrogen is found predominantly in the soil water, both sorbed and chemically bound.   To correlate the number of counts (per second) with the amount of moisture contained in the soil (in pounds per cubic foot), calibration curves should be constructed and used.

The method described requires practically no disturbance of the earth mass.   In another method,[29] electrodes are introduced into a block of plaster of paris or are heavily wrapped with nylon fabric.   The electrical resistivity of the earth material between the electrodes is thus measured.   The moisture content is then determined from the *calibration curves* prepared by artificially wetting the experimental soil and plotting its moisture contents against the corresponding electrical resistivities.

* By T. P. Ahrens, USBR.

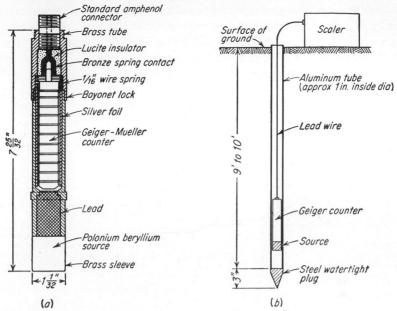

FIG. 5.20. Use of radioactive materials for measuring the water content in soils: (a) moisture probe, (b) sketch of installation.[28b]

5.24. Pore Pressure.[30–32]   Artesian water is a particular case of pore water standing under pressure.   The pore water pressure often is simply termed "pore pressure" (see Sec. 4.16).

There are two basic ways in which the pore pressure may develop.   One way is the increase of pore pressure because of the *increase of hydraulic head* under which water in the aquifer stands.   Figure 5.21 illustrates this situation.   The hydrostatic pressure in a box with saturated soil material provided with a tube filled with water equals the hydraulic head h times the unit weight of water γ_w.   If, for instance, the hydraulic head at a point is $h = 10$ ft, the hydrostatic pressure at that point is $10 \times 62.4 = 624$ psf.   It is obvious that in the case of Fig. 5.21 the hydrostatic pressure increases from top to bottom of the aquifer, and as water is added to the tube, the hydrostatic pressure at all points of the aquifer increases accordingly.

In an artesian aquifer protected by huge aquicludes with considerable shearing strength, the increase in hydraulic head only compresses the pore water but generally cannot destroy the aquifer.   In a saturated clay deposit with a relatively limited shearing strength, the increase in pore pressure may cause a shear failure, often in the form of a landslide (Chap. 17).

The other way in which the pore pressure can be developed is an *application of a load to the aquifer*, particularly to a compressible one. Figure 5.22 shows a composite aquifer consisting of a compressible clay deposit between two pervious sand layers. If the clay is fully consolidated before any load is applied at the ground surface, the original hydrostatic pressure is triangular.   If, however, a load is applied at the ground

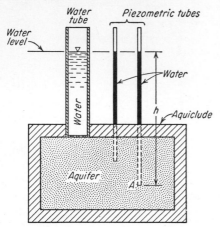

FIG. 5.21. Pore pressure ($h\gamma_w$ at point A).

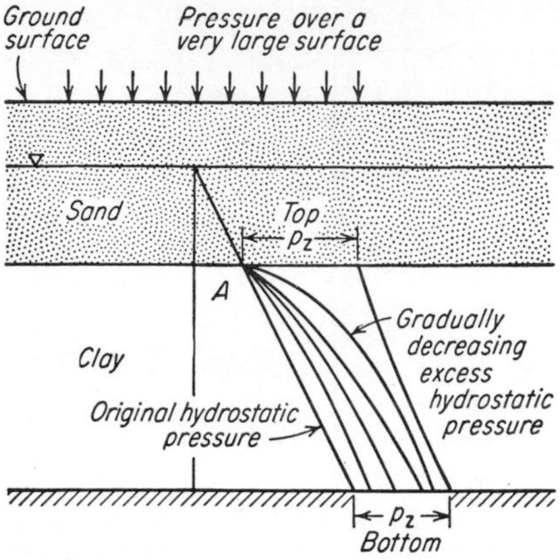

FIG. 5.22. Gradual decrease of pore pressure during consolidation.

surface and vertical pressures are developed through the aquifer, pore water will be easily squeezed out from the sand pores but detained in the

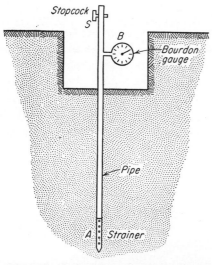

FIG. 5.23. Measuring pore pressure in the field.

clay pores where the familiar *process of consolidation* starts. If at a point A of the clay mass (Fig. 5.22) the vertical pressure from the load is ρ_z, the whole pressure from the load will be taken by water and gradually transmitted to the clay particles (Sec. 4.22). At the end of consolidation, the pore-pressure diagram returns to its original triangular shape.   The gradual decrease in pore pressure is illustrated by a series of curves labeled "gradually decreasing excess hydrostatic pressure" in Fig. 5.22.   Hence, if field measuring of pore pressure does not reveal a triangular pore-pressure diagram, the corresponding clay deposit

(Fig. 5.22) is probably still in the process of consolidation.

When a high embankment such as an earth dam is built, gradual addition of fill material may trap air if it cannot readily escape. The air in the pores thus trapped will stand under high pressure and be gradually sorbed in pore water. In this case, the pore pressure in both air and moisture is controlled by the weight of the overlying earth material.[33,34]

While fills made of rather impervious materials are under construction, pore pressure in them may be revealed by simply driving an open pipe into the body of the fill. Figure 5.23 represents schematically the method of measuring pore pressure in the field (point A). A pipe with a strainer at the tip is driven to reach point A, filled with water to the top of the pipe, and stopcock S closed. Reading R (in pounds per square inch) is taken on the Bourdon gauge (manometer) B. The length of the pipe l is measured in inches. Pore pressure at point A is then (in pounds per square inch)

$$p = R + l\gamma_w \qquad (5.10)$$

Figure 5.19 shows that, in general, pore pressure depends on the configuration of the flow net, particularly on the shape of the equipotential lines. Hence, strictly speaking, the results of the field pore pressure measurements refer to one point only (A in Fig. 5.23).

GROUND-WATER OCCURRENCE AND CONSERVATION

5.25. Principal Groups of Aquifers. The three basic groups of aquifers are discussed hereafter. The *first group of aquifers* consists of coarse-grained, loose sand and gravel surface deposits. These are primarily porous strata in glacial drifts and alluvial deposits. The most outstanding pervious members of *glacial drifts* are (1) glacial outwash, generally most of the terminal moraine in the immediate vicinity of the glacial front or in old melt-water channels, and (2) sand and gravel in eskers and kames (Sec. 3.12). Under certain circumstances, some lake deposits and wind deposits such as dunes or even sandy varieties of loess may be regarded as aquifers. *Alluvial aquifers* may consist of mountain outwash made up of silt, sand, and gravel deposits of considerable thickness with great amounts of water in storage (e.g., Rocky Mountains outwash). Alluvial valleys may be underlain by several hundred feet of alluvium (e.g., the Mississippi Valley) or contain alluvial deposits in the form of alluvial cones and fans (e.g., the Great Valley of California).

The *second group of aquifers* consists of sedimentary rocks, primarily *sandstones* and limestones. The sandstone aquifers may be found in many places in the United States, but they are mostly concentrated in the northern part of the country (the Dakotas and the neighboring states, Illinois, Wisconsin area). Sandstones are often prominent sources of artesian wells.[28,29] *Limestones* are good water reservoirs if they are porous or have openings caused by the solution of calcite, the primary constituent of limestone.　Water stored in a limestone deposit may be discharged in the form of springs (Florida, Ozark region).

The *third group of aquifers* is mostly formed by fissured or broken igneous rocks (e.g., lava in the Columbia River plateau or the Gila River lava field in Arizona and New Mexico).　Sometimes intrusive igneous rocks, such as granite, and crystalline metamorphic rocks are aquifers. Ground water may occur, even at great depths, in the upper partially weathered zone overlying the unweathered rock if considerable weathering *in situ* has taken place.　Occasionally, faults may store considerable amounts of water.

For classification purposes the entire area of the United States is often subdivided by the geologists into ground-water provinces.[26]　These subdivisions have but a limited civil engineering significance.

5.26. Actual Demand on Ground Water.[37]　A comprehensive national survey of water use has never been made in the United States.　Available statistics are based on limited information.　The national average per capita use of water varies from 60 gal/day in small communities to 140 gal/day in cities with populations of more than 10,000.　Thus, 12 to 14 billion gallons a day are required for municipal supplies.　In addition, the rural domestic use of water is estimated at 3 to 5 billion gallons a day. The nation also uses 75 to 100 billion gallons of water a day for irrigation, 35 billion for the generation of steam and electric power, and 35 to 45 billion for other industrial purposes.　All in all, the national use of water is of the magnitude of 200 billion gallons a day, which represents about 15 per cent of the total flow of American streams to the oceans.　It is estimated that about one-sixth of the water for irrigation, industrial, and public use (roughly 30 billion gallons a day) is furnished by ground water, i.e., by wells and springs.　Ground water is unimportant as a direct source of water for power, navigation, or recreation.　It is, however, of great indirect importance in these fields because the minimum flow of streams is sustained chiefly by the ground-water inflow.

The demand on ground water by industry, including air conditioning, is of some importance.　Water for industrial purposes must be of a quality that will not damage the product being manufactured, clog boilers or pipes, or damage water-heating devices.　For air conditioning, it also must be of the right temperature.　Ground-water temperature will undoubtedly play an increasingly important role in the installation of heat pumps, which are being installed in a larger number of homes and plants every year for the purpose of heating and air conditioning.　It should be noted, however, that part of the water used for air conditioning is generally turned back into the ground (recharged), the other part being discharged to sewers or other sumps.　More and more cities, however, are passing legislation requiring that air-conditioning water be returned to the ground and that evaporative-type coolers be eliminated. Where air-conditioning water is obtained from a city supply coming from some distance and water supplies become critical, the users may be required to put in recirculating systems in order to avoid a drain on the supply. This was done in 1955 in Denver, Colorado, where water was obtained from a drainage basin remote from the city.

5.27. Problems of Ground-water Conservation.[37] The American public, and particularly engineers and geologists, should realize that ground water represents a *national resource* of great importance. Ground water is as vital to the life and economy of the country as, for instance, are the national supplies of coal and iron. It should be understood that in order to meet the *national demand* on ground water adequately, the latter must be *properly managed*. Thus, a number of problems arise.

Reservoir Problems. These problems result from a ground-water shortage in a given locality. The areas of serious ground-water shortage are those in which wells, year after year, draw water from the ground-water reservoir in excess of annual recharge. Such problems are current in California, Arizona, New Mexico, and Texas and to a lesser extent in Colorado, Nevada, Utah, the Louisiana rice area, and Grand Prairie, Arkansas.

Pipeline Problems. These problems of ground-water transportation arise when the total capacity of a reservoir is adequate but the ground water does not move rapidly enough to meet the demands of specific wells. The term "pipeline problems" is analogous to the problems of water supply systems afflicted with inadequate pipe lines or distribution mains. Whereas reservoir problems arise in the Western United States, pipeline problems most commonly occur in large cities and in industrial areas, whether they are in humid or arid regions. Local overpumping from an otherwise sufficient reservoir results in cases of general lowering of water levels over broad areas.

The city of Brooklyn, New York, is located on the western tip of Long Island, which is bordered on all sides by salt water. Because of the high humidity of the local climate, precipitation usually provides sufficient natural replenishment. In 1943, however, heavy drafts on ground water for industrial and public use lowered the water table under the northern part of the city to 15 ft below sea level. As a result salt water encroached on many of the pumped wells.

Watercourse Problems. Generally speaking, any stream accompanied by an underflow of ground water may be termed a *watercourse*. An insufficiency in the underflow cross section in a watercourse may cause a ground-water shortage.

A ground-water shortage exists in the so-called "Golden Triangle," Pittsburgh, Pa. This is a 0.4-sq-mile area of flood plain at the junction of the Allegheny and

Monongahela Rivers. It comprises Pittsburgh's downtown district. The sand and gravel aquifer through which the ground water flows is only 40 ft thick; hence, there is a limited supply of water for air conditioning during the hot seasons.

Safe Yield.[38] The withdrawal of ground water from a reservoir should not exceed the value of *safe yield*. This is the annual volume of water which can be extracted from the ground water without (1) exceeding the average ground water recharge, (2) lowering the water table sufficiently to permit encroachment of undesirable water (such as salt water), and (3) lowering the water table to such a depth that the cost of pumping becomes prohibitive. Numerous state statutes and court decisions tend to regulate the safe-yield question. The methods of determining the value of the safe yield are widely discussed in geological and technical literature.

5.28. Ground-water Recharge: Spreading.

Water retained for future use may be stored in either surface or underground reservoirs. Ground-water storage is of importance in localities where periods of intensive rainfall or floods may be followed by periods of subnormal rainfall or droughts and where local topography or seismicity does not offer damsites satisfactory for formation of surface reservoirs. Water stored underground is ensured against evaporation losses and pollution, provided, however, that the water table is kept deep.

Under natural conditions, the aquifers are in an approximate state of dynamic equilibrium, although the water table fluctuates seasonally and from year to year. The recharge or replenishment of the ground-water reservoir may be natural—from precipitation or stream flow—or artificial. One method of recharge is *spreading*. This can be done by (1) ditches that convey water from a surface reservoir or stream to furrowed plots of ground (such a method is used in irrigation); (2) the "basin" method, in which harrowed areas are surrounded by retaining dikes and a shallow sheet of water is maintained over the diked area (surface basin); or (3) feeding water down trenches, shafts, or wells, an operation that requires a preliminary desilting of the water.

In Long Island, New York, water used for cooling is returned to the ground through special wells. The seasonal excess of water coming from the San Gabriel Mountains toward Los Angeles, California, is held by special dams and then allowed to percolate from shallow basins. Configurations of the mound formed above the natural water table by spreading and characteristics of the flow are discussed in detail in Ref. 39.

5.29. Rejected Recharge.

The phenomenon of rejected recharge arises if the amount of water available for recharge exceeds the capacity of the reservoir to accept water because of prior saturation of the recharge terrain or inadequate transmissibility or because the reservoir is already filled. A rejected recharge zone is shown to the left in Fig. 5.24. Here the rejected recharge water is partly taken away by streams and partly consumed by vegetation. Proper location and spacing of wells in this zone may increase the total perennial yield of the aquifer. In fact, each new well in that zone reduces the amount of the rejected recharge. When a cone of depression for one or more wells has been developed (Fig. 5.24, right), however, any new well in the area alters that cone of depression.

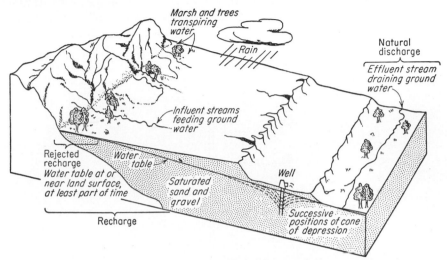

FIG. 5.24. Rejected recharge of an aquifer. (*From Theis, Civil Eng., vol. 10, p. 277, 1940.*)

DRAINAGE

Subsurface drainage is discussed hereafter. It may be simply defined as "the removal of excess water and salts from the soil." This broad definition applies to both engineering and agricultural uses of drainage.

5.30. Dewatering Excavations for Foundations. Prior to the start of an excavation, the locality should be drained using gravity only, i.e., by digging ditches and letting subsurface water move to natural sinks or low spots where the excess water can be stored. This may lower the natural water table somewhat.

An excavation should be dry enough in order to permit emplacement of concrete and other materials forming the foundation of the structure. In general, the original water table, even lowered by ditches as described, is higher than the bottom of the excavation. The problem, then, is to pump enough water so as to keep the depression surface (Sec. 5.14) *below the proposed* bottom of the excavation. In smaller excavations a deep *sump* is dug at a corner of the planned excavation and filled with coarse filter material, such as clean gravel, and pumping is done from that sump. Pumping from open excavations generally involves the danger of *caving*. In modern excavations, this danger is eliminated by using *well points*.

(*Siemens wells* used in Europe are provided with an 8-in. casing and a 6-in. suction tube and spaced 20 ft or more center to center.)

There are several systems of well points. Basically, a well point is a perforated and screened pipe, about $1\frac{1}{2}$ in. in diameter and a few or several feet in length. The well point is fixed to a *riser pipe*, also of the same diameter, and the whole is driven with or without jetting. If the natural soil is not pervious enough, an opening is made around the riser

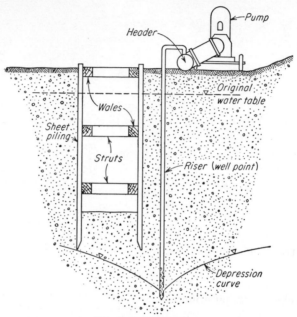

FIG. 5.25. Well points.

and filled with filter material, e.g., coarse sand or fine gravel. The opening can be formed by washing or by reaming up a chain fixed to the point and driven down with it. The individual riser pipes are attached to a header pipe or manifold that leads to a pump (Fig. 5.25). Lines of well points may be set at the edge of the excavation, along one side, or completely surrounding it. In general building construction, the well points are placed 3 to 6 ft center to center (commonly 4 ft). The suction lifts possible with well points range from 15 to about 25 ft decreasing with the altitudes at which the work is being done and depending on the efficiency of the system. Where deep excavations are required, well points are placed on descending steps 15 or 20 ft high ("multiple-stage setup").

Investigations for a well-point installation should cover determination of the thickness of saturated material and the depth to which it must be drained; the coefficients of storage and permeability of the material; the source, direction of flow, and probable quantity of water to be handled;

and the character of the aquifer in regard to the presence of nonpermeable beds and lenses, artesian conditions, and gradation of grain sizes in different strata. In practice, the work is done by specialized companies,[40] which often appraise the situation on the basis of their experience and are satisfied with only a portion of the above data.

Deep Wells. If the aquifer to be drained is thick and of high permeability and deep excavations are planned, deep wells equipped with turbine pumps often are more satisfactory than well points. The same data are required in the design of a deep-well system as for a well-point system. The individual installations of the deep-well turbines are more efficient than the well points, and the lift is limited only by the diameter and depth of the well and the capacity of the pumps. The individual wells may be placed outside the area of operations. The installations are designed to do the entire job; successive steps of well points with consequent interference with construction operations are avoided. The initial cost of the deep-well system may be higher, but the operation cost usually is about the same as for the well-point system. The deep wells operate

more positively and efficiently, and are less subject to breakdown, and in many instances are much more flexible in meeting emergency demands. Submersible motors attached directly to the pumps also are used.

On both systems, i.e., well points and deep wells, once pumping is started, it must be continued until the excavation work is completed. Failure of a pump or power unit can be disastrous, so adequate standbys must be on hand at all times, and a pump man on duty 24 hr a day. It should be remem-

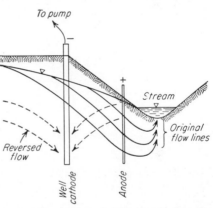

FIG. 5.26. Reversal of natural flow by electroosmosis (dotted arrows).

bered that when either well points or deep-well turbines are used, the maximum drawdown occurs at the well point or well. The zone of saturation is always somewhat higher between individual wells and well points. Designs therefore are based on the maximum drawdown between points of withdrawal of water.

5.31. Drainage by Electroosmosis.[41-46] If two electrodes are driven into a saturated earth mass, there is a general tendency of the water to flow from the anode toward the cathode. This tendency is schematically illustrated in Fig. 5.26 in which solid flow lines depict the flow before switching on the electric current and the dotted lines correspond to the flow as reversed by electroosmotic action.   The cathode in such cases is a perforated tube from which water should be pumped out.   Careful laboratory investigations[46] show, however, that if the whole surface where the electrodes are driven is flooded, there is no noticeable change in the distribution of moisture even in the neighborhood by the electrodes. However, there is a considerable moisture redistribution if "there is free water outside the anode."[46]   The phenomenon of electroosmotic flow may be explained in a simplified way.   There is a double electric layer at the surface of a clay particle, the inner layer being negatively charged whereas

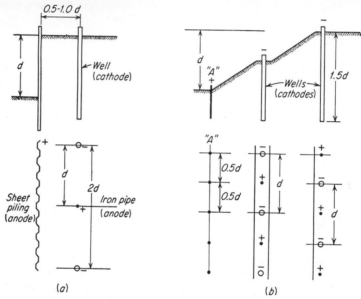

FIG. 5.27. Arrangement of electrodes for electroosmosis: (a) sheeted excavation, (b) cut 20 ft or more deep (anodes A are auxiliary).

the outside layer consists of positive ions.   When an electric field is formed, the positive ions move toward the cathode carrying associated water molecules with them.

The electroosmotic coefficient of permeability measured in centimeters per second under the action of a potential gradient of 1 volt/cm *is much higher* than the coefficient of permeability for water in soils with colloidal content.   Attempts to apply electroosmosis to sands, even to fine ones, are doomed to failure.   The method appears best suited to the drainage of silts and silty clays and is then a good substitute for well points, which are not always successful in such soils.   The high cost of electroosmosis prevents its extensive use, however.   A simple relation between the amount of water discharged by osmosis and the quantity of electricity consumed does not exist.   Data from a few large projects indicate a consumption of energy between 0.4 and 1.0 kwh/cu yd of excavation. On small projects more energy per cubic yard is spent.

In general, a potential of about 100 volts will be satisfactory on an actual project; the actual potential *gradient,* however, should not exceed 0.5 volt/cm in order to prevent loss of energy due to heating of the ground. Sheet piling protecting an excavation (Fig. 5.27*a*) may be used as an anode, though old iron rods or pipes of 1 or 2 in. diameter may serve the purpose. The anodes corrode considerably in the course of a few weeks of treatment and then should be replaced. Figure 5.27*b* shows a case of slope stabilization by drainage. The diameter of the well should be 4 in. or over. In general, both anodes and cathodes should reach to equal depths (Fig. 5.27*b*), though exceptions exist.

Safety measures in using electroosmosis are as follows: Rubber boots should be worn in working close to the electrodes; touching one electrode or the wiring and simultaneously touching the other electrode or the ground may cause serious harm and should be avoided; occurrence of short circuits by the excavation machinery also should be avoided.[46]

5.32. Engineering Subdrainage. For interception and removal of detrimental ground water, *drains* (subdrains) are used. A drain is formed by placing a pipe in a trench, backfilling it with properly graded filter material, and protecting it against silting from the top. A French or blind drain is a drain as described, but without pipe. Before the drain installation starts, it is advisable to lower the local water table by open ditches as advised in the beginning of Sec. 5.30.

In highways[47,48] the *underdrains* are mostly interceptive in character. These intercepting drains should collect water flowing toward the road and discharge it where no damage will result. In the case of a sidehill location the intercepting drain is placed in the gutter on the uphill side of the road. In level areas one underdrain under either gutter or shoulder usually suffices. The pipe should be placed below the pervious saturated soil, thus extending to the underlying impervious soil. Presumably it has to be below the local frost line to avoid freezing of the water in the pipe.

To permit ground water to enter the pipe, it may have open joints (as clay or reinforced concrete pipe) or be perforated (case of metallic pipes). In the latter case the holes may be "up" (i.e., at the top of the emplaced pipe) or "down." The manner of placing the holes is a controversial matter.

The backfill material must be fine enough to keep the adjacent soil from entering the pipe and coarse enough not to enter the pipe perforations or joints. Soil that might enter the drain would "silt" or "clog" it. Assuming the diameter of perforations at $\frac{1}{4}$ to $\frac{3}{8}$ in., a $\frac{3}{8}$-in. stone next to the pipe would be satisfactory. Proper grading of the filter material has been studied by the Army Engineers.[49–51]

Water pockets formed in *railroad ballast* where water collects and is confined also need to be drained.[47]   Once started, a water pocket descends to the wet, unstable subgrade; holds additionally arriving water; and rapidly enlarges under heavy wheel loads and high speeds.   Formation of water pockets can sometimes be prevented by placing drains during construction, especially in cuts.   Where water pockets have already been formed, their exact location is established by test holes, and subdrains are placed at right angles to the track in the form of 6- or 8-in. pipes backfilled with pervious material.   Soft spots under fills and wet cuts are treated in a similar manner.   Grouting (Sec. 15.20) also has been successfully used to force water out from the subgrade.

Areal subdrainage, as in airports or large housing projects, is designed together with the surface drainage on the basis of a grading plan for the project, indicating heights of the fills and depths of the cuts.   Smaller areas such as parks, playgrounds, race tracks, etc., are drained also on the basis of the grading.

There should be drainage behind *retaining walls*, especially if the backfill has some expansive properties.   This is true also of the *bridge abutments*, especially of the so-called U type as discussed in Chap. 14.   As to the *weep holes*, or openings in a retaining wall for discharging water accumulated behind a wall, their usefulness is questionable, since generally they discharge a relatively small portion of drainage water and on expensive structures spoil the appearance because they discolor the concrete or masonry (see also Chap. 13).

All drains must have good unobstructed outlets.

5.33 Agricultural Drainage.   Agricultural drainage in the United States started soon after the first settlement was made.   In 1835, Scotch patterns were brought to the United States, and clay tiles were molded by hand.   Further invention of machines to facilitate drainage, such as a drain scoop for finishing the bottom of the trench to grade, various ditching machines, the dragline, and floating dredges, all served rapidly to increase the installation of drainage controls.   The use of pumps further served to facilitate drainage operations, and by 1950 there were 378 drainage enterprises equipped with pumps serving about 1,700,000 acres.   The total area made cultivable by drainage in the United States is estimated from 50 to 60 million acres.

In 1902 a Department of Agriculture engineer discovered that a deep drain properly located would drain an entire field; thus, Western lands which were being rapidly damaged by seepage and alkali accumulation could start to be reclaimed.   This discovery led to the present general practice of developing drainage plans and irrigation plans at the same time.

One of the most important modern drainage problems is the development of economical methods of maintaining drainage improvement. Weeds, for instance, may reduce ditch capacity by 50 per cent in one year without maintenance. Ditch cross sections, therefore, should be ample enough to permit operation of the equipment controlling vegetation and removing silt. To decrease silting, erosion on the watershed should be controlled; also silt deposits in the outlet ditches can be reduced by proper construction of the lateral outlets.

Agricultural drainage is the removal of both the excess water and detrimental salts in order to make the land cultivable. The soils that require immediate drainage are as follows:

1. *Saline soils* containing highly soluble salts which can be handled by leaching.

2. *Alkaline soils* that cannot be leached because of the dispersed condition of the soil profile but require special chemical treatment such as by gypsum or sulfur.

3. *Saline-alkali* soils which, when irrigated, are initially similar to saline soils but, after the soluble soils are leached, develop characteristics of alkaline soils, resulting in low permeability, puddled texture, and eventual failure of the soil as a "crop grower."

If the soils are good from the agricultural point of view and are neither saline nor alkaline, poor-quality irrigation water may make them unsuitable for agriculture.

The engineer is ready to design a drain system once he has gathered physical and chemical data on the soil, the substrata profile, the water quality, and the physical features of the irrigation development. This system may include pumped wells, inverted wells, open ditches, closed tile lines, artesian taps, or any combination thereof.

The location of the main drain can be controlled by the position of the outlet and the size, shape, and slope of the area to be drained. In a *natural system* layout, the drain lines follow the depressions in the ground surface, and in a *regular system* there is a main line with regularly spaced laterals emptying into it from one or both sides to form a gridiron or herringbone pattern. *Intercepting drains* follow the base of a hill and intercept water as it flows from the high land. In complicated springs or seepage areas, the drain often is located exactly through the source of the water. References 52 through 60 contain additional information on agricultural drainage.

REFERENCES

1. Tolman, C. F.: "Ground Water," McGraw-Hill Book Company, Inc., New York, 1937.
2. U.S. Geological Survey: Over 1,000 *Water Supply Papers* contain a wealth of information on surface and subsurface water, particularly the following:
2a. Fuller, M. L.: Underground Waters of Eastern United States, *USGS Water Supply Paper* 114, 1905.

2b. Meinzer, O. E.: The Occurrence of Ground Water in the United States, with a Discussion of Principles, *USGS Water Supply Paper* 489, 1923.

2c. Meinzer, O. E.: Outline of Methods for Estimating Ground Water Supplies, *USGS Water Supply Paper* 638, 1932.

2d. Wenzel, L. K.: The Thiem Method for Determining Permeability of Water-bearing Materials, *USGS Water Supply Paper* 679, 1937.

2e. Wenzel, L. K.: Methods of Determining Permeability of Water-bearing Materials, *USGS Water Supply Paper* 887, 1942.

2f. USGS *Water Supply Papers* 227, 520E, 597, and 598 are concerned with the Dakota sandstone.

3. Meinzer, O. E., and L. K. Wenzel: "Physics of the Earth," vol. IX, Hydrology, chapter on ground water, McGraw-Hill Book Company, Inc., New York, 1942.

4. Bennison, W. W.: "Ground Water, Its Development, Uses and Conservation," Edward E. Johnson, Inc., St. Paul, Minn., 1947.

5. De Beer, E. E., and H. Raedschelders: Some Results of Water-pressure Measurements in Clay Layers, *Proc. 2d Conf. on Soil Mech. and Foundation Eng.*, vol. I, Rotterdam, paper III, p. 2, 1948.

6. Symposium on Fluctuations of Ground Water Level, *Trans. Am. Geophys. Union*, vol. 17, 1936.

7. Dore, S. M.: Permeability Determinations, Quabbin Dam, *ASCE Trans.*, vol. 102, 1937; also Quabbin Dike Built by Hydraulic-fill Methods, *ibid.*, vol. 103, 1938.

8. Plummer, F. L., and S. M. Dore: "Soil Mechanics and Foundations," chaps. 19, 20, 21, Pitman Publishing Corporation, New York, 1940.

9. Maag, E.: Ueber die Verfestigung und Dichtung des Baugrundes, Report at the Seminar at the Swiss Federal Technical University, Zurich, 1938; also a personal communication by A. von Moos of the same university, dated Nov. 9, 1949.

10. Frevert, R. K., and D. Kirkham: A Field Method for Measuring the Permeability of Soil Below a Water Table, *Highway Research Board Proc.*, vol. 28, 1948.

11. Kirkham, D., and C. H. M. van Bavel: Theory of Seepage into Auger Holes, *Soil Sci. Soc. Amer. Proc.*, vol. 13, 1948.

12. van Bavel, C. H. M., and D. Kirkham: Field Measurement of Soil Permeability Using Auger Holes, *Soil Sci. Soc. Amer. Proc.*, vol. 13, 1948.

13. Ahrens, T. P., and A. C. Barlow: Permeability Tests Using Drill Holes and Wells, *USBR Geol. Rept. G*-97, 1951.

14. Hvorslev, M. J.: Time Lag and Soil Permeability in Ground Water Observations, *U.S. Corps Engrs. Waterways Expt. Sta. Bull.* 36, Vicksburg, Miss., 1951.

15. Slichter, W.S.: Contribution to the 19th Annual Report, U.S. Geological Survey, Washington, D.C., 1899; also *USGS Water Supply Papers* 67, 1902, and 140, 1905.

16. Hazen, A.: Contribution to the Annual Report, Massachusetts State Board of Health for 1892, p. 553; *ibid.*, *ASCE Trans.*, vol. 73, p. 201, 1911.

17. Fair, G. M., and L. P. Hatch: Fundamental Factors Governing Streamline Flow of Water through Sands, *J. Am. Water Works Assoc.*, vol. 25, 1933; Tolman, C. F.: "Ground Water," pp. 204–206, McGraw-Hill Book Company, Inc., New York, 1937.

18. Meinzer, O. E.: Compressibility of Artesian Aquifers, *Econ. Geol.*, vol. 23, 1928.

19. Meinzer, O. E.: Problems of the Perennial Yield of Artesian Aquifers, *Econ. Geol.*, vol. 40, 1945.

20. Theis, C. V.: The Relation between the Lowering of the Piezometric Surface and the Rate and Duration of Discharge of a Well Using Ground Water Storage, *Trans. Am. Geophys. Union*, vol. 16, 1935.

21. Theis, C. V.: The Significance and Nature of the Cone of Depression in Ground Water Bodies, *Econ. Geol.*, vol. 33, 1938.

22. Jacob, C. E.: "Flow of Ground Water," chap. 5, John Wiley & Sons, Inc., New York, 1949.
23. Brown, R. H.: Aquifer Test Analysis, *J. Am. Water Works Assoc.*, vol. 45, 1953.
24. Thiem, G.: "Hydrologische Methoden," J. M. Gebhardt, Leipzig, 1906.
25. Wisler, C. O., and E. F. Brater: "Hydrology," John Wiley & Sons, Inc., New York, 1949.
26. Belcher, D. J., *et al.:* The Measurement of Soil Moisture and Density by Neutron and Gamma-Ray Scattering, *CAA Tech. Develop. Rept.* 127, 1950.
27. Belcher, D. J., *et al.:* Use of Radioactive Material to Measure Soil Moisture and Density, Symposium on the Use of Radioisotopes in Soil Mechanics, *ASTM Spec. Tech. Publ.* 134, 1952.
28. Horonjeff, R., *et al.:* (a) Field Measurements of Soil Moisture and Density with Radioactive Materials, *Highway Research Board Proc.*, vol. 32, 1953. (b) Paper given at the Sixth Annual Street and Highway Conference, Feb. 3–5, 1954, University of California, Los Angeles.
29. Bouyoucos, G. J.: Methods for Measuring the Moisture Content of Soils under Field Conditions, *Highway Research Board Spec. Rept.* 2, 1952.
30. Huizinga, T. K.: Measurement of Pore Water Pressures, *Proc. 2d Conf. on Soil Mech. and Foundation Eng.*, vol. I, Rotterdam, paper III, p. 4, 1948.
31. Casagrande, A.: Soil Mechanics in the Design and Construction of the Logan Airport, *J. Boston Soc. Civil Engrs.*, vol. 36, no. 2, April, 1949.
32. Gould, J. P.: Analysis of Pore Pressure and Settlement Observations on Logan International Airport, *Harvard Univ. Dept. Eng. Publ.* 476, December, 1949.
33. Speedie, M. B.: Experiences Gained in the Measurement of Pore Pressures in a Dam and Its Foundations, *Proc. 2d Conf. on Soil Mech. and Foundation Eng.*, vol. I, Rotterdam, paper 111, p. 1, 1948.
34. Walker, F. C., and W. W. Daehn: Ten Years of Pore Pressure Measurements *Proc. 2d Conf. on Soil Mech. and Foundation Eng.*, vol. III, Rotterdam, paper IV, p. 7, 1948.
35. Lang, J. D.: Australian Water Resources, *J. Inst. Engrs. Australia*, vol. 18, 1946.
36. Werner, P. W.: Note on the Occurrence of Flow Time Effects in the Great Artesian Waters of the Earth, *J. Inst. Engrs. Australia*, vol. 18, 1946.
37. Thomas, H. E.: "The Conservation of Ground Water," McGraw-Hill Book Company, Inc., New York, 1951.
38. Conkling, H.: Utilization of Ground Water Storage and Stream System Development, *ASCE Trans.*, vol. 111, 1946.
39. Baumann, P.: Ground Water Movement Controlled through Spreading, *Proc. ASCE Sep.* 86, August, 1951.
40a. "The Wellpoint System in Principle and Practice," booklet issued by Griffin Wellpoint Corp., 881 East 141st St., New York, 1950.
40b. "Moretrench Wellpoint System," booklet issued by Moretrench Corp., 90 West St., New York, 1949.
40c. "Stang Wellpoint System," booklet issued by Stang Corp., 2 Broadway, New York (also 8221 Atlantic Ave., Bell, Calif.), 1950.
40d. "Complete Wellpoint System," booklet issued by Complete Machinery & Equipment Co., 36–40 11th St., Long Island, N.Y.
41. Reuss: Paper in the *Mem. Imp. Naturalist Soc.*, Moscow, vol. 2, pp. 327–337, 1808.
42. Casagrande, L.: "The Application of Electro-osmosis to Practical Problems in Foundations and Earthworks," Department of Scientific Industrial Research, London, 1947.
43. Markwick, A. H. D., and A. F. Dobson: Application of Electro-Osmosis to Soil Drainage, *Engg.*, vol. 163, 1947.

44. Preece, E. F.: Geotechnics and Geotechnical Research, *Highway Research Board Proc.*, vol. 27, 1947.
45. Casagrande, L.: Electro-Osmosis, *Proc. 2d Conf. on Soil Mech. and Foundation Eng.*, vol. I, Rotterdam, paper II, p. 1, 1948.
46. Casagrande, L.: Electro-Osmosis in Soils, *Géotechnique*, vol. 1, 1949.
47. "Handbooks of Culvert and Drainage Practice," Armco Drainage Product Association, Middletown, Ohio, 1955.
48. Keene, P.: Underdrain Practice in the Connecticut State Highway Department, *Highway Research Board Proc.*, vol. 24, 1944.
49. Subsurface Drainage, *Highway Research Board Bull.* 45, 1951.
50. Investigations of Filter Requirements for Underdrains, *U.S. Corps Engrs. Waterways Expt. Sta.,* Vicksburg, Miss., 1941.
51. Filter Design, Soils and Paving Laboratory, U.S. Engineering Office, Providence, R.I., 1942.
52. Maierhofer, C. R.: The Role of Drainage in Irrigation, paper presented at Houston, Tex., meeting of the American Society of Civil Engineers, Feb. 21, 1951.
53. Jones, L. A.: Drainage as a Tool for Increased Crop Production, *Agr. Eng.*, April, 1953.
54. Aronovici, V. S., and W. W. Donnan: Soil Permeability as a Criterion for Drainage Design, *Trans. Am. Geophys. Union*, vol. 27, 1946.
55. Christianson, J. E.: Ground Water Studies in Relation to Drainage, *Agr. Eng.*, vol. 24, October, 1943.
56. Israelsen, O. W., and W. W. McLaughlin: Drainage of Land Overlying an Artesian Ground Water Reservoir, *Utah Agr. Expt. Sta. Bull.* 259, 1935.
57. Israelsen, O. W., D. F. Petersen, and R. E. Reeve: Effectiveness of Gravity Drains and Experimental Pumping for Drainage—Delta Area, Utah, *Utah Agr. Expt. Sta. Bull.* 345, February, 1950.
58. Kirkham, D.: Flow of Ponded Water into Drain Tubes in Soil Overlying an Impervious Layer, *Trans. Am. Geophys. Union*, vol. 30, pp. 69–88, 1949; also Reduction of Seepage to Soil Underdrains Resulting from Their Partial Embedment or in Proximity to an Impervious Substratum, *Soil Sci. Soc. Amer. Proc.*, vol. 12, 1947; also Artificial Drainage of Land, *Trans. Am. Geophys. Union*, vol. 26, December, 1945.
59. Farm Drainage, *USDA Farmers' Bull.* 2046, 1952.
60. Land Drainage, *Calif. Agr. Expt. Sta. Circ.* 391, 1949.

CHAPTER 6

SUBSURFACE EXPLORATION

Considerable field and design practice is required to obtain proper exploratory data on a project and to give them an adequate engineering interpretation. Such ability cannot come from just a book; therefore, this chapter should be considered as merely an aid in gaining some background knowledge in the vast realm of subsurface explorations. Existing subsurface exploratory methods are not something final or rigid but are subject to development and improvement. To make the most efficient use of available exploratory methods in a given locality, the investigator should be acquainted with geology in general and with local geological conditions in particular.

6.1. General Principles. The most common procedure in making subsurface investigations is to drill holes on the chosen site to extract samples of rock or soil or both for further study. The information on the quality of the subsurface material may be obtained not only from the samples but also from the field observations of the resistance to advancing the hole. These observations constitute an essential element of field explorations and often are more important than sampling alone, especially when investigating soil materials.

The machine used for making bore holes commonly is called a *drill rig*. There is no universal rig available; i.e., there is no one type of rig capable of taking every type of sample in every type of subsurface material. The choice of an exploratory device is based primarily on *economics* as explained in the following chapters, particularly Chap. 13. The smaller and less expensive the proposed structure for which the investigations are being made, the simpler and less expensive should be the investigation. Next to economics, the geological conditions exercise perhaps the most influence upon the selection of the exploratory equipment and the extent of its use. For example, the choice of equipment is controlled by whether the subsurface material is soft or hard. The relative merits of the various exploratory methods also depend upon the type of terrain, accessibility, and local topographic conditions. In extremely rugged canyon country, truck-mounted drill rigs would not be economical whereas small skid-mounted drill rigs could be used (Sec. 6.4). In the early stages of an investigation, passable roads may not be available, thus lightweight, portable rigs or even hand-dug test pits might be used (Sec. 6.17).

It has not been desirable to present in this book an actual cost breakdown of the various subsurface exploratory methods. The costs depend upon all factors previously mentioned. In addition, they vary considerably throughout the world depending upon the vagaries of contract bidding, local labor conditions, and the urgency of the work. For example, penetration test holes have been known to cost from 75¢ to $3 per foot (Secs. 6.1 and 6.11), and diamond-drill core holes (Sec. 6.11) have varied from as little as $2 to as much as $65 per foot.

6.2. Preliminary Exploration by Sounding. The term "sounding" generally

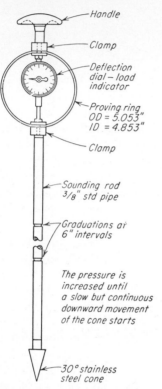

FIG. 6.1. Waterways experiment station penetrometer. (*From. J. Hvorslev, "Subsurface Exploration and Sampling of Soils for Civil Engineering Purposes," U.S. Corps Engrs. Waterways Expt. Sta., Vicksburg, Miss., 1949.*)

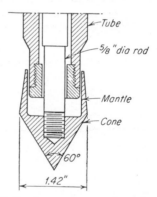

FIG. 6.2. Dutch penetrometer (sounding end).

means a subsurface investigation by observing *penetration resistance* of the subsurface material without drilling holes. This is done by forcing into the ground (1) a rod, (2) a rod enclosed in a sleeve, or (3) a wire and electrical resistor body. Closed pipes, about 1 in. in diameter, with flush points and provided with a driving tip may be used as drive rods.   They may be jacked or driven by a sledge hammer or drive weight.   Sounding may give a rough preliminary idea of soil conditions, but in soft, swampy or loose, sandy soils, the results may have to be considered as final.   Soundings also are helpful in locating soft spots in fills.

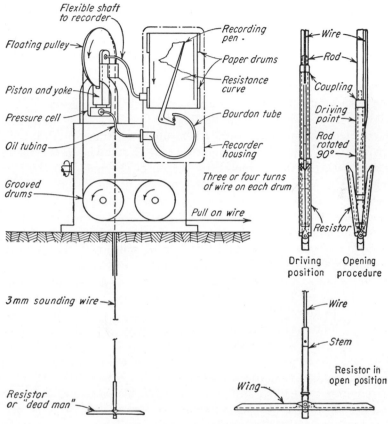

FIG. 6.3. Kjellman "Insitu" apparatus.[2]

Figure 6.1 shows the elaborate type of penetrometer built in the U.S. Corps of Engineers Waterways Experiment Station at Vicksburg, Mississippi.   The value of penetration is measured by a dial combined with the proving ring.   Cone penetrometers originated in Holland give quite satisfactory results in sand sounding.[1]   Figure 6.2 shows the lower end of a Dutch penetrometer, which is essentially a rod with a cone at its end.   The rod is enclosed in a tube and may be pushed down with or without the tube.   When the rod is pushed down without the tube, it has to overcome the point resistance to the penetration of the cone.   When the rod is pushed down with the tube, both point resistance and friction of the tube against the soil have to be overcome.   The path of the cone each time it is pushed down is about 4 in.; the resistance is measured by a Bourdon gauge at the top of the rod.

The Swedish "Insitu" (i.e., in place) apparatus[2,3] shown in Fig. 6.3 consists basically of a small resistor which is pushed down in a folded condition; it is unfolded at a given depth (as shown in the figure) and pulled up. The wire rope of the resistor is attached to a small winch and a recorder; the withdrawal resistance diagram is traced automatically. The height of the apparatus above the ground level is about $2\frac{1}{2}$ ft. The apparatus gives especially good results in soft soils.

DRILLING AND SAMPLING IN SOIL MATERIALS

6.3. Bore Holes. *Bore holes* or *drill holes* may be vertical, oblique ("angle holes"), or horizontal. As the boring (drilling) progresses, it is necessary to loosen the earth material and to extract it from the hole. This loosening is done generally either by a chopping or drilling *bit* acting at the bottom of the hole or by an *auger* which has an action in the hole similar to that of a wood bit drilling into wood. The earth material extracted from a bore hole usually is termed "cuttings." Holes of small diameter above water level generally are stable. The danger of caving increases with (1) the diameter and depth of the hole, (2) the presence of cohesionless sands, and (3) the presence of ground water. Filling bore holes with water sometimes counteracts the tendency toward caving. Uncased bore holes filled with water generally are used in cohesive clays or in soft rocks if a hole has to remain open for a long time. If the walls of the hole contain salt or anhydrite, a strong solution of salt may be used instead of fresh water to prevent dissolving of the salts.

Drilling Fluid. Bore holes may be stabilized by filling them with drilling fluid or "driller's mud." Driller's mud is a slurry prepared by mixing commercial products such as Volclay or Aquagel (essentially bentonite) with water; if fat clay is obtainable locally, it may be used as an admixture. For example, the Muroc Lake (in the Mojave Desert) mud, sometimes used in the Western states, is mixed with bentonite and barium sulfate. The mud stabilizes the walls of the bore hole both by coating them with a relatively impervious film and by exerting on them a lateral (hydrostatic) pressure. Although in civil engineering mud sometimes is used to stabilize small-diameter bore holes, it is more common to oil-well-drilling operations.

When drilling muds are used during sampling operations, their use should be noted on the drill log (Sec. 6.29); otherwise surface contamination of the sample by the mud later may erroneously be regarded as part of the soil sample.

Casing. The most expensive but the safest method of protecting the walls of a bore hole is by the use of casing (or "pipe"). Many types of standard and special pipe are used as casing. Standard or extra strong black pipes are used mostly, as they are readily available. The extra strong withstands repeated use better than the standard type and generally is preferred. The nominal sizes commonly used for casing are shown in Table 6.1 (Sec. 6.21). Casing usually comes in 5- or 10-ft-long sections. Joints of the sections may be either open with recessed couplings or flush. The lower end of the casing is provided with a shoe of hardened steel, especially if the casing has to be driven through a layer of hard material. In flush-jointed casing, the shoe may simply consist

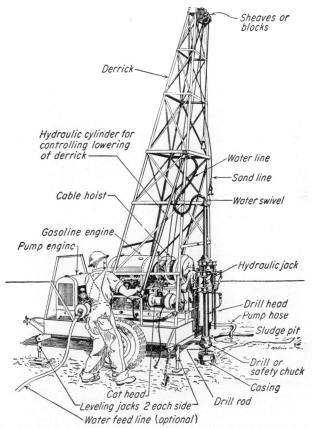

FIG. 6.4. Trailer-mounted rotary drill rig. (Drawing by J. Vitaliano.)

of a short pipe section. If a boulder is encountered during the driving and cannot be pushed aside by the casing, it has to be cored (drilled through by proper bits) or blasted out of the way. In this latter case, to avoid damage, the casing is pulled back 8 or 10 ft before blasting. If the ground is fairly cohesive, borings are made without any casing. Casing usually is sunk into the ground by repeatedly dropping a heavy weight on top of the casing (Fig. 6.6). When the bore hole no longer is needed, the casing is pulled out and may be reused.

6.4. Drilling Methods and Equipment. There are three basic types of drill rigs: (1) the rotary, or core-drill, rig (Fig. 6.4); (2) the cable-tool, or churn, rig (Fig. 6.5); a similar bucket rig (Fig. 6.8); and (3) the auger rig described in Sec. 6.7. All rigs commonly are provided with motors (gasoline, diesel, compressed air, or electric).

In the *rotary* (core-drill) *rig*, the motor is connected to a *drill head* (or table) which actuates a *drill rod* with a bit at the end. In Fig. 6.4, the drill rod is rotated by means of a gear and the bit is forced into the ground by a hydraulic or screw jack. Thus, the bit progresses in depth, and the material acted upon is cut, chipped, and ground. The drill rod in rotary drills generally is thick-walled hollow pipe (see Table 6.1, Sec. 6.21).

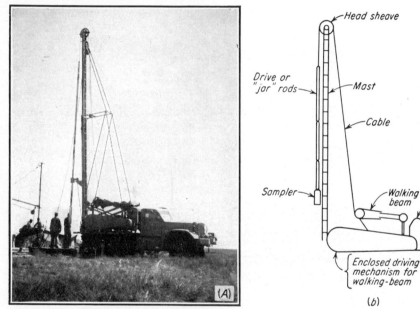

FIG. 6.5. Churn drill.

In the *cable-tool* (or churn) *rig*, a "string of tools" is used. This consists of a chopping *bit* screwed into a *stem* which, in turn, is attached to a set of two *jars* held by a rope or a cable. The chopping bit may be a heavy bar 4 ft or more long, working on the bottom of the hole. It acts together with the string of tools in a hammering, or *percussion*, action. The energy required for this action is provided in several ways, one of which is shown schematically in Fig. 6.5. This figure shows a "walking beam" that moves up and down by means of eccentrically mounted bars driven by a motor. The jars serve as a hammer to drive the sampler after the hole is made or may be used to jerk the bit loose if it sticks at the bottom of the hole. In this method the hole usually is fully or partly filled with water. The slurry formed at the bottom of the hole is periodically *bailed* out by means of *bailers* or *sand pumps* (Sec. 6.8).

Kelly Bars. In a rotary rig the upper part of the drill rod (termed "kelly" or "kelly bar" in large rigs) is connected to the water swivel. The term kelly also is used to designate a heavy, often rectangular bar with rounded edges in cross section and weighing several hundred pounds or more. It may be used to drive the sampler into a harder material after the hole in the upper soil strata has been made by wash boring (Sec. 6.5) or other means. It then is operated by a cable-tool rig, although the latter generally does not have a kelly.

Lifting and Shifting Equipment. Drilling involves a number of auxiliary operations such as placing and pulling casing and drill rods and lifting and pulling samplers. To perform most of these operations, rigs usually are provided with derricks or masts. A *derrick* is a simple crane provided with a block ("crown block") or sheave at the top centered over the drill head, with ropes or cables. The motive power is supplied by hoists. Derricks may be four-legged, e.g., 28 to 34 ft high, or three-legged (tripods). The latter are used mostly with skid-mounted rigs that move on steel runners or skids placed directly on the ground. Whereas a derrick may be used separately from the rig, a lifting *mast* is a constituent part of a rig (e.g., Fig. 6.5) formed of two or four legs close together and solidly interconnected. An *A frame* is another type of mast, of limited height (10 to 20 ft), with two legs connected in the shape of capital letter A.

Lifting devices known as *hoists* may be either of a cable type (Fig. 6.4) with a wire cable permanently fixed to the drum and wound around it or of a friction type usually termed a *cathead*. In the latter, manila rope is wrapped three or four times around the drum and hoisting is accomplished by pulling the rope tight against the drum to develop high frictional forces. Figure 6.13 shows a cathead being used to take samples with a penetrometer-type device (Sec. 6.11). Drilling equipment may be truck- or trailer- or jeep-mounted (e.g., Figs. 6.5 and 6.10) for easy movement in gentle terrain or skid-mounted (Sec. 6.1). Derricks and masts may be collapsible and raised and lowered mechanically.

6.5. Wash Borings.[4,2] To make bore holes by a simple method known as *wash boring* (or wash borings or jet holes), a four-legged derrick (Fig. 6.7) with a block at the top generally is used. Before washing starts, the hard surface material should be broken up and removed, e.g., to a depth of about 1 ft, with a pick or bull point (iron bar with sharp point). If the washable soft material is hardened at the top or covered with other hard layers, it is first necessary to make a regular bore hole by one of the methods described hereafter. The casing is then driven into the uncovered washable material by repeated blows of a drop hammer (Fig. 6.6). The usual weight of a hammer used with a cathead is about 140 to 300 lb, the height of free fall being 2 to 4 ft. With cable hoists, hammers weighing up to 800 lb may be used. The casing encloses a wash pipe, generally 1 in. or more in diameter, with a chopping bit at the end. Thus, the wash pipe in this case is usually what is termed a *drill rod* (Sec. 6.4). After the casing is driven to a reasonable depth, earth material is washed out from inside the casing. The drive

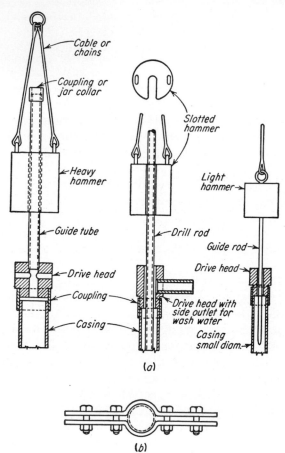

FIG. 6.6. Driving casing: (a) drop-hammer arrangement, (b) casing clamp.[2]

head of the casing is replaced by a T pipe (i.e., a pipe shaped like the letter T). The wash pipe is connected to a swivel (also called a "water swivel"), and the swivel connected to a pump. The wash pipe ejects a jet of water through the bit. The water thus erodes the soil material next to the tip of the casing and returns between the casing and the wash pipe, bringing the broken and eroded soil to the surface as a suspension. The suspension is received in a tub or a sump in the ground under the horizontal portion of the T. (It should be noted that drill mud is applied to borings in the same way, i.e., forcing it under pressure through the wash pipe and having it return between the pipe and the casing.) Except, of course, if special drill muds are used, the return water that flows into a sump or "sludge pit" is constantly returned to the hole.

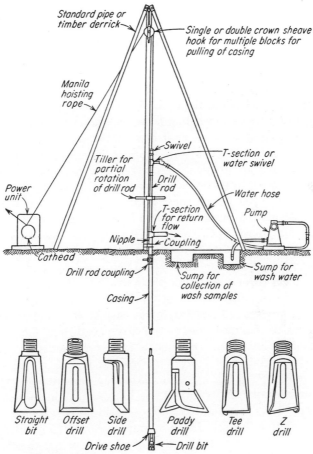

FIG. 6.7. Wash boring rig and typical bits. (USBR.)

The mixture is kept at the correct specific gravity by occasional additions of water or chemicals. If the drilling work is close to a river or other source of water, it is preferable to pump water directly instead of reusing the water from the sump as shown in Fig. 6.7. To facilitate rapid erosion of the material under the bottom of the bit, the wash pipe is churned up and down and twisted. Hoisting, in this case, may be accomplished by wrapping rope around the water swivel and lifting and dropping the pipe by alternately pulling and loosening the rope wraps on the cathead. Hand wrenches are used to twist the pipe.

When the hole is washed out, sampling usually follows. Wash borings, however, may be of value even without subsequent sampling. Such is the case when it is known that rather shallow rock underlies a certain locality and the exact depth of the rock is to be considered in the design. (This application is not practical if the rock erodes so easily under water action that its exact depth cannot be determined.) Wash borings without sampling also may be useful in the preliminary exploration before expensive, deep, undisturbed sampling starts. The cuttings (or "sludge") returning from the hole will give some, though incomplete, information about the general character of the soil and the sequence of the soil strata. Such identifications have to be regarded with caution, as the wash water tends to wash out the fines from the soil. The returning cuttings thus may indicate that the soil is coarser than it actually is. Similarly, large gravel may be broken into fine chips under the bit and thus not show in the cuttings. However, the presence of such coarse material sometimes can be detected by the action of the bit and by the presence of numerous sharp rock fragments in the cuttings. Often, washing out of soft material such as soft clay or organic silt can be done to a considerable depth, such as 60 or 70 ft, by the use of a short length of casing (e.g., 15 ft) at the top of the hole only. Generally, during the washing, the casing should be firmly gripped by a clamp or safety "dog" (Fig. 6.6b) to prevent its possible loss into the hole. In very soft materials, the casing may sink down several (and sometimes many) feet under its own weight. To remove the casing, it is pulled by upward blows of the drive hammer against the drive head (Fig. 6.6a).

A wash boring crew usually consists of a foreman and two helpers. It is the foreman's duty to submit daily progress reports to the engineering office and keep track of the following data: (1) diameter and type of the casing per lineal foot; (2) weight of the hammer, average height of fall, and number of blows required to force the casing through each lineal foot; (3) loss or gain in wash water if any (compare Sec. 6.25); and (4) his own log of the boring (Sec. 6.29) regardless of whether or not there is an engineer or geologist at the site.

6.6. Rotary and Percussion Drilling in Soils. The rotary drilling method as such is very successful in rock boring, but because of the use of wash water and high speed of rotation, many soils are disturbed by this method. In addition, operational costs and high amortization generally do not justify the use of the rotary method in soils.

Percussion or churn drilling is not very successful in predominantly sandy or clay material, but good results are obtained in mixtures of the two. Deposits of coarse gravel are usually handled fairly well with percussion methods, whereas the rotary methods are either severely handicapped or stopped completely by such materials. Similarly, boulders often are more easily penetrated by churn drilling than by rotary drilling, particularly if the boulders are loose and thus tend to roll around the hole. Although percussion methods are slow, they still provide one of the most satisfactory ways to penetrate mixtures of sand, gravel, and boulders such as are found in some alluvial and glacial deposits. Sometimes a combination of the drilling methods is used: The churn drill penetrates the coarse overburden, and the rotary drill cores the underlying rock.

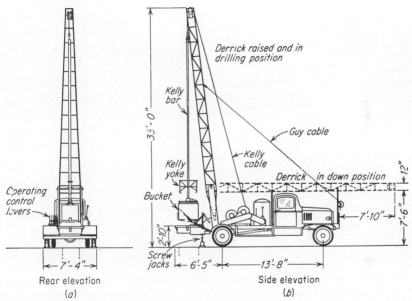

FIG. 6.8a and b. Rotary bucket rig. (The kelly yoke has not yet reached the directing and supporting ring.) (*Calweld Inc., Los Angeles 22, California.*)

Rotary Bucket. A successful combination of the rotary and percussion principles is found in *rotary buckets*. Figure 6.8a and b represents a rotary bucket rig. The buckets are 10 to 72 in. in diameter and are provided with teeth arrangements at the bottom. One variety of such an arrangement is shown in Fig. 6.8c. The bucket is fastened to the end of a kelly bar (e.g., 22 ft long and 3 in. square) and is gradually rotated and churned down. "Telescopic" kelly bars of double length are also available. When the bucket is filled, it is emptied simply by tipping it over. The bucket method for sampling is used if the bore holes are not very deep (30-ft holes are not uncommon) and do not reach the water table. The operation does not require wash water and is very rapid provided large

boulders are not encountered.   The boulders may be blasted out or in
large holes may be pried loose and hoisted out.   Larger diameter bucket
holes (over 3 ft) may be drilled by using a special side reamer attachment.
Wide holes permit visual inspection and direct sampling by a man lowered
into the hole.

FIG. 6.8c. Bottom of rotary bucket (upper part of photograph).   Portion of ring that supports kelly yoke and directs rotation (lower part of photograph).   (*Calweld Inc., Los
Angeles 22, California.*)

A modification of the rotary bucket described is used in California and elsewhere
(1956) for belling-out the holes for concrete piers (Fig. 13.6c).   A regularly shaped,
bell-like surface is obtained in this way and not a cone as in Fig. 13.6b.   Such buckets
are used for drilling deep holes for piers or for ventilation shafts for tunnels.   At
Oahe Dam, South Dakota, for example, a rotary bucket rig was used to drill 30-in.
holes up to 100 ft deep through glacial till and shale to provide ventilation shafts for
tunnels.

6.7. Auger Borings.   An auger is advanced by a rotational movement
combined with application of pressure.   Augers may produce holes 4 in.
to 3 ft or more in diameter.   Hand-operated augers may be used for
exploring soil conditions up to 20 ft.   In this case the auger consists
of a handle and a sampling tip that may be of the spoon or barrel type
(*A* in Fig. 6.9a), of helical surface type (worm auger, *B* in Fig. 6.9a), or

ɔf other varieties. Additional lengths of pipe are screwed onto the tip as the hole is deepened. When the hole goes deeper than 20 ft, it becomes difficult to withdraw the auger from the hole. Although the auger may still be turned by hand in deeper holes, it will be necessary to set up a tripod and hoist the auger from the hole by means of a block or hand winch.

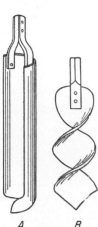

A   B

FIG. 6.9a. Sampling auger tips.

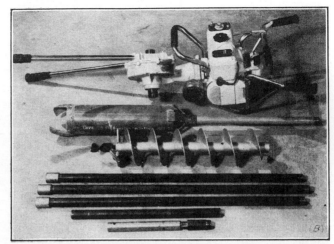

FIG. 6.9b. Hand-held power auger and sampling tools. (*USBR photograph.*)

The small, hand-held power augers (Fig. 6.9b) generally can excavate a hole from 4 to 6 in. in diameter, and the auger is advanced by a helical worm. The sections are each 5 ft long and can be screwed together. Difficulty may occur if this rig is used in hard-packed, dry silt or clay deposits. The worm tends to leave horizontal ridges on the sides of the hole, and as the motion of the auger cannot be reversed, it may be necessary in some cases to excavate it from the hole. Figure 6.10 shows a medium-size auger rig that can be operated by two men. The holes excavated are 6 in. in diameter, and the auger consists of 5-ft-long sections. The rig is self-sufficient and generally does not require additional lifting devices, although a simple hoist and tripod is useful in holes over 10 ft deep. Large-size augers will drill into slightly cohesive soils with appreciable quantities of gravel. Augers will handle cobbles or boulders up to a size slightly less than the vertical distance between the auger blades. Some of the large-size augers are capable of drilling shallow holes only (up to 12 ft), whereas others may drill considerably deeper holes.

Auger drilling does not require wash water; above the water table it leaves a perfectly dry and relatively clean hole. Cuttings brought to the surface, although disturbed, generally are suitable for positive identification of the material.   This constitutes a great advantage of auger boring over wash borings and percussion and rotary drilling (where cores are not obtained).   With power augers, when determining the depth from which the cuttings come, it is necessary to take into account the speed with which the cuttings travel up the hole.   The determination of the ground-water level, generally very difficult and often uncertain in cohesive soils, is easier in the case of auger drilling than in other drilling methods.   Casing generally is not needed in auger drilling except when

FIG. 6.10. Auger rig mounted on truck.   (*Woodward, Clyde & Associates, Oakland, California.*)

drilling through noncohesive sand and gravel and sometimes below the water table.   The disadvantage of the auger method is the impossibility or extreme difficulty of drilling through (1) fluid soils such as super-saturated sands and (2) boulders and hard strata, such as firmly cemented sands, located at a considerable depth.   Drilling practice, however, has shown that where auger drilling is *applicable*, the work progresses so fast that in drilling holes not deeper than 100 ft, it is preferable to other methods.

6.8. Cleaning the Bore Hole.   Successful sampling may be performed only in a *clean* (and primarily stable) bore hole.   Small diameter holes made by wash boring generally are sufficiently clean, but in holes made by rotary or percussion methods, the slurry (or sludge) formed at the bottom of the hole should be removed. This is done by a simple device known as a *bailer*. In one type of bailer, a vertical pipe section is provided with a flap valve at the bottom. The valve is open (i.e., it is vertical) when the bailer is pushed into the slurry and permits the slurry to fill the pipe. On withdrawal of the bailer, the valve drops into a horizontal position under the weight of the slurry in the pipe and prevents most of the slurry from flowing out. A similar but more complicated device used for cleaning bore holes is a sand pump.[2]

Coarse material such as pebbles or small rocks at the bottom of the hole may be detrimental to the stability and integrity of the upper part of the sample to be taken. Washing large holes with water jets directed upward sometimes is done to remove such occasional coarse material or lumps of cohesive soil. When the soil consists of very soft clay, silt, or fine sand with admixtures of coarse materials, it may be advisable to first take a short disturbed sample, using, for example, a barrel auger (A in Fig. 6.9a) or a similar device,[2] and only then proceed to the regular undisturbed sampling (Sec. 6.9 and following). In any case as it is seldom possible to clean the hole thoroughly, it always is advisable to discard the upper 2 or 3 in. of a sample.

6.9. Disturbed and Undisturbed Samples. Samples extracted from the ground may be disturbed or undisturbed. *Undisturbed samples* are taken primarily for all shear and consolidation tests, which generally are essential for the design and analysis of foundations. Second, undisturbed samples are taken in all cases where the properties of rocks and soils *in place* have to be studied. If, however, rocks and soils are to serve as construction materials, e.g., for runways, dams, etc., disturbed samples may be used. In a *disturbed sample*, the material must be representative of the mass from which it has been taken. Any type of soil sample should contain the same ingredients as the mass from which it is extracted and in proper proportion, including the moisture content. It is desirable to preserve in the sample, in so far as possible, the stresses and strains existing in the natural mass, since such stresses and strains may influence the results of the laboratory tests. This is difficult, however, because of the tendency of the sample to swell following its extraction from the ground and the consequent relief in pressure. In fact, in such cases there are elastic rebound of the unloaded solid material and expansion of gases in organic clays and similar soils. Nor can migration (redistribution) of *moisture within the sample* and internal plastic flow be avoided. Careful research, particularly longitudinal cutting of the sample[2] and its careful examination after drying, has shown that the horizontal layers become distorted, especially next to the walls of the sampler. Therefore, efforts should be made *to cut out* samples by hand from natural earth masses if possible and to extract from greater depths, if possible, samples large enough to permit both vertical and horizontal trimming in the laboratory. For example, a laboratory specimen 2 in. in diameter may be trimmed from a field sample 3 in. in diameter. Also, from a thick sample, an adequate laboratory specimen may be obtained by trimming the top and bottom.

So-called "undisturbed" samples (which in reality are disturbed), if taken with due care, generally furnish a satisfactory basis for judging whether or not a material is suitable for the given engineering project. Disturbed or somewhat damaged samples give, in the laboratory, smaller values of shear strength and larger values of probable settlement when compared with their field prototypes. Hence, computations based on tests on slightly damaged samples are always *on the safe side*. If, however, the samples are *badly* damaged, the design will be oversafe; i.e., the factor of safety will be unduly large. Thus, the computations may erroneously indicate that the structure is economically unjustified.

6.10. Drive Samplers: General Data.   Samplers may be divided into two basic groups: (1) drive samplers that are forced into the ground by pushing or jacking (static method) or hammering (dynamic method) and (2) core samplers used to cut a core in the given material, separate it from the rest of the mass, and then extract it.   Core boring is used mostly in rocks but can be applied to hard, cohesive soil types.   Small explosive charges sometimes are used for driving samples.

Dry Sampling.   This term is a misnomer but is often used in practice. A more accurate term would be "drive sampling," to distinguish it from the wash borings and sampling in the form of a suspension (Sec. 6.5).   The procedure is as follows: When a bore hole is made by the wash boring process, the wash pipe is lifted and the chopping bit is replaced with a section of extra heavy pipe (about 1 in. in diameter, 12 to 16 in. in length) or a special spoon and forced into the bottom of the hole.   Such a drive sample is disturbed, but its constituent parts are in natural proportion, and often it furnishes sufficient information for the design of a medium-size structure.   Drive samples generally are preserved in jars with airtight stoppers to prevent moisture loss.   The extra heavy pipe is the simplest type of drive sampler.

General Data on Drive Samplers.   A sample diameter of 2 to 3 in. usually is satisfactory for the samples intended for routine laboratory tests.   Larger diameters (up to 6 in.) are used for special tests and for research.   All lengths of samples from 5 to 12 or 18 in. are found in practice.   Greater lengths, as explained already, permit trimming and besides are less exposed to the danger of sample loss during the drilling (see succeeding paragraphs).

Figure 6.11 depicts schematically *thick-walled* and *thin-walled* drive samplers.   Careful longtime research[2] and numerous field observations have led to the conclusion that samplers for taking undisturbed samples should be *thin-walled*. A thin-walled sampler conventionally has a sampling tube (Fig. 6.11) with a wall thickness less than 2.5 per cent of the diameter. It is obvious that the amount of soil displaced by any sampler should be as small as possible in comparison with the volume of the extracted material. In other words, the ratio of the volume of the soil displaced by the sampler to the total volume of the sample extracted ("area ratio") should be reduced to a minimum. In thin-walled samplers, the area ratio is about 10 per cent. A small area-ratio value obviously means that the damage caused to the sample in the process of sampling is relatively small.

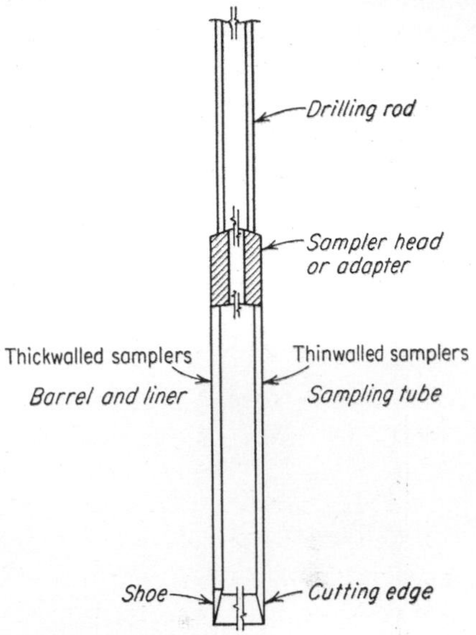

FIG. 6.11. Terminology of a drive sampler.

Conversely, in the thick-walled samplers, the area ratio and consequently the damage caused to the sample are relatively large.

During the extraction of the sampler from the bore hole there are an excess hydrostatic pressure above the sample and suction forces at its bottom. The combination of these two factors may lead to the *loss* of the sample, especially in soft soils. To prevent loss, various devices were developed such as vents at the head of the sampler with check valves to open and close the vents, wire loops at the bottom of the sampler to prevent the sample from falling out, etc. In order to contain these devices, it was necessary to thicken the walls of the samplers. Thick-walled samplers, however, are being gradually abandoned.

The sample is shorter than the column of earth from which it has been extracted. This *shortening* of the sample S (Fig. 6.12) is caused by some compaction of the material, by the downward deflection of the soil layers near the walls of the

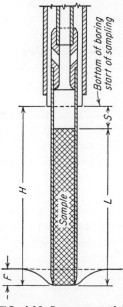

FIG. 6.12. Recovery ratio[2] L/H.

sampler, and by the stretching of these soil layers. The ratio of the final length of sample L to the length of the soil column H from which it has been extracted is the *recovery ratio* L/H. Note that the sampler deflects the strata at its cutting edge (F in Fig. 6.12).

6.11. Sampler Driving Methods. *Blow Count.* The three methods of driving the sampler into the ground are jacking, pushing, and hammering (Sec. 6.10). *Jacking* by levers or commercial screw or hydraulic jacks produces an intermittent slow motion of the sampler. If a steady downward force acts on the sampler without interruption (as in soft materials), this is *pushing*, and a continuous motion of the sampler results. Figure

FIG. 6.13. Driving penetrometer-type sampler. A = penetrometer, B = drive head, C = load, D = cathead (Sec. 6.4), E = gasoline engine.

6.13 illustrates taking a sample, starting at the ground surface, and using a penetrometer, a device similar to that shown in Fig. 6.1. The device in Fig. 6.13, however, may be used for both pushing and hammering. Both jacking and pushing are considered to be preferable to *hammering*, i.e., repeated blows of a drop hammer or a few blows of a heavy kelly, with resulting intermittent fast motion of the sampler. In the hammering method the impact and vibrations of the heavy load applied to the sampler contribute considerably to the damage of the sample structure and a decrease in its shearing and compressive strength (Sec. 6.9). However, the hammering procedure of driving the sampler still is much used, because the *blow count* taken during the hammering provides a rough but easily obtainable, very tangible (and in many cases sufficiently correct) characteristic of the earth material in place. Limitations of the blow count should be clearly understood, however.

Standard Penetration Test: Blow Count. The so-called standard penetration test consists of driving a standard sampling spoon or sampler (Fig. 6.14) using a weight of 140 lb falling *freely* from a height of 30 in. The first 6 or 7 in. of penetration are not considered, but the blows

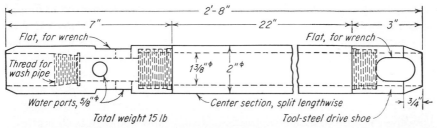

FIG. 6.14. Sampling spoon for standard penetration test. (Raymond Concrete Pile Co.)

required to drive the sampler the ensuing foot into the ground are counted. If some other type of sampler is used in investigations, this sampler should be calibrated against the spoon shown in Fig. 6.14. This can be done by driving both samplers in several places under equal conditions and correlating the results. Such calibration may be unnecessary if the blow count by a sampler is, by virtue of a fairly long practice, correlated with the local soil properties, particularly with known bearing values.

Terzaghi[5] proposed a chart for estimating permissible soil pressures for footings on sand from the blow count in the standard penetration test (Fig. 6.15). This was an excellent start, but each user of the curves should adjust those curves according to his own practice of drilling in sands and other soils. For example, it sometimes is assumed that a count of 10 blows per foot indicates a fair foundation material that may support a *total load* (i.e., dead and live plus seismic, if any) of about 4,000 psf and

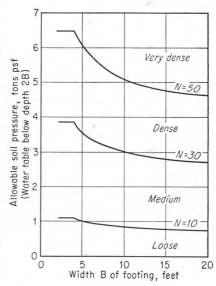

FIG. 6.15. Chart for estimating permissible soil pressure from the results of a standard penetration test. (From K. Terzaghi and R. Peck, "Soil Mechanics in Engineering Practice," John Wiley & Sons, Inc., New York, 1948.)

more, according to the depth of the base of the footing below the ground surface.

Research data by the USBR[6] indicate that besides the relative density and hence the dry density,* the blow count in a uniform and homogeneous sandy material increases with the depth. In other words, to make the sampler penetrate 1 ft into the ground requires more energy at depth than close to the surface. In fact, the standard penetration test is basically a shear test (Sec. 4.18), and the shearing strength of a uniform sand deposit depends on the weight of the overburden or surcharge, i.e., the depth from the surface. Increasing the moisture content in cohesionless sand has only a slight effect on penetration resistance. At high moisture contents, however, there is some trend toward lower penetration resistance.[6]

Fine sands generally are excellent foundation materials if they are prevented from escaping from underneath a foundation or are slightly cohesive because of some clay admixtures. Blow count in such materials may be low, however. Sometimes a decrease in blow count in an apparently uniform deposit may be explained by the proximity of the water table. Saturated fine sands may liquefy under the hammer blow.

Summary. A high blow count is indicative of good foundation characteristics of materials with a *stable* ("conservative") structure, primarily of coarse sands. In materials with alterable structure, such as clays (Sec. 4.6) or loess, the blow count is indicative of the shearing strength only at the time of the test. In this case, the test results may not apply over a long period because of the variations of the moisture content, alteration of the soil structure due to increased moisture, changes in physical-chemical properties of the constituent clay minerals, and other factors. The blow count should be considered only as a crude, semiqualitative method of establishing the suitability of a given soil material for foundation purposes. With all its uncertainties, however, the standard penetration test gives a quick and inexpensive, although rough, estimate of the suitability of a given foundation material. It should be done in every soil foundation study except, of course, in self-evident cases of very coarse, gravelly, or fluid or soft and similar materials.

6.12. Types of Samplers for Soils. Open-drive samplers and piston samplers are used for sampling soft materials, such as most soils.

Open-drive Samplers. The open-drive sampler attached to a drilling head (Fig. 6.11) is essentially a sampler head (or adapter) and a removable sampling tube. An example of a thin-walled open-drive sampler is the *Shelby tube* (Fig. 6.16). This is a trade name for hard-drawn seamless steel tubing, but any other type of thin-walled steel or brass tubing of comparable properties may be used. These samplers usually come in three sizes (2.0, 2.8, and 3.37 in. ID) and in lengths up to 30 in. Very common and convenient are 3-in. thin-walled tubes 36 in. long that per-

---

* The relationship between the dry density of a soil material and its relative density is discussed in Sec. 4.8.

mit samples 30 in. long to be taken; in this type, both ends of the tube have to be sealed as explained in Sec. 6.15.   Also, there may be accumulation of sludge and disturbed material at the top of the sample that has to be trimmed off before using the sample.   As in other thin-walled samplers, the thin-walled tubing is fixed to the sampler head by two setscrews.

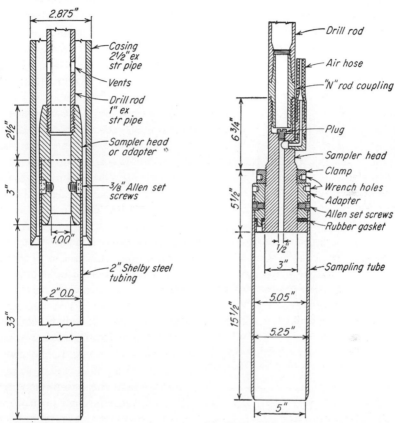

FIG. 6.16. Thin-walled drive sampler with Shelby tubing.[2]

FIG. 6.17. Thin-walled sampler, Waterways Experiment Station, Vicksburg, Miss.[2]

Another type of thin-walled open-drive sampler was developed by the U.S. Corps of Engineers Waterways Experiment Station at Vicksburg, Mississippi (Fig. 6.17).

Two basic disadvantages of open-drive samplers are (1) the entrance of sludge and disturbed soil from the bottom of the hole if the latter is not sufficiently clean and (2) the formation of large excess hydrostatic pressures over the sample.   These disadvantages are, at least partially, eliminated by the inclusion of a reversible pump.   This furnishes compressed air before the actual sampling starts and thus contributes to the removal of sludge and disturbed material from the samples.   During the sampling, the pump produces a vacuum over the sample that tends to balance the excess pressure.

6.13. Piston Samplers.   A piston sampler is a drive sampler in which the lower end may be closed by a piston operated from the ground surface by the piston rod.   Thus, a sampler with a closed bottom may be forced into the ground until the proper depth is reached.   In this way the cleaning of the bore hole may be decreased or completely eliminated.

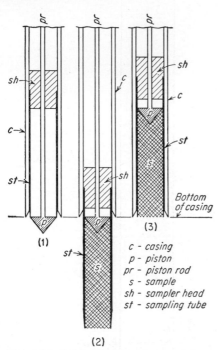

FIG. 6.18. Drive sampling using stationary piston: (1) lowering, (2) sampling, (3) withdrawal.

c - casing
p - piston
pr - piston rod
s - sample
sh - sampler head
st - sampling tube

There are three kinds of pistons: stationary, retractable, and free. In the *stationary* type, the piston rod is clamped to the drill rod in such a way that the piston is flush with the cutting edge.   In this position, the piston reaches the level where the sample has to be taken (position 1, Fig. 6.18).   In a general case, this is the bottom of the casing.   During the whole time of actual sampling the piston remains at this level by releasing the clamp to the drill rod and clamping the piston rod to the casing or some other stable object at the ground surface.   The latter clamp is released when the sampling tube is full (position 2, Fig. 6.18), and withdrawal proceeds until position 3 is reached.   The possibility of an accidental downward movement of the sampler during the withdrawal is prevented by a special cone clamp.

In samplers with a *retractable piston*, the piston is withdrawn to the top of the sampler just before the start of actual sampling.   An example of such a sampler is the California sampler shown in Fig. 6.19.   The piston is held in this position by a nut section in the sampler head and threads on the lower part of the piston rod.   The California sampler takes three to five 12-in.-long consecutive samples, thus covering a continuous sampling range up to 5 ft.

In a modified California thin-walled and pistonless sampler, which is used in conjunction with the rig of Fig. 6.10, four 4-in. consecutive samples, 2 in. in diameter, are taken.   Such an arrangement is very convenient for regions of "young geology," where the soil materials are in reality half-decomposed rocks; thus, sometimes the material in the upper 4-in.-long tube is quite different from that in the lower tube.

The *free-piston*-type sampler has a piston that is free to move with the top of the sample during the actual sampling operation. The sampler with stationary piston will act as a sampler with free piston when the piston rod is not clamped to the casing or ground surface during the drive.

6.14. Cohesionless Soils.

In cohesionless soils, ordinary sampling methods usually result in loss of the sample when the sampler is withdrawn. An attempt to prevent this loss by a purely mechanical method has been made in the Sprague and Henwood (S. & H.) Main-type sampler with flaps or flap doors.[7a] The material in this apparatus is partly disturbed, however, in passing by the entrance flap. More efficient is the impregnation of the sample with chemicals, asphalt, or cement grout. The chemicals are injected through a pipe welded to the outside of the sampler barrel and leading to a groove in the sampler shoe. A plug at the bottom of the sample is thus formed. Before testing, foreign substances introduced into the solidified part of the sample should be removed by evaporation or some other laboratory method.

The lower part of a cohesionless sample may also be stabilized by

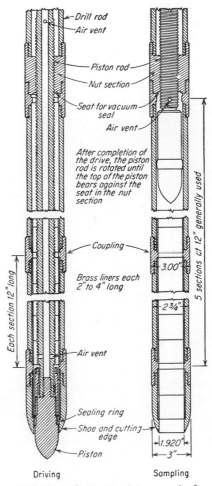

Driving   Sampling

FIG. 6.19. California drive sampler.[2]

freezing. This method is primarily intended for sampling saturated sand and silt and normally requires casing. The latter should be driven gently, without vibrations; hence, hammering is excluded. There is no conclusive evidence as to which type of sampler gives better results in the freezing method; apparently drive samplers with stationery pistons are preferable. The sampler is pushed down, being clamped to an annular auger that keeps the sampler centrally in the bore hole. After the sampler is forced into the ground, the annular auger is withdrawn and replaced by an annular freezing chamber. The freezing chamber designed by the Providence District, Corps of Engineers, U.S. Army, consists of coiled copper tubing in an insulated barrel with ethyl alcohol which is cooled to approximately −80°F by means of dry ice.[2] The main objective of undisturbed sampling in cohesionless soils, such as sand, is somewhat different from sampling in cohesive soils, where samples for consolidation and shear tests are needed. In the case of sand, it is of primary importance to know whether the deposit is dense or loose (i.e., the voids ratio of the material or its relative density) and whether or not it is saturated (i.e., its moisture content).

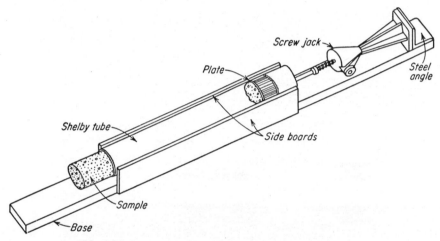

FIG. 6.20. Device for extruding sample from a Shelby tube.

6.15. Some Details of Sample Handling. In large engineering organizations such as the Corps of Engineers or the Bureau of Reclamation, the procedures to be followed in obtaining, preserving, and transporting samples are carefully codified. These instructions have been compiled primarily for the usually large government-built engineering structures and are not always applicable to the letter in the case of medium and small structures dealt with in private consulting practice. Hereafter, the gist of some of these instructions is given.

A "rest period" of 10 to 12 min should be allowed the sample before the sampler is withdrawn from the bore hole; this is considered necessary to ensure proper positioning of the sample in the container. The sampler then is twisted and pulled up. This is a good rule for soft clays and similar materials provided, however, that the position of the sample is not disturbed again during the withdrawal of the sampler from the hole.

Sample containers should be sealed at the ends with hot paraffin or similar material to prevent undue loss of moisture. This should be done if the samples have to be preserved for a long time before examining and testing. If, however, the samples for a minor structure will be tested a few days after extraction, sealing with masking tape may be sufficient.   It is not advisable to store sealed samples for too long except in a cool, humid atmosphere.   It has been found that even thoroughly paraffined samples lose a small but increasing percentage of moisture over a several months' period.   An excellent rule is to weigh the sealed container with sample immediately

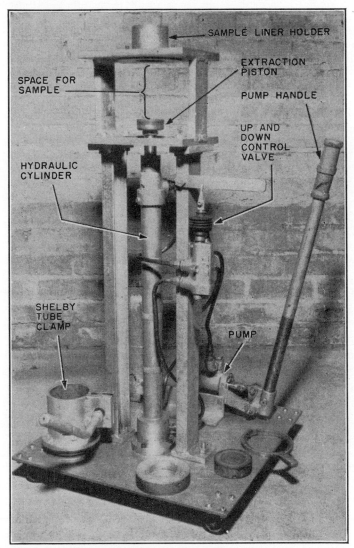

FIG. 6.21. Device for extruding samples from a sampler.   (*Institute of Transportation and Traffic Engineering, University of California, Berkeley.*)

after sealing and just before testing to ascertain the possible moisture loss during this time interval.

Samples should be protected during transportation against rolling and vibration of the containers.   This can be done by careful packing in sturdy wood or metal boxes and enclosing each sample with packing material such as sawdust, excelsior, vermiculite, or wadded paper. (In the far North, samples have been protected by packing them with moss.) Freezing protection is necessary, even in summer, if air transportation is contemplated. In fact, sample boxes may be carried in the unheated part of an aircraft and exposed to the low temperatures that prevail at high altitudes.

Extraction of the Sample from the Tube. A soft sample to be tested often can be removed from its container by simply pulling the latter down while the sample is supported by a plunger of a proper diameter. In other cases, water, air pressure, or simple jacking is applied, preferably at the top end of the sample in order to avoid the reversal of the natural stresses. Figure 6.20 shows a simple device designed to push out soft-clay samples from a thin-walled tube. Prior to pushing, the tube should be clamped or otherwise stabilized in its position and the caps and the paraffin encasing removed. As the sample is pushed out, the protruding part of it can be separated by using a tight wire saw.

Figure 6.21 shows a sample extruder designed and built in the University of California workshops at Berkeley. The sample is pushed out by a piston of interchangeable diameter acted upon by the pressure in light oil filling the vertical cylinder. Pressure is produced by a small hand pump. In many laboratories, long, thin-walled tubes with the sample still inside are cut into small portions for soil extraction. The use of circular saws for this purpose is not advisable because of vibrations.

6.16. Boring and Sampling in Submerged Areas.

For boring in a submerged area, the drilling equipment may be placed on (1) a platform protruding above the water level and supported by piles or other relatively rigid supports; (2) a raft supported by pontoons, boats, or its own buoyancy; or (3) a barge. The raft or barge usually has an opening in the bottom for drilling. Casing is needed, at least in open water. High tides, swift currents, and waves are troublesome if rafts or barges are used; the use of stable platforms is indicated in such cases.

Different arrangements for drilling through water are described in Refs. 2 (pp. 46, 47) and 23. Figure 6.22a shows a steel tower that was used by Dames and Moore (Consulting Engineers, Los Angeles, California) to make borings in the Pacific Ocean through water of variable depth. The tower may be adjusted to any water depth up to 100 ft by cutting its legs off or welding on additional lengths as necessary. Figure 6.22b shows the tower in working conditions, and Fig. 6.22c shows it being lifted to the barge for adjustment.

6.17. Sampling from Exposed Earth Surfaces.

In addition to bore hole sampling in soils, samples may be taken by hand from naturally or artificially exposed earth surfaces found in excavated trenches and pits. Samples of cohesive soils can be obtained by using a galvanized iron cylinder that is gently worked down by hand (Fig. 6.23). Earth around the sampler is removed, and the sample cut off from the rest of the mass. It is then dipped into hot paraffin or covered with cheesecloth and hot paraffin and sealed in a slightly larger container (but not so large that the sample can bounce around in the container). Paraffined samples also can be placed into plastic bags to retard the loss of moisture. If the cohesive

FIG. 6.22. Offshore foundation exploration: (a) diagrammatic sketch of drilling setup, (b) tower in position for drilling, (c) tower raised to deck of barge preparatory to cutting off portion of the legs. (Dames & Moore, Consulting Engineers, Los Angeles, California.)

soil contains gravel and pebbles, a sample may be obtained by placing a box (top and bottom removed) around a column of soil that has been carefully cut out from the earth mass (Fig. 6.24a). Empty space between the sample and the walls of the box is packed with damp sand, the top is screwed into place, and the sample is cut off from the rest of the mass with a spade or knife (Fig. 6.24b). The box is turned over (Fig. 6.24c), and the surplus soil removed. Empty spaces, if any, are again filled with damp sand, and the remaining side of the box is screwed into place.

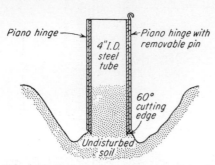

FIG. 6.23. Surface sampling in cohesive soil without gravel or pebbles.

Figure 6.25 represents a brass cylinder driven into sand in an analogous manner to that shown in Fig. 6.23. The excess sand at the top is leveled off flush, and a cap is put on (Fig. 6.25b). The person taking the sample places one hand on the sampler and the other on a trowel introduced

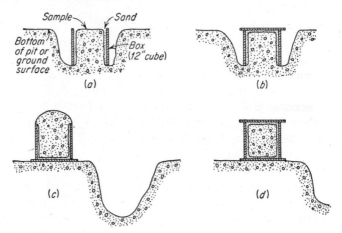

FIG. 6.24. Surface sampling in cohesive soil with gravel and pebbles.

below the sampler. The sampler is taken out and reversed, and the excess sand is again leveled. Since the volume of the container is known, porosity n or the void ratio e of the given sand in the natural state may be easily found by assuming a specific gravity value of sand grains ($G = 2.65$ for quartz sand). In some sands, the primary purpose of the pit or trench is to obtain the relative density in place of the sand. This method is described in Sec. 4.8.

If hand labor is cheap or earth-moving equipment such as bulldozers or draglines is readily available, exploratory trenches are excavated. These usually are from 2 to 8 ft in width. Their depth depends upon the capacity of the digging equipment; e.g., a bulldozer may be able to excavate only 12 or 15 ft from the ground surface, a small dragline may be able to go to 20 or 30 ft, and a hand-dug trench usually is not over 2 or 3 ft. The main advantages of the trench method are the following: (1) It provides a continuous profile of the soil strata; (2) in cases where bedrock is reached, an excellent view is obtainable of the over-burden-bedrock contact and the structural features in the bed-rock, such as jointing and frac-

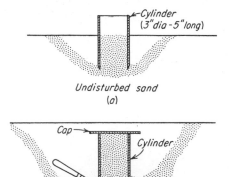

FIG. 6.25. Surface sampling in sand.

turing; and (3) it is easy to obtain in-place soil samples as described in this section. Trenches usually are dug without bracing, and their depth is limited by the stability of their slopes in a given material. It usually is not feasible to excavate a trench below the water table because of the high cost of keeping the water pumped out.

Another way of exposing soil strata in place is to dig *test pits*, usually by hand, although sometimes clamshells or orange-peel buckets are used. The cross-sectional dimensions of test pits vary from 4 by 4 to 6 by 8 ft. Generally they are laterally supported by sheeting (lagging) and cribbing, and they are braced in both vertical and horizontal directions (Fig. 6.26). If sampling has to be done below the water table, well points may be sunk around the pit to lower the water table, or pumps within or outside the pit can be used. (Centrifugal pumps generally used for this purpose can usually *lift* water only about 16 ft at sea level but can *push it up* a considerably greater distance.) Test pits offer the same possibilities for inspection of soil strata as trenches but can be dug considerably deeper; 100-ft-deep test pits are not uncommon. Because of the greater depth of excavation, which makes the use of bracing obligatory for safety, the cost per cubic yard excavated of test pits is higher than that of trenches. In deep pits, provisions should be made to keep fresh air at the bottom of the pit. Some air circulation can be maintained by the use of a long length of stove pipe with the lower end a few feet above the pit floor and the upper end far enough above the ground surface to cause a draft or by installing small blowers and vents. The exhaust of gasoline pump motors in a pit always should be vented above the ground surface. Men have

been asphyxiated in unvented deep test pits owing to the lack of oxygen.
Extreme precaution should be taken when it is necessary to enter a deep
pit that has been boarded over for a long time.   In such cases, the pit
should be adequately vented and a careful watch maintained for poisonous
snakes (which often crawl into the relatively cool interior of the pit).

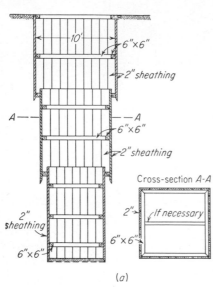

FIG. 6.26a. Test pit timber sheeting and   FIG. 6.26b. Steel pipe frame bracing of a pit.
bracing.

DRILLING AND SAMPLING IN HARD MATERIAL

6.18. Core Boring: Equipment and Methods.   The objective of the
core boring is to obtain samples of a rather hard material, whether rock
or hard soil, in the form of a cylindrically shaped *core*.   This is done by a
rotational process (40 to 1,000 rpm or more) using the rotary rig described
in Sec. 6.4.   Water, drilling mud, or air is used in core boring.   The
samplers used generally are termed *core barrels*.   There are single-tube
and double-tube core barrels (Fig. 6.27).   In the latter type, the inner
tube retains the core and usually does not rotate with the outer tube.
The rotating tubes in both types of core barrels have *drill bits* at their
cutting ends.   The actual grinding or cutting devices, or "media," may
be (1) permanently fixed to the bit, such as the cast-set diamonds used in
diamond core drilling; (2) fixed to the bit but exchangeable or refaced
when worn, such as steel teeth and cutters, tungsten carbide (or "hard
metal") inserts, or hand-set diamonds; or (3) fed between the bit and the
rock, as *chilled shot* (shot drilling).

The core boring process is as follows: (1) The bit makes an annular opening in the rock, (2) the barrel gradually slides down into this opening, (3) the core enclosed in the barrel is separated from the rest of the rock mass as later described, and (4) the barrel is lifted to the ground surface. The measures taken to avoid loss of the core consist of (1) relieving water or air pressure above the sample, for example by a small ball-check valve that opens automatically if the pressure increases above a certain limit, and (2) supporting the core from below by a *core catcher* or *core spring* (Fig. 6.28). This is a circular, springlike device which permits the core to move upward but, by spreading out, prevents it from falling out of the barrel.

In drive sampling, the material displaced by the sampler is pushed laterally and may be pressed into the sample. In core boring, the material around a sample is ground up and removed by the circulating water or drilling fluid. Sometimes the core boring is done in a dry bore hole, and the cuttings are removed by circulating air at high pressure or by using a helical hollow auger. In some types of core barrels, the coarser particles in the cuttings moving up in the drilling fluid tend to fall back because of the decrease in velocity of the fluid. To prevent this, core barrels, especially in the case of shot drilling, are provided with special sludge barrels for collecting coarse cuttings. Such a sludge barrel sometimes is termed a "calyx" barrel.

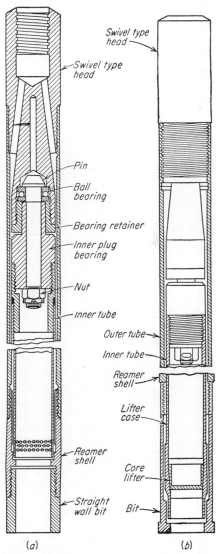

FIG. 6.27. Core barrels: (a) Double-tube core barrel—X series, swivel type; (b) barrel and bit—M series, swivel type. (Joy Mfg. Co.)

Calyx also is used to mean a very large diameter hole (30 in. or greater) excavated by shot drilling; also the actual method of shot core boring is interchangeably called "calyx boring" (Sec. 6.19).

Single-tube barrels are used in sound rock or in large-diameter holes in all kinds of rock. If the diameter of the hole is small, or if the material to be sampled is soil or fissured or soft rock, the core should be protected

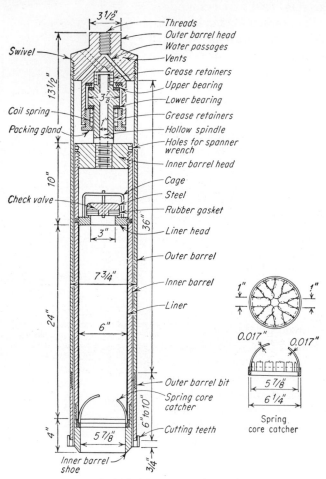

FIG. 6.28. Denison core barrel.[2]

from the erosive action of the drill water. In such cases a double-tube barrel is indicated. The inner tube should be provided with a *liner* if soil cores are to be obtained. The liner with the soil material is then taken out of the sampler, sealed, and sent to the laboratory. The *Denison sampler* is such a double-tube core barrel with a liner in the inner tube (Fig. 6.28). It is used for sampling stiff to hard clays, soft shales, soft and friable sandstone,[8] and also some types of cohesive sands and

silts, particularly if they are located above the water.   The presence of
hard particles such as occasional gravel or rock fragments will seriously
disturb a soft sample or even destroy it, as they are ground against the
sample by the drilling operation.

6.19. Diamond and Calyx (or Shot) Core Boring.   Two types of
diamonds are used for rock boring: black diamonds, or *carbons*, and gray,
yellow, or brownish industrial diamonds known as *bortz* or bort.   Carbons
are of higher quality and much harder than bortz but cost roughly ten
times as much.   Diamond bits may be hand or mechanically set (Fig.
6.29).   In the latter case they are termed "cast" or "cast-set" bits.   In

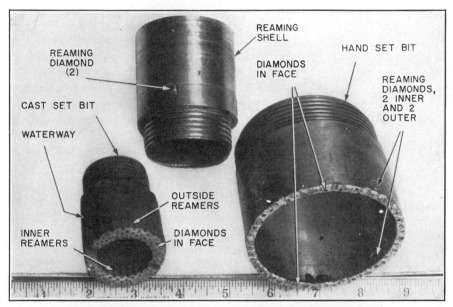

FIG. 6.29. Diamond bits and reaming shell.   (*USBR photograph.*)

the former case, holes are drilled in the inner and outer edges of the bit,
the carbons placed in the holes, and the bit steel is hammered or peened
around the carbon (Fig. 6.29).   Usually a few carbons are set on the
outside of the bit about $\frac{1}{4}$ in. or more above the bit face (i.e., its lower
edge).   These diamonds act as reaming stones (Fig. 6.29) during the
drilling to widen the hole and thus protect the carbons and steel in the bit
face and prevent the cuttings from binding ("freezing") the bit.   Figure
6.29, left, shows the so-called "M" type bits which may be effective if the
material of the core is erodible in water.   These bits permit the drill water
to discharge through ports on the side of the bit and thus minimize the
water contact with the core.   In all cases, the bit is screwed into the
bottom of the core barrel, which is rotated at various speeds depending

upon rock conditions.   As the heavy friction heats the bit, cold water is forced through the hollow drill rod to keep it cool and wash the cuttings upward.   The success of a diamond-coring operation depends upon the skill of the drill operator in maintaining his drill speed, his water pressure, and his drill bit pressure at just the right levels for the particular kind of material that is being drilled.

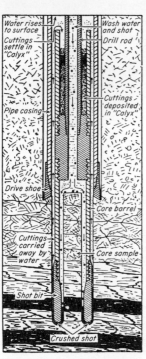

After the *core run* is completed (i.e., the full length of the core barrel has penetrated the rock), the core is separated from the rest of the rock.   In one method the circulating water is shut off and the core is "dry-blocked" or "burned-in"; i.e., the clearance between the barrel and the lower part of the core then is packed with ground materials, and the core may be extracted, though with some damage to its lower part.   In some rocks, a rapid increase in rotation speed may be all that is required to break the core loose.   In hard cores of large diameters, the barrel is withdrawn and replaced with one having a core retainer and special bit that undercuts the core. Unfortunately, sometimes none of these methods work, and the broken-off core at the bottom of the hole is picked up during the next core run (and usually is severely damaged).

FIG. 6.30. Shot core borings (notice calyx or sludge recovery). (*From H. S. Jacoby and R. P. Davis, "Foundations of Bridges and Buildings," 3d ed., McGraw-Hill Book Company, Inc., New York, 1941.*)

Calyx or Shot Core Boring.   In this type of boring, single-tube barrels only are used. Chilled-steel shot is fed into the wash water; it will become partially embedded in the soft steel of the coring bit.   To be effective, the shot must be crushed during the operation; sometimes precrushed shot ("calyxite") is used.   The cuttings are removed from the bit by circulating water and deposited in the calyx above the barrel proper (Fig. 6.30).   The core may be retained by methods similar to those used in connection with diamond drilling.   In rock with numerous open fractures, this method may be unsuccessful, as the shot may be lost in the rock openings.   Large-diameter holes (36 in. and larger) have been drilled by this method.   It also has been used to sink ventilation shafts up to 2,000 ft in depth for mines.   For the extremely large diameter holes (such as 10 ft) the motor operating the bit is placed directly above it, and the entire apparatus follows the bit down the hole by being lowered on cables.   The operator in such cases may ride directly on the top of the motor.   This large-diameter boring requires specially built equipment; even so, the work is difficult if the rock is broken or the ground-water pressures are very high.

Extraction of core in large-diameter holes is accomplished by drilling a small hole in the top of the core, grouting in a steel lifting eye, and then pulling the core to the surface by a hoist.

6.20. Angle or Oblique Holes.   When rock formations dip at steep angles, it often is more economical to drill them at an angle to the ground surface.   The drill head is inclined until its axis is normal to the dip of the beds; thus the hole penetrates a greater footage of different formations with a specific depth of hole than if the hole had been drilled vertically to that same depth.   Angle drilling is possible on most standard rotary drilling rigs if they have a swinging or rotating drill head.   It is not too effective when shot are used because of the difficulty of keeping the shot equally distributed around the bit.   Angle holes can be drilled in any direction from vertical.   Such holes seldom are advisable if the overburden requires casing; in such cases, the casing can be sunk only by underreaming the hole and pushing the casing in by the pressure of the drill head.   If the hole angle is at or near a horizontal plane or in an upward direction, it is difficult to maintain sufficient water circulation to keep the bit cool and wash all the cuttings from the hole.

The initial inclination of an angle hole often tends to vary owing to the centrifugal forces and the force of gravity acting on the drill rods.   The direction of the hole can be ascertained by occasionally inserting into it small glass vials filled with acid.   The vials are allowed to stand at rest in the hole for a sufficient time to permit the acid to etch a line on the glass.   When the vials are retrieved, the etched line shows the dip of the hole from the horizontal.   If necessary to change the direction of the hole once it is started, steel wedges are inserted in the bottom of the hole; when the bit contacts the wedge, it is deflected.   The angle of deflection is controlled by the angle of placement of the wedge.

If the core from an angle hole shows bedding, the observer must realize that the angle of the bedding is not true.   That is, the angle of the beds in the ground is complicated by the angle of the hole.   Similarly, when an angle hole is logged or such a log read, the log should state whether the depths shown are measured along the axis of the hole or are calculated depths projected along a vertical line.

6.21. Core Sizes and Core Recovery.   Rock cores generally come in four sizes (diameters) corresponding to four sets of casing, couplings, and bits as shown in Table 6.1.

The AX and BX cores are most commonly used.   The cores are preserved in special core boxes with narrow long partitions.   They are placed in the boxes in the order of their extraction from the hole. The depths are indicated on small wooden blocks inserted between each *core run*. A core run is the amount of core obtained each time the core barrel is retrieved from the hole and emptied. As in the case of soils, the ratio of the length of the cores to the length drilled (depth of the bore hole) is termed recovery ratio or simply *core recovery* and usually is expressed in per cent lost or per cent recovered. If the recovery ratio is not uniform along the bore hole, it is advisable to record it at various depths. Sound rock usually furnishes high recoveries, often about 100 per cent; seamy rock may furnish low recovery and badly broken cores. Diamond core boring gives smoother and more regular cores than shot boring.

TABLE 6.1. Standard Sizes, in Inches, for Casings, Rods, Core Barrels, and Holes

Size symbol		Casing OD	Casing bit OD	Core barrel bit OD	Drill rod OD	Approx. diam. of core hole	Approx. diam. of core
Casing, core barrel	Drill rod						
EX	E	$1\frac{13}{16}$	$1\frac{27}{32}$	$1\frac{1}{16}$	$1\frac{5}{16}$	$1\frac{1}{2}$	$\frac{7}{8}$
AX	A	$2\frac{1}{4}$	$2\frac{5}{16}$	$1\frac{27}{32}$	$1\frac{5}{8}$	$1\frac{7}{8}$	$1\frac{3}{16}$
BX	B	$2\frac{7}{8}$	$2\frac{15}{16}$	$2\frac{5}{16}$	$1\frac{29}{32}$	$2\frac{3}{8}$	$1\frac{5}{8}$
NX	N	$3\frac{1}{2}$	$3\frac{9}{16}$	$2\frac{15}{16}$	$2\frac{3}{8}$	3	$2\frac{1}{8}$

It is advisable to use as large a diameter of core as economy will permit. The larger the diameter, the more accurate are the geological observations on jointing and fracturing. Also less core loss is apt to occur if fractured rocks are drilled with large diamond bits. Very small diameters such as EX are to be avoided when possible. Often a small-diameter core will give deceptive information on the extent of fracturing, as many widely spaced fractures may be missed entirely. In extremely deep holes, it may be necessary to resort to EX holes because of the weight of the drill rods for the low capacity of the particular rig being used. Also, for ease of drilling deep holes, the size of the hole commonly is reduced with depth; e.g., the hole may start with NX, be reduced to BX at 50 ft, be further reduced to AX at 150 ft, and finally be drilled with EX bits beyond 250 ft. This reduction in size permits easier handling of casing when it must be carried to great depths because of extremely broken rock. In the latter case, the size of the hole usually is reduced two sizes at a time to permit a one-size reduction in casing; i.e., an NX hole with 4-in. casing may be carried to 25 ft, then the remainder of the hole is cased with 2½-in. casing, and AX core barrels are used.

To ensure recovery in fragmentary and seamy rocks, the bore hole may be pre-grouted. For this purpose a small diameter pilot hole, about EX, is drilled, washed, and grouted. After the grout has set, a larger hole, about NX, is drilled directly over the EX hole. A skilled operator may thus obtain a grouted core of fragmentary material (Fig. 6.31). A similar method is to bore an NX pilot hole on a slightly oblique angle, using a steel wedge inserted at a shallow part of the hole. After the hole has been grouted, another NX hole is drilled vertically at the same location.

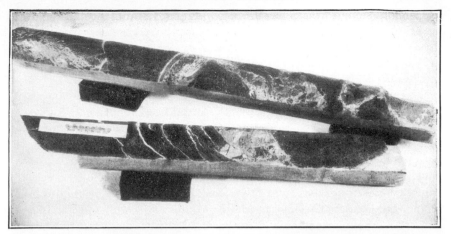

FIG. 6.31. Recovery of fragmental core by pregrouting (the white masses and seams are grout). (*USBR photograph.*)

FIG. 6.32. Exploratory drift. (*Photograph by C. McHuron.*)

6.22. Exploratory Drifts and Tunnels. One method of exposing a rock surface covered by overburden is to excavate a drift into the hillside. Drifts are tunnels with one entrance; two-entrance tunnels are used occasionally. A drift usually is 4 to 6 ft wide and 4 to 8 ft high; its ultimate dimensions depend on the necessity for timber supports. The floor of a drift should slope toward the entrance to facilitate drainage and removal of the excavated material ("muck") by either a wheelbarrow or a small hand-pushed muck car.   From within a drift, core holes can be made in any direction, from vertical downward to vertical upward.   The main purpose of a drift is to permit detailed examination of joints, fractures, and other rock conditions (Fig. 6.32).   Often the dimensions of a drift are increased for special purposes, e.g., for additional borings.   Increased size drifts also are used to study the *capability* of an unsupported rock mass to *carry its own weight without crumbling*—a factor important in the design of underground power plants and, obviously, in all mine excavations.   The cores taken in the drift should be properly marked, i.e., whether it was obtained from the walls, floor, or roof (the tunnel terms invert and crown or back are also used for floor and roof, respectively, see Sec. 9.2).   Such orientation of the cores may prove important for proper interpretation of the results of compression or triaxial tests.   Drifts are expensive, but in the case of a major structure their cost may be fully justified.

6.23. Jackhammer Rock Exploration.   A jackhammer is a hammer-type rock drill operated by compressed air or a self-contained gasoline engine.   During the drilling, the rate of penetration of the jackhammer and the color of the dust issuing from the hole are observed.   These factors, in conjunction with previous boring data, may enable the geologist to establish a rough picture of the underlying strata at a given place.

Jackhammer exploration cannot be substituted for regular core boring, however; this is an auxiliary exploration method which may be useful during construction or in other emergency cases when the available geological information has been proved insufficient and there is no time to set up a core boring rig.

In a particular case, when the excavation for a masonry dam had been carried down to the proposed finished grade (i.e., design elevation shown on plans), the presence of soft spots under that grade was suspected.   A series of jackhammer holes 5 ft deep permitted the location of the soft spots and an estimate of their thickness.   Thus, the soft material was removed and replaced with lean concrete.

6.24. Depth of Core Borings.   Core boring commonly is used for the foundation exploration of heavy structures such as large dams, large bridges, powerhouses, and monumental buildings.   The depth of the boring depends primarily on local geology and the structure load.   Presumably, the excavation for the foundation of a heavy structure should go through the upper weathered or otherwise weak rock strata, for instance, as explained in Sec. 15.19 for masonry dams.   Even if the rock is homogeneous and solid, the foundations for a heavy structure should be placed below its surface.   In fact, the heavier the structure, the larger the shearing stresses in the underlying strata, and the shearing strength of the rock resisting these stresses rapidly increases with the depth. If feasible, the boring should penetrate to that depth in the rock where the stresses from the structure load are still critical. Very often, ledge rock with undercut ledges is mistaken for a continuous depth of solid rock.

Similarly, boulders of large diameter have been confused with bedrock, and these are additional reasons for core boring below the apparent hard rock surface.

6.25. Ground and Wash Water in the Soil and Rock Borings. If a bore hole strikes the water table, the position of the latter should be determined in terms of either its *elevation* above a definite local datum plane or its depth below the ground surface at the bore hole. In pervious materials, the free water table may be seen or its presence established by simply throwing a rock fragment into the bore hole. In impervious soils and rocks the situation is more complicated as explained hereafter. The position of the water level in a bore hole should be recorded after it is *stabilized*, i.e., is not rapidly rising or dropping. In any case, it is advisable to check the position of the water table 24 hr or so after water has been struck. All water-table records must be accompanied by the date

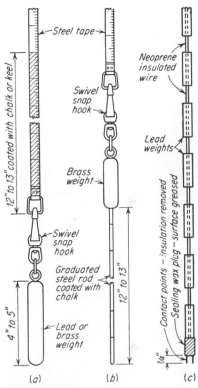

FIG. 6.33. Devices for measuring depth of the water table.[2]

of the final observation, since seasonal and other fluctuations of the ground water may occur (see Sec. 19.5).

It always is advisable to observe the interior of the bore hole by using a strong flashlight or a small mirror reflecting the sun's rays into the hole. A simple device to measure the depth of the water table is a pipe section 2 in. long and 2 in. in diameter. The upper, blind end of the pipe section is provided with a stopper and a hook. The device is fixed to the end of a tape or a rope and dropped into the hole; when it strikes free water, a dull but quite distinguishable sound may be heard. Other free-water-table-measuring devices are shown in Fig. 6.33. The device labeled (*a*) is a tape coated with chalk or painted with a lime solution, and a weight is hung on the end of the tape. This end is dropped into the hole, whereas the other end remains fixed at the ground surface. The boundary between the washed (where the lime or chalk is removed) and the lime-coated portion of the tape indicates the position of the water surface. To account for the volume of water displaced by the weight, the reading can be made on a thin rod attached to the weight (Fig. 6.33b). In another type of measuring device, advantage is taken of the difference between the electrical conductivity of air and water (Fig. 6.33c). As soon as the ends of the two wires touch the water, the reading of the galvonometer to which the wires are connected changes. Note the small weights distributed along the length of the tape. (Metal or metallic tapes should always be used when measuring water tables, as cloth tapes will stretch when wet.)

Difficult and, in many cases, uncertain is the determination of the water-table position in materials of low permeability. Generally, it cannot be done hastily. Since the total volume of the bore hole is larger than the volume of the pores, a certain time is required for water in the bore hole to assume the true level. Also there may be a stratum of relatively high permeability located above the bottom of the bore hole. This stratum, no matter how thin, then will provide water to the bore hole until the water in the latter stands at the level of that stratum. Obviously, the existence of such a stratum, if unnoticed by the observer, will confuse the results of the measurements. Finally, the upper portion of a bore hole may be in impervious material and the lower portion in pervious, fully saturated material. If the moisture in the latter is connected with a reservoir or basin located at a higher level, the water in the bore hole is *under pressure* and will rise. The ultimate water level will be that of the previously mentioned reservoir or basin, minus some friction loss. The time required for the final water level to be attained in the bore hole depends on the permeability of the material through which it is sunk.

The total amount of *wash water* forced into the bore hole during the drilling should return to the ground surface. If it does not, this fact should be reported by the drill foreman. If wash or drill water is lost, it may mean either that at a certain depth there is a porous material with empty voids or that there is a downward hydraulic gradient that pushes the wash water downward together with other water already moving in that direction. Although the former may cause design problems (in dams, Sec. 15.14), the latter case may indicate that serious trouble will be encountered during or after construction. In fact, it may mean that when the excavation for a structure reaches the design depth, it will be difficult to remove the water by pumping. Sometimes, even during the exploration period, the water level in the boring cannot be lowered by bailing—a definite danger sign that should be investigated further by pump-in or pump-out tests or by pressure tests (Sec. 5.14).   In some types of highly impervious materials, the wash water may remain in the drill hole for several days, and its level may be erroneously taken for a high water table.   Therefore, if such a condition is thought to exist, the hole should be thoroughly bailed and time (at least 24 hr) allowed for the water table, if any, to stabilize.   In important cases, some of the bore holes are left open or vertical small-diameter pipes are set in the completed holes; periodic water-table measurements are taken during and after the entire exploration program is completed.

6.26. Drilling Safety Practice.   Like many other operations involving machinery, drilling is dangerous only when the operators become careless.   Most operators wear gloves, which can become entangled in the rapidly rotating hoist drums or drill heads if the operator becomes careless.   When using the cable hoist or cathead, the operator should be wary of having his gloves or his hands pinched by the opening and closing of the cable or rope strands as tension on the cable is released and applied, respectively.   Similarly, the cathead rope may become suddenly entangled around the drum owing to an unusually loose wrap of rope, and the sudden jerk can catch a hand or arm between the rope and the cathead drum.   Loose clothing should always be avoided because it may become entangled in the numerous moving parts—the pump shaft, drill head, chuck, drive shaft from the motor to the drill head, or the drill rods.   Ridiculous though it seems, hands are often crushed against the rig when an unwary driller places a wrench on the drill rod to loosen it and the other driller suddenly throws in the clutch on the motor to spin the rods to assist the uncoupling procedure.

When casing is being driven, caution is always necessary.   The heavy driving hammer may break the casing and swing free against a driller's head, the twisting motion of the hoisting cable applied to the hammer may loosen the drive pipe and the suddenly released weight can drop on a driller's foot, or in an attempt to hold the casing perfectly vertical, the driller may get his fingers between the dropping hammer and the top of the drive head.   A standard safety rule is to wear a "hard hat"; this pressed fiber or metal helmet offers considerable protection from a sudden blow or a part dropped from the mast or derrick.   Often one of the crew may be working on the upper part of the mast and accidentally drop a tool; the common warning cry in such a case is "Headache!"   Many of the drilling tools are very heavy, and lifting injuries are not uncommon.   These can be avoided by using the power hoist when possible. If lifting is done by hand, the driller should assume a squatting position, fix his hands and arms securely around the object to be lifted, and then raise it by straightening his legs.   Heavy objects should never be lifted by bending over and trying to pick them up.

LOGGING AND LOGS

The term "logging" in its broad sense means recording the earth crust materials along a single direction, usually along a vertical line starting at the ground surface.   Such records are represented in the form of a *log*. Borings in soils and rocks, previously described, are direct loggings; several methods of indirect logging also are described in this chapter (Secs. 6.30 and 6.31).

6.27. Contents of a Bore Hole Log. The log of a bore hole may be made in written form or plotted graphically to scale. The character of a log depends on the materials of the bore hole, on the method of drilling, and finally on the purpose for which the bore hole is being drilled. For example, the logs of shallow bore holes drilled in soft materials for medium-size buildings may be much simpler than the logs of drilling in rock for a heavy structure, such as a masonry dam. All records should be made *in the field*, and if the field records are redrawn or rewritten, the field original should be *preserved* for a reasonable time.

There should be a separate log, on a separate sheet or sheets of paper, for each bore hole, test pit, trench, or drift. Date or dates of drilling are indicated on *each log sheet*, also the elevation of the ground surface at the bore hole. The water table, if found, should be recorded as an actual depth or elevation, and the date of such measurement given. (On cross sections the position of the water table may be indicated by a horizontal line with a small equilateral triangle standing on its vertex on that horizontal line.) If no water has been found, this *should be clearly stated* on the log (and the date of observation given).

6.28. Logs of Soil Materials. As stated in Sec. 6.1, a proper appraisal of soil conditions should be based not only on the examination and testing of the samples extracted during the drilling but also on observations on the resistance to the advancement of the hole and on penetration resistance of the material. Penetration resistance is characterized quantitatively by the blow count during the drilling (Sec. 6.11). The resistance to the advancement of the hole may be evaluated by the recovery of the sample (Sec. 6.10) and by the *speed of drilling*. Other features to be observed are largely a matter of judgment and experience of the drilling personnel. Such comments in the log as "drilling becomes easier (or harder)" indicate the local loosening or strengthening of the soil structure, respectively, a fact which generally cannot be detected from the samples.

The simple example of a soil log in Fig. 6.34 refers to an auger-bored hole, though it may be adapted for other types of drilling. Either the location of the samples to be extracted is predetermined by the engineering office, or samples are taken at predetermined intervals (e.g., every 10 ft or at every change in material). Only the engineer or geologist in charge or a very experienced technician may order the extraction of additional samples or change the sampling instructions of the engineering office.

In the first column of the sample log (Fig. 6.34) the field identification of the soils from visual inspection is given. This is checked and corrected by the laboratory when the samples are opened. The primary purpose of the other columns is self-evident. It should be recalled that the blow count on the first 6 or 7 in. is recorded but is not considered in estimating the penetration resistance. Observations on the resistance to the advancement of the hole are recorded in the first or the third columns. Some laboratory test results (such as natural moisture content, m.c.; dry density, D; and the unconfined compressive strength, U.C.) are placed in the fourth column. The location of the water table and the pertinent observation data (e.g., height and duration of the water-table rise) are usually given in the second column.

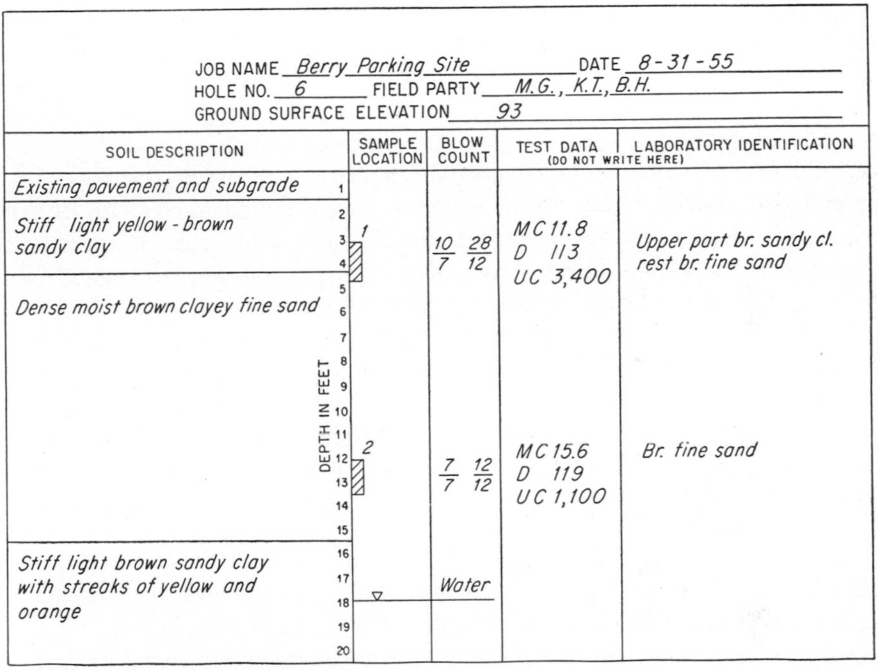

JOB NAME _Berry Parking Site_   DATE _8-31-55_

HOLE NO. _6_   FIELD PARTY _M.G., K.T., B.H._

GROUND SURFACE ELEVATION____ _93_

SOIL DESCRIPTION	SAMPLE LOCATION	BLOW COUNT	TEST DATA (DO NOT WRITE HERE)	LABORATORY IDENTIFICATION
Existing pavement and subgrade 1				
Stiff light yellow - brown 2 sandy clay 3 4	1	$\frac{10}{7}$ $\frac{28}{12}$	MC 11.8 D 113 UC 3,400	Upper part br. sandy cl. rest br. fine sand
5 Dense moist brown clayey fine sand 6				
7 8 9 10 11 12 13 14 15	2	$\frac{7}{7}$ $\frac{12}{12}$	MC 15.6 D 119 UC 1,100	Br. fine sand
16 Stiff light brown sandy clay 17 with streaks of yellow and 18 orange 19 20	▽	Water		

DEPTH IN FEET

FIG. 6.34. Example of shallow boring log.   (*Woodward, Clyde & Associates, California.*)

Presumably more columns for additional information, if any, could be added; other types of log forms also are used.

Sometimes direct shear tests are performed by using the unit weight of the overburden overlying the sample as the vertical unit load. The shearing strength of the sample then may be indicated graphically as a heavy horizontal line across the log. Other test results such as the consolidation test or triaxial compression test usually are recorded separately (i.e., not in the logs).

If there is casing, the blow count should be indicated for each foot of penetration. The diameter of the casing should be given. If the sampler is jacked or pushed, the force required to move it is recorded. In other words, the record shows the work done in moving the sampler 1 ft.

6.29. Logs of Core Borings. If core boring is done for a heavy structure which obviously will be founded on rock, the overburden of soft material over it is not logged in great detail. Its thickness and general description may suffice in this case. The geological origin of the overburden, if known, should be indicated (alluvium, wind-blown, glacial, talus, etc.). If more details are needed, the example of Fig. 6.35 may be followed.

For *rock material*, besides the name of rock in each stratum and the elevations of the top and bottom of each stratum, the following information should be included in the log: (1) color, fabric, texture (such as fine-grained), type of cementation (such as calcite-cement, clay-cement, etc.), weathered or not;* (2) the presence of joints, fractures, or seams and whether they are open or closed or filled with other material and their dip; (3) amount of core recovered and the average lengths of the core; (4) location of the drill hole, by coordinates as explained in Sec. 7.3 or by stations; (5) elevation of either the ground surface or collar (top) of the casing, carefully indicating which of the two is given; (6) angle from vertical and the bearing of the hole (the compass direction in which an angle hole is driven); (7) in rotary drilling, speed of penetration of the bit (e.g., 1 ft in 10 min) and the number of rpm during this period; and (8) unusual occurrences during the boring (e.g., loss of core).

The logs are plotted on cross-section paper as vertical lines or strips spaced on distances proportional to their actual (field) spacing. If the tops of the logs (i.e., the collar elevation) are referred to the same datum plane, strata of similar material located on the various logs may be joined by straight lines or smooth curves. Thus, soil profiles and *geological sections* are obtained (Sec. 7.6). Although it is generally preferable to make the vertical and horizontal scales of the soil profiles and geological sections equal, the vertical scale may have to be increased (e.g., 10 times) for clarity.

An elevation view of the planned structure may be plotted on a geological section (or a soil profile) at the same vertical and horizontal scales as the section. If these scales differ from each other, the structure will appear distorted, but the plot will clearly show the relationship between the foundation of the structure and the supporting strata.

Bore Hole Camera. To record true dips of deep formations, intensity of deep jointing, and general character of the rock *in situ*, "deep-well cameras" are used in order to photograph the walls of a bore hole (see Ref. 9, p. 664). The original idea of such cameras was advanced in the thirties by Reinhold.[10] Further development of this by the Geology and Geophysics Branch of the Corps of Engineers and the Engineering

---

* Negative geologic information is rarely recorded. For example, "no sand" is superfluous, as the mere omission of the word "sand" from the log is self-explanatory.

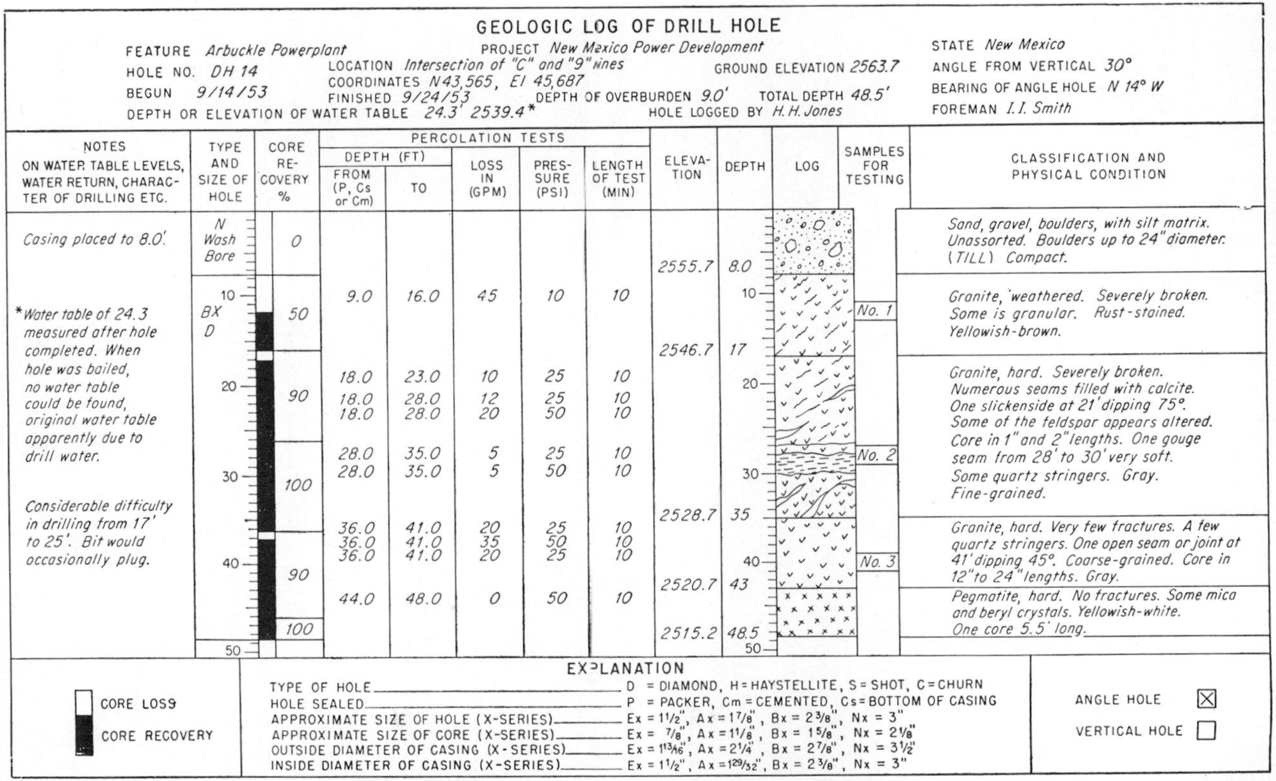

GEOLOGIC LOG OF DRILL HOLE

FEATURE *Arbuckle Powerplant*
HOLE NO. *DH 14*
BEGUN *9/14/53*
DEPTH OR ELEVATION OF WATER TABLE *24.3*

PROJECT *New Mexico Power Development*
LOCATION *Intersection of "C" and "9" lines*
COORDINATES *N43,565, El 45,687*
FINISHED *9/24/55* *DEPTH OF OVERBURDEN *9.0'* *2539.4**

STATE *New Mexico*
ANGLE FROM VERTICAL *30°*
BEARING OF ANGLE HOLE *N 14° W*
GROUND ELEVATION *2563.7*
TOTAL DEPTH *48.5'*
HOLE LOGGED BY *H.H.Jones*
FOREMAN *I.I. Smith*

NOTES ON WATER TABLE LEVELS, WATER RETURN, CHARACTER OF DRILLING ETC.	TYPE AND SIZE OF HOLE	CORE RECOVERY %	DEPTH (FT) FROM (P, Cs or Cm)	DEPTH (FT) TO	PERCOLATION TESTS LOSS IN (GPM)	PRESSURE (PSI)	LENGTH OF TEST (MIN)	ELEVATION	DEPTH	LOG	SAMPLES FOR TESTING	CLASSIFICATION AND PHYSICAL CONDITION
Casing placed to 8.0'.	N Wash Bore	0	0	9.0	45	10	10	2555.7	8.0			Sand, gravel, boulders, with silt matrix. Unsorted. Boulders up to 24" diameter. (TILL) Compact.
*Water table of 24.3 measured after hole completed. When hole was bailed, no water table could be found, original water table apparently due to drill water.	BX D	50		16.0							No. 1	Granite, weathered. Severely broken. Some is granular. Rust-stained. Yellowish-brown.
		90	18.0 18.0 18.0	23.0 28.0 28.0	10 12 20	25 25 50	10 10 10	2546.7	17			Granite, hard. Severely broken. Numerous seams filled with calcite. One slickenside at 21 dipping 75°. Some of the feldspar appears altered. Core in 1" and 2" lengths. One gouge seam from 28' to 30' very soft. Some quartz stringers. Gray. Fine-grained.
		100	28.0 28.0	35.0 35.0	5 5	25 50	10 10	2528.7	35		No. 2	Granite, hard. Very few fractures. A few quartz stringers. One open seam or joint at 41' dipping 45°. Coarse-grained. Core in 12" to 24" lengths. Gray.
Considerable difficulty in drilling from 17' to 25'. Bit would occasionally plug.		90	36.0 36.0 36.0	41.0 41.0 41.0	20 35 20	25 50 25	10 10 10	2520.7	43		No. 3	
		100	44.0	48.0	0	50	10	2515.2	48.5			Pegmatite, hard. No fractures. Some mica and beryl crystals. Yellowish-white. One core 5.5' long.

TYPE OF HOLE
HOLE SEALED
APPROXIMATE SIZE OF HOLE (X-SERIES)
APPROXIMATE SIZE OF CORE (X-SERIES)
OUTSIDE DIAMETER OF CASING (X-SERIES)
INSIDE DIAMETER OF CASING (X-SERIES)

EXPLANATION

D = DIAMOND, H = HAYSTELLITE, S = SHOT, C = CHURN
P = PACKER, Cm = CEMENTED, Cs = BOTTOM OF CASING
Ex = 1½", Ax = 1⅞", Bx = 2⅜", Nx = 3"
Ex = ⅞", Ax = 1⅛", Bx = 1⅝", Nx = 2⅛"
Ex = 1¹³⁄₁₆", Ax = 2¼", Bx = 2⅞", Nx = 3½"
Ex = 1½", Ax = 1²⁹⁄₃₂", Bx = 2⅜", Nx = 3"

☒ ANGLE HOLE
☐ VERTICAL HOLE

CORE LOSS
CORE RECOVERY

FIG. 6.35. Example of log for deep bore holes. (USBR.)

Research Associates Division of Remington Rand, Inc., resulted in the construction of a stainless-steel tube $2\frac{3}{4}$ in. in diameter and $31\frac{1}{2}$ in. long containing an oil-damped compass which supports a hollow and truncated conical mirror (Fig. 6.36).   The position and vertical dimensions of the mirror correspond to those of a continuous cylindrical quartz window in the tube wall and act as the camera lens.   The camera lens (a 15-mm lens) facing down is located a short distance above the mirror.   Illumination is provided by a high-voltage circular flash tube midway between the cone mirror and camera lens and is actuated by a pulsing device on the lowering rig.   Directly above the lens is a conventional 16-mm motion-picture

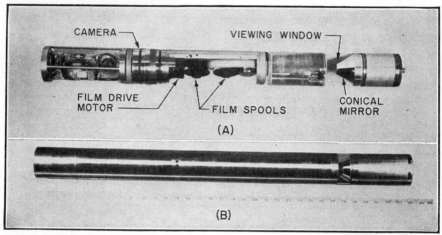

FIG. 6.36. Bore hole camera: (a) interior view, (b) exterior view.   (*Engineering Research Associates Division of Remington Rand, Inc.*)

camera with a motor drive, synchronized by the same pulsing circuit which actuates the flash tube.   The device is run into the bore hole on a three-conductor cable so constructed that twisting is prevented; it may operate in a bore hole as small as 3 in. in diameter.   Pictures usually are taken while the camera is being retrieved from the hole to ensure more constant cable tension.   Seventy-five feet of boring can be photographed at each loading of the camera.   Water in the hole is no problem, as the camera can withstand 500 ft of hydrostatic head without leaking, and pictures in dry and wet holes may be equally successful providing mud does not obscure the lens.   All equipment is portable and thus adaptable to any site where diamond-drill rigs can be taken.

The resultant photographs, which actually are photographic logs, give a 360° picture of the bore hole.   The compass orientation is automatically recorded on the film.   The film is placed in a special projector that projects the image on a circular ground-glass tube.

6.30. Electrical Logging.[11-13]   The two indirect logging methods often used in the oil industry, but very little in civil engineering, are electrical and radioactivity logging.   *No samples are taken* in either method.

In the electrical logging, measurements are made in an *uncased* hole filled with driller's mud, preferably fresh.   A system of electrodes (known as a sonde) is lowered into the hole on a multiconductor cable, and the readings are recorded on photographic film or by other means.   The basic purpose of this method is to obtain the in-place resistance to the flow of electric current, or the *resistivity* of the rock materials at various depths, and, in addition, the natural electrical potential developed in the bore hole.

Rocks are capable of transmitting electric currents only by means of sorbed water; dry rocks are electrically nonconductive.   Fresh water is a poor conductor; hence, fresh-water sands and gravels and dense limestone and strongly cemented sands with very little pore water are poor conductors.   Conversely, low resistivity is characteristic of salt-water sands and gravels and also of shales and clays, since their pore water is generally of high salinity.

Resistivity is measured in ohms meter squared per meter of the depth (m²/m) which amounts to ohm times meter, or *ohm-meter*.   In the following table,[12] the resistivity of various substances is given in increasing order.   Thus, copper listed in this table is the best conductor, whereas quartz is the poorest.

TABLE 6.2. Electrical Resistivity of Different Materials, in Ohm-meters

Copper............................. 10^{-8}
Pyrite (iron sulfide, FeS_2)........... 10^{-6}
Concentrated salt water............. 2×10^{-2}
Clay, wet, plastic................... 1–3
Lignite or gypsum................. 10^3
Oil.............................. 10^4
Calcite or dense limestone.......... 10^7
Quartz........................... 10^{10}

The zigzag curve of an electrical log is traced by plotting resistivity against depth.   From this curve, a skilled interpreter can determine the boundaries of formations.   As most rock types have characteristic curves, long practice will permit the interpreter actually to recognize and thus name the formations.   Also he can detect variations in the fluid content of the rocks and, hence, locate water supply aquifers.

6.31. Radioactivity Logging.   The concept of fast and slow neutrons which is used in the radioactivity logging already has been discussed in Sec. 5.23.   One of the two methods of radioactivity logging, namely, *neutron logging*, has much in common with the neutron scattering method as used for the determination of the moisture content in soils.   In the neutron logging (Fig. 6.37) an ionization chamber filled with an inert gas is lowered into the bore hole.   There is a *source of neutrons* at the bottom of the ionization chamber (crosshatched in the figure).   This source is formed by combining beryllium nine, Be^9, with helium four, He^4.   The fast neutrons leaving the chamber bombard the walls of the bore hole and, thus, release γ rays that ionize the gas in the ionization chamber.   The electric current generated is amplified, and its voltage, which is proportional to the intensity of the γ rays, measured.   An automatic recorder (see Fig. 6.7) traces the log by plotting these measured values as horizontal ordinates from a vertical base line representing the hole.   In a formation containing a considerable amount of fluid (such as gas, oil, or water)

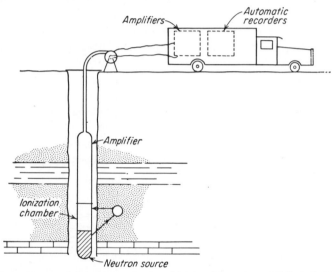

FIG. 6.37. Radioactivity logging methods.   (*Adapted from Lane-Wells Bull. RA-47-B-3.*)

and thus porous, the hydrogen reduces the velocity of the neutrons (Sec. 5.23) so that only a small percentage of neutrons actually bombards the formation.   Thus the recorded curve recedes to the left of the vertical base line, indicating a low γ-ray value.   Conversely, if the formation is very tight and thus contains little fluid, the neutron bombardment is almost unimpeded and the considerable number of γ rays emitted cause the recorded curve to extend to the right.

Besides locating saturated zones in limestones and sandstones, the neutron logs give some indication of the classifications of the formations. Sand, limestone, dolomite, anhydrite, salt, and coal are low in radioactivity, whereas, shales, bentonite, and volcanic ash have the highest values of radioactivity encountered.[9]

The other method of radioactivity logging is the *γ-ray method*.   Both methods require special equipment which in both cases is essentially the same, although there is no source of neutrons at the bottom of the ionization chamber in the γ-ray method.   All rocks contain measurable quantities of radioactive materials which release γ rays.   (These are primary γ rays, whereas the γ rays in the neutron logging are secondary.)   When the γ rays strike the gas in the ionization chamber, it ionizes.   The subsequent phenomena are the same as in the neutron logging.   It appears that the results of the γ-ray logging down to a depth of 70 ft are affected by cosmic rays.   Hence, the method is not applicable in civil engineering unless foundation data are required at depths greater than 70 ft.

For further details on radioactivity logging, see pp. 419–439 of Ref. 9 and Refs. 14 to 16.

GEOPHYSICAL EXPLORATION[17]

Geophysical exploration is a form of field investigation in which physical measurements normally are made at the ground surface by using special instruments to secure subsurface information.   It is a blend of physics and geology because the physical measurements are interpreted in terms of subsurface geological conditions.   Electrical and radioactivity logging are special cases of geophysical subsurface exploration in that they require bore holes for placement of the equipment.   In the methods described hereafter, the exploratory bore holes may be used merely to provide correlative data and thus are not always necessary.   Hence, these methods are appropriate for a rapid though approximate solution of certain geotechnical problems in an area, such as the depth to bedrock at a damsite.   There are four major geophysical methods: seismic, resistivity, magnetic, and gravity.

6.32. Seismic Measurements.[18]   The problem solved in Fig. 6.38 may be formulated thus: Determine the average thickness d of a layer in which seismic waves travel with a velocity V_1.   In this V_1 layer the seismic waves travel at considerably less velocity than in the underlying layer wherein they travel with a velocity V_2.   The upper low-velocity layer may be soil overburden or weathered rock.   The solution of the problem basically is to measure V_1 and V_2; then the thickness d may be easily determined by the formula in Fig. 6.38a.

Procedure.   A number of seismic wave receivers termed *geophones,** or pickups, are set in a line on the ground at measured distances apart (100 ft in Fig. 6.38).   At the *shot point,* a charge of a few pounds of a special explosive (usually nitramon†) is exploded.   Seismic waves sent into the ground by the explosion reach the geophones, are picked up by

* A somewhat different meaning of "geophone" is given in Sec. 2.6.

† Nitramon is an ammonium nitrate explosive.   It is obtained in tin cans and can be easily and safely handled under rough transportation conditions.

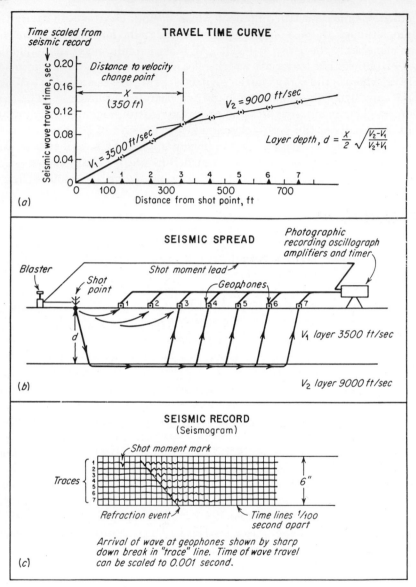

Time scaled from seismic record

TRAVEL TIME CURVE

Distance to velocity change point

X (350 ft)

$V_2 = 9000$ ft/sec

$V_1 = 3500$ ft/sec

Layer depth, $d = \dfrac{X}{2}\sqrt{\dfrac{V_2 - V_1}{V_2 + V_1}}$

Seismic wave travel time, sec

Distance from shot point, ft

(a)

SEISMIC SPREAD

Photographic recording oscillograph amplifiers and timer

Blaster

Shot moment lead

Shot point

Geophones

V_1 layer 3500 ft/sec

d

V_2 layer 9000 ft/sec

(b)

SEISMIC RECORD
(Seismogram)

Shot moment mark

Traces

6"

Refraction event

Time lines $^1\!/_{100}$ second apart

Arrival of wave at geophones shown by sharp down break in "trace" line. Time of wave travel can be scaled to 0.001 second.

(c)

FIG. 6.38. Seismic refraction method.

them, amplified, and transmitted to the recording device, or *oscillograph* (Fig. 6.38b). The seismic record, or *seismogram*, thus produced consists of several lines or *traces* (Fig. 6.38c), one for each geophone. The shot instant is recorded by a special mark on one trace, and the instant of arrival of the *first wave* (i.e., the wave energy) at each geophone can be determined from the seismogram. Some of the waves (direct waves) travel close to the ground surface with a velocity V_1; others cross the upper layer obliquely and are refracted, after which they travel along the top of the underlying material with a velocity V_2. The quickest way for the latter wave to reach a geophone is to cross the upper layer in both down and up traveling under a certain angle ("critical angle of grazing incidence") as shown in Fig. 6.38b. It should be noted that the laws of refraction of the seismic waves are analogous to those of light refraction, hence a similarity in terminology.

To obtain the *travel-time curve* (Fig. 6.38a) the distances between the shot point and all geophones are plotted horizontally; the times between the shot instant and the times of arrival of the first waves causing "breaks" in the traces are plotted vertically. Joining the points thus obtained produces two intersecting *straight lines* (Fig. 6.38a). At all geophones to the left of the point of intersection in that figure, the direct waves arrive first, whereas at the rest of the geophones the refracted waves arrive first. At the point of intersection itself, both a direct and a refracted wave arrive simultaneously. Distance x of this point of intersection can be simply scaled. The distance (in feet) from the shot point to any geophone at which direct waves arrive first is divided by the time (in seconds) the direct wave requires to reach that geophone to obtain velocity V_1 (in feet per second). The distance between any two geophones at which refracted waves arrive first is divided by the time of travel between those two geophones to obtain velocity V_2. The geophones usually are equally spaced from 50 to 100 ft apart; the whole distance covered by the geophones is often made equal to from 3 to 12 times the desired depth of penetration. From measurements of seismic wave speed as described above, it is possible to find the depth to certain types of geological horizons, such as bedrock, at several points in an area. The bedrock at a damsite, for example, may be the horizon which carries seismic waves at a relatively high velocity.

6.33. Continuous Vibrational Measurements.[19-22] Seismic waves may be produced in the ground either by blasting, as in the seismic method, or by an oscillator such as developed in Germany by the engineers of Degebo (Deutsche Gesellschaft für Bodenkunde). The device consists of two eccentrically supported disks which revolve in opposite directions (Fig. 6.39). Small circles in the figure indicate the axes of rotation of those disks. Since the disks rotate in opposite directions, horizontal centrifugal forces are balanced, and the summation of the vertical centrifugal forces may be graphically represented by a sine curve as shown in Fig. 6.39. The vertical ordinates of a sine curve indicate the upward and downward movements of the ground at the source of vibrations, i.e., at the oscillator. Similar vibrations occur at all points of the ground surface around the oscillator, the vertical upward and downward movements decreasing with the distance from the source of vibrations. Though there is no translational movement of rock or soil particles through which a wave propagates, a certain time is required for the wave to move from one point of the ground surface to another. The magnitude of the upward and downward vibrational movements may be detected by seismographs as used in the study of earthquakes, since in reality an oscillator (and the blast in the seismic method) produces a miniature earthquake. Geophones and oscillographs as described in Sec. 6.30 also may be used in this connection. The curves traced by any two seismographs are similar in shape, but one is retarded with respect to the other. Their *phase difference*, or distance between the analogous points of the seismograms (e.g., the highest or the lowest), as measured in terms of time, is the time of wave travel from one seismograph to the other (Fig. 6.40). Velocity of the wave propagation, then, is the distance between the two seismographs divided by the time of the wave travel between them. Loose sand has a low phase speed (250 to 350 mph); the phase speed in clay is about 400 mph and

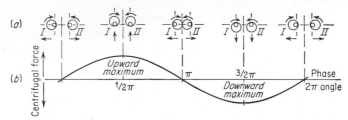

FIG. 6.39. Exciting vibrations: (a) two revolving masses, (b) sine curve.

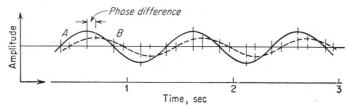

FIG. 6.40. Phase difference.

close to 1,000 mph in dense gravel. Much higher values are obtained in rock. Vibrational measurements may be used for detecting buried channels, subsurface gorges, and similar geotechnical problems. (More information may be found in Ref. 2, pp. 32–34, and items 51 and 52 of the bibliography of Ref. 2.)

6.34. Resistivity Measurements. From a great number of electrical geophysical methods, most of which are used in searching for ore deposits, the *resistivity* method also has proved useful in civil engineering investigations. According to usual practice, the electrical resistivity of a portion of the ground at a site is measured as follows:

Four electrodes are set in the ground in a line and at an equal distance apart (Fig. 6.41a). A set of B batteries and a milliammeter are connected in series with the outer pair of electrodes. They are the current electrodes and consist of sharp-pointed metal rods. A potentiometer for measuring voltage is connected between the inner pair of electrodes, which are the potential electrodes. They are porous pots containing a solution of copper sulfate, some of which seeps into the ground and makes a good electrical contact. The circuit through the pot is completed by a copper bar which passes through the lid of the pot and into the solution.

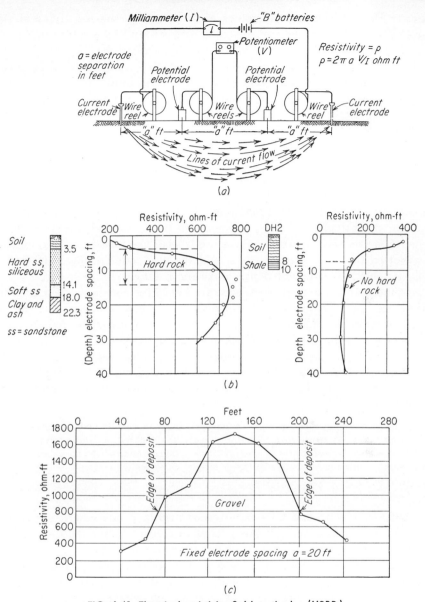

FIG. 6.41. Electrical-resistivity field method.   (USBR.)

To determine the resistivity of the ground, the current I that flows from the batteries and through the ground between the current electrodes is measured on the milliammeter. At the same time the voltage V between the potential electrodes is measured on the potentiometer. The equal spacing of electrodes, which is measured in feet, is designated as a. From the measured values of I, V, and a, the resistivity of the soil or rock between the current electrodes is found by the formula given in Fig. 6.41a. This formula states that resistivity $\rho = 2\pi a\,(V/I)$. The term $2\pi a$ relates to the volume of material measured, whereas the term V/I is electrical resistance in ohms. The units of resistivity are ohm-feet if a is measured in feet. Ohm-centimeters or ohm-meters sometimes are employed.

The resistivity method may be used to make (1) resistivity depth measurements at a selected point or (2) fixed depth resistivity measurements along a traverse line. In method 1, the electrode spacing a is progressively increased to pick up changes in resistivity with depth. As the distance a between the electrodes becomes greater, the current penetrates to a greater depth, related to a. This procedure has been called, figuratively, *electrical drilling*. On the basis of the field measurements the resistivity depth curves are traced, the resistivity in ohm-feet is plotted on the horizontal axis, and the electrode spacing a in feet is plotted on the vertical or depth axis (Fig. 6.41b).

In method 2, the four electrodes are kept at a constant spacing while they are moved along a line, and resistivity measurements are made at various stations. Lateral changes in resistivity of materials are indicated, in contrast to the vertical changes obtained in method 1. The procedure of method 2 has been called, also figuratively, *electrical trenching*. Field measurements are shown graphically as resistivity traverses. The stations are plotted on the horizontal axis, and the resistivity, in ohm-feet, is plotted on the vertical axis.

Figure 6.41b shows the resistivity curves traced in search of a high-resistivity siliceous rock. Progressive increase in resistivity until the maximum resistivity is reached and the curve starts to descend shows the presence of the rock (left curve). Conversely, low resistivity shown by the right curve indicates the absence of rock. A traverse traced at a gravel deposit (Fig. 6.41c) indicates the boundaries of the deposit as the resistivity drops below a certain minimum (about 800 ohm-ft in this case).

6.35. Magnetic and Gravity Measurements.

The magnetic and the gravity methods are somewhat similar in respect to the field operations but distinctive as to the type of physical measurements. In magnetic procedure, a magnetometer is used to measure the vertical component of the earth's magnetic field at closely spaced stations in an area. These measurements are corrected for certain systematic variations. The resulting magnetic values relate to local subsurface conditions. In the gravity method, a gravimeter is used. Measurements are made of the force of gravity at certain field stations and corrected for variations as necessary. The corrected gravity values reflect conditions below the surface.

These two methods are not used in engineering investigations so much

as the seismic and resistivity methods.   The main reason for this infre-
quent application is that gravity and magnetic measurements are less
easily interpreted in quantitative terms than are seismic layer velocities
or the measured resistivity of rocks and soils.   There are cases, however,
where either the magnetic or gravity method may be uniquely suited to
obtain subsurface information.   For example, the magnetometer is an
excellent tool for outlining intrusive dikes, and the gravimeter can be
used in searching for buried solution channels or caverns.

6.36. Accuracy of Geophysical Methods.   For geophysical methods
to supply geotechnically useful answers, there must be contrasts in
physical properties of subsurface layers.   These contrasts must be
related to or affect the physical measurements that are made.   Basically,
therefore, the greater the contrasts, the sharper is the measured response
and the more accurate the results.   Consider the seismic method as an
illustration: If the bedrock layer carries seismic waves at a much higher
velocity than the overlying material, the seismic measurements may be
very accurate because of the great velocity contrast.   Conversely, if
there is a gradual transition from overburden to bedrock, the velocity
contrast is relatively small; thus, the ability of the seismic data to provide
accurate depth determinations is correspondingly reduced.   If there are
no sharp contrasts in velocities, the seismic method may be unsuccessful.
Such lack of velocity contrasts might be found where a semi-indurated
clay shale was overlain by compact clay or where granite bedrock was
overlain by a very compact mass of large granite boulders.

The geophysical methods, as are other methods, are applicable only
where the basic assumptions of the methods used are valid.   The seismic
method, for example, cannot be used for determining the depth of
stratum that carries the seismic waves at a lower velocity than the overlying stratum.   Difficulties in the use of the seismic method also arise
if the surface terrain and/or the interfaces of layers are steeply sloping or
irregular instead of horizontal and smooth.

The resistivity method requires a high resistivity contrast between
the materials being located.   For example, when this method was used
to outline bodies of silicified rock (Fig. 6.41*b*), it was successful because
the rock had a resistivity some five or six times greater than the surrounding shale and clay.

It should be appreciated that the results of geophysical explorations are
interpretations of physical measurements.   These physical measurements are not, in themselves, geological facts relative to the subsurface
at a test locality.   A seismic depth determination, for instance, is a
measure of the distance from the ground surface to a "layer" that carries
seismic waves at relatively high speed.   Such a determination *cannot*
have the degree of accuracy provided by core drilling.   Furthermore,

the geophysical measurements do not necessarily identify the rock type or describe its physical properties. The use of geophysical methods in civil engineering *requires* that frequent comparisons be made between seismic layer depths and core-drilled depths to bedrock. In the practice of the USBR, where drilling was available for checking purposes, it was found that seismic-predicted depths to bedrock showed an average accuracy of 90 to 95 per cent to depths of about 100 ft.

REFERENCES

1. Plantema, G.: Construction and Method of Operating of a New Deep Sounding Apparatus, *Proc. 2nd Conf. on Soil Mech. and Foundation Eng.*, vol. I, Rotterdam, paper III, p. 6, 1948.
2. Hvorslev, J.: Subsurface Exploration and Sampling of Soils for Civil Engineering Purposes, *U.S. Corps Engrs. Waterways Expt. Sta.*, Vicksburg, Miss., 1949. A comprehensive bulletin on exploration methods and equipment.
3. Kjellman, W.: "Diagrammatic Sketches, Photographs and Descriptions of the 'Insitu' Apparatus," Stockholm, Sweden, 1947. See also Ref. 2.
4. Mohr, H. A.: "Exploration of Soil Conditions and Sampling Operations," 3d ed., Harvard School of Engineering, Cambridge, Mass., 1943.
5. Terzaghi, K., and R. Peck: "Soil Mechanics in Engineering Practice," John Wiley & Sons, Inc., New York, 1948.
6. Progress Report of Research on the Penetration Resistance Method of Subsurface Exploration, *USBR Earth Lab. Rept.* EM-314, 1952; rev. 1955.
7a. Catalogue, Bull. 36-A, Sprague and Henwood, Inc., Scranton, Pa., 1937.
7b. Catalogue 15, American Instrument Co., Silver Spring, Md., 1939.
7c. Catalogue on Drilling Equipment, Ingersoll-Rand Co., New York, N. Y.
7d. Catalogue on Drilling Equipment, Joy Mfg. Co., Pittsburgh, Pa.
8. Johnson, H. L.: Improved Sampler and Sampling Technique for Cohesionless Sands, *Civil Eng.*, vol. 10, 1940. See also Ref. 2.
9. LeRoy, L. W. (ed.): "Subsurface Geologic Methods—A Symposium," rev. ed., Colorado School of Mines, Golden, Colo., 1950.
10. Haddock, M. H.: "Deep Bore Hole Surveys and Problems," McGraw-Hill Book Company, Inc., New York, 1931.
11. Review of Schlumberger Well Logging and Auxiliary Methods, Document Number 2, Schlumberger Well Surveying Corp., Houston, Texas, 1949.
12. Jones, P. H.: Electric Logging Methods, Principles of Interpretation, and Applications in Ground-water Studies, *USGS*, July, 1952.
13. Guyod, H.: Electric Log Intrepretation, a reprint of articles from *Oil Weekly*, Dec. 3, 10, 17, 24, 1945, published by Halliburton Oil Well Cementing Co.
14. Downing, R. B., and J. M. Terry: Introduction to Radioactivity Well Logging, reprinted from *Petroleum Engr.* 1950 Reference Annual, published by Lane-Wells Co., Los Angeles, Calif., 1950.
15. Lane-Wells Radioactivity Well Logging, Bull. RA-47-B-3, rev., Lane-Wells Co., Los Angeles, Calif., 1952.
16. Tittle, C. W., H. Faul, and C. Goodman: Neutron Logging of Drill Holes: The Neutron-Neutron Method, *Geophysics*, vol. 16, no. 4, October, 1951.
17. Jakosky, J. J.: "Exploration Geophysics," 2d ed., Trija Publishing Co., Los Angeles, Calif., 1950.

18. Wantland, D.: The Application of Geophysical Methods to Problems in Civil Engineering, *Trans. Can. Inst. Mining Met.*, vol. 56, pp. 124–132, 1953.

19. Timoshenko, S.: "Vibration Problems in Engineering," D. Van Nostrand Company, Inc., New York, 1928.

20. Den Hartog, J. P.: "Mechanical Vibrations," 3d ed., McGraw-Hill Book Company, Inc., New York, 1947.

21. Bernhard, R. K.: "Mechanical Vibrations," Pitman Publishing Corporation, New York, 1943.

22. Heiland, C. A.: Geophysical Investigations Concerning the Seismic Resistance of Earth Dams, *AIME Tech. Publ.* 1054, February, 1939.

23. Gross, H. E.: A Method of Drilling in Deep Water, *AIME Tech. Publ.* 1722, October, 1943.

CHAPTER 7

MAPS AND AIRPHOTOS

Planning for the use of land and water to meet the needs of man is one of the basic features of civil engineering. Before intelligent planning can be done, advanced knowledge must be obtained of the following features: (1) topography of the site, i.e., the configuration of the land surface; (2) geology and soil conditions of the site, i.e., the deposits that compose the land and its weathered surface; and (3) hydrology of the site, i.e., the occurrence of surface and subsurface waters. These three kinds of features can be graphically represented on maps and airphotos.

TOPOGRAPHIC MAPS

7.1. General Description. A topographic map (Fig. **7.1**) is the representation of natural and man-made features of an area by means of conventional signs upon a plane surface. A topographic map shows both the horizontal distances between the features and their elevations above a certain level called the *datum plane* (which for most maps is taken at sea level, elevation 0.00). The configuration of the land surface, which is known as *relief*, can be represented on the map by means of contour lines, hachures (minute crosshatching to show slope of terrain), or hypsometric tints (colors indicating various elevations). According to the variations in elevations there is slight, moderate, or strong relief. All topographic maps show surface water such as streams, springs, lakes, and swamps (conventionally termed "drainage") and the so-called *culture*, involving roads, railroads, airports, cities, and groups of buildings such as isolated hospitals or housing projects. Culture is shown on maps by symbols (Fig. 7.2). This information sometimes is supplemented with symbols indicating vegetation (trees, shrubs, etc.). Fence lines, power lines, telephone lines, etc., are also indicated on many maps. This information is especially useful in planning preliminary investigations for structures.

Scale. The scale can vary according to the terrain and to the possible use of the map. Topographic maps specially prepared for surveys of

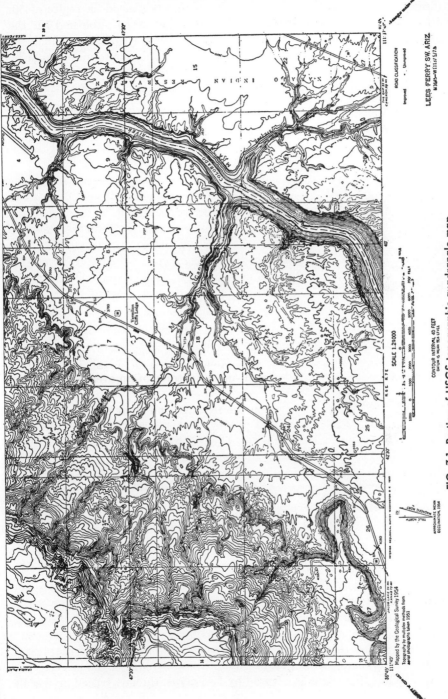

FIG. 7.1. Portion of USGS topographic quadrangle map.

large structures such as tunnels or dams may be traced on scales of 50, 100, or 200 ft to an inch, whereas topographic maps published for the general public may be on a scale approximately 1 mile to an inch, or more accurately 1:63,360. This means that 1 in. on the map would be equal to 63,360 in. on the ground. The latter method of designating scales is used in the topographic maps prepared by the U.S. Geological

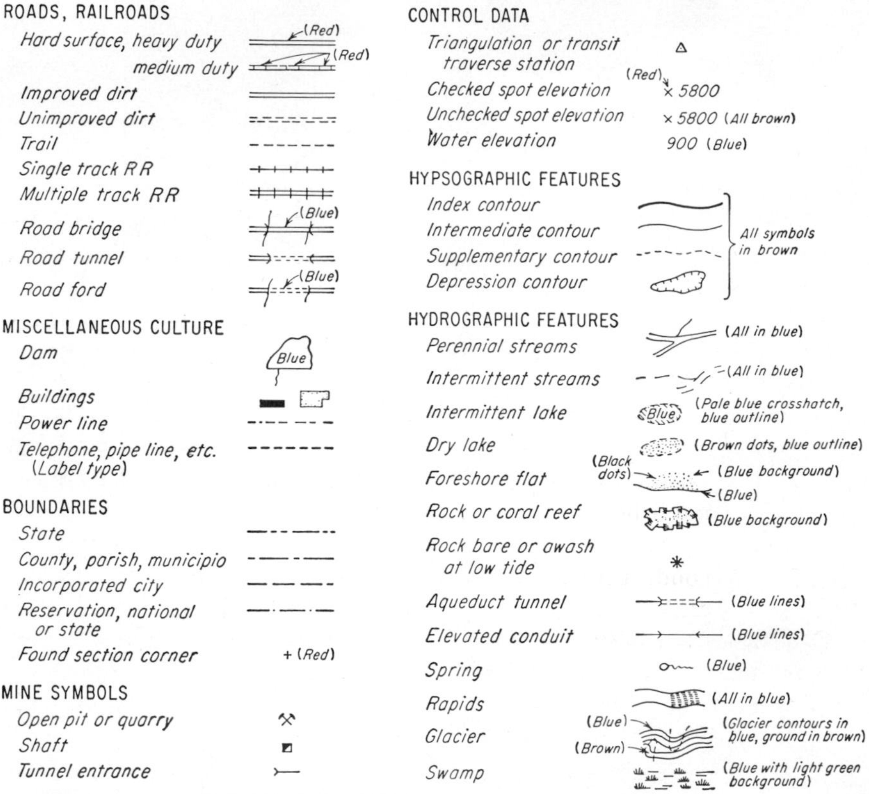

ROADS, RAILROADS
  Hard surface, heavy duty   (Red)
              medium duty   (Red)
  Improved dirt
  Unimproved dirt
  Trail
  Single track R R
  Multiple track R R
  Road bridge   (Blue)
  Road tunnel
  Road ford   (Blue)

MISCELLANEOUS CULTURE
  Dam   (Blue)
  Buildings
  Power line
  Telephone, pipe line, etc.
    (Label type)

BOUNDARIES
  State
  County, parish, municipio
  Incorporated city
  Reservation, national
    or state
  Found section corner   + (Red)

MINE SYMBOLS
  Open pit or quarry
  Shaft
  Tunnel entrance

CONTROL DATA
  Triangulation or transit
    traverse station
                                  (Red)
  Checked spot elevation   × 5800
  Unchecked spot elevation   × 5800 (All brown)
  Water elevation   900 (Blue)

HYPSOGRAPHIC FEATURES
  Index contour
  Intermediate contour   All symbols
  Supplementary contour   in brown
  Depression contour

HYDROGRAPHIC FEATURES
  Perennial streams   (All in blue)
  Intermittent streams   (All in blue)
  Intermittent lake   (Blue) (Pale blue crosshatch, blue outline)
  Dry lake   (Brown dots, blue outline)
  Foreshore flat   (Black dots) (Blue background)
                                  (Blue)
  Rock or coral reef   (Blue background)
  Rock bare or awash
    at low tide
  Aqueduct tunnel   (Blue lines)
  Elevated conduit   (Blue lines)
  Spring   (Blue)
  Rapids   (All in blue)
  Glacier   (Blue) (Glacier contours in blue, ground in brown)
                                  (Brown)
  Swamp   (Blue with light green background)

FIG. 7.2. Topographic map symbols.   (USGS.)

Survey (USGS) and other government agencies, and it also is used to indicate the scale of airphotos. However, for engineering purposes (for design drawings), these scales are more commonly expressed in so many feet to an inch or so many miles to an inch, using round figures, such as 1 in. = 1,000 ft or 1 in. = 2 miles, etc.

If the actual distances are considerably decreased on the map, it is said to be a "small-scale" map. The opposite would be a "large-scale" map. Thus, if an area is covered by two maps, the one traced on a scale of 1 in. = 1 mile is a small-scale map and the one drawn to a scale of 1 in. = 50 ft is a large-scale map.

Contour Interval. Points with the same elevation are joined with *contour lines* (contours) on the map. The contour interval is the vertical distance between the contours. In low, rolling country, contour intervals may be as small as 2 or 5 ft, whereas in rugged, mountainous terrain, the interval may be 50 or 100 ft or more. The contour interval also may vary according to the intended use of the map. On special maps for large structures it is not uncommon to use a contour interval of 2 ft, regardless of relief, since a detailed knowledge of minor variations in relief may be important. For instance, the location and size of boulders are of significance in determining the amount and kind of stripping for a structure.

7.2. Control.[1] The topographic survey of an area requires the establishment of a system of key stations interconnected by precise measurements; this system generally is termed the *control*. The *horizontal*, or *ground, control* consists primarily of determining the latitudes and longitudes of certain points on the ground. The *vertical* control is the determination of precise elevations of the key stations. These are marked on the ground by *bench marks* (special metallic plates firmly and safely emplaced in concrete blocks in the ground). The control net connecting these points should be closed; i.e., the survey is done in the form of a circuit, from point to point, ending either at the starting point or at a previously established point with known location and elevation. According to the magnitude of error in the closure, the control net may be of the first, second, third, or fourth (the least accurate) order. First-order horizontal control generally is established by triangulation, which is the precise measurement of the mutual location and elevations of three widely spaced points on the ground. A triangle on the ground is surveyed, and other somewhat similar triangles are gradually added. The type of control depends on the ultimate purpose of the survey, which thus predetermines the selection of surveying instruments and the character of the surveying operations.

As an example, the triangulation over the top of a mountain where a tunnel is to be located should be done with first-order accuracy. In fact, if the tunnel is constructed from two headings (i.e., from both sides), both parts of the tunnel should meet accurately *on line* and *on grade*. On the other hand, a reservoir survey does not require so much accuracy, and thus the third- or even the fourth-order control can be used for mapping.

7.3. Grid Systems. Engineers ordinarily need maps on large scales which are referred to orthogonal axes of coordinates for convenience in measurement. For this purpose, maps (not all, however) are provided with *grid systems*.

A grid consists of two sets of parallel lines intersecting at right angles to form a network consisting of squares with sides that may be as large as 3,000 yd or more or as small as 500 ft.   A grid is applied to a plane area that for mapping purposes replaces an actual area at the earth's surface. This replacement, made according to certain rules, incurs some negligible distortion of the area, but this is unavoidable, since the curved surface of the earth cannot be developed on a plane.   The central meridian of the area (passing approximately through its center) is the Y axis of the grid.   The X axis of the grid is perpendicular to the Y axis and is located to the south of a given area.   Intersection of these axes is the origin of the coordinates.   Its X coordinate is not zero (0) but a value greater than half the width of the given area to avoid negative X coordinates. The X coordinates increase numerically from west to east, and the Y coordinates increase from south to north (mnemonic rule: "read right and up").   The system of numbers used for the X coordinates should be considerably different from that used for the Y coordinates to avoid confusion during plotting.   For example, the north, or Y, coordinates can be given a system of numbers starting with 500,000 and increasing to the north, whereas the east, or X, coordinates can be given a system of numbers starting with 100,000 and increasing progressively to the east.   Thus the map would locate a point, e.g., north 550,000; east 125,000.   Unless very large areas are mapped, it is not desirable to use coordinate numbers in the millions because of the inconvenience in plotting.

There are several grid systems.   The *United States military grid system*, initially in yards, since 1947 has been expressed in meters (UTM or universal transverse Mercator).   This system subdivides the continental United States into seven zones labeled A to G in the east-west direction.   Thus Connecticut is located in the A zone, and California is mostly within the G zone.[2]   The world grid is practically an extension of the United States and British military grid systems.   The *state coordinate systems*[3] are based on the fact that in a strip of the earth's surface 158 statute miles wide (measuring along a meridian) the change in the actual length of a line on the grid constructed for this strip is negligible (below 1 per cent).   Areas of the states thus are subdivided into grid areas not exceeding 158 statute miles in the south-north direction. In this connection California has seven grid areas, whereas Connecticut has only one.

There is little difficulty in reading a topographic map based on a grid system, since the coordinates are shown by small marks, or "ticks," with numerals at the margin of the map.   The preparation of such a map is a complicated process, however, requiring the application of formulas and tables which in the United States are published by the USCGS.[3]

The coordinate system should be established either before or during the explorations for a structure.   Only then can all exploration operations, particularly the drilling of the holes, be referred to that system.   If an investigation program is likely to continue for several years, it is desirable to refer a few of the initial exploration holes to some permanent land-marks or existing bench marks in the area. It is not uncommon to find that more than one coordinate system has been used at a site, and in the passing years, references between systems are lost. Thus, logs of holes referenced to earlier systems may not be accurately relocated, if at all.

7.4. Published Topographic Maps and Similar Information. The majority of topographic maps published in the United States are pre-pared by the U.S. Geological Survey. The USCGS established the primary horizontal and vertical control nets. The USGS, in its work, subdivides these larger areas into the third- and fourth-order control nets required as a base for preparation of topographic maps.

Since 1942, the USGS has adopted certain accuracy requirements for all their topographic mapping: (1) *Horizontally*, 90 per cent of the well-defined planimetric features have to be plotted in correct position on the published map within a tolerance of 40 ft on the ground for 1:24,000 and 1:62,500 scale maps, and (2) vertically, 90 per cent of the elevations interpolated from the contours shall be correct within a tolerance of one-half contour interval.

Topographic maps based on the United States military grid are pub-lished by the various agencies of the U.S. defense establishment, par-ticularly the Army Map Service. Highways and roads are drawn on these maps, and their traffic conditions specified. Another important source of topographic maps, although on a small scale, is the USCGS, who publish the aeronautical charts (or "flight maps") of the United States and its territories. These charts commonly show topography by hypsometric tints and contours, with a large contour interval (as much as 500 to 1,000 ft). They are useful in determining access to proposed sites of engineering structures and often are the only available maps showing roads and topography for certain areas in the United States (as not all of the United States has been mapped on large-scale topographic maps). Another source of information, particularly in the Western states, are the maps issued by the U.S. Forest Service and the National Park Service. These maps seldom show topography but usually show drainage, trails, and culture in national forest and national park areas.

County road maps, available from practically all state highway depart-ments, are another excellent source of information on access to a site. These maps are based on county, range, township, and section lines (Sec. 7.5) and show major streams and all roads, from major paved highways to country roads or mountain trails.

River and Shore Survey Charts. The river survey information of the USGS gives the course and gradients of streams, configuration of the valley floor and adjacent slopes (by contours), and location of towns, scattered buildings, irrigation ditches, and roads.[4] River topography also can be obtained for some streams from the Corps of Engineers, U.S. Army. If a topographic map of some other type is not available, the river survey map may contain sufficient data for initial planning stages in the investigation of an engineering site, particularly a damsite.

Engineers interested in shore and harbor structures can consult charts prepared by the USCGS.[5] These charts cover the Atlantic, Gulf, and Pacific Coasts and also Alaska, Puerto Rico, Hawaii, and the Philippine Islands. Practically all charts are drawn on a decimal scale, generally from 1:10,000 to 1:80,000.

The American Geographical Society, Broadway at 156th St., New York 32, New York, has prepared excellent maps covering most of South and Central America. These can be obtained in sections, at scales up to 1:1,000,000. Topography is shown by contours and hypsometric tints. All trails, highways, railroads, airports, etc., are shown.

7.5. Subdivisions of Topographic Maps. Topographic maps are published on sheets, each covering what is known as a quadrangle, i.e., a certain area limited by meridians and latitude parallels. The over-all dimensions of these sheets may vary considerably, although 16½ by 20 and 22 by 27 in. are the most common.

Most of the mapped area of the United States was subdivided into areas 24 miles on a side. These areas, which are bounded on the east and west by meridians and north and south by parallels, are subdivided into townships and ranges. *Townships* are squares 6 miles on a side, contain 36 sections, and are numbered in increasing order to the north or south from the base line of the survey, e.g., Township 1 North (T1N), Township 2 North (T2N), etc. Townships also are numbered in increasing order to the west or to the east of the principal meridian of the survey in strips 6 miles long called *ranges*, e.g., Range 1 West (R1W), Range 2 West (R2W), etc. Each *section* is 1 sq mile and contains 640 acres. The sections occasionally are subivided into halves, quarters, and even smaller portions. Most real estate in mapped zones may be located according to township, range, section, and portion of a section, using certain conventional symbols.

The USGS, the USCGS, and the U.S. Land Office (GLO) place bench marks at some township, range, and section corners. Sometimes additional markers are placed at the one-quarter-section points ("quarter corners") and at the one-eighth-section points ("eighth corners"). (Such markers should not be confused with "claim stakes," which are markers, often permanent, placed on the property lines of mineral claims.) Bench marks are used as starting or reference points in surveying properties.

GEOLOGIC MAPS AND MODELS

7.6. Types of Geologic Maps. There are two basic types of geologic maps: (1) surface maps and (2) subsurface maps. The former are compiled from surface geologic data, and the latter from borings, well logs, geophysical surveys, and extrapolation of surface data. The surface maps often are subdivided into (1) surficial maps and (2) areal maps (Fig. 7.3). The *surficial* maps show the character and distribution of surface materials such as outcrops, i.e., places where bedrock is

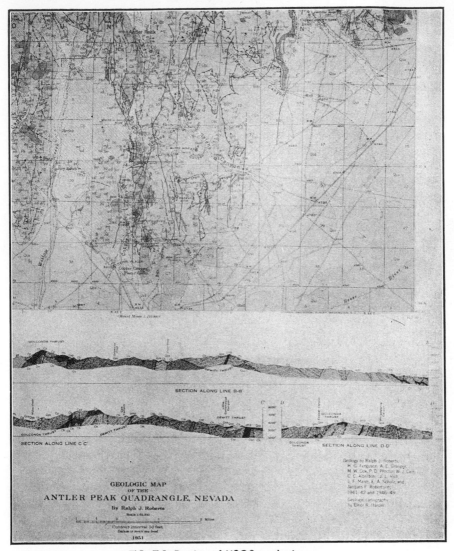

FIG. 7.3. Portion of USGS geologic map.

exposed, or both outcrops and soils. Soils on such maps are identified according to origin, such as glacial, alluvial, etc. (Chap. 3). These maps can be very useful to engineers; e.g., the construction materials maps in the Missouri Basin states prepared by the USGS show the locations of sand and gravel deposits and riprap sources. Geologic *areal* maps (Fig. 7.3) show the earth's surface as it would appear if the overburden materials were stripped from it. Outcrops may also be indicated on such maps by certain symbols. Structural details, such as faults (the solid **and** dashed lines trending north and south on Fig. 7.3), strikes and dips of formations, axes of anticlines and synclines, etc., also are shown on such maps by symbols. The contacts between various formations also are indicated. Each geologic map carries a *legend*, or *explanation* of symbols

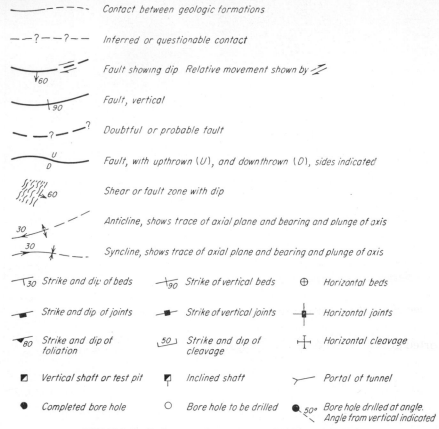

FIG. 7.4. Symbols on geologic maps.   (USGS and USBR.)

used in its compilation (Fig. 7.4), and rock stratigraphic symbols are included in it as shown below.

Geologic periods or ages are identified or symbolized by using the following letter system:

Q	Pleistocene and Recent	T	Triassic	S	Silurian
T	Tertiary	P	Permian	O	Ordovician
K	Cretaceous	C	Carboniferous	$\mathcal{C}$	Cambrian
J	Jurassic	D	Devonian	$p\mathcal{C}$	Pre-Cambrian

The symbols used for each of the various formations on a particular map are listed in the "Explanation" in descending order, with the most recent (by age) formations at the top of the list and the oldest formations at the bottom. Small (lower case) letters are used to indicate specific formations, e.g., K_d, which means Cretaceous Dakota formation, or D_g, which means Devonian granite (wherein that particular granite has no specific formation name).

Subsurface rock structure, being three-dimensional, cannot be adequately shown on a two-dimensional surface map, which essentially is nothing more than a horizontal projection. Therefore, to complete the information given by the surface geologic map, geologic sections or columnar information are commonly added to geologic maps. The *geologic section* (Fig. 7.3 has an illustration of such a section) presents the strata as they would appear on a vertical plane if the locality was divided by such a plane. Geologic sections usually are highly interpretative if correctly constructed. In preparing a geologic section, the geologist is aided by the information obtained from bore holes, well logs, mines, road cuts, geophysical surveys, or some type of excavation at the site. A geologic section might be considered analogous to a section or "elevation" of the structure on an engineering drawing. The geologic section will have a vertical and horizontal scale which should be equal to each other. However, because of the limitations imposed by the size of a workable sheet of paper or by reason of clarity, it often is necessary to *exaggerate* the vertical scale, i.e., to make the vertical scale several times larger than the horizontal scale (or vice versa under some circumstances). For example, a geologic section could have a horizontal scale of 200 ft to the inch and a vertical scale of 20 ft to the inch, in which case the distorted section has a vertical exaggeration of 10:1. When distorted geological sections are used, correct interpretation should be given to any slopes shown. If the vertical scale is exaggerated, the dips of the strata in the distorted section, as compared with actual ones, will be considerably increased and the oblique exploration holes ("angle" holes) will appear closer to vertical than they are in actuality (compare Sec. 6.20).

Figure 7.5a shows how to plot a distorted dip. Plot the actual dip (5° in this case) to intersect any vertical line BD at point C. Then plot distance BC from point B, up as many times as indicated by the vertical exaggeration ratio (10 times in this case). Then AD is the distorted dip.

Figure 7.5b shows how to plot the distorted direction of a bore hole from the actual angle made by the hole with the vertical (40° in this case) and the vertical exaggeration (5:1 in this example). Trace any horizontal line BD, and *divide* it in as many parts as indicated by the vertical exaggeration ratio (five times in this case). Then AC is the distorted direction of the bore hole.

Columnar information (or the *geologic column*) is given in the form of a vertical strip about ½ to 1 in. wide on one side of the map. The geologic column shows the sequence, thickness, and identity of the various strata encountered in the area to which the column refers. It roughly resembles a boring log but usually has no vertical scale, as the strata thicknesses indicated are relative only.

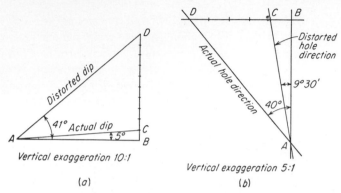

Vertical exaggeration 10:1

(a)

Vertical exaggeration 5:1

(b)

FIG. 7.5. Distorted geological features.

7.7. Block Diagrams. Sometimes, surface maps and geologic sections, both being two-dimensional, cannot adequately depict an existing complex geologic condition. In such cases, *block diagrams* may be used. There are two types of block diagrams: isometric and perspective. In the former, the parallel sides of the block also are parallel on the drawing representing this block, and measurements can be scaled directly along any of the three axes (Fig. 7.6a). In the simplest case of a perspective block, the sides of the block intersect at a "vanishing" point A as shown in Fig. 7.6b. The vanishing point is located in the horizontal plane that

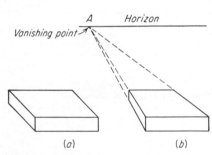

FIG. 7.6. Block diagrams: (a) isometric, (b) perspective.

passes through the observer's eyes, and distances can be scaled only along the front horizontal axes.[6] Considerable time is needed to draw a block diagram accurately, and in some cases the use of peg models (Sec. 7.8) is more convenient.

7.8. Peg Models. Figure 7.7 shows a peg model that depicts geologic conditions under a portion of a damsite. The base sheet can be either a topographic map (as shown in Fig. 7.7) or a large-scale airphoto. The drill holes and test pits used on the project are plotted on this base sheet. The latter is given a certain elevation and thus becomes the datum plane to which all other elevations are referred. At each bore hole or pit location, a hole in which to insert the peg is drilled; if there are oblique holes, they can be adequately represented by drilling the hole at an angle. The pegs are rods made of any sturdy material such as wood dowels, welding rods, copper tubing, etc.   The log of the bore hole is represented on the peg by painting the peg in various colors; each color corresponds to a definite formation.   Pieces of string, sometimes of different colors, are used to show the contact planes between formations.   The string can be glued to the peg or threaded through a small hole drilled in the

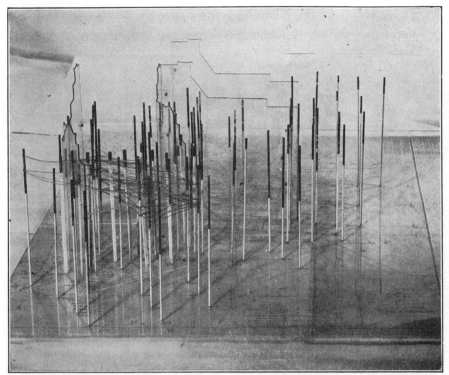

FIG. 7.7. Geologic peg model of portion of Davis Dam, Arizona.   (USBR.)

peg.   After all strings are in place (or sometimes beforehand, for convenience), the models of the proposed engineering structures are prepared by cutting them from plastic sheets and gluing them to the base sheet (as was done in the model in Fig. 7.7).   Each model must have a vertical scale, generally made of cardboard or wood; the horizontal scale is shown on the base sheet.

The time required for preparation of a peg model will depend upon the size and complexity of the model and the degree of accuracy desired.   The model in Fig. 7.7 was prepared in about 4 man-days, using drilled holes in the pegs for insertion of the strings.

7.9. Sources of Geologic Maps.   The best source of geologic maps and other geologic information is the USGS.[4]   Their maps are published in several forms: (1) *Folios*, which contain topographic quadrangle sheets with superimposed geologic features and a discussion of the detailed geologic history of the area [since 1945, individual geologic quadrangle sheets (Fig. 7.3) have been published instead of the folios]; (2) *Bulletins* and *Professional Papers*, which generally discuss in considerable detail a specific geologic subject or locale; and (3) *Water Supply Papers* (Ref. 2, Chap. 5), which provide data on surface and subsurface water of a given area, a brief description of the geology of the area, and a map.

In 1951 the USGS initiated the preparation of engineering geologic maps and folios[7] which contain the general geology of an area and data of interest to a specialist in engineering geology, e.g., suitability of sites for structures, location of construction materials, etc.

Occasionally the USGS also publishes oil and gas maps for the areas where such deposits are prevalent.   However, most of these maps are published by individual oil companies or state geological surveys.   The latter agencies are generally very active in publishing local geologic information but do not exist in all states.   Notable work has been done by the Mississippi River Commission of the Corps of Engineers in studying and representing in graphic form the geology of the Mississippi River Valley.[8]   A less common source of geological data, but one that sometimes proves fruitful, is the library of a local college.   For example, all graduate students' theses are filed in these libraries,[9a] and often a geological thesis discusses the area in which the engineer desires to work. Occasionally state highway departments prepare bulletins which contain geological information.   Some large maps, but drawn to a very small scale, depict geology and certain geologic features of the United States as a whole: for example, "The Geologic Map of the United States" (USGS), "The Glacial Map of the United States" and "The Map of Loessial Deposits or Windblown Soils in the United States" (both published by GSA), and "The Landforms of the United States" (prepared by E. Raisz and obtained from Harvard University).

7.10. Use of Geologic Maps in Engineering.   The average civil engineer, with his knowledge of surveying and descriptive geometry, should have little difficulty in reading geologic maps and cross sections. Before using such data, however, he should be aware of the degree of accuracy with which the particular map has been prepared.   The small-scale geologic maps, such as those found in the USGS bulletins and folios, are intended to give a *generalized* picture of the geology in the area, and whereas they may include all structural geologic features present, they usually are not of sufficient accuracy to be used directly in engineering practice.   For example, when designing a canal, the engineer has to determine the lined and unlined portions of it.   By using the available maps, he can discover the various types of material the canal will traverse, but he may make a serious error if he attempts to scale the corresponding lengths of formations from a published map.   In large engineering organizations this is remedied by preparing special geologic maps for proposed structures directly in the field.

There usually are two stages in the preparation of geologic maps for specific engineering structures.   In the first stage, a reconnaissance-type survey is made.   In this survey, the geologist usually uses a Brunton

FIG. 7.8. Geological mapping methods: (a) use of Brunton compass to measure dip, (b) use of plane table and alidade.   (*USBR photograph.*)

compass, or so-called "pocket transit," to measure horizontal angles, slope inclinations, and strikes and dips (Fig. 7.8*a*).   In the second stage, for more details, the geologist usually uses a plane table and alidade (Fig. 7.8*b*).   With these, he is able to establish, practically with the same degree of accuracy as the published topographic maps, the location of formation contacts and structural features in the geology of the area. The basis for either one of these stages can be a portion of a topographic map, an enlarged airphoto, or merely a sheet of paper upon which the ground control has been plotted.   In the third case, the geologic map is superimposed at a later time on the topographic map of the given site. Where topographic maps or ground control are not readily available, elevations may be taken by barometric readings or a pocket altimeter. In all cases, the geological surveys are closed, either at the starting point or at a definitely located point on the ground control net.

The amount of detail on a special geologic map depends upon the purpose of the map. Normally, the following information should be given: (1) the *lithology* or stratigraphy of a given locality, i.e., the rock types, their mode of occurrence, grain size, color, and mineral constituents; (2) the *structure* of the rock, such as stratification, lamination, dip and strike of beds, cleavage, fractures, joints, etc.; and (3) the depth and characteristics of *overburden* and *weathered rock*. If some of the above details are inconvenient for placing on the map, they should be recorded

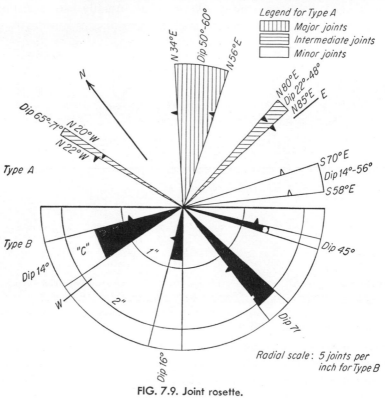

FIG. 7.9. Joint rosette.

in an accompanying geologic report. In any case, all detailed field observations should be carefully entered into a field notebook for later record. The geologist's field operations in the detailed mapping stage are very similar to those of the topographer with some variations. The plane table is set up at a certain station, the rodman moves from point to point, and each point is given a number and plotted on the map. The rodman is usually a geologist (Fig. 7.8*b*), and under the number of a point he enters into the notebook the geological data observed at that point. Later in the office, the pertinent notebook data are plotted on the geologic map.

Joint rose or *joint rosette* on the geologic map (Fig. 7.9) presents diagrammatically the direction and intensity of the jointing in an area. This diagram permits the engineer readily to ascertain the direction of the most intensely jointed rock in relation to his structure. The rosette is particularly useful in the design of tunnels, arch dams, and powerhouses. To trace it, the geologist simply records as many joints as he can observe in a given area and groups them according to their strike. The rosette also can be used to depict the degree and severity of fractures if they follow any determinable pattern.

There are two types of rosettes (Fig. 7.9). In both, the strikes of the joint bearings and dips are indicated. In type *A* the order of joint magnitude is indicated by crosshatching. In type *B* the number of joints in a given direction can be measured along the radius from the center. Thus in the direction *C* there are eight joints. Their strike varies from N 65° W to N 85° W, the dip being about 14°.

The final geologic maps prepared after the subsurface exploration is completed contain the results of such explorations and often the outlines of proposed structures. Additional information on the use of geologic maps and interpretation may be found in Refs. 9 to 11.

7.11. Use of Agricultural Soil Maps in Engineering. The agricultural soil maps are prepared by the U.S. Department of Agriculture (USDA) and by the states. The soil surveys on which these maps are based ordinarily extend to a depth of from 3 to 6 ft only. Because of this restricted depth, the use of such maps is limited to those branches of engineering activities where shallow soil data are sufficient for design purposes, such as highway, runway, and irrigation projects.

The basic agricultural soil literature, consisting of reports and maps, is twofold. Voluminous, but necessarily generalized, information on soils is contained in the third part of the "Atlas of American Agriculture."[12] This atlas, similar to the "Geologic Atlas of the United States,"[4] is published in large "library edition" size and contains information on land relief, climate, soils, and natural vegetation. A relief map of the whole country (1:1,800,000) is given, and a large number of soil maps prepared on the basis either of detailed or only of reconnaissance surveys is available. Also, there are a great number of papers and maps generally prepared by the USDA in cooperation with state agricultural experiment stations. These bulletins describe soil conditions in different areas of individual states. As a rule, a paper of this class contains a general description of the soils of the area, following, of course, the pedological soil classification.

The agricultural maps are based on topography, the same as are geological maps. The soil surveyor collecting field data for the preparation of soil maps takes notes on the soil texture and structure; its plasticity, horizon thicknesses, color, and organic matter content;

character of parent material; and other pertinent data, if necessary. According to this information, the soils of a given area can then be classified in series and types, a series containing several types as explained in Sec. 4.1. *Soil profiles of the soils of the same type are similar in all respects.* Soils belonging to a certain type are not necessarily concentrated in a single spot but may be spread in the form of isolated islands in the midst of larger territories occupied by other soil types. The simplest way to use an agricultural map, for example, in highway engineering, is to plot the proposed highway or the relocation of an existing highway on a soil map and study a few samples of the soil types involved. This procedure will save considerable time and energy for the engineer, because it makes continuous sampling along the whole project unnecessary.[13]

In planning irrigation projects, soil maps are used for *land classification;* i.e., the land is classified according to its suitability for various types of crop cultivation and for its drainage characteristics. Soil maps are also used in hydrologic studies, as many soil data published by the USDA give infiltration rates for various types of soil, i.e., the amount of water that will filter into the soil during a prescribed period. From these data, the runoff characteristics of the soil can be determined. Also soil maps may be used for locating suitable borrow areas for large embankments.

A notable example of the use of pedological soil information in engineering is found in the soil manuals for highway engineers prepared by the Michigan and Missouri State Highway Departments. The Michigan manual[14] contains graphically represented soil profiles of the pedological soil types accompanied by practical "construction information" on the peculiarities of earthwork for various soil types.

AIRPHOTOS

7.12. Aerial Photography.[15–20] The Committee on Nomenclature of the American Society of Photogrammetry defined *photogrammetry* as [20a] "The science or art of obtaining reliable measurements by means of photography." This definition has been amplified to include the "interpretation of photographs."

There are two types of photogrammetry: (1) ground photogrammetry, wherein the photographs are taken from the ground, and (2) aerial photogrammetry, wherein the photographs are taken from an airplane. The human eye sees perspectively, and a photograph is a perspective projection, whereas a map is a horizontal orthogonal projection of a locality or, in the case of engineering structures, a "plan." The principles of map preparation from photographs are basically the same in both aerial and ground surveys and are explained in various textbooks and manuals on photogrammetry.

MAIN DAM

INTAKE GATES

POWERHOUSE

CUTOFF TRENCH FOR SPILLWAY

1/2 MILE

Example of airphoto. (Construction of Fort Randall Dam, *USBR* photo.)

7.13. Airphotos. The term "airphoto" has been adopted by many to include all types of aerial photographs. Airphotos can be taken very rapidly and in any locality, whereas a corresponding ground survey may require many days of labor and in some cases, where the terrain is inaccessible, cannot be done at all. Because of the instability of the camera on the plane and possible inaccuracy in determining the flight altitude, the results of an aerial survey are, as a rule, less accurate than those of a good ground survey. For the majority of practical applications, this inaccuracy is not important, however, and the USGS is using airphotos almost exclusively as a base (in conjunction with a highly accurate triangulation net) to prepare topographic quadrangle sheets.

A *vertical photograph* is one taken with an aerial camera leveled as perfectly as practicable to keep the optical axis vertical. An *oblique photograph* is one taken with the optical axis of the camera deviating considerably from the vertical. In the *trimetrogon system*, both oblique and vertical photographs are taken simultaneously. The oblique photographs may or may not include the visible horizon.

The *scale of an airphoto* depends on the ratio of the flying height H (Fig. 7.10a) to the focal distance of the photographic camera (F_1 and F_2), as expressed in the same length units (e.g., inches). The higher the flight, the smaller the scale but the larger the territory covered by the survey. If a camera with larger focal distance is used, a larger scale photograph is obtained (photograph 2 in Fig. 7.10a).

To measure the flying height (altitude of the flight), airplanes are equipped with an altimeter. The performance of the altimeter depends on the changes in air pressure and is affected by meteorological variations,

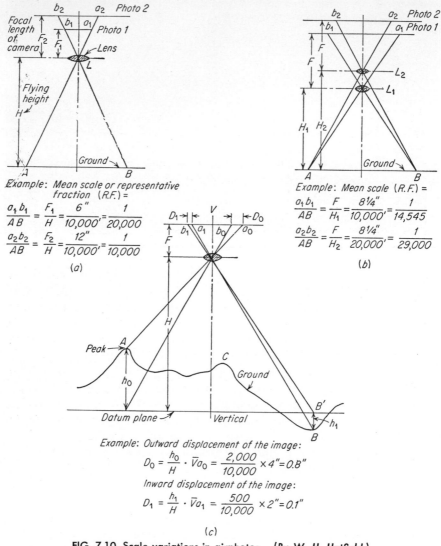

Example: Mean scale or representative fraction (R.F.) =

$$\frac{a_1 b_1}{AB} = \frac{F_1}{H} = \frac{6''}{10,000'} = \frac{1}{20,000}$$

$$\frac{a_2 b_2}{AB} = \frac{F_2}{H} = \frac{12''}{10,000'} = \frac{1}{10,000}$$

(a)

Example: Mean scale (R.F.) =

$$\frac{a_1 b_1}{AB} = \frac{F}{H_1} = \frac{8\frac{1}{4}''}{10,000'} = \frac{1}{14,545}$$

$$\frac{a_2 b_2}{AB} = \frac{F}{H_2} = \frac{8\frac{1}{4}''}{20,000'} = \frac{1}{29,000}$$

(b)

Example: Outward displacement of the image:

$$D_0 = \frac{h_0}{H} \cdot \overline{V a_0} = \frac{2,000}{10,000} \times 4'' = 0.8''$$

Inward displacement of the image:

$$D_1 = \frac{h_1}{H} \cdot \overline{V a_1} = \frac{500}{10,000} \times 2'' = 0.1''$$

(c)

FIG. 7.10. Scale variations in airphotos. (By W. H. Hatfield.)

including temperature and humidity. In their turn, erroneous altimeter readings can greatly affect the computation of the airphoto scale.

The actual scales of airphotos vary according to the purpose for which the photographs were made; the greater part of the United States has been mapped by photos with scales of 1:20,000 and 1:40,000. If the scale is not shown on the photograph, it can be obtained from the agency for which the photograph was made or from the aerial survey company which did the flying.

Figure 7.10c shows the effect of relief on the linear distortion of a vertical airphoto. The peaks (A in the figure) are displaced outward, and the low points (such as B) inward. The distortion is greater the farther the given point from the vertical line passing through the lens. The image of point C is in true position.

The term *flight* means the run of the plane over a strip of land in one direction. The flights usually are made approximately parallel to each other with a certain prescribed percentage of overlap. Contracts for airphoto flying usually specify the percentage of overlap desired; e.g., the USGS requires 60 per cent or more overlap for photographs taken for mapping purposes. The higher the percentage of overlap, the greater the degree of accuracy which can be obtained when the photographs are combined into a mosaic (Sec. 7.15).

7.14. Airphoto Code Symbols. Various letters and numbers usually appear on the margin of airphotos. Each of these has a particular meaning. Sometimes the scale is recorded on the first photograph in a flight, simply as 1:20,000. On some photographs, particularly those taken by the U.S. Air Force, a small photograph of the altimeter taken during the flight is printed on the photograph so that the flying height can be read directly. Most photographs have numbers in one corner that indicate the date flown, e.g., 5 5 54 indicates that the photograph was taken on May 5, 1954. The airphotos are kept in rolls, and there may be photographs of several flights in one roll. The rolls and photos are numbered; e.g., the number 7 82 indicates that the photograph is the eighty-second photograph in roll 7. These symbols showing the sequence of photographic operations in a flight are very important in the preparation of mosaics (Sec. 7.15). The name of the agency for which the flying is done usually is stamped on the back of each photograph; e.g., PMA means that the photograph was taken for the Production Marketing Administration of the USDA, USFS indicates the U.S. Forest Service, and USBR indicates the U.S. Bureau of Reclamation. Other letters on the photograph indicate codes by which the agency identifies the surveyed areas.

7.15. Mosaics. A mosaic is a collection of photographs of a given area which are joined together to form a continuous over-all photograph of that area. Usually, in the preparation of such mosaics, the 60 per cent overlap photographs are used. The edges of the photograph are torn off so that only its center portion is used; this eliminates, to a great extent, the distorted relief usually found near the edges of individual airphotos (Sec. 7.13). For the planning of large structures, it usually is necessary to make a mosaic, as one photograph rarely covers the entire structure area unless the photograph is on a very small scale. Mosaics are particularly useful in laying out railroads, highways, canal lines, transmission lines, and irrigable lands. Also mosaics can be used as base sheets for geologic mapping, although usually a selected airphoto of the particular area is enlarged for this purpose. The preparation of mosaics often requires the services of an expert in order that the least amount of error will be present.

In the *controlled mosaic*, the ground control is plotted on a board prior to laying out the airphotos. The photographs then are arranged in such a way that the mutual position of two points and the distance between them as plotted to scale will check those given by the ground control. This permits measuring distances on a controlled mosaic with a high degree of accuracy. In an *uncontrolled mosaic*, ground control normally is not used, and the photographs are arranged by matching conspicuous features, generally landforms such as rivers, mountains, etc. Though in this case the relative position of objects on the adjacent photographs will be more or less accurate, their mutual spacing will not be measurable with a very high degree of accuracy.

The use of a stereoscope in examining airphotos permits the viewing of a large section of terrain in perspective. As a result, landforms are clearly visible. However, the relief is exaggerated (more pronounced), which should be considered in appraising the results of such study. The airphotos are placed on a table and viewed through a hand stereoscope or a specially built stereoscopic device. According to the type of stereoscope used, the photographs should overlap each other; 60 per cent overlapping photographs are recommended. The photographs should always be placed with the shadows in them falling toward the observer. Incorrect positioning will result in a reversal of relief; e.g., ridges will appear as rivers, and trees as holes in the ground! The perspective view can, after a little practice, be obtained by what is known as "stereoscopic vision"; the eyes achieve the perspective view without the use of artificial devices.[17]

7.16. Airphoto Interpretation.   Before discussing the use of airphotos in geotechnics, some general principles of the airphoto interpretation should be given. To identify any object on the airphoto, the following features of that object as represented by its image should be given attention:[20b] (1) shape, (2) size, (3) pattern, (4) shadow, (5) tone, (6) texture, and (7) relationships.

The *shape* of an object, such as a building, may readily identify the object. Many man-made objects will have characteristic shapes, but some of nature's forms will be more difficult to recognize. The *size* of the object can sometimes be identified by relating it to known objects on the photograph or by studying the shadows cast by the object in question. The *pattern* is particularly important, as it is often said that nature never makes a straight line or a regular spacing. Thus, from the pattern or peculiar arrangements of objects in the photograph, it can be determined whether the object is natural or man-made. The *shadow* helps to determine size, and the shape of the object can often be detected by the profile of the shadow it casts. The *tone* designates the amount of light that is reflected by the object. For example, a soil type having a very dark tone may be indicative of very wet, impervious soils or high ground water, whereas light-toned soils may indicate an arid condition (Fig. 3.13). *Texture* of a photographic image* can be defined as "the frequency of tone change within the image. Texture is produced by an aggregate of unit features too small to be clearly discerned individually on the photograph."[20b] The *relationship* of the object to other objects in the photograph may assist in identification; e.g., a large body of water with a straight line across one end of the body would be indicative of a dam, or as in Fig. 7.11, the parallel spacing of the ridges and the location of a piece of construction equipment (in this case, dredges *a*) at the end of some of the ridges indicate that the ridges are man-made and that the picture represents dredge-placering operations.

After the preliminary examination of the airphoto as described, the airphoto is analyzed from the geotechnical point of view. In this connection the following features should be studied: (1) landforms, (2) drainage patterns, (3) gully shape and gradients, (4) vegetative cover, and (5) culture, or land used by man (Sec. 7.1). In identifying these features, the graphical identification charts in Ref. 21 and the detailed discussion of drainage patterns in Ref. 22 are very helpful. *Landforms* often permit the identification of soils and their parent materials. For example, the ground surface formed by loess is a repetition of identical or similar hills with a series of parallel ridges, whereas sand dunes form a characteristic wavy surface. The configuration of the ground surface and the drainage pattern in Fig. 3.27 indicate loess, whereas the peculiar forms in Fig. 3.32 are dunes.

A *horizontal ground surface* alone is not indicative of the parent material. The latter may be a sedimentary rock; if sinkholes are seen on the ground surface, further study of the airphoto may show limestone present. In regions that have undergone considerable erosion, isolated islands of sedimentary rock (Fig. 3.1), often of irregular shape and fairly high, are left standing on the otherwise horizontal ground surface ("mesas" or "buttes" in arid regions, "monadnocks" in humid climates). A horizontal ground surface also is found in plains areas such as the Great Plains and Coastal Plains (alluvial soils) or the till and outwash plains in glacial regions (Illinois, North Dakota, and others). If the sedimentary strata are tilted, there is a rugged ground surface (Fig. 7.15).

The *slope gradient*, preferably in the vicinity of streams where the slopes tend to be higher and eroded by accumulated runoff, may indicate the character of soil materials. Coarse materials, such as gravel and sand

* Not to be confused with the texture of rocks or soils.

(and obviously rock), tend to assume steep slopes; in the case of flatter slopes, a relatively high clay content can be expected in the soil. When river terraces are examined as a source of gravel and sand materials, the

FIG. 7.11. Airphoto showing dredging operations (gold placer mining), California. (*USBR photograph.*)

sections with rolling and flat slopes ordinarily should be excluded from consideration.

As previously mentioned, the moisture content of the soil often can be determined by the color tones in the photograph. Very dark colors in a relatively flat ground surface may indicate that water is close to the surface, and thus vegetative growth may be rather heavy (Fig. 3.13).

Conversely, very light or even white areas in a photograph and a sparsity of vegetation may indicate that low moisture content or very deep ground-water tables can be expected. (Sometimes, however, white spots may be due to the reflection of sunlight on water at the time the photograph was taken.) The type of vegetation, as well as its density, also is indicative of moisture content of the underlying soils. Certain types of vegetation such as willows (known as "indicator plants") are known to grow only where moisture content is high, and conversely, other types such as sage and mesquite are known to grow only where moisture content is low. The plants usually can be identified by their pattern, color tones, and relationships. For further discussion see Ref. 20*b*.

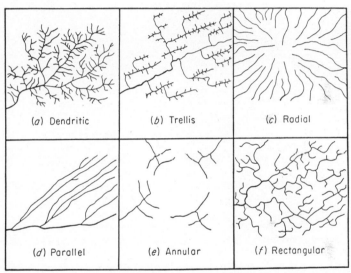

FIG. 7.12. Basic drainage patterns. (*From Parvis, Highway Research Board Bull.* 28, 1949.)

7.17. Interpretation of Drainage and Erosion Patterns. Running water erodes the ground surface, producing *drainage patterns*, i.e., systems of streams or streams and gullies, large and small. Drainage patterns are characteristic of a given soil or rock or of a complex of several materials, and a change of soil or rock type generally is accompanied by a change of drainage pattern. Of the many stream patterns formed on the ground surface, six discussed hereafter have been classified as *basic drainage patterns*.

Dendritic (arborescent or treelike) drainage commonly develops over horizontal homogeneous rocks and exhibits uniformity in all directions. The pattern may be considerably or only slightly developed according to the rock type; e.g., dendritic pattern on granite is much simpler than that on shale. Figure 7.12*a* shows dendritic pattern schematically, whereas Fig. 7.13 gives a comparison between an airphoto of a limestone-shale area and the drainage map of that area traced from the airphoto. Small circles on the map indicate sinkholes. *Trellis* patterns (Fig. 7.12b) may be compared with a garden trellis (grapevine pattern) and are usually developed in folded or dipping rocks with a series of parallel faults. The primary tributaries of the main stream are long and straight and often parallel to each other and to the main stream. Secondary

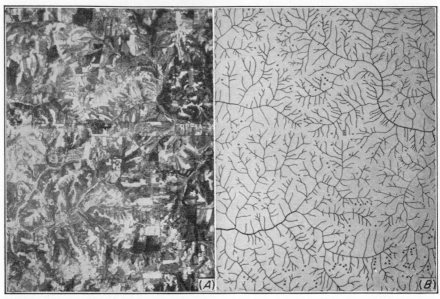

FIG. 7.13. Drainage pattern of limestone-shale. Small circles are sinkholes. (*From Parvis, Highway Research Board Bull.* 28, 1949.)

tributaries are short and stubby and join the primary tributaries approximately at right angles. *Radial drainage patterns* (Fig. 7.12c) consist of streams that flow radially, either from a center (e.g., a volcanic cone) or toward a center, e.g., to a basin. In a *parallel drainage pattern* (Fig. 7.12d), the streams are nearly parallel to each other (this pattern has been termed *cauda equina*, or "horse's tail"). Such patterns may develop on more or less uniform, rather loose deposits, e.g., valley fills. In an *annular drainage pattern* (Fig. 7.12e), the main streams are radial and the tributaries roughly annular, e.g., running around a dome. Streams following the faults and cracks in jointed rocks produce *rectangular drainage patterns* (Fig. 7.12f). The so-called *angulate* pattern (which is not basic) is a variation of the trellis pattern, but like the rectangular pattern, it reflects the influence of rock joints.

In a drainage pattern, streams and their tributaries are easily distinguished. If the tributaries are *closely spaced*, local soils and bedrocks have poor resistance to erosion (clay shales, silts, or sandy clays).   Conversely, if the tributaries are *widely spaced*, the bedrock or soil mantle is resistant to erosion and consists, for example, of sandstones, granular deposits, or unconsolidated glacial till.   In any case, any criteria based on spacing of streams have to take into account the over-all topography of the area and the quantity of precipitation that occurs in the area. Very rugged relief with high precipitation would tend to give a close spacing to the tributaries even though they were flowing over highly resistant rocks.   Conversely, very flat relief, combined with a low precipitation (e.g., in deserts), would result in wide spacing of the tributaries even if the soils are of low erosion resistance.   As soon as the relationship of a pattern (regardless of its scientific name) and the bedrock or soil mantle is established, similar pat-

terns in the same region indicate similar bedrocks or soil mantles.

The cross section of *gullies* is not the same in different soils. The gullies in *granular* (sandy) soils are short and steep and of a rather triangular cross section.   In *silts* and *loess*, the gullies are rectangular, with compound gradients, first steep and later flat.   The *clay* gullies are

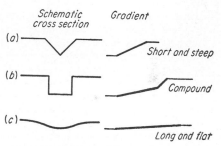

FIG. 7.14. Gullies in (a) granular materials, (b) silt and loess, (c) clay.

long and flat with flat, curved cross sections (Fig. 7.14).   Here again, as in the case of drainage patterns, precipitation quantity and general topography can influence the shape of gullies sufficiently to reverse the above criteria.

Except for those gullies shown in Fig. 7.14, no systematic drainage pattern develops on sand surfaces, including kames and eskers.   The absence of a regular drainage pattern usually means a high degree of permeability in a given environment (or possibly almost complete lack of precipitation).

Such is the case with red or yellowish-red silty clays developed on limestones when the clay particles are segregated into small lumps with ample space between them for the passage of water (pervious clays).   Fissured and otherwise pervious parent material, i.e., limestone, readily sorbs the infiltrating water, and the clay lumps do not swell.   This structure, however, is destroyed either in the laboratory by swelling in the permeameter or in the field under a roller; the material then performs as any other impervious clay.

7.18. Examples of Geologic Airphoto Interpretation.   The identification of various landforms and drainage patterns and their relation to various soil and rock types often enables the geologist rapidly to develop a picture of geologic conditions on a site and construct a rough geologic map without even setting foot on the ground. As it is difficult to set forth definite criteria for complete geologic airphoto interpretation, several examples are discussed hereafter.

Figure 7.15. The drainage patterns in area *A* of this photograph indicate resistant rocks underlain by softer ones. Because of the apparent tilting of the beds and the characteristic erosion patterns, the rocks are probably sandstones overlying softer shales, all tilting or dipping toward the upper part of the photograph. When the beds are traced to the left across the photograph, a sharp displacement in their regularity or continuity may be found; this displacement is repeated near the center and lower portion of the photograph. This indicates some geologic anomaly, probably a severe fold or possibly a synclinal structure. The sharpness of the folding indicates that there are probably one or more large faults trending from the upper left-hand corner of the photograph toward the lower right-hand corner. This is a warning to the geologist and calls for a detailed ground inspection. Besides the preceding fault, there is another one starting at the river at the center of the photograph.

Judging from the pattern and scarcity of vegetation, the gullylike developments in the area indicating cloudburst-type erosion normally found in arid climates, and the predominance of the outcrops, the area should be semiarid. If prospecting for a dam is being done, the dark color tones along the river (through the approximate center of the photograph) indicate high moisture contents in the soils, which possibly are clay materials. They are not too abundant, however, at that point. Furthermore, as the land has been cultivated (as can be seen from the regular rectangular patterns at the left of the photograph), it would appear that the material was suitable for crop production and thus would be silty or clayey. From these deductions it can be assumed that there is some impervious embankment material at the area, but not in very large quantities. A further study also shows that sand and gravel are likely to be scarce. In fact, the size of the buildings and cultivated ground dimensions as compared with the stream would indicate that the stream is not very wide, and thus large surface alluvial deposits are not to be expected, although a few sand and gravel bars (light spots along the stream course) can be seen on the photograph.

Figure 7.16. The jointing and fracture characteristics of bedrock often are easily identified in airphotos. In this photograph, the general joint systems are seen as dark, almost straight lines cutting across the river in the center of the photograph (letter *a*), and the intensity of the lines would indicate highly resistant rock (in this case, granite). The main river *d* is deeply incised into what appears to be very competent rocks (granite) with the formation of a deep canyon and cliffs. Two drainage patterns, both of dendritic type but of different degrees of development (*b* and *c*), would indicate two rock types, very soft (siltstone in this case) and very hard (granite). A further study shows that the soft rocks seem to fill the old drainage hollows in the hard rocks, which suggests that the soft materials were deposited much later than the granite. The dark streak indicated by *e* is either a dike or a fault zone. (Later ground examination disclosed that it was an intrusion of dolerite or basaltic rock.)

Figure 7.17. This photograph depicts typical alpine glaciation. The dark, roughly circular spots are glacial lakes, and the lack of vegetation indicates that they are above timber line. The rugged terrain and large quantities of water (as indicated by the lakes) denote a fairly high precipitation. As glacial lakes usually are left after the glaciers have melted, large deposits of talus and coarse rock can be expected to occur near the lakes. (This is true of alpine glaciation where the glaciers have not traveled very far from their source, and thus fine-grained materials are likely to be

FIG. 7.15. Airphoto illustrating faults, folds, hogbacks, and tilted sedimentary beds, Wyoming. (USBR photograph.)

HOGBACKS

A

CULTIVATED LAND

FAULT (?)

CULTIVATED LAND

SCALE
1 MILE

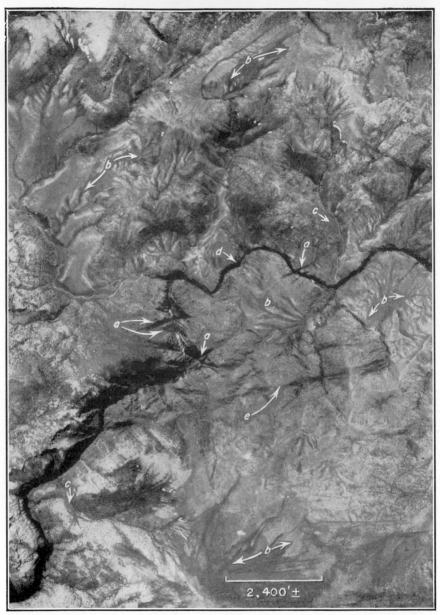

FIG. 7.16. Jointing system in granite and sedimentary beds overlying granite, Wyoming. Note fault trace. (*USFS photograph.*)

absent.)   Because of the rugged relief, the underlying rock is likely to be very resistant to erosion (particularly since a fairly high annual precipitation has already been
assumed) and thus quite hard.   The latter fact would be important to a search for
suitable riprap materials for dams (Chap. 8).

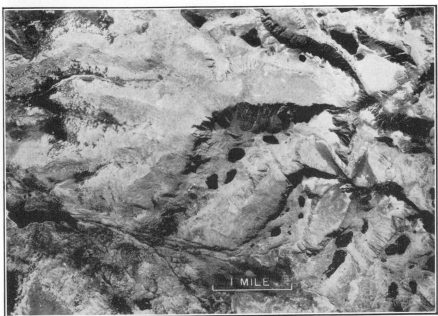

FIG. 7.17. Airphoto illustrating alpine glaciation, Wyoming.   (*USBR photograph.*)

FIG. 7.18. Old beach lines, Indiana (airphoto).

Airphotos are particularly valuable in interpreting the geology of regions that have
undergone continental glaciation.   Landforms as a result of such glaciation may have
such a great magnitude that ground recognition may be difficult, but in an airphoto,
such landforms exhibit very characteristic outlines to the experienced observer.
In Fig. 3.13, for example, the roughly ovate-shaped form near the center of the photograph and the surrounding horizontal ground surface indicate to the trained photo-

grammetrist that the landform is an esker and the surrounding terrain is glaciated. The nonuniform and often erratic permeability often found in glacial materials is very well illustrated in this photograph by the alternation of dark and light tones in the horizontal ground surface.   The dark tones represent high moisture contents, and the light tones, low moisture contents.   The various types of moraines also are usually easily identified, as an airphoto will show the relationships between the moraines and the general drainage, and thus it can be determined whether the moraine is lateral, recessional, or terminal (Chap. 3).   Glaciers exhibit characteristic flow lines because of included sediment and thus are easily identified (Fig. 3.19).

Figure 7.18.   Another important element that can be determined from airphotos is the former geologic history of an area that has later been modified by erosion or man-made culture.   In this photograph, ancient beach lines *a* are clearly seen, although a ground observer rarely would note them.   Similarly, ancient drainage courses that may affect the location of a proposed housing project often can be distinguished on an airphoto, although not apparent on the ground.

REFERENCES

1. Swainson, O. W.: Topographic Manual, *USCGS Spec. Publ.* 144, 1928.
2. Lobeck, A. K., and W. J. Tellington: "Military Maps and Air Photographs," McGraw-Hill Book Company, Inc., New York, 1944.
3. Mitchell, H. C., and L. G. Simmons: The State Coordinate Systems, *USCGS Spec. Publ.* 235, 1945.
4. Publications of the U.S. Geological Survey, Department of the Interior, a comprehensive catalogue published annually.
5. Catalogue of Nautical Charts and Related Publications, *USCGS Serial* N665, October, 1949.
6. Lobeck, A. K.: "Block Diagrams," John Wiley & Sons, Inc., New York, 1924.
7. Eckel, E. B.: Interpreting Geologic Maps for Engineers, *ASTM Spec. Tech. Publ.* 122, 1951.
8. Fisk, H. N.: Geological Investigation of the Alluvial Valley of the Lower Mississippi River, *U.S. Corps Engrs. Mississippi River Commission*, Vicksburg, Miss., 1944.
9. Chalmers, R. M.: "Geologic Maps," Oxford University Press, New York, 1926.
9a. "Bibliography of Geology Theses on File at Colleges and Universities of the United States," compiled by D. S. Turner for Petroleum Research Co., Micro-SORT-Card Petroleum Geology Library, Denver, Colo.   This lists over 6,000 theses at more than 80 colleges.   It is available in most college geology libraries and is being kept current.
10. Dake, C. L., and J. S. Brown: "Interpretation of Topographic and Geologic Maps," McGraw-Hill Book Company, Inc., New York, 1925.
11. Interpreting Ground Conditions from Geologic Maps, *USGS Circ.* 46, May, 1949.
12. "Atlas of American Agriculture, Physical Bases Including Land Relief, Climate, Soils and Natural Vegetations of the United States," Government Printing Office, Washington, D.C., 1936.
13. Engineering Use of Agricultural Soil Maps: A Symposium, *Highway Research Board Bull.* N. 22, Washington, D.C., 1949.
14. "Field Manual of Soil Engineering," rev. ed., Michigan State Highway Department, Lansing, Mich., 1946.
15. Bagley, J. W.: "Aerophotography and Aerosurveying," McGraw-Hill Book Company, Inc., New York, 1941.

16. Eardly, A. J.: "Aerial Photographs: Their Use and Interpretation," Harper & Brothers, New York, 1942.

17. Smith, H. T. U.: "Aerial Photographs and Their Interpretation," Appleton-Century-Crofts, Inc., New York, 1943.

18. Aerial Photography, *War Dept. Tech. Manual* TM 5-240, Washington, D.C., 1944.

19. Trorey, L. G.: "Handbook of Aerial Mapping and Photogrammetry," Cambridge University Press, London, 1950.

20. American Society of Photogrammetry: "Manual of Photogrammetry," 2d ed., George Banta Publishing Company, Menasha, Wis., 1952: (*a*) Chap. 1, The Development of Photogrammetry by G. D. Whitmore; (*b*) Chap. 12, Photographic Interpretation for Civil Engineers by R. N. Colwell.

21. Jenkins, D. S., *et al.:* The Origin, Distribution, and Airphoto Identification of U.S. Soils, *CAA Publ.*, Washington, D.C., 1946.

22. Parvis, M.: Drainage Pattern Significance in Airphoto Identification of Soils and Bedrocks, *Highway Research Board Bull.* 28, 1950.   Good bibliography.

CHAPTER 8

ROCK AS A CONSTRUCTION MATERIAL

Building facing, protective blanketing of earth dams, base and subbase of airport runways, and concrete constituents are only a few uses of rock and rock products in engineering as covered in this chapter. Though in practice the strict distinction between "rock" and "stone" (Chap. 2) is not always observed, it is maintained through this chapter.

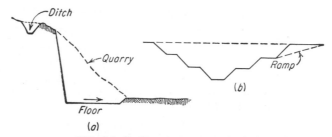

FIG. 8.1. Shelf and pit open quarries.

8.1. Terminology. *Quarries* are places where rock is separated from its natural beds and processed for use in construction. There are open and underground quarries. Open quarries may be *shelf quarries*, where the rock is extracted from a hillside (Fig. 8.1a), or *pit quarries*, where rock is excavated from a certain depth in the ground (Fig. 8.1b). Quarry products are dimension stone, crushed stone, and broken stone. *Dimension stones* are blocks with more or less even surfaces of specified shape and usually of specified size. Dimension stones are rapidly being replaced in building construction by reinforced concrete and baked-clay products and therefore are generally used only for *facing* of expensive buildings (granite, sandstone, limestone, including the well-known Indiana limestone, and others). Stone and rock for pavements also are gradually being replaced by concrete and asphalt, though pavement blocks still are moderately used. As the production of dimension stone has decreased, the *crushed-stone* industry has progressed. Based on 1952 figures[1] the annual production of crushed stone in the United States amounts to about 300 million tons, of which about two-thirds is used in construction, the rest being consumed in the cement and lime industry.   Most of the crushed stone comes from limestone, although small amounts are obtained from basalt (traprock), granite, sandstone, and quartzite.   Crushed stone is sold by the ton or cubic yard at relatively low prices ($1.37 per ton average in 1952),[1] whereas dimension stone is comparatively costly and is sold on a per-cubic-foot basis.   As a result of this differential, the cost of long transportation hauls is a minor part of the cost of dimension stone but a very important factor in the cost of crushed stone.   Generally a quarry produces both dimension and crushed stone, although there are some that specialize in one or the other.

Riprap is broken stone or boulders used as a protective layer on the upstream face of an earth embankment to protect it from wave action and general erosion by water.   As in crushed stone, the transportation factor is a very important component of the riprap cost.   A riprap quarry is generally opened for a given project.   It should be as close as possible to the project it serves and is satisfactory even if it contains only the amount of rock required for that project without any reserve for the future.

FACTORS AFFECTING ENGINEERING SERVICE OF ROCK

8.2. Frost Action.   If a rock is saturated, i.e., its pores are completely filled with water, freezing of the water induces tensile stresses in the rock and cracking may result.   However, the degree of saturation of rocks generally is less than 100 per cent.   Frost action, therefore, will cause cracking only when there is *not* enough space within the pores to permit the freezing water to expand (about 9 per cent by volume of water).   Also, frost action depends as much on the diameter of the pores as on the degree of saturation.   Rock with wide pores that give up water readily (i.e., by evaporation) is not particularly susceptible to frost damage even in cold climates.   Furthermore, because the freezing point of water in fine capillaries is below 32°F,[2] rocks with fine capillaries, such as granites, are frost-sensitive only at very low temperature.   In freezing-thawing tests on granites[3] there were no signs of disintegration after 5,000 experimental cycles; each cycle consisted of 6 hr of freezing and 1 hr of thawing in water at 68°F (20°C).

Most freshly quarried rocks, especially limestones and sandstones, possess a considerable percentage of "quarry water" (geologically known as "connate water," see Chap. 5) and should be seasoned, i.e., reasonably dried, before being placed into structures (but not overdried, see Sec. 8.30).   Since laminated rock may scale badly if water freezes in open bedding planes, it is particularly inadvisable to place it with the cleavage or bedding planes vertical.   Limestones with shaly layers or any rock with seams that sorb water readily are not recommended for use in cold climates.

Frost action on riprap is most severe where the rock is subjected to wetting and alternate freezing and thawing, i.e., at and above the contact of the water surface with the riprap. The same factors of porosity affect durability of riprap as affect building stones. Rocks with high capillarity or with numerous fissures (although nonporous) are rapidly destroyed by frost action.

FIG. 8.2. Scaling of a tooled sandstone surface (average thickness of scaling about 0.09 in.).

8.3. Chemical Destruction. Building stones in large cities are perhaps more severely affected by atmospheric gases than by any other destructive agent. Carbon dioxide, CO_2, sulfur dioxide, SO_2, and sulfur trioxide, SO_3, are the main detrimental gases. The latter forms sulfuric acid when combined with air moisture. (The well-known Los Angeles, California, "smog" is partially composed of sulfuric acid.) Calcite, if acted upon by sulfurous gases, may be transformed into sulfate, and scaling will result. Consequently, marbles and limestones may not be highly resistant to weathering.

Riprap sometimes is destroyed by the effects of alternate wetting and drying or by dissolution of water-soluble substances in the rock. The latter type of failure is particularly common to rocks originating in arid regions, where natural leaching is very slow or nonexistent. Where the content of bentonites or similar clays in rock is high, destruction caused by cyclical wetting and drying processes is common.

8.4. Physical Destruction. Although the physical agents discussed in this section more commonly affect building stones, occasionally riprap and even concrete are destroyed by their action. In sandstones, these agents may cause scaling, cracking, or the destruction of the cementing material (Fig. 8.2). Crystallization of salts (sulfates and chlorides) at the outside surface of building stone sometimes occurs in the form of white spots known as "efflorescence" or "bloom." During a continuous rain, the pores in a stone surface may be filled with a supersaturated salt solution. Crystal growth then will occur which may cause an excess of hydrostatic pressure and a consequent disintegration of the stone. Owing to continuous hydration, the crystals themselves may considerably increase in volume and thus contribute to the increase of the hydrostatic pressure. For example, because of hydration, sodium sulfate in the form of thenardite, Na_2SO_4, may recrystallize into glauber salt, $Na_2SO_4 \cdot 10H_2O$, with a resultant 400 per cent increase in molecular volume.

Cyclical heating to a high temperature and subsequent cooling (as may occur during a fire because of the action of cold water) also may destroy a rock, particularly if the constituent minerals are calcite and feldspar.[4,5] When heated, a calcite crystal will expand along the direction of one crystallographic axis (the c axis) and contract in a direction perpendicular to that axis (along the a or b axis, Fig. 1.1). The expansion in an orthoclase crystal will be more than 12 times greater in one direction (along the a axis) than in another direction (along the b axis),[6] though no contraction occurs. Similarly, both plagioclase (sodic feldspar) and quartz have different coefficients of expansion in two mutually perpendicular directions. These differences cause irreversible

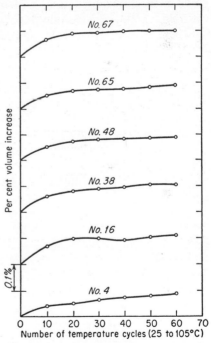

FIG. 8.3. Volume-expansion curves for six different granites. (*After Kessler, National Bureau of Standards.*)

sliding of the crystals and increase in volume, or "bloating," of granite after cyclical heating and cooling (Fig. 8.3). In the same class of phenomena is the "sugaring," or granulation, of marble surfaces due to the differential thermal expansion of tightly knit calcite crystals. As a result of sugaring, up to 1 per cent increases in the volume of marble have been observed.[6]

Temperature variations also may cause spalling or exfoliation of a rock. Extensive temperature changes induce internal stresses, with a resultant "peeling" of the rock surface. This type of weathering is common in granites and other coarse-grained igneous rocks. Exfoliation may, in combination with other factors, cause rejection of the rock source for use as a riprap.[7]

QUARRIES: EXPLORATIONS AND PROCEDURES

8.5. Explorations for Quarries.   The search for rock material suitable for building stone, crushed rock, or riprap is controlled by the same basic factors.[8]   They are (1) quality, (2) supply of the material, and (3) economics of production and delivery.   Usually the investigation can be done singly or by a combination of surface geologic mapping, subsurface exploration with drills, and geophysics.

1. *Quality.*   A competent geologist usually can establish the quality of the material by examining rock specimens that he considers representative of the entire quarry.   If the rock is to be used in an industry where the chemical properties only are essential, sufficient data usually can be obtained from churn-drill samples.   However, for dimension and most crushed stones and riprap, the physical properties should be determined from diamond-drilled cores, usually 3 in. in diameter (the larger the diameter, the more accurate is the determination of fracture and joint patterns and their influence on the eventual quarry operations).   Rock for dimension stone must be free of cracks, of uniform texture, of attractive color, and in some cases, capable of taking a polish.   Horizontal bedding in such stones as limestone or sandstone is desirable.   Crushed stone and riprap must have satisfactory strength, soundness, and low water sorption.   Particularly, rock selected for riprap should be roughly squared and reasonably flat faced.   The specific gravity is of utmost importance.   Values of 2.6 and higher are preferred, as the rock must be heavy enough to resist displacement by wave action, ice pressure, and thrust from drifting objects.[7]   Flat or stratified rock can be used as riprap only if carefully placed on the structure.   The gradation of the riprap should provide a reasonably smooth layer of uniform thickness. Fine material should be avoided, as the pressure of water trapped behind the riprap might induce slumping, although spalls, gravel, or crushed rock can be used to fill voids between rocks.

The breaking characteristics of a rock material earmarked for riprap can be determined by a blast test.   Usually small experimental charges of high explosives are placed into fissures on the outcrop face and exploded.   The results of the blast test are considered satisfactory if a low percentage of small (about 1 in.) sizes and a high percentage of large (about ½ cu yd) sizes are obtained. Fragments should be angular to ensure proper interlocking in the riprap layer, as explained. Hence the following rocks are unsatisfactory for riprap: (a) rocks shattering in the blast and producing a large percentage of fines, (b) if the blast produces flat fragments such as may occur from rocks containing platy materials like mica or from some sandstones with thin bedding and a scarcity of smaller sizes (Fig. 8.4), (c) conglomerates producing rounded (and not angular) fragments in the blast.

FIG. 8.4. Unsatisfactory riprap produced by a blasting test (sandstone). (*Photograph by E. A. Abdun-nur.*)

When the quality is studied, full attention should be paid to the geological features of the locality that may control the rock processing. The strike and dip of the strata and the jointing of the rock are of major importance. In a shelf quarry in sedimentary rock, it may be advantageous to work the quarry floor parallel to the bedding of the strata. If the strata have a steep dip, the quarry may be worked as a steep-walled cut, extracting good strata only. Joints may facilitate the removal of blocks, but if too many, they may limit the size of dimension stone. Lenticular strata occasionally are worked for riprap providing the cover of overburden is not too deep and there are no other sources of riprap within economical haul.

2. *Supply.* If the rock appears in outcrops, especially in a cliff along a stream, the volume of a quarry may be readily estimated from simple visual inspection and a few measurements. If outcrops are not available, the volume can be estimated on the basis of borings or geo-

physical work.   Care should be exercised in such estimates, as lenticular rock bodies may appear as thick, continuous ledges in outcrops whereas back under the hill they may thin out to only a few inches (Fig. 8.5). The detection and estimation of quantity of rock in such deposits can be accomplished by geophysical surveys combined with bore holes.

In the zones that have been heavily glaciated (e.g., Montana, the Dakotas, etc.) advantage often can be taken of abundant boulders left by the ice.   They may be used for riprap, though this type of deposit is expensive to process.

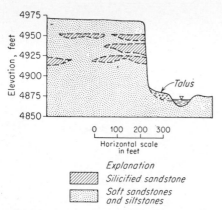

FIG. 8.5. Geologic cross section of lenticular riprap deposit (occurrence near Ogallala, Nebraska).

The rock supply of a quarry generally is estimated in tons. For dimension- and crushed-stone operations, the supply should be sufficient for about 20 years if initial expenses and amortization costs are to be justified.   For riprap quarries, as already explained, an economically feasible operation usually is possible even if the supply is only sufficient for the immediate use on the structure.

3. *Economic Factors.*   One of the most important factors, which may make the operation economically prohibitive, is the *cost of transportation* of the product to the place of consumption.   Local deliveries are made by trucks, and if possible, long-distance hauls are made by water, since water rates usually are lower than railroad rates (e.g., large tonnages of limestone are carried on the Great Lakes).   Other factors that may increase unit prices are high wages, labor shortages, poor drainage conditions in the quarry, and expensive stripping (the removal of clay, sand, gravel, and unsuitable rock from the surface of the desirable rock, see Sec. 8.6).

Geophysical Methods of Explorations.   Both seismic and resistivity methods can be used to determine the volume of rock in potential quarry sites.   The selection of the proper method will depend upon geological conditions and should be left to the judgment of the geophysicist (Sec. 6.37).

A deposit of siliceous (partly cemented with silica) sandstone known to lie in lens-shaped, interconnected bodies in northeastern Colorado was investigated by the USBR for riprap.   Where siliceous rock was present, the resistivity was high (from 400 to 1,200 ohm-ft), as compared with the low resistivity elsewhere in the area being investigated (from 50 to 200 ohm-ft).   Examples of the two types of curves are shown in Fig. 6.41.   The fixed depth resistivity traverses permitted the edge of the rock to be accurately located, as it was marked by the appearance of low resistivity on the traverse line (Fig. 8.6).   The resistivity findings also assisted in determining the quality of the sandstone, as, in general, the more silicified (and thus harder) the rock, the higher the resistivity.

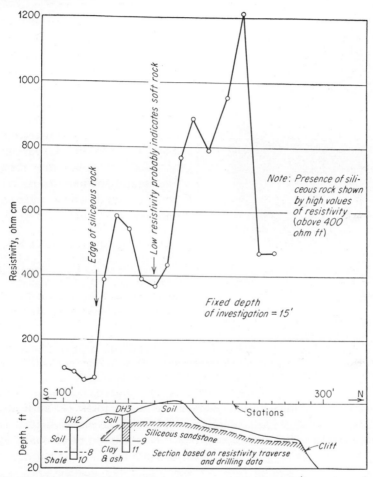

FIG. 8.6. Fixed depth resistivity traverse across a quartzite deposit (traverse run between DH2 and DH3).   (USBR data.)

8.6. Stripping and Drilling.   A better understanding of quarry economics can be obtained by a study of the methods used in excavating and processing rock in a quarry.

Stripping.   The depth of overburden in actual quarries ranges from zero to perhaps 30 ft.   The limits of a quarry with a widespread rock supply are continuously extended, hence continuous stripping.   In some specialized quarries, the rock is obtained by actual mining methods, and thus stripping becomes unnecessary.   In small quarries, simple handwork with a pick and shovel is used; in large quarries, excavating equipment such as a power shovel or dragline is used.   If the overburden is friable enough to be eroded by water, the hydraulic method is used and the overburden is simply washed off by high-pressure streams of water. Clean stripping is essential only if the rock is to be used in the chemical industries.   Dirt left on dimension stones during initial quarrying is immaterial, since the stone is cleaned during subsequent processing.   The spoil material (the dirt) obtained in excavation may be used for filling low places or for making embankments or simply dumped into abandoned quarries.

FIG. 8.7. Airhammer bit with tungsten carbide inserts, 0.8 ✕.

In the Indiana quarry districts, a "red" clay covers the limestone beds to a depth of 1 to 20 ft or more.   Stripping is done partly by power shovels and partly by the hydraulic method, wherein the clay is washed into abandoned pits or other low spots.   Below the clay in Indiana, there is from 5 to more than 15 ft of shattered and broken noncommercial limestone also to be removed.   Poor rock separated from the usable rock by a layer of clay or shale is blasted with light charges of black blasting powder (Sec. 8.7).   If poor rock is intermixed with good, the separation process is expensive.

Drilling.[9]   Drilling is done either by hand tools or with drilling machines.   In either case, steel drill rods with specially shaped tips— bits—are forced ("fed") into the rock.   Bits may be hexagonal, octagonal, or cross-shaped.   The basic functions of a bit are penetrating, crushing, reaming, and mixing the rock material.   For deeper holes and for machine work, detachable bits are used.   One of the efficient developments in this field is the use of "carbide inserts," or "slugs" of tungsten carbide mounted in slots machined in the steel body of the bit (Fig. 8.7). In hard rock "jet piercing" may be used.   This consists of directing a flame produced by burning oxygen and a petroleum-base fuel in a special blowpipe against the rock face.   Hand drilling, i.e., driving a hand drill into rock with blows from a sledge hammer (often called a "double jack" or "single jack" depending upon its weight), is used in soft rocks and shallow bore holes or where mechanical equipment is not available.

Hammer drills (or jackhammers) are hand held and operated by compressed air furnished by portable compressors. Compressed-air or pneumatic hammers also may be mounted on a frame on wheels to form a wagon drill (Fig. 15.14 shows wagon drills in operation). Churn drills, generally used in well drilling, also are used in quarries.

Diameter, spacing, and depth of holes and presumably the amount of explosive are factors influencing the output of excavated rock. The diameter of bore holes varies from 1 in. in handwork to 12 in. in mechanical drilling. There is a strong tendency to use a larger hole diameter, such as 6 to 9 in., especially in the crushed-stone industry. Among other numerous advantages, large holes need not be sprung, i.e., widened at the bottom, for the placement of explosives. Required spacing for holes sometimes is determined experimentally, though tables exist for their determination.[10,11] Occasionally in excavating rock to "neat" lines (i.e., exact and specified lines) for the foundations of a structure, line drilling is employed. *Line drilling* is the drilling of closely spaced holes along a straight line at the predetermined, specified edge of the excavation. Weak explosive charges are detonated in these holes, leaving a fairly even and straight rock wall. This method also is used in rocks that are known to be so badly shattered that they might break in very uneven lines and thus cause considerable overexcavation.

8.7. Explosives and Blasting.[10,66] The following discussion is presented in connection with quarry operations; however, it is highly pertinent to all geotechnical explorations.

The two basic kinds of explosives are *black blasting powder* and *high explosives*. The latter detonate with a sharp jar, shatter the rock, and produce fissures in it. Black powder tends to "push" the rock and thus acts more gently to break the material along a few well-pronounced shear surfaces. Therefore, high explosives are used in crushed-stone quarries and most civil engineering excavation operations, whereas black powder is used in dimension-stone quarries (and also in coal mining).

Black blasting powder may be either *"A" blasting powder* (charcoal, potassium nitrate, and sulfur in proportions of about 15:75:10) or *"B" blasting powder* (charcoal, sodium nitrate, and sulfur in proportions of about 16:72:12). B is slower and less expensive than A and therefore is used in producing broken stone (riprap).

High explosives may be (1) those that contain mainly nitroglycerin and nitroglycol and (2) those that do not (mostly military-type explosives). Nitroglycerin-nitroglycol types (symbol NG) are the main types used in civil engineering. The two types are very similar in explosive properties, and the purpose of the mixture is to provide (1) compositions that will not freeze and (2) economy, since nitroglycol is less expensive than nitroglycerin. They commonly are referred to as *dynamite*, and they may be of either a granular or gelatinous nature. In addition to NG, the principal ingredients are ammonium nitrate, sodium nitrate, and absorbents such as woodmeal, rye flour, etc. They are packed in a paper cartridge. Gelatin dynamite employs nitrocotton to thicken the NG into a gelatinlike substance. Because they are plastic, cohesive, and practically waterproof, "gelatins" are excellent for underwater work or for small bore holes in which they have to be molded tightly in place. Dynamites usually are rated according to their *strength* which indicates the amount of energy the dynamite will develop when exploded.

Proper *storage* of dynamite involves the construction of a magazine (a small house or underground cave) that has a constant cool temperature, humidity control, and proper ventilation. If excellent storage conditions are not maintained, dynamite may deteriorate to the extent that it is hazardous to use. It then should be examined and, if necessary, destroyed *by an expert*. This generally is done by burning it in small quantities. At the present time, all dynamites are "nonfreezing," i.e., with freezing points well under the lowest temperatures observed in the continental United States.

Detonation. A *primer* is a high-explosive type of cartridge ("stick dynamite") with a detonating device inserted into the cartridge. The *safety fuse* is the medium through which a flame is conveyed from outside the bore hole to the primer. This fuse is a powder core wrapped in some textile; it resembles an insulated electric wire and is placed in an ordinary blasting cap. Explosion also may be caused electrically by providing the primer with an *electric blasting cap* (a small cylindrical brass object resembling a small-caliber rifle shell and containing an explosive compound such as lead azide that explodes upon application of an electrical current). When the primer explodes, the rest of the dynamite in the bore hole is set off by the detonation. After the charge (including the primer) is in place, the bore hole is stemmed (with clay, sand, loam, or even water) or plugged with a special plug. *Firing,*[10,11] i.e., setting off the blasts, should be done with precautions as explained in Refs. 10 and 11. The common warning cry that a blast is about to be detonated is "Fire in the hole!"

The *breaking of the rock mass* by an explosion is caused by the shearing and tensile stresses that develop in the rock owing to the pressures exerted in all directions by explosion gases. Figure 8.8a shows the closed failure surface developed by the explosion of a weak charge C in a sprung hole. Failure surfaces CF' and CF'' in Fig. 8.8b, caused by a stronger charge, reach the free surface of the rock. If a vertical face OV (Fig. 8.8c) is far from the charge, a conical "crater" may be formed instead of far-reaching failure surfaces. A "corner break" with two rows of holes

(Fig. 8.9) is not recommended because of the diagonal resultant of the forces and hence the waste of the possible energy of the charges. A corner break with one row of holes is considered economical.

Excavating a mixture of solid and loose rock 1 cu yd in volume measured in place makes on the average about 1.4 cu yd in fill. To blast rock in large areas, such as road cuts, from 1.0 to 1.25 lb of explosive per cubic yard of rock is used. In confined places such as ditches or foundations, the quantities of explosives necessary may double. The amount of explosive required and the type used will depend upon the strength

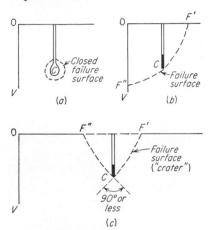

FIG. 8.8. Rock excavation.

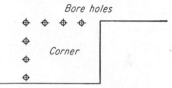

FIG. 8.9. Corner break (two rows of bore holes—not recommended).

of the explosive, the density and fracturing in the rock, and the possible damage[12] by the blast to nearby buildings or other structures.

8.8. Crushed-stone Quarrying. Crushed stone may be obtained by (1) crushing rock or (2) crushing oversized gravel or cobbles and small boulders from gravel pits. In many parts of the country, demand for crushed gravel exceeds even that for natural pebbles.

Crushing Characteristics. A micro- and megascopic study of the rock fabric usually discloses the crushing characteristics of the rock. For example, rocks with a weakly interlocked grain structure may show excessive granulation in crushing (Fig. 8.10*a*), whereas dense rocks with tightly interlocked grains will produce satisfactory crushed stone (Fig. 8.10*b*). Dense, pure limestone is a good material for crushing but may produce an excess of fines if closely fractured or clayey. In sandstones, the cementing material, as well as the interlocking of the grains, closely controls the crushing characteristics. Hard, dense sandstones generally produce good crushed stone.

Quarry Methods. In crushed-stone quarrying, high explosives and deep bore holes are used, whereas in dimension-stone quarries, black powder and relatively shallow bore holes are used. In a crushed-stone quarry the bore holes must reach the bottom elevation of the rock to be quarried and are even "subdrilled" from 3 to 5 ft below that elevation

FIG. 8.10. Two types of crystal interlocking: (a) weak, rounded crystal contact, granitic texture, 14 $\times$; (b) strong, felting of lathlike crystals into very fine matrix, diabasic texture, 25 $\times$. (*Photograph by W. Y. Holland.*)

(Fig. 8.11). Subdrilling is eliminated, however, if there is natural horizontal parting at the quarry floor. Also, the diameter of bore holes is generally larger than in dimension-stone quarries.

A crushed-stone quarry may be operated either by the *bank method* (Fig. 8.11), the height of the bank being up to 100 ft or more, or by the *coyote-tunnel* (gopher-hole) method. The cross section of a coyote tunnel is 4 to 5 ft square and its plan resembles the capital letter T, the stem of the T being perpendicular to the quarry face. The wings of the tunnel (corresponding to the bar of the T) and a part of the stem are piled with high explosives, and the rest of the tunnel is stemmed or tamped with broken stone. The product obtained is broken stone, which may be used as riprap or jetty stone (see Sec. 11.10).

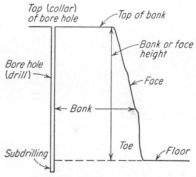

FIG. 8.11. Quarry terminology.

It should be noted that the unprocessed, natural rock product obtained by blasting or otherwise is termed "quarry run" or "pit run."

Crushers. Gyratory and jaw crushers commonly are used for reasonably hard rock such as granite, traprock, or limestone. The *gyratory crusher* consists of a concave hopper in which a stout vertical shaft is suspended at the top. The shaft operates as a kind of pendulum which describes a small circle at the bottom of the hopper. The shaft carries a renewable crushing head, and the hopper is lined with renewable chilled-iron or -steel staves. The size of the final product is from 1½ in. in small machines to 7 in. in large ones. In *cone crushers* a swinging cone is substituted for the shaft. A *jaw crusher* consists of an immovable and a movable jaw lined with interchangeable, generally corrugated, wear-resistant lining plates. The movable jaw rotates about a horizontal axis in a pendulum action and crushes the stone material by striking the immovable jaw (Blake, Dodge, and other systems). In a large operation, the crusher usually does not reduce all the material to the desired size, so the coarser material is passed through *secondary crushers*—gyratory and jaw crushers with some modifications. The rock material released by a crusher is the "crusher run."

Screening. This is the process of separating the crusher-run material into a series of different-sized products, the product in each series being approximately of the same size. Particles that remain on a screen are said to be the oversize of that screen, while the materials passing through the screen apertures are the undersize. Openings in a screen are either square or round. Although there are many types, the simplest kind of screen is made of wire mesh fixed within a rectangular frame which is placed at a moderate angle to the horizontal. The rotary or revolving screen is a cylindrical frame with an envelope of wire cloth or perforated plate, its longitudinal axis being set at a small angle (5 to 7°) with the horizontal. Screened material below 1 in. in size generally is termed *screenings* or *chats*.

Instead of the square or round openings used in a screen, a *grizzly* has parallel bars spaced at a fixed distance and set at angles of 20 to 50° with the horizontal (Fig. 8.12). There are grizzlies of more complicated systems, e.g., rotary grizzlies.

FIG. 8.12. Grizzly.

Washing crushed stone is not ordinarily done but may be necessary in order to remove a clay coating and thus recover good material from what otherwise would be quarry waste.

Quarry Hazards. Accidents in quarries mostly are due to (1) falls of the overburden or slides of the rock slopes, (2) mishandling of explosives, and (3) poorly organized haulage of the rock material together with carelessness and/or fatigue of the drivers. A serious occupational disease of stonecutters is *silicosis*, a lung infection that develops as a result of inhaling quartz powder during working hours.[13,14] For protection against the disease, dust respirators are used.

SAND AND GRAVEL

8.9. Sources of Sand and Gravel.[7] Sand and gravel are used in engineering mostly to make concrete and for placement in those parts of earth and other structures where permeability is required, e.g., in highway subgrades, drains, railroad ballast, etc. In housing projects and in warehouses, sand and gravel cushions are often placed under floor slabs to create a "capillary break" and prevent humidity in the building. The places (or, speaking scientifically, "geological terrains") wherein sand and gravel may be found are discussed hereafter. For terminology, consult Chap. 3.

Alluvial Fans. These deposits, found usually in the valleys at the base of mountains (Sec. 3.18), range from steeply sloping deposits which contain large rock fragments and boulders to gently sloping, almost plains-like deposits which usually contain fine-grained materials. Usually the coarser material is near the steep outer slopes at the head of the fan (the upstream end) and the finer materials are near the outer edges (downstream end and sides). The material in fans commonly tends to be angular to subrounded owing to the short transportation distance involved before deposition; poor stratification and lenticularity are characteristic.

Terrace deposits form benches in the topography of a region and commonly occur along the sides of a stream or in the flood plains of rivers (Sec. 3.8). The sands and gravels in such deposits usually are well stratified, well graded, and well rounded. However, because of the age* of the terrace deposits, care has to be exercised in their use, as many of the gravels may be severely weathered and soft or certain chemicals such as iron oxides or silicates may have been deposited in and on the gravels, rendering them useless for some purposes such as concrete (Sec. 8.16).

Flood-plain materials occur within the boundaries of a stream valley and may be many miles in width normal to the course of the stream. Usually such materials are fine-grained, and this is particularly true near meandering streams. Coarser materials such as sand and gravel may occur adjacent to braided streams. If flood-plain deposits are to be used, considerable processing (such as washing) to remove fine materials may be required.

Glacial outwash is deposited from streams flowing out from a glacier and may be many miles from the point where ice actually occurred. The deposits tend to be stratified and many square miles in areal extent unless they are concentrated in narrow valleys. Glacial outwash may contain all sizes from clay to boulders. Generally the grain sizes tend to decrease toward the lower end of the deposit, although due to alter-

---

* The relative age of any sand and gravel deposit may give the geologist a clue as to the degree of weathering. For example, Tertiary terraces, being several million years older than a Pleistocene terrace, could be expected to contain considerably more weathered rock fragments.

nating retreats and advances of the ice sheet during the glacial epoch, considerable variation in grading of various layers may occur. Outwash deposited by alpine glacier streams usually is more erratic in stratification and may not contain so high a percentage of fine materials as outwash from continental glaciers.

Glacial Till. These deposits are very heterogeneous and generally are poor sources of sand and gravel. Usually all gradations of materials from rock flour to large boulders are found, but sorting is erratic. Considerable amounts of fines, such as rock flour, generally are present, although occasional lenses of sand or boulders may occur.

Moraines. These deposits remain wherever ice has once been. They may occur along the sides of valleys, across valley floors, or as hummocks in plains-like terrain. The deposits usually are erratic mixtures of soil, gravel, and boulders. The alpine deposits tend to contain a higher percentage of large-size material, whereas the continental deposits may contain much rock flour.

Sand dunes may be found in semiarid plains, along ancient alluvial flood plains where the wind has sorted the ancient alluvial deposits, or along beaches. Dunes are also very common in arid regions, where vegetation is sparse or lacking. The sand generally is fine to medium in size but poorly graded.

Residual Deposits. Occasionally sand (and rarely, gravels) may be obtained from deposits resulting from the destruction of outcrops by weathering. Typical sources are granites (under semiarid conditions, Sec. 3.3), sandstones, and, under certain conditions, quartzites. In all such occurrences the deposits usually are very limited in extent, and careful petrographic analysis is necessary to assure that the grains are not too weathered to be used for engineering purposes.

Other Sources. Occasionally, a *talus* deposit (Fig. 8.13) may be used as a sand and gravel (usually the latter) source if grading permits. Usually, some crushing is required to obtain a sufficient proportion of small sizes. *Delta* deposits formed at the mouth of very large rivers or streams contain some sand but may be high in silt and clay content (Fig. 3.9). *Lacustrine* deposits (lake deposits), which generally consist of clay or silt, sometimes contain coarse materials as the result of deposition by ancient streams which have flowed into those lakes. For example, the extinct Lake Lahonton in Carson Sink, Nevada, was the sand and gravel source for the construction and rehabilitation of Fallon Airfield.

Blending is the process of improving the poor grading of a sand material by adding to it the missing grade sizes. Violent and prolonged water and wind action contributes to the production of hard, well-rounded sand or gravel material concentrated in a few grade sizes. Upon proper blending, such sands become an excellent engineering material. Among cther sources of blending materials for concrete, *sand dunes* should be particularly mentioned.

Sand and Gravel in Glaciated Areas.   Abundant sand and gravel deposits exist in New England, in the state of New York, in portions of the middle western and northern Great Plain states, and in some parts of Idaho and Washington.   *Terraces* left by outwash glacial streams are found near the following rivers: Mohawk, South Platte,

FIG. 8.13. Talus—potential crushed-rock source.

Susquehanna, Wabash, Illinois, Minnesota, Wisconsin, Missouri, and parts of the Mississippi.   *Morainal* sources of sand and gravel are common in Illinois, Iowa, the Dakotas, Montana, Washington, Colorado, and other states that were subjected to the direct action of either alpine or continental glaciers.   The southern portion of so-called "greater New York" is on a terminal moraine.   In Alaska, the most abundant sources of sand and gravel are glacial outwash and moraines.

8.10. Investigations for Sand and Gravel.   A competent geologist or engineer should perform the investigation.   His report should include the following information:

1. Geographic location and the type of the structure, where the material from the deposit is to be used, and the amount of material required.

2. The geographic location of the deposit, including the county and township; the distance from the deposit to the structure and the type of transportation proposed; and data on the cost of various transportation methods available in the locality.

3. The topography of the deposit area (whether it is in a valley, on a high terrace, or near the ocean) and the topographic relief of the deposit. A contour map should accompany the report. The scale of the map should be about 1,000 ft = 1 in.; the contours should be at 5- or 10-ft intervals. In exceptional cases for extremely large deposits, smaller scale maps may be used. Photographs of the deposit may also be useful.

4. The geology of the deposit. The geologic origin and the type of rocks in the vicinity, the approximate dip and strike of rock strata adjacent to the deposit, any geological differences in the deposit itself (e.g., part of it may be alluvial and part glacial), and the composition of any overburden that may cover the deposit.

5. Ground-water conditions. The water-table elevations in the deposit at various seasons and also the quality of the water. (Whether local water can be used for processing or water has to be imported, distance and cost of transportation in the latter case). Knowledge of the water-table depth is necessary to establish the type of equipment to be used to develop the deposit. Should excavation be done above the water table, dry-land equipment such as draglines (Fig. 8.14), power shovels, scrapers, or similar equipment may be used.* Excavations under water can be done by dredging or pumping. If raising or lowering the water table would improve the situation, investigations should be made to see if this is possible and, if so, how to do it. Disposal of waste water also should be studied; if adjacent streams and water basins cannot be used for this purpose, *spill ponds or settling basins* can be planned.

6. Volume available and physical properties, especially the gradation of the material. The volumes available above and below the water table should be separately estimated. Roughly 1 cu yd of material in place will provide 1 cu yd when placed in the structure, providing there is no excess waste material.† The amount of stripping should be estimated (e.g., 1,000 cu yd of overburden about 2 ft thick has to be stripped to expose 25,000 cu yd of suitable sand and gravel). The gradation of

---

* For information about excavating equipment, see Ref. 15. Dredging is discussed in detail in Refs. 16 and 17 and in Ref. 11, which also contains data on the organization of a sand and gravel plant.

† Compare also table, Sec. 16.4.

the material as disclosed by screen analyses should be given. If all sizes required in the concrete-mix design are not available, construction information should be given on possible sources of blending materials or other methods to make up the deficiencies.

FIG. 8.14. Dragline obtaining sand and gravel.   (*USBR photograph.*)

7. The methods used by the writer of the report in the exploration and recommendations for further exploration if required. The deposit can be sampled by shovel cuts, test pits, trenches excavated by a bulldozer, churn drills, large-diameter rotary drills, or drive sampling (Chap. 6). Other investigative methods such as airphotos, vegetative-cover interpretations, surface mapping, logging of test holes, etc., are discussed in detail in Chaps. 6 and 7. Figure 8.15 shows a typical cross section through one part of a potential sand and gravel source; the number of

such sections given in the report depends upon the irregularity of the stratification of the material.

8. If the aggregate is to be used for the manufacture of concrete, it is advisable to obtain service histories of nearby concrete structures that were made using the same aggregate. These records should incorporate, if possible, the age of the structure and the types of cement and mixing water used.

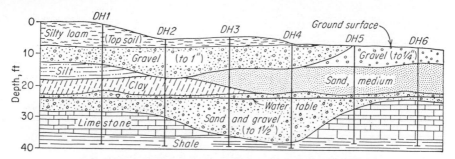

FIG. 8.15. Geologic cross section of sand and gravel deposit.

CONCRETE AGGREGATES

8.11. Terminology. The American Society for Testing Materials defines the term "aggregate" as follows:[18] "Aggregate—In the case of a material of construction, designates inert materials which when bound together into a conglomerated mass by a matrix forms concrete, mastic, mortar, plaster, etc."

In reality aggregates are not inert but are chemically and physically active and in many ways control the properties and the behavior of the mass into which they are incorporated. The term "mineral aggregate" is used to denote a rock product used as aggregate; hereafter, mineral aggregates used in the fabrication of portland cement concrete are discussed. *Coarse aggregate* used in concrete is crushed rock or natural rock fragments, such as gravel or pebbles, that are retained on a screen with $\frac{1}{4}$-in. holes. *Fine* aggregates are uniformly graded sand or screenings of crushed stone or gravel that pass through a $\frac{1}{4}$-in. screen. As discussed in subsequent sections, the engineering properties of natural rock fragments depend on their shape, size, surface texture, and the coatings of the particles.

8.12. Shape of Aggregates. The shape of natural rock fragments largely depends upon the presence and spacing of natural partings such as cleavage planes in minerals or joints in rocks. If natural partings in a mineral or rock are absent or few, the probability of breaking is equal in all directions, and equidimensional particles are produced. Quartz, for example, has no easy cleavage, and thus grains of quartz sand, both angular and rounded, are equidimensional. Rocks such as granite, marble, and quartzite originate more or less equidimensional, rounded pebbles (Fig. 8.16). Feldspars have two cleavage planes and develop tabular particles when broken. Pronounced stratification of a rock also results in the production of planar or elongated pieces in the aggregate.

FIG. 8.16. Typical round aggregate, 1 ×. (*Photograph by W. Y. Holland.*)

Thus, schists, slates, any rock containing a high percentage of mica, and shales commonly produce slabby or platy forms. Chiplike particles may be found in crushed, fine-grained, rather brittle, massive rocks such as chert or quartzite. Basalts tend to produce sharp, angular particles. Flat or elongated aggregate particles tend to decrease the workability of concrete, and thus more cement, water, and sand are needed. The workability of fresh concrete has been defined as "that property of the concrete mixture that determines the ease with which it can be placed, and the degree to which it resists segregation."[19] Sharp, angular fragments (Fig. 8.17) make a "harsh" mix, which increases the amount of cement required in concrete.

FIG. 8.17. Typical harsh aggregate, 1 ✕. (*Photograph by W. Y. Holland.*)

8.13. Size and Gradation of Aggregates.

The size of rock products, including sands and gravel, ranges from the finest particles to large boulders. The size used is regulated by specifications that establish the maximum and minimum aggregate sizes (Fig. 8.18). The natural material, if poorly graded, has to be processed (usually screened and sometimes washed), and all oversize and undersize particles are rejected. If certain sizes are lacking but are necessary, the aggregate should be blended as explained in Sec. 8.9.

The grading of the aggregate has a direct influence on the workability of the concrete mix. For example, a mix that contains a predominance of one size of sand or gravel will tend to have numerous voids between the aggregate particles. These

voids decrease the workability, because insufficient cement mortar may be present
to fill the void space. If the cement content is increased to compensate for this
difficulty, the water content also will have to be increased. Thus the final concrete
may be expensive or unsatisfactory from a strength standpoint. Occasionally, the
original source may have good grading, but owing to breakage during handling or to
excessive wear of the separation screens, undue amounts of both undersize and oversize material, respectively, may occur in the final mix.

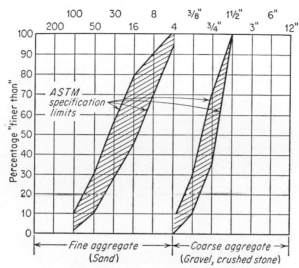

FIG. 8.18. Gradation of concrete aggregate.

8.14. Surface Texture of the Aggregate.

The periphery of a pebble
differs more or less from its interior. It may be rather soft and porous
because of leaching and alteration of original mineral constituents.
The surface also may be smooth and impervious to liquids. In a concrete
mix, a smooth pebble surface is not conducive to a good bond between
the pebble and the cement mortar. Figure 8.19 shows the comparatively
rough surface texture required for good bond in concrete.

The number of pores in an aggregate particle directly influences the
freezing and thawing durability, strength, elasticity, abrasion resistance,
specific gravity, rate of chemical alteration, and, as already mentioned,
the bond characteristics. If the pores are abundant and mostly less
than 0.004 mm in diameter,[21] reduced freezing and thawing durability
can be expected of the finished concrete. Because of capillarity, these
small pores readily sorb water and tend to remain saturated after being
enclosed by the mortar. Freezing will cause the enclosed water to
expand with a possible failure of the enclosing mortar. Scarcity of pores
contributes to the density of the rock, thus making it highly resistant to
abrasion, a factor which is particularly important if the concrete is used
for the construction of highway pavements. Quartz, quartzite, and

many dense basalts are used for making wear-resistant concrete. The specific gravity of the rock material is important if the specifications require that the concrete have a certain minimum or maximum weight. Specific gravity also serves as a quick identifier of poor aggregates. Weak and absorptive aggregate usually possess a low specific gravity.

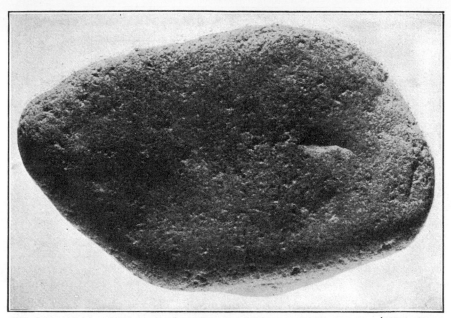

FIG. 8.19. Good surface texture for concrete aggregate—basalt, $1\frac{1}{2}$ X. (*USBR photograph.*)

In some atomic energy installations it was necessary to design dense, heavy-weight concrete.   For this purpose aggregates containing a high percentage of barium mineral were used (specific gravity of barite, the barium sulfate mineral of barium, is 4.5).

8.15. Soundness of the Aggregate. It is very important that the constituent particles of a concrete mix resist weathering—both chemical and physical.   Highly porous aggregates, particles that are easily broken, or those which may swell when saturated all indicate an aggregate that is easily deteriorated by weathering.   Such materials produce weak bond in concrete or induce cracking or spalling or even "popouts" (small blisters or holes in the surface) in the concrete.   Micaceous rocks, shales, friable sandstones, and clayey rocks are examples of such physically unsound aggregates.   Some very coarse crystalline rocks and some cherts also may be unsound for similar reasons.   The destruction of concrete by freezing and thawing of unsound aggregate, combined with poor concrete-mix design, is illustrated in Fig. 8.20.

8.16. Coatings on Aggregate. Subsequent to the deposition of a sand-gravel mass, its upper part, if exposed to the air, is subject to weathering just as an outcrop. At the same time, that part of the deposit below the water level is, because of precipitation, gradually covered with mineral matter. In this way, *coatings* at the pebble surface are formed. The effects of weathering and coating formation generally are very pronounced in terrace deposits. The mineral matter in the coating may be clay, silt, calcium carbonate, and, to a lesser extent, oxides, opals, chalcedony, and gypsum. The thickness of a

FIG. 8.20. Deterioration by freezing and thawing of concrete due to poor aggregate and mix design. (*USBR photograph.*)

coating ranges from a fraction of a millimeter to several millimeters. A coating rarely covers more than half of the surface of a pebble (Fig. 8.21) and generally is located on the bottom of the pebble (in its natural position in the pit). The coatings may be hard or soft and loosely or firmly bonded to the pebble surface. In the latter case, abundant coating material deposited by water may cause cementation of particles and may even form conglomerates such as caliche.

Gypsum and clay coatings detrimental to concrete often may be easily removed by screening and washing. Detrimental coating formation and cementation may affect only the top or the bottom layer (or both) of a sand and gravel deposit. This possibility should be considered when inspecting a sand and gravel deposit. Siliceous coatings (e.g., these are

FIG. 8.21. Coatings (white material) on aggregate, 0.8 ×.   (*Photograph by W. Y. Holland.*)

opal or chalcedony) are as detrimental to concrete as any other admixture of that nature (Sec. 8.18).

8.17. Physical Properties of Aggregate in a Concrete Mix.   These properties are discussed in many engineering books and articles,[22] and only some of the most important are briefly discussed hereafter.

Tenacity of the bond between the aggregate and the cement paste primarily depends on the structure of an aggregate particle and also on the substances filling the pores of the particle.   If these substances are soluble, they may be leached from the exposed surfaces of the concrete mass with a resulting decrease in concrete strength.

Volume changes in the concrete may be due either to high thermal expansivity of the aggregate or to water sorption.   In the former case, when the temperature decreases, tensile stresses may cause fracturing of the mortar or breakage of the bond, whichever is less resistant.

Clay minerals, especially those of the montmorillonite and illite groups, constituting admixtures or impurities in the aggregate structure expand if permitted to sorb water. Thus stresses in concrete far in excess of its strength may be created. Other expandable impurities would act similarly.

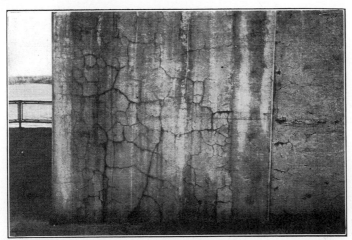

FIG. 8.22. Pattern cracking at Chickamauga Dam. (*Photograph by L. H. Tuthill.*)

FIG. 8.23. Deterioration of limestone riprap, Chickamauga Dam (pencil at center of photograph indicates scale). (*Photograph by L. H. Tuthill.*)

It is believed that expansion of the aggregate caused cracking of the concrete (Fig. 8.22) in some of the structures of the Chickamauga Dam near Chattanooga, Tennessee. The aggregate used in these structures was a carboniferous limestone containing as much as 13 per cent clay minerals. The rapid deterioration of this limestone under normal weathering also has been observed in some of the upstream riprap on the dam (Fig. 8.23).

A well-known instance of surprisingly rapid deterioration of concrete because of thermal expansivity is associated with the use of "sand-gravel" aggregate from the

Republican River, Kansas.[24] This aggregate contains high proportions of orthoclase and microcline feldspars with exceedingly low coefficients of thermal expansion in one crystallographic axis as compared with the other axes (Sec. 8.14).

8.18. Cement-aggregate Reactions. When concrete is setting and hardening, hydration of the cement takes place and *alkalies* (oxides of sodium and potash) are released. All *silicate and silica minerals* are attacked by these alkalies.

The results of this attack become very serious if the so-called "reactive" minerals (as listed in Table 8.1[25]) are present in the aggregate.

TABLE 8.1. Rocks Which Are Deleteriously Reactive with High-alkali Cements*

Reactive Rocks	*Reactive Component*
Siliceous Rocks:	
Opaline cherts.....................	Opal, $SiO_2 \cdot nH_2O$
Chalcedonic cherts.................	Chalcedony, SiO_2
Siliceous limestones................	Chalcedony and/or opal
Volcanic rocks:	
Rhyolites and rhyolite tuffs	
Dacites and dacite tuffs	Glass, devitrified glass, and tridymite, SiO_2
Andesites	
Metamorphic rocks:	
Phyllites........................	Hydromica (illite)
Miscellaneous rocks: any rocks containing veinlets, inclusions, coatings, or detrital grains of opal, chalcedony, or tridymite. Apparently, also quartz highly fractured by natural processes.	

* From D. McConnell *et al.*, Petrology of Concrete Affected by Cement-aggregate Reaction, "Berkey Volume."

If aggregate is manufactured from one of these reactive rocks, there may be expansion, cracking, and decrease in strength of the concrete. The modulus of elasticity also declines but fully or partially recovers. The mechanism of concrete deterioration from alkali aggregate reaction may be briefly explained as follows.[25] During the mixing of the aggregate, cement, and water and during the following several hours, the water acquires alkalies from the cement because of their solubility. The concentration of the alkalies in water gradually increases because some of the water is being used up in the hydration processes. The caustic liquid thus produced attacks susceptible aggregate particles, and *alkali silica gels* begin to form. The reaction proceeds rapidly for highly reactive materials, such as opal, and more slowly for less reactive ones, such as glassy volcanic rocks. The formation of gels and their imbibition of water produce *osmotic pressures* that distend and ultimately may rupture the paste near the reactive particle. The alkali silica gels fill the fractures.   The fissures gradually extend until they reach the surface of the concrete, and the cracks commonly exude gel.   Among the explanations, there is also an opinion[26] that sufficient lime has to be produced in the reaction to liberate the alkali from the alkali silicates in the aggregate.

Figure 8.24 indicates locations in the United States where deterioration of concrete as caused by the cement-aggregate reactions has been observed.   Figure 8.25 shows an open crack in the concrete of the

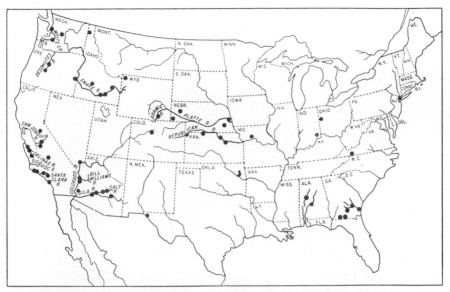

FIG. 8.24. Location of observed deterioration of concrete due to alkali aggregate reaction. (*USBR data.*)

Tuscaloosa locks, Alabama.   A considerable percentage of chalcedonic chert was used in this aggregate.[27]   Where the cement-aggregate reaction is severe, pattern cracking develops (Fig. 8.22) and white to colorless, extremely hard gel appears on the surface of the concrete.   Microscopic examination will then disclose a network of fine white lines intersecting the concrete in all directions, and occasional blobs of gel may be noted.

Research on alkali reaction in cement was initiated by Stanton in California[28] and has been successfully continued by others.[19-30]   It has been proved that troubles from the use of deleterious and reactive rocks for concrete aggregate may be predicted and possibly avoided if *preliminary petrographic analyses* of the aggregate and of the cement are made.   In this examination the number of particles containing reactive substances are determined by actual visual count.   The USBR has found[20] that aggregates containing more than 0.25 per cent by weight of opal, more than 5 per cent by weight of chalcedony, or more than 3 per cent of glassy or cryptocrystalline acidic to intermediate volcanic rocks or tuffs will cause deleterious reactions in the concrete unless cements low in alkali content are used. If any of these reactive minerals are enclosed or admixed with innocuous substances, the aggregate can probably be safely used. For instance, it may be safe to use a chalcedonic chert containing a considerable amount of calcium carbonate.

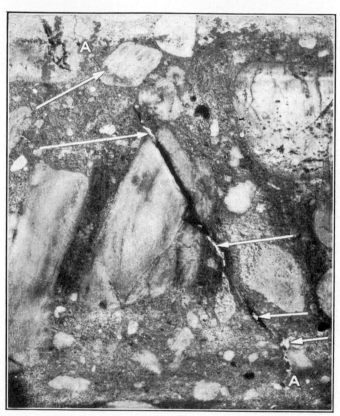

FIG. 8.25. Open crack (A–A) in concrete caused by the use of chalcedonic chert, Tuscaloosa locks, Alabama. (*Photograph by Waterways Experimental Station.*)

Besides the petrographic analysis, there are other methods to establish the reactivity of an aggregate. These are (1) a chemical test which involves dissolving the crushed aggregate in sodium hydroxide solution, NaOH, and determining the reduction in alkalinity of the solution; (2) measuring the change in length of small mortar bars made from the suspicious aggregate that have been stored for several months under moist condition, as reactivity, if any, should increase the length of the bars; (3) measuring the length change in concrete beams subjected to alternate wetting and drying with simultaneous heating and cooling; and (4) examination of concrete structures built using suspicious aggregate.[31]

Reactivity in concrete may be controlled by the use of *low-alkali cements*, containing less than 0.6 per cent of sodium and potassium oxides (American Federal specifications). The use of pozzolanic materials (Sec. 8.23) may prevent or reduce the alkali reaction. Occasionally, the incorporation of an "air-entraining" chemical compound into the concrete may increase the percentage of voids and thus provide space available for expansion.

8.19. Sulfides and Organic Substances in Concrete. The inclusion of sulfide minerals, such as pyrite and marcasite (both FeS_2), in an aggregate results in concrete deterioration. The sulfides are oxidized and then hydrated with an increase in volume when incorporated in concrete. The process is particularly detrimental in warm, humid regions[21] but may have only a minor effect in cold or dry climates.[29] The final results of this chemical deterioration are unsightly staining of the concrete surface and occasional popouts.

Organic substances are particularly detrimental because of the presence in them of organic acids. These acids inhibit the hydration of the cement, with consequent decrease in strength and durability of the concrete.[20] Coal is objectionable, as it decreases the freezing and thawing resistance of concrete and its strength is inherently low.

8.20. Sulfates and Sea Water in Concrete. The action of various chemical salts such as sulfates, chlorides, carbonates, etc., depends upon the amount of these salts present and the type of cement used. Sulfates may cause expansion and disintegration of the concrete by their reaction with the cement, but generally their action can be inhibited by the use of special cement.[20] Many of the contaminating agents listed may be removed by washing; however, care has to be taken that some of them do not dissolve in the mixing water during the manufacture of the concrete.

The use of sea water in making concrete and its effect on concrete structures are somewhat controversial. Some believe that the use of sea water for mixing water has no detrimental effect other than accelerating the setting of the concrete.[32] Conversely, others believe that mixing water containing high percentages of chlorides and sulfates (constituents of sea water) is very detrimental to concrete.[20] The action of sea water on concrete structures in some cases has caused deterioration owing to the reaction between hydrated cement and the sulfates and/or other chemicals in the sea water.[33] On the other hand, it is believed that much of the deterioration observed in concrete in contact with sea water was initially caused by alkali aggregate or some other reaction in the concrete, the resulting cracks providing access for attack of the concrete by the sea water.[34] In fact, with dense concrete free of cracks, sea water practically has no detrimental effect.

8.21. Thermal Effects on Aggregate in Concrete. *Small* temperature variations may cause expansion and occasional fissuring of concrete (Sec. 8.17), which are purely physical phenomena. *Intense* heat causes both physical and chemical deterioration of concrete. Because of the variability of the coefficients of expansion in different directions, concrete may fail purely in tension. Intense heat may actually decompose the aggregate, creating new chemical compounds. For example, it appears that the intense heat exerted by jet engines on concrete pavements of airfields decomposes the limestone in the aggregate.[35]

This same problem arises in fireproof building construction.[36] In tests by the Bureau of Standards, it has been found that the type of aggregate has a definite effect on the fire resistance of the finished concrete or of finished plaster. Siliceous aggregates which contain large quantities of such siliceous minerals as quartz and chert result in a concrete or a building block that is subject to severe damage at high temperature. At temperatures as low as 410°F (210°C) for chert and 1063°F (573°C) for quartz, abrupt volume changes occur that lead to destructive high stresses in the finished wall or floor.

Aggregates composed mainly of calcareous minerals are less subject to damage. Under intense heat, calcination occurs which produces a material that is a good insulator. The latter retards the transfer of heat from the side of the wall that is exposed to the fire to the unexposed side.

Tests showed that a wall composed primarily of siliceous aggregates collapsed 58½ min after exposure to fire, whereas the calcareous aggregate wall withstood the fire for 1 hr and 51 min.[36] In these tests, the calcareous gravels contained over 80 per cent calcareous minerals, the sands contained over 50 per cent calcareous minerals, and the siliceous aggregate contained over 90 per cent quartz with small amounts of feldspar, mica, and clay.

8.22. Artificial Aggregates. Man-made aggregates may be composed of such materials as air-cooled blast-furnace slag or fused loess. The latter material is produced by fusing silty and clayey loesses in a high-temperature electrical kiln.[37] This process, although still in the pilot-plant stage, promises low-cost aggregates (and possibly riprap, Fig. 8.26) for those areas short in natural aggregates. The blast-furnace slags are used mainly for road and railroad construction.

8.23. Pozzolanic Materials.[38] Pozzolans are a very fine-grained natural and sometimes artificial material used in combination with portland cement to make concrete. Originally volcanic ashes near the town of Pozzuoli in Italy were used for this purpose, hence the name. Natural pozzolanic materials are volcanic tuffs, ashes, siliceous sedimentary rocks (particularly opaline shales, cherts, and diatomaceous earth), and clays and shales calcined at temperatures of 900 to 1800°F. Industrial wastes such as fly ash (collected from smoke residue in manufacturing plant chimneys) also are used. All these materials possess forms of active silica or alumina which, though not cement in themselves, combine with lime or portland cement to make a stable cemented compound.

Pozzolans may retard or prevent alkali aggregate reaction,[39] reduce the generation of heat caused by the hydration (setting) of the concrete in a massive structure, increase the tensile strength of the concrete, improve the workability of the mix,[39] and lower the cost of the concrete in many cases (as the pozzolan may be much cheaper than the cement

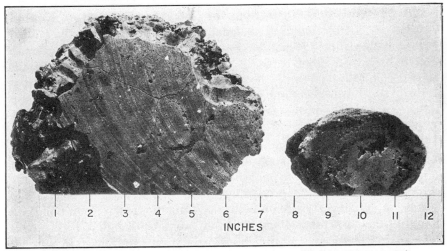

FIG. 8.26. Aggregate made by fusing loessial soils.   (Photograph by USBR.)

for which it partially substitutes). The material has some disadvantages such as increase in the volume of water required for concrete or relatively high drying shrinkage. Before a pozzolan is used, laboratory tests and field observations should be performed and the economics of its use considered. Search for pozzolans is simplified if the aggregate to be used is nonreactive or has a low alkali reaction or if low-alkali cements are used. As with other types of quarries, it rarely proves economical to develop a pozzolan deposit that is sufficient to supply only one engineering structure under construction.

8.24. Lightweight Aggregates. These materials may be natural or artificial, and their use provides concrete that is lighter in weight than when the usual rock and gravel aggregates are used. Usually it is not economically desirable to use lightweight aggregates unless normal bulk weight of rock and gravel aggregates (about 100 pcf) can be reduced to about 50 pcf by the addition of the lightweight material.[40]

Lightweight aggregate materials are shale, clay, slag, pumice, diatomaceous earth, perlite (a volcanic glass), expanded vermiculite (a form of mica that expands to many times its original volume when heated because of its high water content), slate, and scoria (a volcanic deposit). Consultation of Refs. 40 to 42 will aid the search for these materials.

Lightweight concrete ranges from 35 to 115 pcf, and its strength ranges from 200 to 5,000 psi depending upon the type of aggregate used.[41] In most cases, lightweight aggregate concrete is more expensive than normal-weight concrete because more cement is needed in the mix, the initial cost is higher as compared with normal weight aggregates, and special care is needed in placing the concrete. Some of the uses for lightweight aggregate concrete are (1) to reduce the over-all weight of a reinforced concrete structure and thus achieve economy in design of the steel and columns and (2) to provide a concrete with special insulating qualities that will reduce frost penetration from the outside and loss of heat from the inside of a heated structure.

8.25. Sampling and Prospecting for Concrete Aggregate. Natural deposits of concrete aggregates seldom exist in a conveniently graded condition; therefore the sampler must be a keen observer and exercise good judgment in order to select a *representative* sample of a deposit. In performing the field work, the sampler must clearly realize the particular tests which will be run upon the sample. The practice of sending any convenient laborer to obtain an aggregate sample for testing should be discouraged.

Two kinds of sampling are done: (1) sampling to investigate if the material will meet certain specifications, which is done when acquisition of aggregate or a property containing aggregate is considered, and (2) periodic sampling in the processing plant to check the quality of the product at all stages of manufacturing. Instructions for sampling are changed from time to time and therefore are not presented in this book. It is advisable to consult the latest ASTM publications (see Supplemental Reading at end of this chapter). Sampling for the *quality of the aggregate* should be done in place, i.e., in the quarry or in the sand-gravel bank. The walls and the floor of the quarry are examined for different varieties of rock, and their percentages estimated. In a natural sand-gravel bank, sampling should be done at different points spread in plan and at different depths at each of these points. In all samples, tags should be firmly attached to the sample container and a duplicate of the tag placed inside. These tags should give the sample location, both geographically and within the quarry, and the depth from the ground surface (or better, the elevation) at which the sample was secured. Before a prospective aggregates source is rejected, it should be remembered that some deficiencies of a deposit may be corrected by blending or washing the aggregate and sometimes by careful selective excavation.

8.26. Terazzo Aggregates.[43] Terazzo floors usually are made by mixing one part of portland cement with two parts of crushed rock by volume. The hardened concrete, or terazzo, is then ground by polishing wheels to a high polish. The aggregate usually varies from $\frac{1}{8}$ to $\frac{1}{2}$ in. in size. The color is of particular importance, as terazzo commonly is used for decorative interior and exterior floors. The various types of rocks used for terazzo aggregate are marble, dolomite, and mixtures of marble and dolomite called dolomitic marbles in construction practice.[43] Green floors, highly resistant to abrasion, are made from serpentine. Serpentine varies in abrasion resistance, however; the material located close to the ground surface or damaged by water penetrating through open joints is weaker.

HIGHWAY AND RAILROAD AGGREGATES

8.27. Highway and Runway Aggregates. The rock used in a highway or runway pavement is subject to the *abrasion* caused by horizontal forces at the leading wheels of an automobile or the landing wheels of an aircraft. The weight of standing vehicles produces *compression* in the material of the pavement, and additional *impact* stresses produced by moving vehicles may be estimated at from 25 per cent and up of the static weight. The rougher the pavement, the greater the impact. The resistance of rock to impact is a measure of its *toughness*.

The upper portion of a pavement made generally of portland cement concrete or asphaltic concrete (symbol often used is "AC") may be impervious, whereas the lower portion made of aggregate only has to satisfy the requirements of permeability; otherwise, pore pressures may develop within the pavements. These pressures tend to heave and crack the pavement, especially if the subgrade or the ground surface on which the pavement is placed is impervious. The aggregate should be tough enough not only where it is exposed to the traffic but also in the lower portion of the pavement. This is necessary to prevent dust formation and hence an increase in capillarity, which may result in severe frost damage in cold climates (Sec. 10.2).

Classification. The aggregates used in highway-runway construction generally are classified as (1) load-bearing aggregates, (2) binders, and (3) void-filling aggregates.[45] The *load-bearing aggregates* carry the traffic loads and transmit them to lower courses or to the subgrade materials. The *binders* serve merely to bind together the load-bearing materials. These are powderlike materials possessing binding properties if mixed with water. The mass is made cohesive by thin water films around each particle.[45] Clay, crusher dust, limestone dust, various chemicals, etc., may serve as binders.

Void-filling materials are also powderlike substances which may or may not have binding properties. Limestone dust, ground silica, hydrated lime, slate dust, soapstone dust, etc., may be used as void-filling materials.

Load-bearing aggregates may be composed of crushed rock, crushed or uncrushed gravel, sandstone fragments, sand, slag, cinders, caliche, shell, iron ore, etc. The best rock types are usually known in practice as "traprocks"—basalts, andesites, diabase, diorite, etc.[45] Granites are generally not desirable because their abrasion resistance is relatively low and they are difficult to crush (because of the interlocking texture).[49] The traprocks, on the other hand, are naturally blocky and brittle enough to crush easily in mechanical crushers (but not under road traffic) conditions.

Besides the general requirements of abrasive resistance and toughness, the load-bearing aggregates should be stable in fill without addition of cement paste. For this purpose, the aggregate should be angular, without an excess of flat, thin pieces of shale or slate, and resistant to alternate freezing and thawing; should have a low coefficient of expansion; and should be chemically inert.[46] In each particular case, gradation of the aggregate is regulated by specifications.[47] If *limestone* is used as load-bearing aggregate,[48] it should contain at least 70 per cent by weight of calcium and magnesium carbonates, the remainder being silica and iron and aluminum oxides. Sandstone generally has a low wearing value but occasionally may be used in the lower courses of the pavement, where it is far from the immediate effect of the abrasive forces. If crushed volcanic rocks, such as basalt, are used for highway aggregate, they should not be chemically weathered. This condition often can be identified by the presence of an oillike, blue-purple stain across a brown rock. Basalts extracted from below water are seldom usable because the high percentage of water-soluble alteration products they contain rapidly contributes to their deterioration.

Aggregates for Bituminous Mixes.[49–52] Rocks containing a high percentage of silica minerals (acid-type rocks) generally are unsuitable for mixing with asphalt or other bitumens in a wearing course. Such rocks are "hydrophilic," i.e., have a high water-sorption ability but a low bitumen sorption. Rocks high in the ferromagnesian minerals (e.g., basic rocks such as basalt) have a high affinity for bitumen and are termed "hydrophobic." Those rocks which do not stay mixed with the bitumens cause a condition in the pavement known as "stripping," i.e., peeling away of the bitumen from the aggregate. Basalts are generally excellent aggregates to be used in bituminous mixes, but if they contain minerals that have deteriorated into clay, stripping may occur.[49] Although, as previously mentioned, sandstones generally are not suitable for wearing courses, under certain conditions of gradation they have performed satisfactorily in bituminous pavement.[53]

8.28. Testing Highway-Runway Aggregates. Field examination of these aggregates should include determination of the rock type; the presence of accessory minerals such as clays, calcium carbonate, and micas; the structure and texture in order to predict crushing properties and resistance to abrasion; the amount of alteration; and the susceptibility of the rock to cracking, shrinking, expansion, and parting.[54] Petrographic studies are invaluable, as they detect the presence of such detrimental minerals as feldspars, which under proper climatic conditions will deteriorate into impervious clays. Laboratory tests are performed to determine the abrasion, impact and toughness, freezing and thawing resistance, compression strength, and sorption characteristics.

Abrasive resistance is tested by the Los Angeles abrasion test (ASTM Designation C131—51). A charge of 5 kg of aggregate together with a specified number of cast-iron spheres, varying according to gradation, is loaded in a steel drum with an interior projecting shelf. The drum is rotated for 500 revolutions at 30 to 33 rpm. The wear

TABLE 8.2. Hardness, Abrasive Coefficient, Impact Toughness*

Kind of rock	Sclero-scope hardness	Standard deviation, %	Abrasive coeff.	Avg. toughness, in./sq in.	Avg. standard deviation, %
Granite................	95	7	18.0	4.9	19
Greenstone............	81	10	20.0	6.5	17
Limestone.............	27	22	2.6	1.9	13
Marble................	56	9	7.6	2.7	17
Sandstone.............	31	23	1.5	1.8	9
Slate.................	56	9	3.3	3.7	17

* L. Obert, S. L. Windes, and W. I. Duvall, Standardized Tests for Determining Physical Properties of Mine Rock, *USBM Rept. Invest.* 3891, 1946.

is expressed in per cent of the original weight. In the *Deval abrasion* test (ASTM Designation D289—46 and D2—33), a 5-kg charge is placed in a sealed bucket inclined at 30° to the horizontal shaft and rotated for 10,000 revolutions at about 30 rpm. The percentage of wear is computed as in the preceding test.

Toughness (ASTM Designation D3—18) is measured on samples about 1 in. in height and 1 in. in diameter. The test consists in dropping a standard hammer on the sample. The height of fall starts at 1 cm and is gradually increased by 1 cm per blow until a fracture occurs. The height of the blow at failure, as measured in centimeters, is the measure of the toughness.

Considerable research work on abrasion was done by the Bureau of Public Roads, Bureau of Standards, and Bureau of Mines[55] with the Dorry machine, initially built by the Ecole des Ponts et Chaussées in Paris. Other research to be noted is that done on hardness measurement. As is well known (Sec. 1.2), hardness is measured by scratch resistance according to Mohs scale. A number of devices have been proposed and used in industry for the same purpose,[55] including the Shore scleroscope. This instrument measures hardness by the height of rebound of a diamond-pointed hammer dropped vertically on the test surface.

The data given in Table 8.2 show the comparative durabilities of various rock types: The higher the number in the hardness, abrasive coefficient, or toughness column, the more durable is the rock.

8.29. Railroad Ballast. Ballast is the natural or artificial material that supports the ties under a railroad track and transfers the train loads to the subgrade. Generally ballast is composed of crushed stone, crushed air-cooled blast-furnace slag, crushed or uncrushed gravel, or cinders.[56] The type of the ballast and the thickness of the ballast layer depend very much on the type and amount of the traffic and, of course, on economic considerations. In addition, the drainage conditions of the subgrade, climatic conditions (e.g., protection of the subgrade from freezing and heaving), and the bearing properties of the subgrade are to be considered. Ballast must be elastic in order that the ties and rails can return back to true line and grade after the passage of the train. Ballast also has to reduce dust and weed growth.[57] The gradation of ballast should permit satisfactory cleaning. This is done by special machines that pick up the ballast from under the ties; run it through screens to remove fines, dust, and other injurious matter; and dump it back near the ties, where it is hand-tamped in place. Crushed stone is the best ballast type from this point of view. Because of the heavy, suddenly applied loads imposed by engines, ballast aggregate must be particularly resistant to impact. Under a moving series of cars, the ballast particles move back and forth and "work" against one another. To resist the wear, a good abrasive resistance is needed. Glassy or brittle rocks, e.g., those containing considerable amounts of quartz and mica, are not suitable for ballast because of their poor impact resistance. Dolomites and generally rocks containing soluble material are susceptible to erosion by rain water and also are not suitable for ballast. (Magnesium carbonite or dolomite is four times more soluble by rain water than calcium carbonate or calcite.) Sandstones containing quartz cemented with silica (of the quartzite type, Sec. 1.10) and a few mica flakes are excellent sources of ballast. *Cinders* generally are used only on branch lines with light traffic, on sidings, on yard tracks, and on new embankments that are still settling. On wet, spongy subgrades, a subballast of cinders, 12 in. or more thick, may prevent mud and similar materials from working up into the top ballast.[56] Cinders also may be used for replacing frost-susceptible soil material.

The dry density (dry unit weight) of ballast should be not less than 70 pcf after compaction, shrinkage due to compaction being from 10 to 15 per cent by volume.[57] For good gradation the material should be between No. 8 sieve and 4-in. screen sizes. These numerical data should be considered as general only.

FACING STONE

8.30. Stability of Rock Facing. Concrete and reinforced concrete walls are often faced with stone, making them more or less pleasant to the eye and safe against rain-water and atmospheric-gas action. The most widely used facing-stone materials are granite and limestone.

Granite. The rock should be split into veneer stones (slabs) about 2 to 4 in. thick. Generally granites cannot be split to a thickness less than about one-third of the smallest face dimension. Granite that is to be used as a veneer should not be overdried but should retain some quarry water; otherwise it becomes very tough and hard to fabricate. When masonry or steel is faced, cracks may appear in the facing owing to the tensile stresses caused by the different coefficients of expansion of the different materials. Therefore, rigid fixing of veneer stones to the structure being faced should be avoided, and a certain play of the facing allowed by using so-called "relieving joints."

Limestone. Weathering resistance of limestone facing in some buildings often has been questioned. In the majority of cases, such misgivings are unfounded, particularly in the case of the best limestones such as the well-known Indiana limestone. According to Loughlin[58,59] laboratory experiments and computations based on them show that in an atmosphere uncontaminated by excessive smoke, the surface limestone exposed to a rainfall of 40 in./year would be dissolved to a depth of 0.4 in. in 450 years. Weathering conditions in large cities are somewhat comparable, as observations in New York City indicate that limestone weathers about $\frac{1}{16}$ in. in 50 years (more specifically about 0.57 in. in 450 years). Limestones with uneven grain surfaces were affected more by weathering than those with even surfaces.

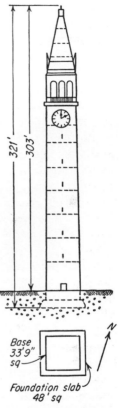

FIG. 8.27. Sather Tower (Campanile), University of California at Berkeley.

If city atmospheres are known to contain excessive quantities of acid, such as sulfur compounds in smoke, it is advisable to use granite or a sandstone that does not contain calcite instead of limestone. Capillary rise of water from the ground affects limestones more than other stones; therefore, direct contact of limestone with the earth surface should be avoided.

Porous stones such as limestone or sandstone have been used in the submerged parts of bridge supports. The use of limestone is objectionable if water in the stream has the capacity of dissolving calcium carbonate (e.g., streams flowing toward the Atlantic usually have this capacity). The situation is improved if the stream already is saturated with lime and does not flow excessively fast (e.g., many streams in the Mississippi Valley[58,59]). Occasionally stones are used to face bridge piers. In such cases, because of their great weight, they have to be affixed to the underlying concrete pier with steel rods.

FIG. 8.28. Campanile shown in Fig. 8.27 under construction in 1914.

8.31. Failure in Building Facing. Distress in the facing stones of a building may, without cause, be attributed to foundation settlement. The factors discussed in Secs. 8.2 through 8.4 should be carefully considered when analyzing cracks in facing stones. The following example illustrates such an analysis.

Sather Tower (Campanile), University of California, Berkeley campus,[60] was built in 1914. The basic dimensions and orientation of this structure are shown in Fig. 8.27. The granite-facing stones are, in general, 20 in. high, 6 to 8 in. thick, and 5 to 7 ft long. They are placed on the 6- to 8-in. edge and set in ¼-in.-thick cement mortar joints. The facing is anchored to a reinforced concrete backing at least 6 in. thick, and the walls are carried by a substantial steel frame (Fig. 8.28). The foundation slab is 45 in. thick and lies on a grillage of 24-in. 80-lb I beams. This grillage rests on a stratum of boulders and hard, gravelly clay. The cracking that gradually developed cannot be attributed to the foundation because of its obvious competence. The tower is designed as an earthquake-proof structure, and in fact, earth shocks did not produce any cracks in it. The general crack pattern is shown in Fig. 8.29.

It is the opinion of many stonecutters that granite, like marble and limestone, becomes harder and more brittle if exposed to air; this perhaps would explain why cracks were not observed in the first decade of the existence of the tower.

Perfect verticality of many of the cracks suggests the action of horizontal tensile stresses. These probably are caused by the restraint imposed on the expansion and contraction of the granite by the "dogs," i.e., the anchors fixing the facing to the structural frame. This anchoring, during construction, was exceptionally snug in every direction, the anchors being hot-bent on the job to fit into holes drilled into the

FIG. 8.29. Cracks on the Campanile shown in Fig. 8.27.

granite.[61] Thus, the differences in the coefficients of expansion of the quartz and feldspar in the granite (as caused by daily thermal gradients) exerted a tensile force that could be relieved only by destruction of the stone.

REFERENCES

1. "Mineral Yearbook," U.S. Bureau of Mines, 1952; also Crushed and Broken Stone in 1952, *USBM Mineral Market Rept.* MMS. No. 2285, May 20, 1954.
2. Buoyoucos, C. M.: *Mich. Agr. Coll. Expt. Sta. Rept.* 31, 1916, and 36, 1917.
3. Kessler, D. W., H. Insley, and W. H. Sligh: Physical, Mineralogical, and Durability Studies on the Building and Monumental Granites of the Unites States, *Natl. Bur. Standards, J. Research*, paper 1320, vol. 25, 1940.

4. "Johnson's Materials of Construction," 8th ed., John Wiley & Sons, Inc., New York, 1939.

5. von Moos, A., and F. de Quervain: "Technische Gesteinkunde," Birkhauser Verlag, Basel, 1948.

6. Birch, F.: Handbook of Physical Constants, *GSA Spec. Paper* 36, 1942.

7. "Earth Manual," U.S. Bureau of Reclamation, Denver, Colo., 1953, rev. 1956.

8. Bowles, O.: "The Stone Industries," McGraw-Hill Book Company, Inc., New York, 1939.

9. The U.S. Bureau of Mines, at irregular intervals, issues *Information Circulars* and *Reports of Investigations* containing information on current methods of drilling in quarries, including costs, types of bits, types of drills, spacing of holes, etc.

10. "Blasters Handbook," 13th ed., E. I. du Pont de Nemours & Company, Inc., Wilmington, Del., 1953.

11. "Pit and Quarry Yearbook," Complete Service Publication Co., Chicago Ill. Contains full information on stone and aggregate industry.

12. Seismic Effects of Quarry Blasting, *USBM Tech. Bull.* 442, 1942.

13. Davies, G. G.: "The Pneumonokonioses (Silicosis)—Bibliography and Laws," Industrial Medicine, Inc., Chicago, 1934.

14. Drinker, P., and T. Hatch: "Industrial Dust," 2d ed., McGraw-Hill Book Company, Inc., New York, 1936.

15. Catalogues of Manufacturing Firms:

15a. Caterpillar Tractor Co., B. G. Le Tourneau, Marion Co., Bucyrus-Erie, Allis-Chalmers, and others.

15b. Advertisements in *Engineering News-Record, Construction Methods, Public Works, Roads and Streets, Civil Engineering*, and others.

15c. Technical manuals of the War Department, especially Use of Road and Airdrome Construction Equipment, TM 5-25, 1945; also Grading, Excavating, and Earth Moving Equipment, TB 5-9720-1, 1944; also Aviation Engineers, TM 5-255, 1944.

16. Shankland, E. C.: "Dredging of Harbors and Rivers," Brown, Son & Ferguson Ltd., Nautical Publishers, Glasgow, 1949.

17. Bailey, S. C.: Types of Modern Dredging Plants, *The Dock and Harbor Authority* (British), April, 1951.

18. Book of ASTM Standards, part II, p. 1408, C58—28T, 1946.

19. Powers, T. C.: Workability of Concrete, Report on Significance of Tests on Concrete and Concrete Aggregates, 2d ed., *ASTM Spec. Publ.* STP 22A, 1943.

20. "Concrete Manual," 6th ed., U.S. Bureau of Reclamation, Denver, Colo., 1955.

21. Litehiser, R. R.: The Effect of Deleterious Materials in Aggregates for Concrete, *Natl. Sand and Gravel Assoc. Circ.* 16, 1938.

22. Blanks, R. F.: Modern Concepts Applied to Concrete Aggregate, *ASCE Trans.*, vol. 115, 1950; also with H. F. Meissner: Deterioration of Concrete Dams due to Alkali-aggregate Reaction, *ibid.*, vol. 111, 1946.

23. Scholer, C. H., and W. E. Gibson: Effect of Various Coarse Aggregates upon the Cement-aggregate Reaction, *Proc. ACI*, vol. 44, 1948.

24. Tuthill, L. H.: Inspection of Dams in Service, *USBR Concrete Lab. Rept.* C-356, June, 1947.

25. McConnell, D., *et al.*: Petrology of Concrete Affected by Cement-aggregate Reaction, "Berkey Volume."

26. Hester, J. A., and O. F. Smith: Alkali-aggregate Phase of Chemical Reactivity in Concrete, *Highway Research Board Proc. 32d Annual Meeting*, pp. 306–317, 1953.

27. Mather, B.: Cracking of Concrete in the Tuscaloosa Lock, Alabama, *Highway Research Board Proc.*, vol. 31, 1951.

28. Stanton, T. E., *et al.:* California Experience with the Expansion of Concrete through Reaction between Cement and Aggregate, *Proc. ACI*, vol. 38, 1942.
29. Rhoades, R., and R. C. Mielenz: Petrography of Concrete Aggregate, *Proc. ACI*, vol. 42, 1946; also Mielenz, R. C.: Petrographic Examination of Concrete Aggregates, *Bull. GSA*, vol. 57, 1946.
30. Folk, R. L.: Petrology of Lower Ordovician Chert in Central Pennsylvania, *Bull. GSA*, vol. 60, 1949.
31. Mielenz, R. C., and L. P. Witte: Tests Used by the Bureau of Reclamation for Identifying Reactive Concrete Aggregates, *ASTM Proc.*, vol. 48, 1948.
32. Dempsey, J. G.: Coral and Salt Water as Concrete Materials, *Proc. ACI*, vol. 23, no. 2, pp 157–166, October, 1951.
33. Terzaghi, R. D.: Concrete Deterioration in a Shipway, *J. ACI*, vol. 19, June, 1948.
34. Stanton, T. E.: Durability of Concrete Exposed to Sea Water and Alkali Soils, etc., *J. ACI*, vol. 19, no. 9, May, 1948.
35. Mellinger, F. M., and B. V. Duvall: The Corps of Engineers Approach to Pavements for Jets, Symposium on Airfield Pavements for Jet Aircraft, U.S. Navy Bureau of Yards and Docks, Point Hueneme, Calif., 1952.
36. Foster, H. D., E. R. Pinkston, and S. H. Ingberg: Fire Resistance of Walls of Gravel-aggregate Concrete Masonry Units, *U.S. Dept. Comm. Bldg. Material and Structural Rept.* 120, March, 1951.
37. Plummer, N., and W. B. Hladik: The Manufacture of Ceramic Railroad Ballast and Constructional Aggregates from Kansas Clays and Silts, *Kansas State Geol. Survey Bull.* 76, part 4, Lawrence, Kans., 1948.
38. Mielenz, R. C.: Materials for Pozzolan: A Report for the Engineering Geologist, *USBR Lab. Rept.* Pet-90, September, 1948.
39. Bollen, R. E., and C. A. Sutton: Pozzolans in Sand-Gravel Aggregate Concrete, *Highway Research Board Proc.* 32d *Annual Meeting*, 1953.
40. Conley, J. E., *et al.:* Production of Lightweight Concrete Aggregates from Clays, Shales, Slates, and Other Materials, *USBM Rept. Invest.* 4401, November, 1948.
41. Price, W. H., and W. A. Cordon: Tests of Lightweight-aggregate Concrete Designed for Monolithic Construction, *J. ACI*, vol. 20, no. 8, 1949.
42. Lightweight Aggregate Concrete, Housing and Home Finance Agency Report, Government Printing Office, August, 1949.
43. Kessler, D. W., A. Hockman, and R. E. Anderson: Physical Properties of Terazzo Aggregates, *U.S. Dept. Comm. Bldg. Material and Structural Rept.* BMS 98, May, 1943.
44. Compaction of Embankments, Subgrades, and Bases, *Highway Research Board Bull.* 58, 1952.
45. "Manual of Highway Construction Practices and Methods," American Association of State Highway Officials, 1950.
46. Woolf, D. O.: Results of Physical Tests of Road-building Aggregate, *Bur. Public Roads Publ.*, 1953.
47. Chastain, W. E., Sr.: Performance of Concrete Pavement on Granular Subbase, *Highway Research Board Bull.* 52, 1952.
48. Thoréen, R. C.: Road Building Limerocks, *Public Roads.*, vol. 16, 1935.
49. Scott, L. E.: Source and Selection of Mineral Aggregates, *Pacific Bldr. and Engr.*, vol. 59, no. 5, 1953.
50. "Specifications for Constructing Bituminous Surfacing," American Association of State Highway Officials, 1947.
51. Tyler, O. R.: Adhesion of Bituminous Films to Aggregates, *Purdue Univ. Eng. Bull.*, vol. 22, no. 5, 1938.

52. Endersby, V. A., R. L. Griffin, and H. J. Sommer: Adhesion between Asphalts and Aggregates in the Presence of Water, *Proc. Assoc. Asphalt Paving Technol.*, vol. 16, 1947.
53. Young, J. L., Jr., and L. E. Gregg: Geologic Considerations in Relation to a Materials Survey, *Highway Research Board Bull.* 62, 1952.
54. Maddalena, L.: Choix des matériaux les plus aptes pour le ballast, *Congr. intern. mines, mét. et géol. appl.*, Paris, vol. II, 1935; trans. in *USGS Trans.* 1, 1946.
55. Obert, L., S. L. Windes, and W. I. Duvall: Standardized Tests for Determining Physical Properties of Mine Rock, *USBM Rept. Invest.* 3891, 1946.
56. "Manual for Railway Engineering," American Railway Engineering Association, Chicago, Ill. Loose-leaf form with regular revisions; 1953–1954 copy was used.
57. Raymond, W. G., H. E. Riggs, and W. C. Sadler: "Elements of Railroad Engineering," John Wiley & Sons, Inc., New York, 5th ed., 1937.
58. Laughlin, G. F.: Indiana Oolitic Limestones—Relation of Its Natural Features to Its Commercial Grading, *Contrib. Econ. Geol.*, pt. 1, 1929; also *USGS Bull.* 811, 1930.
59. Laughlin, G. F.: Notes on the Weathering of Natural Building Stones, *ASTM Proc.*, vol. 31, 1931.
60. Cape, E. L.: An Earthquake-proof Tower, *Eng. News-Record*, vol. 69, 1914.
61. Steilberg, W. T., consulting architect, University of California, personal communication.

Supplemental Reading on Concrete Aggregate

62. Symposium on Mineral Aggregates, *ASTM Spec. Tech. Publ.* 83, chap. 8, 1948.
63. Rhoades, R.: Influence of Sedimentation on Concrete Aggregates, "Applied Sedimentation," John Wiley & Sons, Inc., New York, 1950.
64. 1946 Book of ASTM Standards, part II, pp. 466, 467, 470, 1400, 1437; and part IIIA, p. 11.
65. Supplement to Book of ASTM Standards, part II, pp. 114, 117, 272.

Comprehensive Reference on Explosives and Blasting

66. K. H. Fraenkel (ed.): "Manual on Rock Blasting," 2 vols., Aktiebolaget Atlas Diesel, Stockholm, and Sandvikens Jernverks Aktiebolag, Sandviken, Sweden. In English, French, German, and Swedish, loose-leaf form with annual supplements, 1952 through 1956.

CHAPTER 9

TUNNELS

Tunnels, other than mine tunnels, are essentially an element of transportation. In the country, instead of deep cuts being constructed, tunnels often are used to conduct the line under a natural obstacle such as a hill or ridge. In cities, tunnels carry underground railroads and highways which, because of the complexity of traffic problems, cannot be built above the ground. Even city streets or portions of streets are sometimes underground. Besides passengers and freight, tunnels also carry fluids. For example, there are water supply and sewage disposal tunnels and water tunnels used in producing hydroelectric energy. Subaqueous tunnels (constructed under water) replace bridges, e.g., the Hudson Tunnel and the Lincoln Tunnel in New York City. Another form of underground construction that utilizes many of the tunneling techniques, although technically not classified as tunnels, is the construction of underground rooms and passageways (Sec. 13.17).

TUNNEL CLASSIFICATION AND NOMENCLATURE

9.1. Technical Classification. Tunnels can be driven, or "holed," through a rock or earth mass by methods used in *mining*, including blasting. This can be done either in normal air or using compressed air. In rare cases, mining operations in soft ground such as quicksand may be carried on by applying artificial freezing or chemical stabilization of the ground. If mining with an open front becomes exceedingly difficult or uneconomical, the *shield method* can be used (Sec. 9.18). Figure 9.1 represents a case of tunneling with open front in plastic clay.

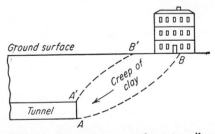

FIG. 9.1. Case when open-front tunneling becomes inadequate.

The clay creeps into the tunnel and causes settlement of the earth surface

347

and thus endangers adjacent structures (*B* in Fig. 9.1). In such cases, it usually is advisable to change from the open-front to the shield method.

In shallow cover and soft ground, the tunnel may be constructed by the *cut-and-cover method*. In this case, an open trench is excavated, the tunnel section (concrete, steel, or timber) is constructed, and the tunnel then is covered or backfilled. Most of the London subway (the so-called "underground"), the Toronto subway, and part of the Buenos Aires subway were constructed in this manner. In some cases with shallow cover, the ground surface may be submerged; therefore the tunnel can be constructed by using a *cofferdam* (Sec. 14.5) and then covered and back-filled. In particularly critical underground conditions, vertical *caissons* (Sec. 14.6) are sunk and the tunnel is built in short sections.

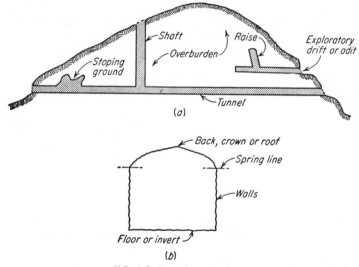

FIG. 9.2. Tunnel terminology.

9.2. Terminology. Tunnel terms were originally borrowed (by civil engineers) from mining practice. In the course of time, the meaning of some of these terms has been modified. In this chapter, the tunnel terms as actually used in civil engineering practice and in engineering geology are given. There is some difference in terminology between these two fields, however. To the engineer, a *consolidated* deposit is one that came to equilibrium under the action of applied forces, e.g., consolidated clay. To the geologist, a clay is consolidated only when it is converted into shale or claystone, i.e., generally into some hard rock. To the engineer, the *overburden* of a tunnel or some other buried structure is any and all material overlying the tunnel, regardless of whether or not the material is consolidated. To the geologist, the term overburden implies unconsolidated, or mostly soil, material (Fig. 9.2).

A *tunnel* is a horizontal or near-horizontal excavation that is open to the ground surface at each end. When an excavation proceeds in a vertical or near-vertical direction and is open to the surface only at the top, it is called a *shaft*. A *drift* (or *adit*) is similar to a tunnel, except that a drift is open to the surface at one end only. The term *stope* or *raise* sometimes is used as a noun indicating an inclined excavation driven from the main tunnel or drift in an upward direction, usually for exploratory purposes. Stope also is used as a descriptive term to indicate that during tunnel excavation, the rock in the tunnel roof keeps dropping out; thus the roof of the tunnel is said to be "stoping" upward (Fig. 9.2).

There are four terms commonly used to describe the location of parts of the tunnel cross section (Fig. 9.2*b*). There is the floor, or *invert* (although this term more properly describes a masonry or concrete inverted arch). At the bottom of the cross section, the flat strut that sometimes is used instead of an arch is called an *invert strut*. The top of the cross section is the roof, also referred to as the *back* or *crown*. The sides of the tunnel section are referred to as the tunnel walls, and the *spring line* is the point where the curved portion of the roof intersects the top of the wall. In a circular tunnel, the spring line is arbitrarily set by the engineer, if used at all. Most geologic tunnel logs are plotted on horizontal projections of the roof and walls and only occasionally on the invert.

9.3. Ground. The material through which the tunnel has to be driven is termed *ground*, and the excavated material is *muck* (or less commonly "tailings"). There is *soft* ground (soil) and *hard* or *rocky* ground (rock).

Soft-ground Terminology. In *raveling* ground, chunks or flakes of material drop from the exposed surfaces. This process may start soon after exposure, especially under water. Or in some instances, raveling may be slow and considerable time is required for it to start. *Running ground* generally is clean, loose gravel or coarse sand. Medium to fine sand also may run if dry enough. *Flowing ground* is wet soil that moves like a viscous liquid and tends to enter the tunnel through every gap in the lining. *Squeezing ground* has a lower moisture content than flowing ground. Squeezing ground flows into the tunnel plastically, i.e., without fracturing, and often tends to raise, or heave, the invert. *Swelling ground* also tends to move into the tunnel, its movement being associated with a considerable increase in volume and moisture content (Sec. 4.6).

Rocky-ground Terminology. A great number of self-descriptive terms are used. Ground that permits advancing a drift without roof support is termed *firm ground* or *intact ground*. Other terms are stratified rock, moderately jointed rock, blocky and seamy rocks, crushed or shattered rock, and others. Presumably rocks can squeeze and swell in the same way as soft ground, and thus the terms squeezing and swelling ground also apply to rocky ground.

SUPPORTING THE OVERBURDEN

9.4. Supports. Tunnels in intact rock may be constructed without *lining*. Lining, if any, generally is plain or reinforced concrete. During the construction of a tunnel that is to be lined, the ground within the tunnel should be supported by wood (timber), steel, or suspension roof supports (Sec. 9.5). The timber or steel supports may be removed prior to lining or left within or behind the lining. Timber supports generally are frames, or *sets*, of heavy round or square timbers (such as 8 by 8 in. or 12 by 12 in.) placed at variable intervals of 2 to 5 ft or more. The terms *cap, post,* and *sill* are illustrated in Fig. 9.3. Behind the sets,

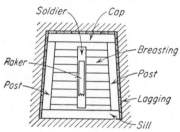

FIG. 9.3. Square-set heading in dry, soft ground.

boards or planks are placed to form *sheathing* (or sheeting) or *lagging*. The *face* (the front vertical end) of a drift (Fig. 9.3) is protected from sloughing by boards that generally are termed *breasting*, though the term *lagging* also is used in this connection. The breasting is supported by one or several vertical *soldiers* (or soldier beams), each held in place by a *raker*, or strut sloping back to the floor. The term *strut* generally is used to designate a member of the set taking up a horizontal force, particularly the horizontal pressure of the earth mass acting on the drift. Thus, cap and sill are struts in this sense.

Formerly, American and European tunnels all were constructed with timber supports; in modern American tunnel practice, however, steel has become almost a standard material for supporting rock in vehicular or water supply tunnels. Timber may be used as a temporary measure until steel supports can be placed. Tunnel steel supports may be of the following five types: continuous rib (Fig. 9.4*a*); rib and post (Fig. 9.4*b*); rib and wall plate or liner plate (Fig. 9.4*c*); rib, wall plate, and post (Fig. 9.4*d*); full-circle rib (Fig. 9.4*e*); and continuous rib with invert strut (Fig. 9.4*f*). In Fig. 9.4, rib, post, and invert strut are elements forming a set. The wall plates serve as sills to the ribs; in their turn wall plates should be placed on firm ground or solid rock. Steel supports may be placed flange to flange or a few or several feet apart. The spacing is guided by actual rock conditions and, in some cases, by safety requirements if occasional roof falls may be expected.

9.5. Roof Bolting.[1-3] In the suspension roof support method (some miners call it the "sky-hook" method) the roof of the tunnel is bolted to a firm stratum located at a certain height above the roof. At the same time, the bolts hold together the strata above the roof and prevent their mutual sliding caused by excavation. Holes are drilled into the back of the tunnel section (Figs. 9.2b and 9.5), and special bolts with a

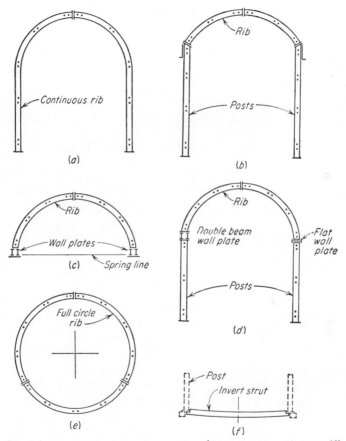

FIG. 9.4. Types of steel tunnel supports. (*After Proctor and White.*[13])

wedge at the end are inserted (detail, Fig. 9.6) and tapped with a hammer. A small steel plate is placed over the outer end of the bolt, and a hexagonal nut is screwed on and seated against the plate (slightly tightened in position). The nut is then tightened, and an impact wrench is used to drive the bolt into rock. Because of the wedge action, the bolt ends spread apart and thus high friction develops between the bolt and the rock. Thus the roof is suspended and prevented from caving down. Furthermore, the bolts tie together the horizontal or near-horizontal strata above the roof to form a "rock beam." The bolts hold the strata together, prevent slippage, and even permit arch action of the rock beam. The method is efficient in fissile or platy rock such as shales, slates, some sandstones, and coal. In slabby or blocky rock, the bolts actually anchor

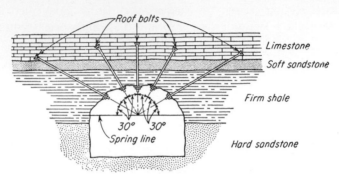

FIG. 9.5. Typical roof bolt installation.

the slabs or individual blocks to firm strata above them. Soft shales, soft friable sandstone, and similar materials do not provide adequate anchorage for bolts. In particularly slabby materials, a lightweight steel channel is placed across the roof and the bolts inserted through it.

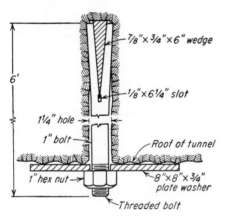

FIG. 9.6. Typical roof bolt detail.

The normal diameter of the bolts used in roof bolting is $1\frac{3}{4}$ in. or more. Bolts of smaller diameters may not have sufficient resistance to driving. In considerably fractured rock, squeezing ground, and similar ground defects, roof bolts may be used to reinforce ordinary steel supports, particularly in large vehicular-type tunnels. Ordinarily the cost of roof bolting is less than that of conventional supports; the difference varies, however, with rock conditions, labor costs, and geographic location.

Some of the advantages of roof bolting are as follows: (1) The roof *immediately* adjacent to the working face can be supported; (2) ventilation of the tunnel is better, since the air flows are not hampered by conventional steel supports; (3) the bolts are not so easily dislodged by blasting or traffic as are ordinary supports; and (4) clearances in the cross section of the tunnel are improved.

It is of interest to note that the use of roof bolting sometimes permits construction of large underground rooms; e.g., in one mine, an experimental room 200 ft long and 110 ft wide was thus stabilized.[1] In the outlet tunnel for Keyhole Dam, Wyoming, roof bolts were used primarily to support loose rock blocks in the roof to prevent large overbreaks (Sec. 9.15). At the start of the tunnel operations, four 6-ft-long bolts were installed for each 4 ft of the tunnel length. It was found by trial and error that adequate safety was provided by fewer bolts, and the number was finally reduced to two bolts for each 6 ft of tunnel length. Better rock also was encountered. Wherever seams or prominent joints intersected the tunnel, additional bolts perpendicular to the seam or joint were installed. A small economy in concrete lining also was achieved.

TUNNELS IN ROCK

9.6. Pressure-relief Phenomena. The rocks in nature, especially the deep ones, are affected by the weight of the overlying strata and by their own weight. Stresses develop in the rock mass because of these factors. In general, every stress produces a strain and displaces the individual rock particles. To be displaced, a rock particle needs freedom to do so; in other words it needs to have space available for movement. If the rock is *confined* and its motion thus prevented, there will be only a partial rock displacement, if any. The stress that could not produce displacement because of the confinement of the rock mass still remains in the rock and is said to be in *storage* in the rock. A more appropriate technical term for such stored stresses is *residual stresses*. As soon as a rock particle acted upon by a residual stress is permitted to move, a displacement occurs. The amount of movement depends upon the magnitude of the stored stress. There may be only a small displacement with an insignificant rupture of the rock, or the displacement may be very large and involve the violent movement of a considerable yardage. The rate at which the released rock mass moves is not necessarily great, and there may be very quiet evidence of *pressure relief*, as such cases are termed. Summarizing, a pressure relief is a decrease of a residual stress or a system of stresses, instantaneous or slow in character, accompanied by the movement of the rock mass with variable degrees of violence. An opening in the rock mass, such as an excavation for the tunnel, causes the adjacent rock to flow into the opening; this in turn may liberate the previously confined rock masses and relieve the pressure.

In deep tunnels, i.e., driven at great depths below the surface, *rock bursts* may occur. This generally is manifested by the popping, or "blowing out," of slabs from seemingly sound rock in the tunnel section. In platy or fissile rock such as shale, the platy beds may slowly deform and "bow" into the tunnel. In this case the rock is not necessarily detached from the main mass, but the deformation may cause fissures and hollows in the rock surrounding the tunnel.

Another pressure-relief phenomenon is that of *bumping ground*, or "bumps."[4] The term refers to a sudden and somewhat violent earth tremor which disturbs deep underground strata. There are several classifications of bumps, all local in nature. In general, bumps are evidenced by (1) a sudden heaving of the tunnel floor and, in some extreme cases, the roof; (2) the sudden release of large masses of rock from the tunnel walls; or (3) shock waves transmitted through the tunnel floor that some-

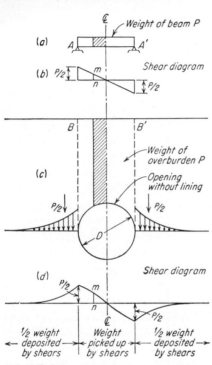

FIG. 9.7. Shear stress action and arching.

times are of sufficient intensity to kill a man. Bumping ground sometimes occurs in conjunction with general seismic (earthquake) movements.

9.7. Arching around a Tunnel. Even if there are no residual stresses in a rock mass, the construction of an opening makes the adjacent rock move into it. This rock is then pulled away by tensile stresses from the rest of the mass, which remains in place. In many cases, this tendency can be stopped only by providing adequate supports during construction and lining afterward. Sometimes the opening can stand safe without any support, both in rock and in firm clay. In any case, construction of the opening destroys the existing state of equilibrium in the material around the opening, and a new state of equilibrium is established. Nature does it by developing self-balanced systems of shearing stresses that produce the phenomenon known as *arching* around the tunnel. An analogy between a simple beam on two supports loaded uniformly with its weight and a tunnel without lining is shown in Fig. 9.7. At any vertical plane *mn* of the beam, an upward shearing stress balances the weight of the beam to the center line of the beam (crosshatched in the figure) and a downward shearing stress transmits this load to the left portion of the beam. Exactly the same occurs at the supports *A* or *A'*, where a half of the weight of the beam $P/2$ is transmitted to the support. The shear diagram of the beam is shown in Fig. 9.7b. Note that because of the symmetry there is no shear at the center line. In the case of a circular opening the shear diagram (Fig. 9.7d) between the hips of the tunnel *B* and *B'* is somewhat curved because of the nonuniform loading of the opening, and the left weight $P/2$ of the overburden is distributed, also by shearing stresses, along a certain distance on either side of the opening. This distance, although theoretically infinite, in practice is limited perhaps to $2D$, where D is the diameter of the opening.

The process of picking up the weight of the overburden by the shearing forces requires that the shearing strength of the material above the opening be not less than the corresponding shearing stress. Otherwise, if no lining is provided, the overburden will fail and crumble down into the opening. In the case of a tunnel with a lining, the weight of the overburden is divided between the lining and the systems of shears around

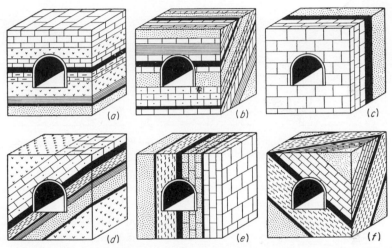

FIG. 9.8. Influence of rock stratification on the tunnel lining. (*After Prof. A. Desio, "Geologia applicata all' ingegneria," Hoepli, Milan, Italy.*)

the tunnel. As already stated, the systems of shearing stresses around an opening should be and are self-balanced, since they do not create or eliminate pressures but simply *redistribute* them.

An appraisal of the arching capacity of the rocks around a proposed tunnel is an important item in the geological study preceding construction. In badly fissured rocks, arch patterns cannot be sufficiently well developed. Massive igneous rocks generally offer favorable arching possibilities. The same also is true of the following layered formations: (1) horizontal or slightly dipping ones with the strike parallel to the axis of the tunnel and (2) steeply dipping formations with the strike perpendicular to the axis of the tunnel. These two types of layered formations in their limiting positions, i.e., with horizontal and vertical layers, respectively, are shown in Fig. 9.8*a* and *c*.

9.8. Influence of Rock Stratification on Lining Pressure. The total
amount of pressure on the tunnel lining and the way in which the pressure
is distributed along the lining depend primarily on the *stratification* of the
rock in which the tunnel is built.   Figure 9.8 shows different tunnel locations in rock strata.[5]   Figure 9.8*a*, *b*, and *c* shows more or less uniform vertical pressure on the lining, whereas Fig. 9.8*d* and *f* shows the
pressure concentration at one side of the tunnel caused by oblique strata.
Figure 9.8*e* shows a case of heavy pressure at the key of the arch.   The
locations shown in Fig. 9.8*a*, *d*, and *e* are favorable for office study
of pressures, since in these cases, a two-dimensional stress distribution
may be assumed.

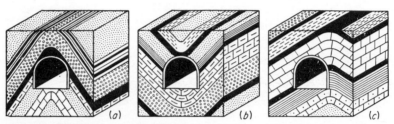

FIG. 9.9. Tunnel located in anticline and syncline.   (*After Prof. A. Desio, "Geologia applicata all' ingegneria," Hoepli, Milan, Italy.*)

Anticlines and Synclines.   Location of the tunnel in an anticline tends
to relieve the vertical pressure on the lining (Fig. 9.9*a*), whereas in a

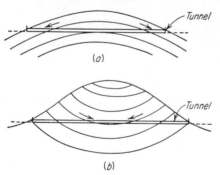

FIG. 9.10. Tunnel crossing (*a*) an anticline
and (*b*) a syncline.

tunnel located in a syncline, there
will be an increase in pressure.
Furthermore, if located in water-
bearing strata, the "anticlinal"
tunnel (Fig. 9.9*a*) will have water
flowing from it, whereas water
tends to flow into the tunnel
located in a syncline (Fig. 9.9*b*).
Consideration also should be given
to the fact that in an anticline,
the upper strata are more bent
and hence more fissured by tensile
stresses than the lower strata.
Therefore, it is advisable to locate the tunnel at a depth where this fissuring will not be consequential.   A deep tunnel in this case also will be
less subject to the percolation of meteoric water from above.   Figure
9.10*a* and *b* shows cases where a tunnel *intersects* an anticline and syncline,
respectively.   In the anticline, lateral pressure on the tunnel is greater

close to the portals (entrances) than at the middle of the tunnel, whereas in the syncline, the converse is true.[6]

Tunnels near Steep Slopes.   In some cases, the stability of the entire tunnel is endangered by the unfavorable stratification of the surrounding rock.   Figure 9.11 shows a situation which often occurs when a railroad or highway follows steep, rocky slopes dipping toward a river or a lake. The line in such cases is located alternately in tunnels and on viaducts or retaining walls.   To decrease the cost of the line, tunnels should be located

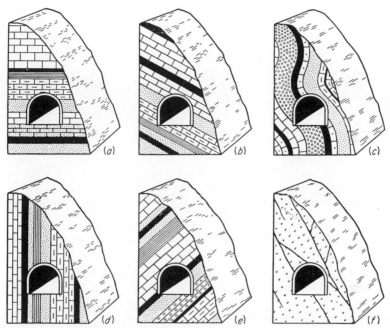

FIG. 9.11. Tunnels close to a steep slope.   (*After Prof. A. Desio, "Geologia applicata all' ingegneria," Hoepli, Milan, Italy.*)

as close as possible to the steep rock slopes; however, such locations demand that their stability be thoroughly investigated.   Figure 9.11*a*, *d*, and *e* shows stable tunnels, whereas Fig. 9.11*b* corresponds to an unstable structure.   Fissuring of rock as in Fig. 9.11*f* also impairs the safety of the tunnel.   The structure in Fig. 9.11*c* is only mediocre from the stability point of view.   It must be kept in mind, however, that the above examples are rather simplified.   Usually there are other circumstances that must be taken into account in locating a tunnel, e.g., water pressure in an unlined water supply tunnel.   In the latter case if the overburden is fissured or fractured as in Fig. 9.11*f*, a *blowout* may occur. Because of the leakage of water through fissures or fractures, the resistance of the overburden to lifting is gradually decreased until the overlying beds are actually pushed or blown off the surface.

9.9. Tunnels in Faulted Zones. Figure 9.12 illustrates different positions of the tunnel with respect to faults. Figure 9.12*a* shows a tunnel located within the faulted zone, whereas Fig. 9.12*b* and *c* corresponds to the location of the tunnel in the foot wall and in the hanging wall, respectively. The tunnel in Fig. 9.12*d* and *f* crosses the fault, obliquely in the latter case. In Fig. 9.12*e* the tunnel is outside the fault.

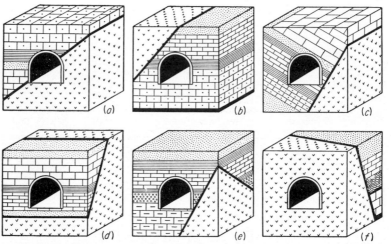

FIG. 9.12. Different positions of a tunnel with respect to a fault. (*After Prof. A. Desio, "Geologia applicata all' ingegneria," Hoepli, Milan, Italy.*)

The geologist should determine if the faulting has occurred in recent times, i.e., within the recorded history of the area, and therefore can be considered "active." If the tunnel is to be built far from a fault but in a region intersected by numerous faults, the possibility of new faulting action should be considered. Generally speaking, if a tunnel intersects an active fault, little can be done to protect the structure. In this case it is best to shift the alignment to avoid the fault or to use open cut within the active fault if possible. Whether the fault is active or inactive, the rock material within a faulted zone generally is shattered and unstable, and there may be considerable water inflow, as in the case of the syncline tunnel location (Fig. 9.10*b*). The space between the fault walls may be filled with gouge (Sec. 2.18). If a section of a tunnel follows the gouge zone, swelling of this material may occur and cause displacement or breakage of tunnel supports during construction. Sometimes the space between the fault walls is filled with sand-sized crushed rock that has a tendency to flow into the tunnel and thus is often mistaken for sand.[6] If, in addition, the tunnel is located under the water table, a kind of sandlike suspension may rush into the tunnel.   An accident of this kind took place in a tunnel forming part of the Hetch-Hetchy aqueduct in California.[7]   When the tunnel crosses an active fault, as in Fig. 9.12d and f, the tunnel may be acted upon suddenly by a shearing force so great as to disrupt any lining completely.   The shear displacement may be in any plane, according to the relative position of a tunnel to the fault. The railroad tunnel near Wright Station in the Santa Cruz Mountains, California, was offset 5 ft, not far from the northeastern portal, by the movement of the active San Andreas fault which crosses the tunnel at an angle of 80°.   This offset was reduced to zero at the opposite portal, about 5,100 ft from the maximum offset section.[8]   It is obvious that shifting of the tunnel away from the faulted zone may save many of the inconveniences; detailed geological information is required, however, before a relocation can be designed and safely accomplished.

9.10. Temperature in Tunnels, Geothermal Gradient.   In deep tunnels, high temperatures can handicap the physical work, but the tunnel temperature usually is not of concern unless the tunnel is more than 500 ft below the surface, except in the case of water tunnels.   In water tunnels, the periodic measurement of tunnel temperature, regardless of the tunnel depth, is of considerable importance.   The reduction of the water temperature by the cooling effect of the surrounding rock, sometimes by only a few degrees, may cause the formation of ice and thus impair the passage of water. Also reduced temperature or greatly increased temperature can affect the performance of the heating or

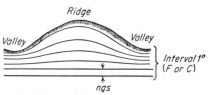

FIG. 9.13. Geoisotherms (symbol ngs means "normal geothermal step") under a plain. (After Andrea.)

cooling system of the plant if the water is supplying energy to drive hydroelectric turbines (Sec. 13.17).

Tunnel temperature largely depends on the value of the *geothermal gradient*.   This gradient shows the change of temperature in degrees centigrade or Fahrenheit for every 100 ft or 1 ft (or meters, respectively) of depth, or, which is the same, the number of feet or meters of depth corresponding to a 1°F or 1°C change in temperature, respectively. (In the latter case the term "geothermal step" sometimes is used.)   The value of the geothermal gradient is not constant in a given locality, however, and is influenced by several factors such as the relief of the locality or difference in thermal conductivity of various rocks.

Figure 9.13 shows schematically a set of *geoisotherms*, i.e., curves connecting points with equal earth temperatures. This diagram illustrates the opinion[6,9] that the geothermal step under the mountains or ridges is numerically larger than under the plains because of the greater exposure of the ground surface; in the case of valleys, the situation is reversed. Air is from 25 to 100 times less conductive of heat than water, the latter being as conductive of heat as some rocks (e.g., marble). It follows, therefore, that the presence of fissured or highly porous rocks increases the value of the geothermal gradient, especially if the quantity and speed of penetrating cool meteoric water are appreciable. Irregularities in the values of geothermal gradient may indicate the proximity of cold or hot water. Stini[6] gives the following average values of geothermal step for long European tunnels as expressed in feet per degree Fahrenheit:

Simplon (depth about 7,004 ft)   = 121 ft
St. Gotthard (depth about 5,748 ft) = 154 ft
Mont-Cenis (depth about 5,282 ft) = 192 ft

Earth temperatures can be measured by sinking thermometers (either wrapped or placed in containers) into bore holes. The thermometers then are left in the hole until they acquire the temperature of the surrounding rock, e.g., about 5 hr in dry borings and 30 min in water.[6] Thermal measurements in either a cased or open hole also may be obtained by means of a continuously recording electronic thermometer (for further details see Ref. 10). The probable temperatures in the tunnel to be constructed, as found from the values of the geothermic step, should be plotted on the longitudinal profile of the tunnel in the form of geoisotherms for guidance during construction. Possible occurrence of hot-water flows during construction should also be indicated.

9.11. Water and Moisture in Tunnels. Construction of a tunnel may vitally change the water regime of a locality. The concept "water regime" involves, in this case, location of water within the rock, direction and velocity of its movement, and changes in both location and motion in time, seasonal and otherwise. The tunnel introduces changes into the water regime in the same way as it changes the local stress conditions (Sec. 9.6). Generally speaking, a tunnel acts as a drain. (In mining areas, tunnels may be driven for the sole purpose of draining adjacent workings.) Firm rocks almost without fissures, such as igneous rocks and some sandstones, may be considered impervious, whereas fissured rock generally represents a good storage reservoir for both stagnant and moving water.[11] In some rocks, such as limestone, water may collect in the caverns formed by erosion. Faults, anticlines, synclines, and certain other geologic features also collect water (Secs. 9.8 and 9.9).[11] Tunnels driven under lakes, rivers, and other surface-water bodies may tap a considerable volume of flow.[12d]

Besides free or gravitational water in fissures and hollows, rocks contain capillary water film or hygroscopic water and water vapor. These kinds of water may move, and water vapor may condense, thus increasing the inflow of water to the tunnel. Also under proper conditions, water in rock may freeze.

Water may enter a tunnel in different ways. It may drip from the top of the drift; this dripping may be of variable intensity, sometimes turning into a veritable rain. Water may penetrate through the walls of the drift in the form of drops or continuous currents. Water under heavy pressure may break in as a gusher at any place on the drift periphery as explained in Sec. 9.9. When a drift which is driven through impervious rock passes over an accumulation of water under heavy pressure, the slab of impervious rock separating the drift from the water may become so thin as to offer no resistance to the pressure. In this case, water will break upward into the tunnel. This situation is illustrated in Fig. 9.14, in which the cross-hatched area signifies impervious rock; letter p designates a conduit or aquifer filled with water under pressure. Water conduits intercepted by a tunnel in rock may be completely or partially filled with water; in time they may become empty or filled with gouge (Fig. 9.14, letters e and g). Blasting may open new water conduits in the proximity of

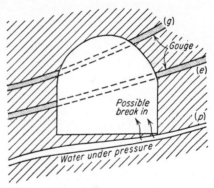

FIG. 9.14. Water channels in rock. (*Adapted from Stini.*[6])

the tunnel, shift the direction of the flow, or even cause a partial flooding of the tunnel.

Water Tables and Tunnels. It is well known (Sec. 5.3) that the ground-water table in the rock may be either local and bounding a limited water accumulation or continuous over a certain distance. It is obvious that a tunnel located above the water table is safe from water invasion (unless thermal water enters from below through cracks in the tunnel floor); hence it is desirable to locate the tunnel above the water table. This is not always possible, however. Figure 9.15*a* shows a tunnel a part of which is invaded by water during construction whereas the other part is dry. The whole excavation will be affected by water, however, if the grade is reversed as in Fig. 9.15*b*.

Construction of a tunnel not only modifies the subsurface water regime, as shown in Fig. 9.15, but also influences the surface flow of water in the locality. Figure 9.16 shows a tunnel constructed in a limestone syncline with a corresponding lowering of the water table. In this case, the ground-water overflow at point A vanishes, and the brook which was fed by the overflow dries out. Also cases have occurred where wells or springs over the tunnel were drained dry by the tunnel excavation.

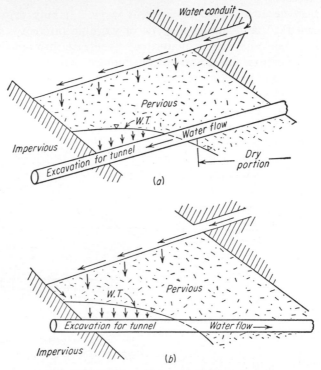

FIG. 9.15. Water table as related to a tunnel.

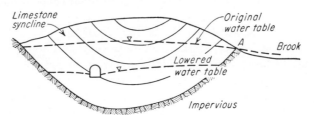

FIG. 9.16. Change in water regime as caused by a tunnel. (*Adapted from Stini.*[6])

Amount of Inflow. As a rule, the amount of water flowing in the tunnel decreases as the construction progresses. This phenomenon is caused by the gradual exhaustion of water at the source of flow and especially by the decrease in hydraulic gradient and hence in velocity of flow. As the flow progresses, the level of water at the source drops with a consequent decrease of hydraulic head. In exceptional cases there may be increases in flow as the construction progresses.[6] If rock around the opening has but few fissures at the beginning of construction and does not conduct much water, additional fissuring and additional inflow may develop if gradually increasing tensile stresses are not offered adequate resistance by the rock material.

Correct estimate of water inflow into the tunnel to be constructed is of great importance, as it influences the organization of the construction plant. In such estimates, the maximum inflow value (e.g., in gallons per day per lineal foot of tunnel) is required, as well as the redistribution of inflow between various sections of the tunnel and its changes with time. An incidental inflow of short duration as, for instance, might be caused by snow melt at the surface may increase the maximum inflow value considerably, but owing to its short duration, it is of no consequence in a long-time water-removal construction operation. The estimate of water inflow, the location of the water table, and its possible fluctuations are a combined responsibility of the geological and engineering staffs. Unfortunately, the geological situation is rarely so easily interpreted as to make accurate quantitative estimates possible.[11] About all the geologists and engineers can do prior to construction is to give rough estimates as to quantity. It should be borne in mind, however, that crude overestimation or underestimation of water inflow can result in severe cost increases, namely, overestimation usually causes the contractor to bid higher than he should, and underestimation will result in enforced construction delays that cause extra payments to the contractor.

Contamination of Tunnel Water. Most detrimental to a concrete tunnel lining are solutions of such sulfuric salts as calcium sulfate (anhydrite, $CaSO_4$); sodium sulfate, Na_2SO_4; or magnesium sulfate, $MgSO_4$. The geologist should be particularly on the watch for tunnel water flowing over gypsum or other sulfates. Such waters may leach the calcium from the cement and ruin limestone aggregate by base-exchange action. Acids diluted in tunnel water act similarly. Solutions of hydrogen sulfide, H_2S, even when weak, are poisonous to humans. Chemical analyses of tunnel waters should be performed, and if pH values of over 7 are found, high alkalinity is indicated. In city tunnels, chemical analysis may indicate whether or not a suspiciously large water inflow is due to a broken water supply or sewer main. Detrimental salt solutions and diluted free acids will attack concrete more severely if they are in warm rather than cold water and if the water is flowing rather than stagnant.[6] As tunnel waters ultimately drain into neighboring streams, if the tunnel water is highly mineralized or contaminated, precautions should be taken as required by local health authorities.

9.12. Gases in Tunnels. Air in a tunnel under construction becomes foul through the workers' respiration and principally through blasting.

Ventilation is absolutely necessary during construction. In large tunnels this may consist of large pipes and blowers leading directly to the working face. In small tunnels, all activities are suspended until *natural* drafts ventilate the tunnel. Some authorities also believe that the use of natural drafts may be more economical for larger tunnels, even though, in the latter case, long delays in working time would be necessary. Artificial ventilation during maintenance is provided in all large vehicular tunnels as shown in Fig. 9.26.

Gas accumulations in rock, and generally under pressure, enter the drift through fissures or openings produced by blasting. If the flow of gas appears to be fairly continuous, the entrance of the flow may be packed with concrete or excelsior. Often the supply of gas is quickly exhausted; however, cases have been reported where the "blowing" of poisonous gas lasted two or three weeks.[6] The possibility of gas hazards should be indicated to the engineers before construction, although this is one of the most difficult tunnel hazards to predict. Gases may be expected in regions of volcanic or hot-spring activities. A dangerous gas, methane, CH_4, also called explosive gas, marsh gas, or "firedamp," may be encountered in coal regions and often is associated with shales. Similarly this gas may be expected in the neighborhood of oil fields and rock-salt deposits.[6] (Methane is the major ingredient of so-called "natural" gas.) Carbon dioxide, CO_2; carbon monoxide, CO; and hydrogen sulfide, H_2S, are other gases sometimes encountered in tunnel construction.

9.13. Bridging Capacity of Rocks: Excavating Cycle. The roof of a horizontal drift driven through rock may stay unsupported for a certain length of time. This bridging, or "stand-up" capacity* of rock primarily depends on the magnitude of shearing and tensile stresses within the unsupported rock mass, which in their turn depend on the span of the unsupported rock mass. Furthermore, the bridging capacity depends on the properties of the rock material, especially on its shearing strength. The bridging capacity also may be related to the toughness of the rock or difficulty of its excavation. The tougher the rock, the higher its bridging capacity. Although this generally is true, there are exceptions; e.g., quartziferous phyllite possesses a very limited bridging capacity though it is hard to excavate.[6] Fissility and piercing of rock with numerous bore holes have an adverse effect on the bridging capacity.

Excavating Cycle. In hard rock, the sequence of tunnel operations in advancing a drift constitutes an *excavating cycle.* This consists of the following stages: (1) *drilling*—in large tunnels this is often done by using a large movable installation containing several drills and referred to as a

---

* In German treatises on tunnel geology, the bridging capacity is expressed by the term *Standfestigkeit*. The closest English equivalent is "stand-up capacity."

jumbo; (2) *charging* the drill holes with explosives; (3) *shooting rounds* (firing), i.e., several rows of drill holes are blasted simultaneously; (4) *ventilating* the tunnel to remove the explosion gases; (5) *mucking* or removing the blasted rock material, which may be done either automatically by a mucking machine containing a large scoop or a series of scoops at the front or by hand; the material is dumped into "muck" cars; and (6) *erecting supports*, or timbering (the latter phrase often is used even though the supports may be of steel). As an excavating cycle is completed, another one starts. (A device commonly used in large tunnels is the "cherry picker." This is a cranelike apparatus which picks up the muck cars to facilitate their switching operations within the tunnel.)

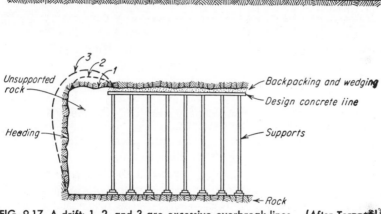

FIG. 9.17. A drift: 1, 2, and 3 are excessive overbreak lines. (After Terzaghi.)

Figure 9.17 shows the longitudinal section of a drift after the last round has been fired. Next to the end of the drift (heading) the roof is supported by the two walls and by the rock material in the heading. If the bridging capacity of the rock is high, the rock material next to the heading will stay in place for a considerable time, as indicated by line 1 in Fig. 9.17. If, however, bridging capacity is low, the rock will start to fall down soon after the last round is fired as shown by lines 2 and 3 in Fig. 9.17. In this case, supports have to be erected as soon as possible. It should be noted that after the erection of supports, the pressure on them generally increases until it reaches a certain ultimate value.

The time during which unsupported rock remains in equilibrium also is termed *bridge-action time*, stand-up time, or bridge-action period (Terzaghi).[13] Obviously, even for a given rock, this time interval is a variable value, and it should be estimated from experiment or observation.

9.14. Methods of Tunnel Excavation in Rock.   The method of building
a tunnel depends very much on the bridging capacity of the rock.   In the
full-face method, the whole area of the tunnel cross section is blasted out
at each round.   Small tunnels with continuous rib-type support generally
are built by this method.   In larger tunnels, this method is applicable
if the bridge-action period allows time for ventilation and mucking before
the roof starts to crumble down.

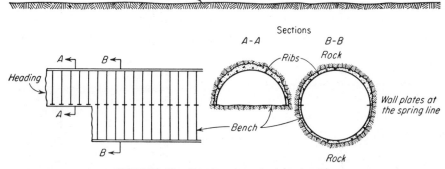

FIG. 9.18. Heading-bench method (a sketch).

In the case of a short bridge-action period, the *heading and bench
method* is used, as shown schematically in Fig. 9.18 for a circular water
supply tunnel under construction.
The top heading drift is carried
ahead of the bench; the bench acts
as a working platform about 15 ft
long in the direction of the drift.
The heading is full width of the
tunnel and is carried down to the
spring line where the wall plates are
placed (see also Fig. 9.4c and d).
The heading and the bench are both
shot at one round, the bench charges
being shot first.   A variation of

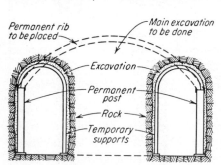

FIG. 9.19. Side-drift method (a sketch).

this method is the *top heading method*, in which the heading is driven clear
through as one operation and the bench is removed in another operation.
To shoot the bench, vertical holes may be drilled instead of the generally
used horizontal ones, with a resulting economy in explosives.

The *side-drift method*, which can be used in poor rock, is shown in Fig.
9.19.   Drifts are driven ahead of the main excavation, and their supports
are removed just before the shooting of the central part of the tunnel
starts.   In the *multiple-drift method*, more than two drifts are used.

In every method, the space between the supports and the rock surface left by blasting should be fully packed with stone or concrete and wedged. Otherwise, as experience has shown, the length of the load increasing period and the amount of the load itself are unduly increased.[13]   After the construction of the permanent concrete lining, all possible voids in the back packing should be grouted with cement mortar under pressure. (In some of the Pennsylvania Turnpike tunnels, the voids were blown full of powdered slag and special chemicals were pumped in which reacted to form a hard mass.)

9.15. Pay Line and Overbreak.   From a stability point of view, a tunnel cross section in intact rock may be any desired shape.   In shattered and unstable rocks, the shape approaching a full circle is most suitable.   The inside shape of the tunnel cross section ("inside" or "neat concrete" line) is determined from the performance requirements of the tunnel.   Then the concrete-lining thickness is designed for the given stress conditions for sections in disturbed or deficient rock (greater thickness) and for sections in intact rock (less thickness).   Generally, plain (nonreinforced) concrete is used for lining except water tunnels under pressure.

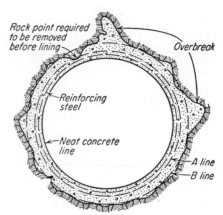

FIG. 9.20. Unsupported tunnel section: payment lines.

Payment for construction of the tunnel and placement of concrete lining is based on the following lines: the inside line, the minimum concrete line (or A line), and the pay line (or B line) (Fig. 9.20).   In tunnels through soft materials, the A and B lines generally coincide.   The A line marks the minimum thickness of concrete required by the design criteria.   Often this thickness is determined by allowing ½ in. of concrete for each 1 ft of inside tunnel diameter; e.g., a 9-ft-diameter tunnel would require a 4½-in. thickness of concrete lining.   In any case, the A line must be even with or clear of the bottom flange of any supports, and rock points are not permitted to project between the A line and the inside line.   If they do, the contractor must remove them without cost to the owner.   The thickness of concrete between the inside line and the B line generally is equal to 1 in. of concrete for every 1 ft of tunnel diameter (USBR practice).   Rock excavated beyond the B line is called *overbreak* (or "overprofile" in European terminology), but payment is made only for the quantities enclosed by the B line, regardless of whether the contractor overshoots or undershoots.

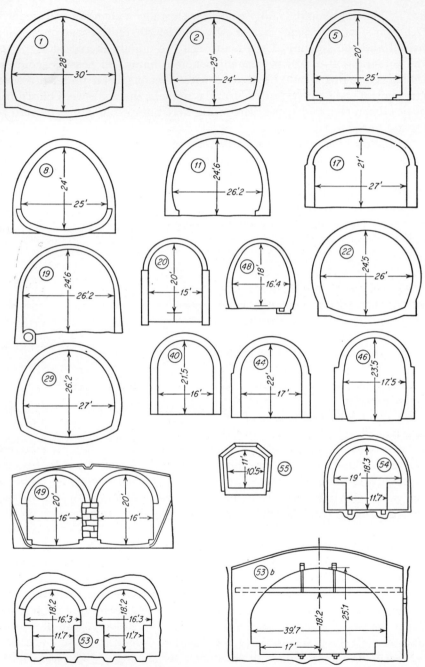

FIG. 9.21. Tunnel cross sections. (*From Merriman, "American Civil Engineer's Handbook," 5th ed., John Wiley & Sons, Inc., New York.*)

In recent research done by W. R. Judd for the USBR, the overbreak records in about 100 tunnels have been analyzed (the research is still in progress at the time of this writing). The results, so far, indicate that the rock types and tunnel sizes (cross-sectional area) have but minor influence on the percentage of overbreak; the construction procedures apparently are more important in this respect. Further detailed research along these lines is required, however. The most frequent overbreak percentage for all tunnels studied has been estimated at from 9 to 15 per cent of the area defined by the pay line (Fig. 9.20), although the overbreak ranges from 0 to 56 per cent. The most frequent overbreak values for various rock classifications have been as follows: sedimentary rocks, 9 to 11 per cent; of metamorphic rocks, only gneiss frequently shows a low overbreak percentage of 8 to 11 per cent, other metamorphic rocks being erratic but generally on the high side; numerous records in granite give a range from 7 to 31 per cent, the greatest frequency being 10 or 12 per cent; the most frequent in basalt is 8 to 15 per cent.

9.16. Rock-tunnel Cross Sections. Figure 9.21 illustrates types of tunnel cross sections used in the United States and Europe. British tunnels are shown in cross sections 1, 2, 5, 8, 22, and 29. Tunnels through the Alps are (1) the Mont-Cenis Tunnel (cross section 11) built in 1871 which joins France and Italy and is about 8.5 miles long, (2) the Swiss St. Gotthard 9.3-mile-long tunnel (cross section 19) built in 1881, and (3) the Simplon Tunnel (cross section 48) built in 1905 which joins Switzerland and Italy and is 12.3 miles long. There are two parallel headings 56 ft apart in this tunnel. Hot-water inflows of 15,000 gpm (up to 127°F) and cold water (55°F) were a few of the difficulties encountered during construction. Note that both the Mont-Cenis and Simplon tunnels and also several others in Fig. 9.21 have cross sections of the so-called *horseshoe shape*.

The longest American railroad tunnel (7.8 miles) is the Cascade Tunnel of the Great Northern Railroad in Washington (cross section 40). The well-known Moffat Tunnel in Colorado is 6.1 miles long and was built in 1930. There are two parallel headings in this tunnel, 75 ft apart. One of the headings is 16 by 24 ft in cross section for the single-track railroad, and the other is 8 by 8 ft for transporting water through the continental divide to the city of Denver. The Adams Tunnel, an integral part of the Colorado–Big Thompson Transmountain Diversion Project, is 13.1 miles long and of a circular cross section. It is reputed to be the longest water supply tunnel driven from only two headings. In the Catskill water supply system for the city of New York there are 25 tunnels in a distance of 160 miles, their accumulated length being about 32 miles.

TUNNELS IN SOFT GROUND

9.17. Rock and Soft-ground Tunnels Compared. Both tensile and shearing strength of soft ground is far below that of hard, firm rocks. It is natural, therefore, that the stand-up time (Sec. 9.13) of soft ground is shorter than that of hard, firm rocks. In stiff clays, however, this time is approximately similar to that in decomposed rocks, i.e., about one day. In other kinds of soft ground, there is a wide range in this time, and in running sands or similar materials, it is zero.

The basic difference between rock and soft-ground tunnels is the influence of the water table on the tunnel construction. In rock tunnels, violent inflows of water or sand suspension (Sec. 9.11) should be considered as temporary hazards; however, in soft-ground tunnels being driven under the water table, the fight with water may be a continuous construction feature.   In soft ground, the water table may be lowered below the tunnel bottom by sinking well points or using some other drainage system.   Water may be more or less completely removed from the tunnel by using compressed air, in which case the work should be done in a closed space to keep pressure, e.g., as is done in the shield method described in Sec. 9.19.   Isolated attempts have been made to grout the water channels that bring water into a tunnel.

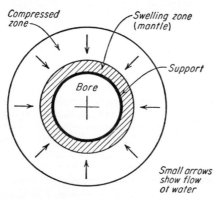

FIG. 9.22. Tunnel in swelling ground.   Circular cross section is recommended.

When the bore is made, water enclosed in the mass surrounding the tunnel starts to move toward the opening because of the difference in pressure.   In a tunnel in clay this causes swelling of the material.   Figure 9.22 illustrates this phenomenon.   Clay surrounding the support swells and exerts pressure on both the support and the outside material.   Since the force of compression remains constant and the area on which the force acts gradually increases (concentric circles around the bore in Fig. 9.22), the compressive stress at a certain distance from the bore becomes negligible.

9.18. Soft-ground Tunneling.   In firm, soft ground (firm clay, sometimes cemented sand) of sufficient tensile and shearing strength, no supports may be needed.   If, for safety considerations, supports are considered, they may be easily placed during the stand-up period.   The firm material may be excavated by using a jumbo machine provided with special pneumatic "clay shovels," such as is done in a tunnel in rock with long stretches of firm soil material.

In *raveling and squeezing ground*, the sides of the tunnel may be supported by timber beams and lagging or by steel "liner plates."   The latter are flat plates that are set in, one by one, and assembled into rings, or "courses," along the periphery of the tunnel until a continuously enclosed tunnel section results.[14]   In *cohesive running ground* it is necessary to support the face by breasting.   As the excavation proceeds, the breasting is gradually moved forward; each move is for a distance equal to the width of the liner plate course.   In *flowing ground* the roof and the sides of a drift are supported by *poling boards* (Fig. 9.23), the face is breasted, and the tunnel bottom reinforced against possible heave of the ground.   The work proceeds by shifting poling boards from position 1 to position 2.   The procedure is slow and dangerous.

9.19. Shield Method.   A shield consists of a circular steel box or ring (bulkhead) generally provided with a transverse diaphragm.   The forward end of the ring is equipped with a cutting edge (Fig. 9.24), and the rear end (tail) projects backward over the completed lining, which usually consists of rings of cast iron.   The shield is pushed forward by hydraulic jacks that react against completed lining at the rear of the shield.   In *firm materials* such as stiff clay, the work is done by steps or jerks.   The shield is advanced a short distance, the jack plungers are withdrawn, one more lining ring is erected, and then the whole cycle is repeated. There is a working chamber in front of the diaphragm, and if the stand-up period of the material is sufficiently long, the excavation is done wholly or partly by the workers. In this case the cutting edge is allowed to break the undisturbed soil.   In *soft materials* such as organic silt under rivers, the shield

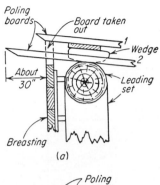
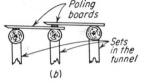

FIG. 9.23. Use of poling boards in flowing ground.

is shoved against the soil face before it, a part of the soil material flows into the tunnel through the openings in the diaphragm (Fig. 9.25), and the rest of the soil material is pushed upward.

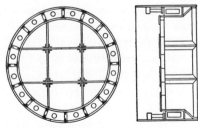
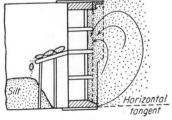

FIG. 9.24. Shield method, Hudson Tunnel, New York City.

FIG. 9.25. Shield method in organic silt, Lincoln Tunnel, New York City.

As the shield is larger than the tunnel lining, the annular space thus formed should be immediately filled in, which is done with grout, i.e., usually a dense suspension of equal parts of cement and sand.   To keep water out of the tunnel during construction, compressed air is used. This method of working water-bearing materials has the disadvantage that the variable water pressure along the top of the tunnel cannot be balanced by a uniform air pressure.   Generally the greater water pressure at the bottom of the tunnel is balanced, and this permits the unbalanced compressed air at the top of the tunnel to escape in the form of bubbles.   A sudden outrush of the unbalanced air ("blow") may cause flooding of the tunnel.[15]

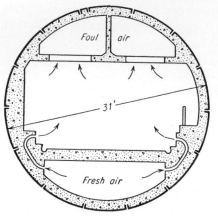

FIG. 9.26. Cross section of the Brooklyn-Battery Tunnel (New York City Standard).

There are numerous well-known shield tunnels in New York City, such as the Holland and Lincoln tunnels under the Hudson River, the Sixth Avenue and other subway tunnels, and tunnels of the Pennsylvania Railroad.   Many tunnels in England and France, including part of the Paris subway, were built by the shield method.   Large tunnels built by the shield method in recent years in the United States are the Queens-Midtown and Brooklyn-Battery tunnels under the East River, Greater New York (Fig. 9.26).   The cross section in this figure has been used for many subaqueous tunnels in New York City.   The arrows in the figure show the path of the air used in ventilation.

PRESSURE ON TUNNEL LINING

9.20. Requirements to Be Satisfied by Tunnel Lining.   It was anticipated in Sec. 9.7 that the vertical pressure on tunnel lining is *less* than the weight of the overburden because a portion of this weight is taken up by the shearing stresses around the tunnel; an exception, however, is swelling ground (Sec. 9.21).   In order that shearing stresses in the rock or soil around the tunnel may develop, small displacements in the overburden must occur (Sec. 4.24).   Hence, if a tunnel is lined, the lining should be *flexible enough* to permit some small displacements in the overburden.   Thus the shearing stresses around the tunnel are "mobilized."   If the lining is *too flexible*, however, the integrity of the rock or soil mass would be destroyed, the shearing stresses would vanish, and the lining would be loaded with the whole weight of the overburden. Though these requirements are clear and simple qualitatively, a quantitative solution of the lining problem does not exist and lining design is based mostly on practice and engineering judgment.

9.21. Vertical and Horizontal Pressures on the Lining.   Estimation of the *vertical pressure* on a tunnel lining may be facilitated by observing actual displacements and displacement tendencies above and around a tunnel *without lining* or on a drift.   In all tunnels in more or less homogeneous rock, the material tends to cave or to fall down into the excavation. In cohesive rocks, such as granite or basalt, the height of the caving zone may be from one to two widths of the drift; its shape often approaches that of a Gothic arch except where distorted by faults and other discontinuities in the rock. The vertical pressure may be symmetrical or nonsymmetrical (compare Figs. 9.8 through 9.12). If the shearing strength of a cohesive rock mass is great enough to withstand the action of considerable vertical shearing stresses, and if there are no serious cracks in the rock mass, the tunnel can be built without lining. At the other extreme, a caving zone in sand or other fragmental material may reach the ground surface,[16] and therefore lining must be used.

Fissured and stratified rocks generally (but not necessarily) need lining. Practice shows that the vertical pressure on the roof of a tunnel in rock is generally independent of the *depth*, except for shallow tunnels, i.e., those constructed near the ground surface. In other words, the shape and size of the caving zone are practically the same regardless of the depth at which the tunnel is built. Shallow tunnels are rare, as deep open cuts are generally used instead.

A considerable number of theories that assume uniformity (and in some cases elasticity) in the material surrounding the tunnel are discussed in detail in Ref. 6. In practice, such theories are seldom used. For convenience in design computations, the weight of the actual rock body exerting vertical pressure on the lining is replaced by an equivalent uniformly distributed load, which is expressed in feet of overburden. This is known as *rock load*. If the unit weight of rock is assumed to be 165 pcf, the expression that the rock load is 20 ft of overburden means a load of $165 \times 20 = 3,300$ psf on the horizontal projection of the tunnel. The most practical way to estimate the equivalent rock load is by using Terzaghi's empirical formulas,[13] which are discussed hereafter in a somewhat simplified form. Symbol B as used hereafter means the over-all width of the tunnel, and symbol H means its over-all height, both values being expressed in feet (Fig. 9.27).

Lining in stratified or moderately jointed rock has to support only some loose rock from overbreak (Fig. 9.28). The height in feet of the corresponding uniformly distributed rock load may be estimated at no more than 0.25 of the width B (in feet). In other cohesive rocks (and firm clays) the actual caving body is replaced by a conventional *ground arch* (Fig. 9.27a), the width of which is variable. It depends on the character of rock and is roughly proportional to the sum of $B + H$. These assumptions being used, the thickness of the uniformly distributed overburden is estimated as the equivalent of the sum $B + H$ multiplied by a coefficient K as given in Table 9.1. The roof of the tunnel in the case of the first three ground materials listed in the table is assumed to

TABLE 9.1. Rock Load in Feet at a Depth of More Than 1.5$(B + H)$*

Moderately blocky and seamy................	$0.25B$ to $0.35(B + H)$
Very blocky and seamy.....................	$(0.35$ to $1.10)(B + H)$
Completely crushed, but chemically intact.....	$1.10 (B + H)$
Squeezing rock, moderate depth..............	$(1.10$ to $2.10)(B + H)$
Squeezing rock, great depth.................	$(2.10$ to $4.50)(B + H)$
Swelling rock..............................	Up to 250 ft

* After Terzaghi.

be located below the water table. If it is located permanently above the water table, the values given for those three materials can be reduced by 50 per cent.[13]

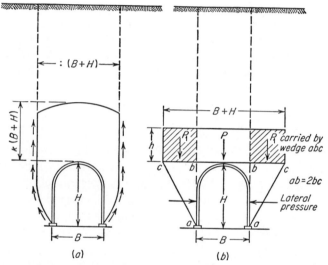

FIG. 9.27. Pressure on a tunnel lining. (After Terzaghi.)

It is obvious that in a general case some sections of a tunnel should be lined whereas others may be left without lining. As an example, Fig. 9.29 shows a tunnel through a granite ridge in which lining is necessary only close to the portals. Of course, if a fault crossed the tunnel somewhere under the mountain, that portion of the tunnel also would have to be lined.

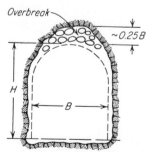

FIG. 9.28. Overbreak to be carried by lining. (After Terzaghi.)

Figure 9.27b shows how to estimate the pressure on the lining exerted by sand or other cohesionless material such as crushed rock (after Terzaghi).[13] The thickness h in that figure depends on the density of the material and the yield of the lining and may be as high as $0.6(B + H)$, as expressed in feet. Attention should be given to the degree of cementation and of compactness of sand deposits along the direction of a proposed tunnel.

Horizontal, lateral, or *side pressure* (on the tunnel lining) is usually small in the case of hard, intact rocks and in blocky and seamy ground. It may be large in other kinds of ground; e.g., a heavy lateral pressure should be expected in all cases of squeezing and swelling ground and in cohesionless deposits. Concrete or steel inverted arches all along a section of the tunnel or struts placed at intervals may stop the swelling ground (e.g., cross sections 2, 8, 22, 29 in Fig. 9.21). Similarly, inside walls of the tunnel supporting the arch may be made curved instead of

vertical to withstand swelling (cross sections 22 and 29, Fig. 9.21). Geological advice in determining lateral pressure generally is of help. It is difficult to expect precise numerical estimates of lateral pressure from geological surveys, but a competent geologist can give a detailed quali-

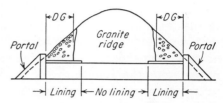

FIG. 9.29. Tunnel through a granite ridge (DG means "decomposed granite").

tative estimate of the capacity of different rocks along the tunnel to produce lateral pressure on the lining.

Pressures on the Tunnel Lining in Soft Ground. The vertical pressure on the lining in this case is estimated by using the rules of soil mechanics. Pertinent information may be found on pages 198 to 201 of Ref. 17. Sometimes it is assumed that the mass around the tunnel acts as a fluid. In this case the vertical pressure on the roof equals the total weight of overburden. The horizontal pressure on the sides of the tunnel generally is assumed to be a certain portion of the vertical. This portion was assumed at between one-third and two-thirds in the design of the Chicago subway, which passed mostly through plastic blue clay, and about one-half in the design of the subaqueous tunnels in organic silt under the Hudson River, New York City.

Change in Pressures with Time. Pressures on the lining in all kinds of tunnels tend to increase with time until they reach a certain value. This process is, of course, tantamount to the gradual vanishing or extinguishing of shearing stresses acting around the tunnel. The load increase immediately after installation of a permanent tunnel support (Sec. 9.13) is just the beginning of this "healing" process, which apparently may last for years.[18] It should be noted, however, that the *rate* at which the pressure on the lining increases decreases rapidly with time.

The vertical pressure on a support of the Broadway Tunnel in San Francisco, California (the type shown in Fig. 9.4d), was, in round figures, 24,000, 29,000, and 30,000 psi, 2 weeks, 1 month, and 2 months, respectively, after the installation of the support. The posts were 10 in. H beams spaced 2 ft center to center. Pressures were measured with SR-4 gauges.

9.22. Pressure Problems in Water Tunnels.

In water supply tunnels, particularly those in which the water is under pressure, the evaluation of lining pressures depends primarily on the internal stresses caused by water. If the lining is designed to resist high internal hydrostatic pressures, it normally is of sufficient strength to resist any external rock pressures. The voids behind the lining should be completely filled with lean concrete or grout to make the rock participate in resisting the internal hydrostatic pressures. The better the rock conditions (completely unfractured, or nearly so, very strong rock), the less the thickness of the lining or the amount of reinforcing steel in it. The evaluation of rock conditions in this case has to be done by both geologists and engineers of long experience, since the amount of pressure likely to be transferred from the tunnel to the surrounding rock is a matter of some controversy among tunnel experts.

One of the major problems in designing water supply tunnels is the possibility of leakage. The best concrete lining ultimately will develop some cracks, and water seepage into the surrounding rock will occur. In this way, external water pressures are built up around the tunnel, and if the water is suddenly withdrawn from the tunnel (by a break in the supply line or for maintenance reasons), the unbalanced outside water pressure may be sufficient to cause the collapse of the lining. Some successful attempts have been made to minimize the initial cracking by injecting a bentonite slurry behind the lining; the resultant expansion of the bentonite places the concrete in compression and reduces the cracking. The seepage problem often is solved by placing drains outside the lining to permit the seepage water to drain away and by constructing the lining in a circular shape of sufficiently strong concrete, reinforced if necessary. Another suggested method is to place flap valves in holes drilled through the lining. These valves are designed to open only toward the inside of the tunnel. Thus, if external water pressure exceeds internal pressures in the tunnel, the valves open and thus relieve the external pressures. In the case of high internal hydrostatic pressures it often proves more economical to construct a full steel liner inside of the concrete lining; thus the tunnel is made perfectly watertight. This latter method generally is employed in the design of high-head hydroelectric power-plant supply tunnels and penstocks. Sometimes the water pressure problem may be solved by constructing the main supply tunnel with only a slight grade (to minimize internal water pressure) and connecting it to a short but steep tunnel or steel pipe for the final run to the power plant. This latter portion of the tunnel may be lined with heavy steel plates, which, however, is more economical than placing steel lining in the entire length of tunnel.

The following physical characteristics of the rock surrounding a water tunnel should be carefully studied: (1) intensity of the fracturing and openness of the fractures, (2) effect of water on rock (whether it will dissolve the rock, deteriorate it, or produce no detrimental effect), and (3) possibility of water percolating into the rock and building up hydraulic pressures.

GEOLOGICAL SURVEY AND REPORT FOR A TUNNEL

9.23. Geological Survey Prior to Tunneling.
In the majority of cases, the alignment, the size of the bore, and the shape of the tunnel cross section are established prior to the geological survey. Usually this is true of city tunnels, railroad and highway tunnels, and also tunnels for the transportation of water in which hydraulic requirements determine the size and shape of the bore. Particularly unfavorable geological conditions as disclosed by preliminary surveys may result in the relocation of the tunnel, however. Some geological considerations on the selection of the route for haulage and drainage tunnels for deep mines may be found in Ref. 12. This latter case, however, is rarely dealt with by a civil engineer.

The basic geological document used in tunnel design is a geological profile along the center line of the tunnel. This profile should show the different types of rocks and soils along the tunnel line, boundaries between them (contacts), and geological defects such as faults. If sufficient trustworthy data can be obtained, the water table also should be shown on that profile. This document usually is accompanied by a geological surface map. The map should show the various formations; structural features such as faults, joints, dips, and strikes of the various beds; contacts between formations; and any unusual water occurrences such as seeps or hot springs. The degree of accuracy with which the surface geology can be projected to tunnel level is left to the judgment of the geologist. In sedimentary rock terrain, such projection usually can be done with some degree of accuracy providing the beds are not too disturbed by folding and faulting. In igneous and metamorphic rock terrain, however, such projection is exceedingly difficult because of the usually highly irregular contacts between formations or rock types, and faults and folds in such terrain usually have a very irregular subsurface trace that is impossible to project accurately without extensive subsurface exploration.

The rock and soil material is explored by pits, drifts, and core drilling. If economically possible, all such exploration holes should penetrate at least to the level of the invert. In the portal zones of tunnels in soft ground, the holes should go to a sufficient depth beyond the invert to provide data for a rational design of the portal shoring and foundations.

The drift (Fig. 9.2a) or pilot tunnel is probably the best method of exploring tunnel locations and should be used if a major-sized tunnel (such as a vehicular one) is to be constructed in a ground that is known to have critical geological conditions. Such drifts should be large enough to permit a man to enter and work (4 ft wide by 6 ft high usually is satisfactory). Temporary shoring constructed of local timber is used

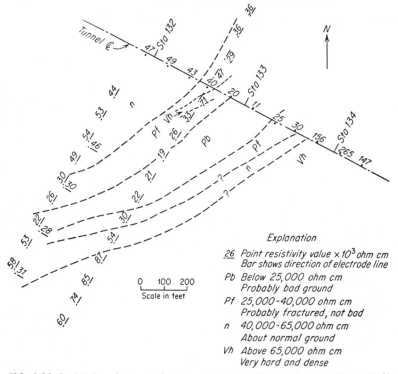

Explanation

26   Point resistivity value × 10³ ohm cm
      Bar shows direction of electrode line
Pb   Below 25,000 ohm cm
      Probably bad ground
Pf   25,000–40,000 ohm cm
      Probably fractured, not bad
n   40,000–65,000 ohm cm
      About normal ground
Vh   Above 65,000 ohm cm
      Very hard and dense

FIG. 9.30. Resistivity plan map for a portion of a tunnel.   See also Fig. 9.31.[19]

for the safety of the workmen. The drift may be hand excavated with hand- or air-driven drills and explosives; in this case the muck is removed on wheelbarrows or small mine cars.

A thick overburden and dense vegetation may make rock and soil investigations very difficult for both shallow and deep tunnels. In such cases, concealed fractures or fault zones, water-bearing horizons, or rock surfaces often can be located by geophysical methods of investigation.

9.24. Application of Geophysics in Tunnel Investigations. The following information, necessary in the tunnel design, has been obtained by geophysical methods: (1) the extent of faulting and fractures of major magnitude, (2) depth of bedrock under deep soil cover, and (3) location of particular rock formations and their possible intersections with the tunnel invert grade.

In the investigation of a transmountain diversion tunnel site in Colorado,[19] the resistivity method was used to determine, among other things, the existence and nature of faults or fracture zones suspected of running parallel to the projected tunnel line (Fig. 9.30). Resistivity field measurements established the classes of ground given in Table 9.2 as related to tunneling. The assumption made in a resistivity

TABLE 9.2. Types of Ground along Proposed Tunnel Line Classified'on the Basis of Electrical-resistivity Measurements

Type of ground	Resistivity, ohm-cm	Map symbol and description
Wet ground, probably caving............	Less than 25,000	P_b (probably bad)
Fractured, partially wet ground; not necessarily caving	25,000–40,000	P_f (probably fractured, not bad)
More or less normal ground or average ground	40,000–65,000	n (about normal)
Dense rock, more or less dry.............	Over 65,000	vh (very hard and dense)

field investigation of this kind is that quantitative geophysical data on the character of the rocks at the surface can be projected downward to tunnel grade. In this way, conditions that will be encountered in driving the tunnel can be predicted. The typical results of this survey are given in Fig. 9.30, the resistivity plan map for an area near tunnel, stations 132 to 134. (This was the suspected fault or fracture zone.) The zone was outlined by fixed depth resistivity measurements at an electrode spacing of 100 ft. Field examination indicated that the zone dipped northward at an angle greater than 45° and less than 80°. Figure 9.31 shows a resistivity profile along the tunnel center line in this area which was plotted from these resistivity measurements. The low resistivity of the fault and fracture zone and the higher resistivity of the rock of better quality on each side of it are clearly evident.

The results of the investigation indicated that the faults parallel to the tunnel would not be likely to cause difficulty in construction. The geophysical survey did disclose apparent zoning in the underlying rocks with one zone of badly fractured (probably bad ground) at an obtuse angle to the tunnel line. This tunnel has not been bored. However, in similar cases, it has been possible to check the geophysical results by drilling and surface geologic mapping. Such checks have shown that the resistivity method is useful if geological conditions are favorable for its application.

Seismic methods frequently have been used to determine depth to bedrock along proposed tunnel lines. For example, when the USBR was endeavoring to find a suitable location for the Low Gap Tunnel of the Columbia Basin Project, core drilling in detail would have proved costly and time-consuming, as some holes had to be over 200 ft deep. Seismic methods were used between widely spaced drill holes to discover thickness of overburden and basalt over tunnel grade. A total of 73 bedrock depth points were secured by seismic methods in the time that would have been required to drill two deep core holes.

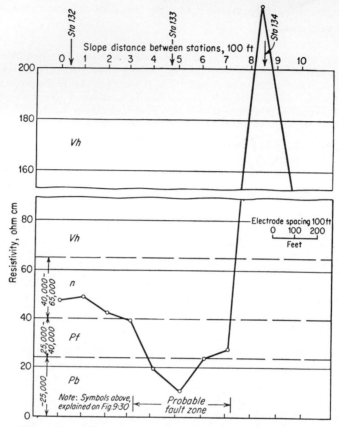

FIG. 9.31. Resistivity profile along a portion of a tunnel line.[19]

9.25. General Geological Comments on Tunnel Design. Three basic points as to the construction methods and corresponding costs should be clarified during the preliminary geologic survey for a tunnel:

1. Is the material hard or easy for tunneling, and will explosives be needed in construction? (A consultation with an explosives expert is essential in this connection.)

2. What part of the tunnel needs supports, and what type of supports will be needed?

3. Is water likely to be encountered, and if so, in what portion of the tunnel and in what quantities?

No preliminary geological survey can give conclusive answers to these vital questions, but in each case the most favorable and the most unfavorable possibilities should be estimated. Nearly all hard, massive rocks excavate readily and may stand without support unless severely disturbed by previous geological mountain-making or orogenic movements.

With diminished hardness, the workability of rocks improves but their stability decreases. Such is the case of many shales, clays, and similar materials, including decomposed rock. Fissility of the rock combined with the presence of nearby water basins located at a higher elevation may indicate a possibility of considerable water inflow into the tunnel. Similarly, the presence of surface springs from which warm water is issuing may indicate deep-seated water flows. Gas bubbles in nearby wells are a warning of possible gas at tunnel level.

Before the construction of a tunnel, it is desirable to have an idea as to the probability of excessive overbreakage and caving. Joints and especially faults are a major cause of severe caving ground in tunnels. Blocky ground may cause trouble if the contact surfaces between blocks are smooth. Surface reconnaissance or airphoto studies often are helpful in locating major fault systems and joint systems (see Sec. 7.18). Sometimes geological data can be obtained from nearby road cuts; e.g., in the preliminary survey for one tunnel, it was found that badly slacked material in an adjacent road cut was the same formation through which the tunnel was to be driven. This led to the correct conclusion that the overbreak in the tunnel would be excessive.[12d] The term "slacking" as used here refers to the surface change of the rock due to oxidation, carbonization, and hydration.

The presence of swelling rock often may be detected in the preliminary geological survey. Whether or not a rock is of the swelling type can be established only by tests on rock samples in their *natural* condition, with their *natural* moisture content fully preserved. Every shale or any other material known to be capable of swelling should be thoroughly investigated. The difference in the density of the rock upon its removal from the core barrel and after it has been exposed to the air for several days indicates the possible volume of swelling. The immersion test also may give indications of swelling ground.

Careful geological investigations prior to tunnel construction are of importance. Many tunnels, particularly those with rather small cross sections, are known to have been successfully built on the basis of an amazingly limited amount of geological information or practically none at all. However, these cases usually occur only when the builders are well acquainted with local geology through previous experience.

9.26. Geological Report. In this report, the results of the preliminary survey are outlined, along with all geologic features along the proposed tunnel line that may be helpful to the design and construction engineers. Complicated geological discussions should be avoided. It is essential to use clear, simple language. The items discussed in this chapter that are pertinent in the given case should be included in the report. Besides the purely technical aspects of the tunneling, the tunnel

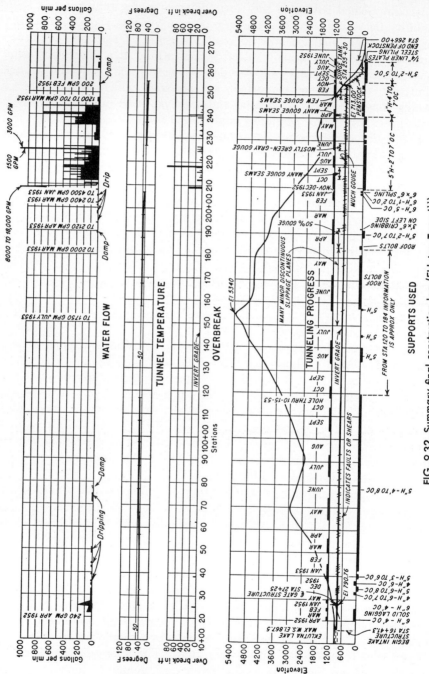

FIG. 9.32. Summary final construction log (Eklutna Tunnel[11]).

owners are interested in avoiding possible legal claims that may arise during and after construction. Therefore, the report should discuss whether or not the tunnel excavation is likely to cause damage to nearby buildings and structures, particularly possible settlements of city buildings and the possible tapping of springs or wells used for water supply. The report should include photographs of adjacent structures that might be damaged by the tunnel. If possible, nearby tunnels having similar geological conditions should be inspected, and the geology and the difficulties during their construction described. Sometimes it is necessary to report on the accessibility of the tunnel to construction equipment (roads, etc.), the availability of concrete aggregate for the tunnel lining, and the general climatic conditions in the area. (For example, in a region of severe electrical storms, lightning discharges hitting the ground above have been known to travel through cracks and fissures to tunnel level and prematurely set off explosives within the tunnel.)

Another feature that the geologist must make clear is the possible influence of the tunnel on the surface and subsurface water regime in the ground over the tunnel. This is particularly true in regions where water is scarce and thus highly prized for domestic and industrial use. Tunnels driven under lakes, rivers, and other bodies of surface water may tap a considerable volume of flow.[12d] (For example, during the construction of the previously mentioned Moffat Tunnel, a surface lake over 2,000 ft above it was completely drained into the tunnel.)

Final Construction Report. If a geologist is retained as a consultant for the whole construction period, he should submit a final report at the conclusion of the work. The basic differences between the predictions of the preliminary (preconstruction) report and what was actually found during construction should be thoroughly explained.[20] The tunnel should be carefully logged during excavation or immediately after the excavation is completed. An example of a summarized final geologic record is given in Fig. 9.32. This log may prove of considerable value if unusual maintenance problems develop in the given tunnel, e.g., severe cracking of the lining, or if another tunnel in the vicinity is planned. A complete log is of great value to the geologist and the engineer in their study of the relationship between surface geologic conditions and possible construction difficulties in future tunnel excavations.

REFERENCES

1. Suspension Roof Support, *USBM Inform. Circ.* 7533, September, 1949.
2. Thomas, E. N., A. J. Barry, and J. A. Metcalfe: Suspension Roof Supports, *Mining Congr. J.*, August and September, 1949.
3. Woodruff, S. D.: Rock Bolts: Theory and Practice in Tunnels, *Western Construction*, vol. 29, July and August, 1954.

4. Miard, H. E.: Sudden Release of Ground Stresses in the Coal Mines of Western Canada, paper given at the meeting of the Canadian Institute of Mining and Metallurgy, Edmonton, Alberta, April, 1953.
5. Desio, Ardito: "Geologia applicata all' ingegneria," Ulrico Hoepli, Milan, 1949.
6. Stini, Joseph: "Tunnelbaugeologie," Springer-Verlag OHG, Berlin, 1950.
7. O'Shaughnessy, M. M.: Construction Progress of the Hetch-Hetchy Water Supply of San Francisco, Calif., *ASCE Trans.*, vol. 85, 1922.
8. Louderback, G. D.: Faults and Engineering Geology, "Berkey Volume," 1950; also Lawson, A. C.: The California Earthquake of April 18, 1906, *Carnegie Inst. Wash. Publ.* 187, vol. 1, 1908.
9. Andreae, Charles: "Les grands souterrains transalpins," Leeman Bros., Zurich, 1948; also "Der Bau langer tiefliegender Gebirgstunnel," Springer-Verlag OHG, Berlin, 1926.
10. LeRoy, L. W. (ed.): "Subsurface Geologic Methods: A Symposium," 2d ed., Colorado School of Mines, Golden, Colo., 1950.
11. Judd, W. R.: Foundation Problems of the Eklutna Project, *Proc. ASCE Sep.* 445, June, 1954.
12. Wahlstrom, E. E.: Application of Geology to Tunneling Problems, with discussions by (a) B. C. Moneymaker, (b) F. A. Nickel, (c) Jacob Feld, (d) Hyde Forbes, and others, *ASCE Trans.*, vol. 113, 1948.
13. Proctor, R. V., and T. L. White: "Rock Tunneling with Steel Supports," with introduction by K. Terzaghi on tunnel geology, Commercial Shearing and Stamping Co., Youngstown, Ohio, 1946.
14. Terzaghi, K.: Liner Plate Tunnels in the Chicago (Ill.) Subway, *ASCE Trans.*, vol. 69, 1943; also Geologic Aspects of Soft-ground Tunneling, "Applied Sedimentation," John Wiley & Sons, Inc., New York, 1950.
15. Terzaghi, K.: Shield Tunnels of the Chicago Subway, *J. Boston Soc. Civil Engrs.*, vol. 29, 1942.
16. Loos, W., and H. Breth: Kritische Betrachtung des Tunnel-und Stollenbaues und der Berechnung des Gebirgdruckes, *Ingenieur*, vol. 24, 1949.
17. Terzaghi, K.: "Theoretical Soil Mechanics," John Wiley & Sons, Inc., New York, 1943.
18. Housel, W. S.: Earth Pressure on Tunnels, *ASCE Trans.*, vol. 108, 1943.
19. Wilson, J. H., and J. Boyd: Personal communication.
20. Zanaskar, W.: "Stollen und Tunnelbau," Springer-Verlag OHG, Berlin, 1950.

The following references are comprehensive treatises on tunnel engineering but are partially obsolete:

21. Rziha: "Lehrbuch der gesammten Tunnelbaukunst," Berlin, 1874.
22. Drinker, H. S.: "Tunneling, Explosive Compounds and Rock Drills," 3d ed., John Wiley & Sons, Inc., New York, 1888.

CHAPTER 10

FROST AND PERMAFROST

In many regions in the world, the engineer has to found his structures and make cuts and fills in ground that is frozen in winter only or the whole year around. Geological features of frozen and perennially frozen ground, known as "permafrost," and the engineering approach peculiar to work in such ground are described in this chapter.

FROZEN GROUND IN TEMPERATE ZONES

10.1. Frost Action in Temperate Zones.[1-3] Moisture within soil pores starts to freeze if the temperature drops to or below the freezing point (0°C = 32°F). Moist earth material when frozen may be compared with concrete; it loses its hardness, however, as soon as it thaws. Soil hardens when frozen because the moisture changes into a solid state. Completely dry sands isolated from any source of water do not freeze. For example, there is dust on earth roads in snowless countries, even during severely cold weather. Similarly it is relatively easy to open a gravel or sand pit during a cold winter, whereas opening a clay pit in frozen ground presents serious difficulties.

The depth of freezing varies according to climate. In Central United States it is between 3 and 4 ft, the building code of the city of New York sets it at 4 ft (though the actual frost depth may be less in many parts of the city), and in northern Maine it is from 4 to $4\frac{1}{2}$ ft. Winter temperature is lower on the outskirts of a city, and therefore the freezing depth is greater than at the center of the city; large, well-heated buildings in the downtown sections may substantially modify the temperature regime of the surrounding earth.

Frost action in temperate zones, particularly on roads and highways, consists of the following: (1) In winter and early spring, there are *frost heaves* spread along a section of the road, and (2) later in spring, the ground at the same places back-settles and softens (sometimes termed "boils"). The cause of frost heaves was formerly ascribed only to the freezing of pore moisture and corresponding increase in volume of the latter (about 9 per cent). Heaves corresponding to 20 or 30 per cent of the thickness of the frozen layer are not uncommon, however. This circumstance and the excess of water at thawed places suggest that during the freezing process, some additional water is drawn from a lower layer.

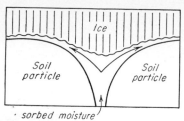

FIG. 10.1. Growth of ice crystals. (After Beskow.[2])

Laboratory and field investigations have proved this point.[3]

10.2. Freezing Point, Ice Crystals. The actual freezing point of water depends on the pressure on and in it. At a pressure of 1 atm it equals 0°C and is lower at greater pressures. In narrow soil pores, water is strongly compressed by attraction to the pore walls. Hence, while in wide soil pores water freezes at 0°C, the actual freezing point in narrower pores is below 0°C. Consequently at the 0° isotherm, i.e., the surface within the earth at which the temperature is 0°C, only a part of the water is frozen. Thus water in narrow soil pores is *overcooled*.

The freezing earth mass expands and exerts a heavy expansion pressure if confined; these phenomena are caused by the growing of *ice crystals*. For the formation of the ice crystals and ice lenses, water is required, and in addition to the interstitial or pore moisture, water is lifted up ("pumped") from the lower strata. There is a difference of opinion about the physical nature of that lifting agency. The growth of ice crystals may be explained in a simplified manner, however (Fig. 10.1). Before freezing, soil particles develop moisture films at their surface by capillary attraction from the water table (the latter is not shown in Fig. 10.1). When the ice crystal starts to grow, suction developed in this connection is stronger than the capillary attraction of moisture by the particle. In this way a soil particle loses its moisture film to the ice crystals which are being formed. The force of capillary attraction of the particles still persists, however, and moisture lost by the particle to the ice crystal is replenished by capillary attraction from the water table.

Excavation in frozen soils often shows the presence of ice lenses, or layers. In clean sands or gravels, the formation of such lenses, or layers, is simply the freezing of small inner pools of water. In silts or silty sands, ice lenses, or layers, are formed if freezing proceeds gradually, because sufficient water can be pulled up and frozen. In the case of rapid cooling, however, there is no such pull of water from lower strata, and only available pore moisture freezes. Figure 10.2 shows ice lenses in frozen soil cylinders obtained in Taber's experiments.[1] In freezing soils, ice also appears in the form of thin layers ("fibrous ice").

The capacity of rapidly changing temperature, known as *diffusivity*, is about eight times greater in ice than in water. At the boundary of frozen and unfrozen soil there is a zone (or curtain) where freezing and melting occur somewhat simultaneously before the soil definitely freezes. In fact, when ice is formed, heat is liberated, and before adjacent warmer water also freezes, ice melts. But when ice melts, heat is consumed (latent heat of fusion) and water freezes.

FIG. 10.2. Ice lenses in frozen soil cylinders. (*After Taber.*[3])

10.3. Frost Heaves in Temperate Zones. The phenomenon of heaves can be explained by (1) the increase in volume of the available pore moisture during freezing and (2) the drawing of additional water from deeper strata and its subsequent freezing to form ice lenses within the mass. This additional water may reach the freezing layer at a high rate if the following conditions are satisfied: (1) There is a nearby source of water supply from which water may be pulled, e.g., *a high water table* (i.e., a water table close to the ground surface); (2) the soil is capable of pulling water upward from that high water table; i.e., it contains a certain percentage of *fine particles*, as coarse soils can pull moisture through a very short distance only; and (3) the path between the water table and the freezing layer is unobstructed; i.e., the soil possesses *good permeability.*

Conditions (2) and (3) are satisfied by silts and clays, whereas condition (3) is best satisfied by silts. Hence, it may be concluded that silts are more apt to frost heave than clays.   This conclusion is widely corroborated by observations.

In the majority of cases a high water table is accompanied by a complete or partial saturation of the soils due to capillary rise and condensation of water vapor in the soil pores.   This condition is favorable for the formation of heaves if it persists when the frost period starts.   Furthermore, a gradual dropping of the seasonal temperature is a factor favorable for heave formation, since, in such a case, ice lenses and layers are being formed in the earth mass.   Conversely, if a heavy frost occurs suddenly, the pores become sealed and thus prevent further access of water.   Mild temperatures somewhat below the freezing point do not stop this access, however, because a part of the soil moisture is still in a liquid state.

According to A. Casagrande[3] the particle size critical to heave formation is 0.02 mm.   If the amount of such particles is less than 1 per cent, no heave is to be expected, but considerable heaving may take place if this amount is over 3 per cent in nonuniform soils and over 10 per cent in very uniform soils.   In rare cases, coarse sands with an excess of moisture have been reported to heave.

Pressure Developed during Heaving.   Information on such pressure is rather fragmental.   The following data are from Ref. 4.   A bridge pier 8 ft square and weighing 31,000 lb was heaved 2¾ in.   A brick wall weighing 2,000 psf (14 psi) was raised ¾ in., and several piers supporting columns and roof trusses were raised ½ to 2¾ in.   Taber[1c] measured expansion pressures of 104 psi and more during heaving in clay.

10.4. Technical Measures against Frost Action in Temperate Zones.

Moisture in the uppermost layers of a frozen soil melts in the spring, and since it is prevented from percolating downward by the underlying, still frozen soil, the melt water softens the earth mass.   On roads, this situation may be alleviated by underdrains (Sec. 5.32) that facilitate the removal of the excess water from the subgrade.   It cannot be overemphasized that underdrains, as any ditch or drainage, must have a certain minimum gradient (slope) and an adequate exit to carry the water away from the road.

Moisture may be drawn from the lower soil strata, not only by natural frost action, but also by *artificial frost*, as in cold-storage warehouses and freezing lockers.   The working temperatures in such buildings may be 14°F or lower almost continuously for a period of years.   As a result the frost depth may reach several feet, and on one occasion a depth of 9½ ft below the ground surface was reported.   The freezing isoline is deeper at the center of such a cold store than it is at the perimeter or near open corridors where normal temperature prevails.   Lightly loaded floors present small resistance to heaving; hence, they heave and crack before the heavier parts of the buildings do.   Thawing a damaged cold store should be done *very gradually;* otherwise, the excess of moisture that has been pulled from deeper strata for a considerable time may suddenly be released and convert the soil supporting the building into mud.   A similar reaction occurs under roads in the spring.[5]

Insulators tend to decrease the downward penetration of the freezing isotherm.   Snow is a natural insulator, and sand layers, tree leaves, and straw also may be used for the purpose.   The freezing point of the soil itself may be lowered by mixing it with solutions of calcium chloride, $CaCl_2$, or sodium chloride, $NaCl$, in concentrations of $\frac{1}{2}$ to 3 per cent by weight of the soil mixture.   Silty material susceptible to heaving may be partially or totally removed; this often is done in runway construction.

Building codes establish an official frost depth, and the base of the footings of the structure should be located *below this depth* to avoid possible heave.   The earth material located above the frost depth adheres to the footings, but this force of adhesion is negligible and cannot produce any uplift of the structure (except in permafrost areas, as later described).   Water supply lines and sometimes sewers also should be located below the official frost depth.

Spillways for dams (Sec. 15.6) also must be protected from frost-heave damage.   If the foundation materials for such structures (particularly the large but relatively thin slabs in the spillway chute) are susceptible to frost heave, the slabs may crack and heave out of place.   In the case of spillways not only critical soils but also critical rock types must be studied as to their susceptibility to frost heave.   Some of the fine-grained, porous rocks, such as chalk, have been reported to attract capillary water to distances up to 9 or 10 ft; as a result ice lenses may form in the rock and heave the overlying concrete slabs.   A thick sand and gravel blanket under the spillway slab relieves the situation, since this coarse-grained material acts as an insulator and also stops the upward water flow. Consideration also should be given to the use of certain lightweight aggregates (Sec. 8.24) in the concrete of the spillway slab; these aggregates can change the thermal characteristics of the concrete sufficiently to provide some insulating action.

PERMAFROST AND ITS GEOTECHNICAL PROPERTIES

10.5. Definition and Origin of Permafrost.   Permanently frozen ground, or "permafrost," is defined as a deposit of soil or rock of variable thickness and depth in which subfreezing temperatures exist throughout the year and probably have existed continuously for a long time (presumably for 2,000 years or more).

According to many theories, the origin of permafrost can be traced to the period of refrigeration of a large portion of the earth's surface at the beginning of the Pleistocene or Ice Age, perhaps a million years ago. Validity of this glacial origin theory has not yet been proved, however. If the last glaciation was responsible for originating the permafrost, its deposits should be decreasing by now. Observations show that apparently because of climatic changes, permafrost deposits vary in their thickness and areal extent. Where the mean annual temperature is below freezing, permafrost may be forming even at the present time. Generally, an increase in thickness of a permafrost deposit is termed *permafrost aggradation;* and conversely, decrease in thickness is *permafrost degradation.*

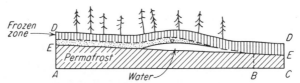

FIG. 10.3. "Drunken Forest." *DE* is the thickness of the active zone; frost zone between *B* and *C* coincides with the active zone.

Permafrost underlies approximately one-fifth of the land area of our globe. Permafrost areas include about 80 per cent of Alaska, half of Canada, a considerable part of Siberia, and some areas in China. Presumably, there is permafrost in the antarctic. General information on permafrost, both geological and technical, may be found in Refs. 6 through 10.

10.6. Basic Features of Permafrost. The permafrost terminology contains many terms borrowed from Russian, though an attempt has been made to replace these terms by those of pseudo-Greek origin.[11]

The top and bottom surfaces of permafrost deposits are not horizontal. The irregular top surface of a permafrost deposit is the *permafrost table.* All ground above the permafrost table is designated as the *active zone.* The upper part of the active zone is subject to intermittent freezing and thawing and is called the *frost zone.* Where seasonal freezing penetrates to the permafrost table, both frost zone and active zone coincide (Fig. 10.3); otherwise, between the bottom of the frost zone and the permafrost table there is unfrozen soil known as *talik* (Fig. 10.4). If a cold summer is followed by a cold winter, a frozen layer may develop at the bottom of the active zone. The product of this upward permafrost degradation that may remain unthawed during one or more summers is called *pereletok.* Water contained in permafrost is mostly in the form of ice, though fine capillaries may contain fluid water and there may be groundwater flow in permafrost (Sec. 10.11). Ground perennially frozen but containing no ice, as in the case of some sandy materials or some bedrock, is *dry permafrost,* or dry ground. Dry ground is not subject to heaving or settling, and it does not change appreciably in volume. The opposite is *saturated ground*—silt, clay, and gravel containing considerable amounts of binding ice or ice in the form of crystals, lenses, veins, or layers. Thawed saturated ground is soft and sometimes reaches the consistency of mire. Finally, permafrost containing but a limited amount of ice is *moist* ground. There may be intermittent layers of permafrost and unfrozen ground, and there may be frozen lenses or kidneys in nonfrozen soils, sometimes very large.

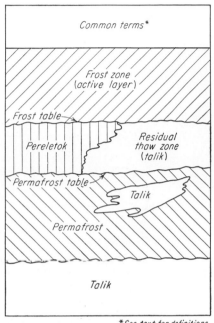

The Army Engineers organization known as SIPRE (Snow, Ice, and Permafrost Research Establishment)[24] classifies permafrost according to the continuity of the frozen mass: (1) *continuous*, where the ground is frozen to full depth without unfrozen inclusions, and (2) *discontinuous*, alternate layers of frozen and unfrozen material. A third type, occasionally referred to, is *sporadic* permafrost, wherein only occasional bodies of permafrost are found in what otherwise appears to be a permafrost-free area. Another type of classification divides permafrost into (1) dry permafrost and (2) detrimentally frozen permafrost wherein the soil contains a large percentage of ice both as a cementing material and as randomly oriented crystals. This classification is just a systematization of the information previously given in this section.

*See text for definitions

FIG. 10.4. Permafrost terminology.

Seasonal thaw in the frost zone penetrates from 1 to 10 ft, depending on insulation, insolation, drainage, and type of soil or rock material. The active zone may vary from 2 to 14 ft or more. The perennially frozen layer itself may be from several inches to about 2,000 ft thick, as in some parts of northern Siberia (unofficial data). In Alaska the greatest thickness is about 1,300 ft, south of Point Barrow.[8] These figures are merely indicative, since continual boring investigations in the arctic may give still greater values.

Permafrost is found in both glaciated and nonglaciated areas. Apparently permafrost is thicker in nonglaciated areas (Klondike about 200 ft, Fairbanks over 300 ft). The thicker permafrost sections are continuous except beneath big river valleys such as that of the Yenisei River in Siberia. Investigations by the Army Engineers[18] have shown that in general, permafrost is deepest beneath hills and shallowest in valleys. Also, permafrost may be absent under sources of either natural or man-made heat supply such as buildings. Permafrost thins abruptly to the north under the Arctic Ocean, and it gradually thins to the south, forming discontinuities.[8]

Perennially frozen bodies of mammals and also bacteria are found in permafrost deposits; the mammals are bison, the prehistoric horse, the mammoth, rodents, and others.

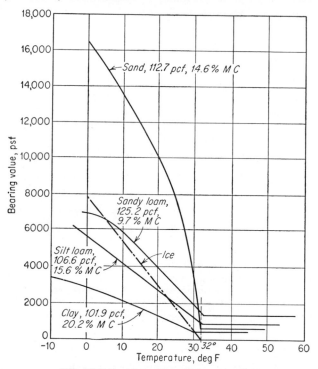

FIG. 10.5. Bearing value of frozen soils.[24]

10.7. Soils in Permafrost. All kinds of frozen soils are found in permafrost,[22] particularly in Alaska. Frozen gravels often contain sticky, micaceous clay, especially near bedrock. Frozen silt is termed "muck" by the miners. (As well known, this term also is applied to the excavated material in tunnels.) "Slud" is frozen, soft, wet mud or mire.

The following two types of soil may cause considerable settlement and even failures of structures if present in the foundation. One is gravel embedded in large quantities of ice. Practicing engineers prefer to remove this material completely[26] because of its inherent instability if thawed. The other critical type is a silty frozen soil with a high percentage of ice that generally becomes waterlogged and spongy when thawed.[26] These silt deposits usually are very thick, and the material has practically no shear strength when thawed.

Soil Strengths and Bearing Values. As a matter of comparison, frozen clays have the lowest shear strengths and frozen clean, cohesionless materials have the highest shear strengths. In any case the shear strength is dependent upon the amount of ice in the soil, its distribution in the soil, the rate of loading, and the outside temperature. For example, the ultimate shearing strength of frozen soils has been found to increase with a decrease in temperature below 32°F. Figure 10.5 illustrates the influence of the temperature on the shear strength of ice and the various soils tested. It appears that the changes in shearing

TABLE 10.1. Indicative Permafrost Vegetation*

Vegetal type	*Characteristics of underlying ground*
Drunken forest (Fig. 10.3).........	Frost mounds and swelling ground
Willows........................	Ground water that freezes for only a short time
Peat and moss (tundra)...........	Thin active layer
Pine and fir.....................	Permafrost absent or very deep. Granular soil
Larch and birch.................	Shallow permafrost
Spruce.........................	Wet ground and high permafrost table
Tree-root depth.................	Indicates thickness of active layer
Curved tree trunks..............	Indicates a creep of the slope over the permafrost table

* From D. J. Cederstrom, P. M. Johnston, and S. Subitzky, Occurrence and Development of Groundwater in Permafrost Regions, *USGS Circ.* 275, 1953, and U.S. Engineer Office, A Test Study of Foundation Design for Permafrost Conditions, *Eng. News-Record,* Sept. 18, 1947.

strength of frozen soils, observed when the temperature changes, are controlled by the respective changes in the shearing strength of the ice contained in the frozen ground.

Information on the safe bearing values of permafrost is insufficient. It may be assumed, however, that the conservative bearing values for frozen soils are as follows: clays and silts, 2 tons/sq ft; fine sand, 3 tons/sq ft; coarse sand, 4 tons/sq ft; and gravel about 5 tons/sq ft. Generally the bearing values increase with a decrease in temperature below freezing and with an increase in dry density.

10.8. Permafrost Vegetation. Forests cover about half the areas of perennially frozen ground in Alaska. Outside this timber area, most (but not all) of the remaining territory is flat and covered with a dense vegetal mat, chiefly composed of lichens, mosses, sedge, berries, and dwarf species of a few trees. This dense mat is called *tundra.* The tundra vegetation grades downward into a layer of peat, and because the flatness of the country prevents surface drainage and there is a lack of subsurface drainage, this peat layer often is saturated like a wet sponge. This circumstance explains the abundance of swamps in the tundra. The typical tundra is very difficult to traverse in the summer because of the swampy ground, shallow ponds, and tussocks of stiff cotton sedge (locally known as niggerheads).

The term "tundra" is also applied, particularly in the Russian literature, to the arctic regions with the natural conditions described.

10.9. Landforms and Surface Features in the Arctic. Mounds of many types can be observed throughout the arctic and subarctic. Some of them are locally known as "pingo," and their origin may be explained either by erosion (dome-shaped hills) or by upward flow of earth suspensions, forming mud cones, locally known as "mud volcanoes." Series of mounds resembling haystacks also may be observed; some of them are as much as 100 to 300 ft in height and $\frac{1}{2}$ miles in diameter.

Some mounds are associated with ground water, such as a seasonal mound variety known as "soil blister." In Fig. 10.3 an intermediate stage of formation of a blister is shown, when the laterally flowing water becomes confined because of the adfreezing of the active zone to the permafrost (distance BE in Fig. 10.3). As the frost penetrates through the whole depth of the active zone all around the ground-water stream, pore pressure increases and the blister may finally explode. Some water then flows out from the blister and freezes, and some water freezes inside the blister. The ice inside the blister does not always melt in the summer, however. Generally, blisters change their position from year to year; their height seldom exceeds 20 ft.

Oversize heaves associated with ground water are termed "hydrolaccoliths" by the geologists. As in the case of a simple blister, the basic cause of hydrolaccolith formation is probably lateral (i.e., more or less horizontal) movement of the ground water to the mound in addition to the vertical pumping of water from lower strata. (The latter phenomenon is similar to that which occurs during freezing in temperate zones; see Sec. 10.3.) It has been suggested[14] that this lateral motion may be caused by freezing of water in water-bearing layers and pushing it to a point of weakness in the overburden, where a heave is formed. Another cause of lateral motion of the ground water, namely, trapping of the ground water between the impervious active zone and the permafrost, is explained in more detail in Sec. 10.11. If the ground water moves laterally, obstructions may contribute to the formation of mounds. According to a hypothesis,[15] local freezing of laterally moving moisture may block further drainage and create heaves ("ice-blocked drainage"). Heaves in the arctic may form in both soils and rocks (Fig. 10.6).

In many places of the arctic, surface cracking forms *soil polygons* of a variety of sizes, shapes, and types (Fig. 10.7). The two basic

FIG. 10.6. Rock massives heaved by ground ice.   (*From D. J. Belcher, Highway Research Board Bull.* 13, 1948.)

FIG. 10.7. Polygonal ground.

polygon types in permafrost areas are (1) those with depressed centers or pans that are enclosed on their perimeters by raised dikes and (2) those with raised centers that sometimes outline very deep channels. A polygon may vary from 15 or 20 to about 200 ft across. The number of sides generally is five or six, and a pattern of smaller polygons often develops within the larger polygon.[16]

Elongated lakes are another feature of the arctic landscape. The lakes are nearly ellipsoidal in shape, the long axis being located somewhat west from the meridian. The elongated lakes occur in sands and silts. Perpendicularly to the major direction of those lakes there often are elongated dunes. The term *thermokarst* refers to subsidences of ancient lakes or basins caused by thawing of large ice masses. These lakes appear to be sunken below the surrounding land surface and do not have outlets.

Surface features (microrelief) on gentle slopes are of especial interest to the arctic engineer, since roads often are built along such slopes to avoid wet, low places. Unstability of gentle slopes is often caused by *solifluction*, or flow of saturated earth material downslope. In some instances solifluction may be started simply by vibrations.[14] *Slow creep* and *viscous flow* of the soil material also are responsible for the formation of surface features on gentle slopes in the arctic (compare Sec. 17.12). These features are soil terraces, lobate terraces, lobes, and tundra mudflows.

Soil terraces (or solifluction terraces) are formed on slightly sloping ground and may be visualized as excavations or slides parallel to the contour lines of the locality, 300 to 4,000 ft long and 100 to 600 ft wide. The surface of a soil terrace generally, is gently sloped, being bounded on the upper side by a steep slope 3 to 20 ft high. Most of the terrace surface is covered by turf and peat. The terrace soil generally consists of coarse sandy silt, with rock fragments and even boulders; its thickness is from a few feet to 20 ft. The substratum may or may not be frozen.

Lobate terraces are characterized by festooned, practically vertical slopes, 1 to 5 ft high, forming individual lobes 20 to 100 ft wide. Most of the terrace surfaces are covered with turf and peat. Soil consisting of sandy silt with rock fragments is slightly movable in wet seasons. The tract of a small railroad north of Nome, Alaska, constantly moves out of alignment where it crosses lobate terraces. *Soil lobes* are tonguelike microrelief features 1 to 5 ft high, 10 to 30 ft wide, and 20 to 150 ft long, formed mostly by viscous flows of the material. Drainage of lobes is very poor. *Tundra mudflows* result from a sudden spewing of fluid soil down steep slopes.

10.10. Temperatures in Permafrost.

The knowledge of ground temperature in permafrost is helpful in determining the bearing capacity of the ground, the thickness of the active layer, the presence of *taliks*, seasonal fluctuations in the permafrost table, the areal extent of permafrost, the effect of buildings on permafrost, and whether or not piles have frozen in enough to be loaded by a building. Up to 1950 the minimum temperature of permafrost found in Alaska[8] was $-9.6°C$ (about $14°F$). According to an informal communication, the minimum temperature of permafrost measured in Siberia was $-12°C$ (about $8°F$).

The precise determination of permafrost temperatures should be made in the beginning of autumn, when the active zone is at its lowest level.[17] To perform these measurements, thermocouples or resistance thermometers may be used. The *thermocouple* is the union of two wires of dissimilar metals, joined at their extremities, for producing an electric current; the voltage of the current is proportional to the difference of temperatures at the two extremities. In order to measure the earth temperature, one extremity of the thermocouple is placed within the drill hole at the given depth and the temperature of the other extremity, that above ground, is measured with a standard mercury thermometer.

From this temperature and the voltage of the electric current, the temperature at a given point within the drill hole may be computed. Thermocouples generally are effective only to about 50 ft in depth.[17] *Resistance thermometers* operate on the principle that the resistivity of certain metals to the flow of electric current depends upon the temperature. The change in resistance is measured by a galvanometer placed at the ground surface. Resistance thermometers may be used successfully at great depths.

For all temperature measurements it is advisable to keep the thermometer in the hole about 6 hr. If the thermometer is placed inside an iron pipe (drill casing), an error of ± 0.2 to $0.3°C$ is likely to occur. Because permafrost can exist when the temperature is only slightly below 32°F or 0°C, all thermometers used should be precise to 0.01°C.

Taber[18,19] measured temperature near Chatanika in a 93-ft hole where the ground remained undisturbed, except for the drilling of the hole several years before. The temperature of the air was 68°F. At a depth of 10 and 80 ft the temperature was 31.5°F. The minimum reading (31.2°F) was at a depth of 40 ft.

The expression "thermal regime" of the permafrost in a given locality means the state of certain equilibrium of the active zone under given environmental conditions including a constant position of the permafrost table. The thermal regime may be affected not only by natural factors but also (and perhaps mostly) by the activities of man, such as stripping vegetation, building cuts and fills, and drainage. When the thermal regime is upset, the permafrost table may be shifted upward or downward, often with undesirable consequences.

The algebraic difference between 32°F and the mean daily temperature in degrees Fahrenheit is known as the *degree-day* for a given date. It is positive $(+)$ when the mean daily temperature is below 32°F and negative $(-)$ when above 32°F. If, starting with a certain date, e.g., the beginning of November, the cumulative number of degree-days is plotted against time for the whole year, a curve similar to that shown in Fig. 10.8 is obtained. The vertical difference between the highest and the lowest point of this curve as expressed in degrees Fahrenheit is the *freezing index* of the locality. The values of the freezing indexes for Alaska are from about 500 at the Arctic Ocean to about 3,000 at the middle of the peninsula, after which, approaching the Gulf of Alaska (Pacific Ocean), it drops again. The general direction of the freezing index isolines in Alaska is from west to east. It should be noted that the isotherm of 30°F average yearly temperature passes very close to the south coast of Alaska. The depth of thaw at various points in Alaska has been correlated with practical observations[20] by the Army Engineers; they have prepared a diagram showing the total thickness of pavement required to prevent freezing of runway pavement (Fig. 10.9).

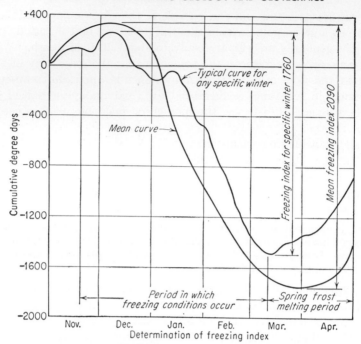

FIG. 10.8. Freezing index.[20]

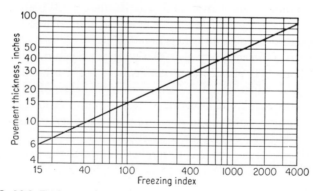

FIG. 10.9. Thickness of pavement as a function of the freezing index.[20]

10.11. Ground Water in Permafrost Areas.[12]   Ground water may be (1) above, (2) within, and (3) below the permafrost.

1. *Water above the Permafrost.* In warm months this may be a source of limited water supply. However, because of its shallowness, contamination may be a problem, and the water usually disappears in winter. If the active layer consists of impervious materials, ground water may be trapped between the active zone and the permafrost. In such occurrences the trapped water under pressure may move horizontally

and contribute to the formation of hydrolaccoliths. Sometimes conduits are formed in the active zone by this water, or it may appear in thawed zones beneath heated buildings in the vicinity of streams or abandoned meander channels.

2. *Water within the Permafrost.* Such water generally occurs in alluvium near rivers, abandoned river channels, or thawed gravel beds. It may also occur in small thawed areas lying between masses of permafrost. Other common occurrences are in or near standing bodies of water, south-facing hillsides, and places where insulation has been removed, e.g., vegetation stripped.

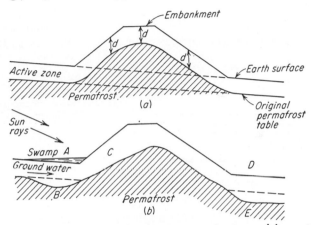

FIG. 10.10. Disturbance of the thermal regime by an embankment: (a) growing of permafrost into the body of the embankment, (b) formation of swamps and deepening of the permafrost table at the sides of the embankment. Nonsymmetrical permafrost table (point C).

3. *Water below the Permafrost.* This occurs in large quantities and generally satisfies the sanitary requirements. It is located mostly in alluvium under the permafrost but also may be in joints, channels, and other available space in the bedrock. In hilly regions, ground water under the impervious permafrost may stand under high pressure, and flowing wells (Sec. 5.12) may result (Dawson, Yukon, Fairbanks). Wells in alluvium 40 to 200 ft deep have been drilled near Fairbanks.

10.12. Ice Fields, Icing.[7,21,22] Figure 10.10a illustrates the disturbance of the thermal regime by embankment (highway fill or earth dam). If the active zone is thin, in other words, the permafrost table is high, the body of the embankment may freeze down to permafrost. It will thaw up, however, only to a certain depth marked d in Fig. 10.10a. Thus permafrost will grow into the body of an embankment. If the embankment is parallel to the direction of the subterranean flow, this ingrow does not influence the behavior of the embankment, but if the embankment is perpendicular to the direction of flow and in addition one of the slopes faces south, the situation will be as in Fig. 10.10*b*. In this case, the ground water, stopped in its movement, softens the soil and forms swamp *A*, depression *B* is formed in the permafrost table, and the ingrowth of the permafrost is not symmetrical with respect to the embankment (point *C*).   Furthermore, if the locality at *D* was swampy, it will dry out, and permafrost at point *E* will thaw up.   Thus the stability of the embankment slopes will be threatened.

As the winter freezing of the soil around point *A* sets in (Fig. 10.10*b*), the arriving ground water is placed under pressure. The pressure gradually increases, since the amount of ground water grows up but the space available for it shrinks because of the ice formation.   Finally the hydrostatic pressure causes the water to force its way to the surface

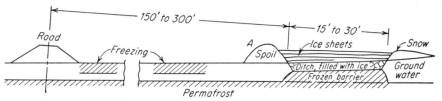

FIG. 10.11. Measures against ice fields, frost-belt method.

of the ice, where it spreads and freezes in successive sheets of ice.   The "ice fields" or "icings" thus formed (the Russian term often used is *nahledee*) are parallel to the embankment often along a considerable distance and attain a thickness of 3 to 10 ft and more.   A similar phenomenon occurs on shallow rivers and brooks when, because of the ice formation, the water in a stream becomes constricted and is forced up through the fissures and weak spots of the ice cover.   Thus river icing is formed, which is distinguished from the ground icings previously described.   Ground icing is very troublesome in road maintenance in the arctic, as the ice sheets may cover a highway for thousands of feet.[21]

Icing control may consist of drainage (swamp *A* in Fig. 10.10*b*), or in shifting the ice field upslope, i.e., against the down-flowing ground water. The "frost-belt method" for such shifting is shown in Fig. 10.11.   The vegetation is stripped and a shallow ditch is dug 150 to 300 ft away from the road upslope.   The ditch may be trapezoidal or otherwise, about 18 in. or more deep, and from 15 to 30 ft wide.   The depth of freezing thus will be greater at the ditch than in the adjacent ground.   If the ditch depth is correctly chosen, the frozen material will reach the permafrost table and thus form a barrier for downward-flowing ground water. Earth material from the ditch ("spoil") should be piled on the downslope of the ditch (point *A* in Fig. 10.11).   In this way, an induced ice field is formed (marked "ice sheets" in Fig. 10.11, right).

Ice fences and heating also have been used to prevent the encroachment of ice sheets upon the highways. In the heating method large oil heaters are placed along the face of the ice sheet before it starts to encroach upon the roadway. This method, although expensive, has proved effective. Provision has to be made for draining the melt water by means of a culvert under the road.

GEOTECHNICAL INVESTIGATIONS IN PERMAFROST AREAS

10.13. Surface and Subsurface Explorations. Because of the variety of factors that may influence the thermal regime under a structure, geotechnical investigations in the arctic generally are more complicated than in the temperate zones. Such studies generally must be confined to the summer or fall months, as snow cover and rigorous temperatures inhibit investigations in the winter. Furthermore for isolated areas, summer studies are almost a necessity, as winter conditions may completely prevent access to these areas. One of the most important considerations in arctic soils investigations is the *self-sufficiency* of the investigation party. Because of the remoteness of much of the arctic from normal transportation routes and more particularly from normal sources of supplies, geologists and engineers engaged in surface or subsurface exploration should have on the project from its very start everything in the way of machinery and instruments required to complete the work. Aerial transportation has been a great assistance in investigations, but the variability of weather, and thus flying conditions, may make airplane communications undependable when most needed. Also, considerable time must be allowed for the purchase of supplies and their transportation to the project site. Unlike most areas in the United States, a driller in the arctic cannot telephone or telegraph a manufacturer for a drill part and expect to have it in a day or two.

Surface Explorations. Among the most valuable assets to the surface studies in the arctic are airphotos. Much of Alaska and Canada has been covered by airphotos flown by the U.S. Air Force, the U.S. Navy, or their Canadian counterparts. From such photographs, permafrost often can be identified[23] by vegetal types (Sec. 10.7), polygonal cracking, and other characteristic landforms. Rock outcrops can be located for further ground study, and accessibility for surface parties can be established.

When rock outcrops are mapped from the ground, the degree of weathering and the amount of jointing should be recorded. If the joints are sealed, the material filling the joints should be identified as to whether it is mud, ice, or a mineral deposit. The size and orientation of the joints also are to be recorded. As joints are opened by frost action, they may most easily be studied in the late autumn or early spring. In such projects as highway route studies, variations in rock character in a lateral direction should be mapped. If the geologist is searching for ground water, he should note if bleaching or encrustation is present on the rock; either of these is an indication of moisture action. Polygonal frost cracks also are important in locating ground water, because they permit infiltration. Also in ground-water studies, the porosity of the ground and the incidence of caverns should be noted.[12]

Topographic features, such as the slope of the locality and its orientation with respect to the sun's rays and prevailing winds, and drainage conditions (in the case of large structures the drainage patterns should be noted) should be studied.[12]   Besides the data normally required in soil studies in temperate zones, in the arctic and subarctic, the thickness of the active and frost zones and the local thickness of permafrost (estimated in difficult cases) should be recorded. Botanical studies are especially indicative and in important cases should be done by a plant ecologist familiar with permafrost conditions.

Subsurface Explorations.   In the simplest case, exploratory holes are sunk; sampling is done by breaking the material with a chopping bit and recovering the soil and rock fragments. Crushing the material between the fingers may give an idea on the degree the material is saturated with ice. A high crushing strength similar to that of dry stiff clay indicates a high ice content, whereas a low crushing strength similar to that of dried silty sand is indicative of low ice saturation. There are, of course, a number of intermediate gradations. Another simple method of subsurface investigations is driving heavy pipes or rails into the permafrost and counting the number of blows per foot. It may be argued that in such a case the permafrost table may be confused, for example, with an unfrozen cemented sand-gravel bed. However, in the case of a large area and a great number of tests, such an occurrence is improbable. In the preliminary subsurface studies, the permafrost table may be located by continuous temperature measurements in a drill hole, provided the hole does not contain free water.

In the case of *important structures*, samples are taken by using rotary samplers, churn drills, or jet-drive rigs. The cable-tool rigs (churn drills) usually are very good for drilling, although cores cannot be recovered. The diamond or rotary drill is satisfactory where cores have to be obtained at great depths. In the jet-drive rig, a wash pipe with perforations at the bottom is forced into the ground by the washing action of the water jet, a small weight being used to drive the pipe. A high water head on the pipe is used. Though this rig is economical for drilling in permafrost, holes deeper than 200 ft cannot be drilled, and the presence of boulders in the ground hampers its action.

Drilling in the arctic requires heating the drilling fluid.[12]   The temperature of the fluid, however, should not exceed 40°F to avoid excessive thawing of the permafrost and consequent caving.   Where caving is not a problem because the ground is stable, hot water at temperatures between 90 and 100°F (or higher) may be used.   A sufficient supply of hot water (estimated at about 400 gph or more per rig) should be available, the drilling generally being done around the clock.

Where it is difficult or costly to heat water, salt brine can be used. The brine is usually a suspension of 35 lb of sodium or calcium chloride in 53 gal (a barrel) of water.   In all cases where drilling fluid is used, the drilling rods should be prevented from freezing to the walls of the hole; otherwise it may prove impossible to retrieve them.

In air-circulation drill rigs, no drilling fluid is necessary.   The cuttings produced by the bit action are simply blown out by compressed air blown through the rods. With such rigs, thawing of the permafrost is held to a minimum, and in most instances no casing is required.   However, in 1955, there were practically no air-circulation drill rigs on actual projects in the arctic.

Practice has shown that in the extremely frigid and windy arctic zones, where temperatures may be as low as minus 25°F, it is necessary to protect the drill motors from freezing.   This can be done with large canvas coverings or wood shacks (wanigans).   The latter also serve to protect the operators of the rigs.   At the same time, measures should be taken to prevent thawing of the ground under the rig and the subsequent sinking of the rig into the ground.

Geophysical Investigations in Permafrost.[25]   The resistivity method of geophysical investigation can be used to determine the areal extent and approximate depth of permanently frozen unconsolidated deposits.   The resistivity values of frozen silt and gravel are from 20 to 50 times greater than those of their thawed counterparts. The resistivity traverses permit the edges of frozen deposits to be located and mapped; the resistivity depth measurements permit the thickness of the frozen layers to be ascertained.

CONSTRUCTION IN PERMAFROST AREAS

10.14. General Criteria.   As in the case of geotechnical investigations (Sec. 10.13), construction in the arctic should be well planned on the basis of careful advance reconnaissance.   The work should be scheduled according to seasons, and supplying of equipment at the proper time should be well organized.   Mechanical equipment cannot be depended upon entirely, since in extreme cold its efficiency rapidly decreases and breakdowns increase.   The human element should be taken good care of in order to maintain satisfactory work standards, although the latter tend to be lowered owing to the rigors of the climate.

Fire protection of the work, the construction camps, and the housing projects to be built requires special attention.   It is impossible to fight fires effectively in subzero temperatures with the usual water supplies available, and once equipment is destroyed by the fire, it is difficult to replace. Furthermore, fires can be fatal to the working force if they become exposed to the weather by the destruction of their habitations.

Water Lines. Though wells can be sunk in permafrost without particular difficulties, distribution of the water to houses and other buildings is difficult. If the pipelines are buried in the permafrost, they must be insulated to protect the line against freezing and also to prevent thawing of the permafrost and subsequent breakage of the lines due to settlement. In some instances the lines are placed on the surface in covered timber or concrete boxes (utilidors) which are heated by steam lines running parallel to the pipes. Here, also, insulation of the boxes is necessary to prevent thawing of the underlying ground. In some cases, the lines have been supported on boxes which in turn are supported by short piles driven into the permafrost. Presumably, if water and sewer lines are placed in the same utilidor, precautions must be taken to prevent infiltration of surface water and flooding of the box with subsequent contamination of the water supply lines.[28] Another method frequently used to assure an unfrozen water supply is to maintain a continuous flow in the water lines. The water is heated prior to placement in the system and then is continuously pumped through the pipes by means of a double circulation system.

Sewage. Disposal of sewage in the arctic is a difficult problem.[27] Some areas use septic tanks enclosed in special heated buildings. In the construction of any such buildings for housing septic tanks, filters, or other processing devices, considerable attention should be given to the possible thawing of the foundations due to the heat generated by the processing. Disposal of sewage by the means used in temperate climates is unsatisfactory; the normal natural processes of organic decay and subsequent harmless rendering of the bacterial matter do not function properly in the arctic. The actual disposal of the sewage waste products may be done in primitive ways such as placing it on ice floes and permitting it to drift to sea and later sink.

Construction Time Table. According to the practice of the U.S. Armed Forces[28a] a generalized planning timetable for the construction of a large permanent installation in an undeveloped arctic area would be

1. *Reconnaissance*—spring through midsummer.

2. *Geotechnical investigations*—late summer and early winter.

3. *Preliminary construction phase* (involving movement of heavy equipment overland, organization of construction camps, clearing, opening of borrow pits, excavation, and stockpiling)—winter through the next summer.

4. *Construction proper* will then start the following spring (after about 2 years of preparation).

In the tundra regions, movement of equipment usually is confined to the winter. This permits haulage by tractor "trains" over frozen lakes and ground. During the thawing months the boggy nature of tundra makes ground transportation difficult, if not impossible.

Two Methods of Construction in Permafrost. There are two known methods of construction in permafrost: (1) the *passive method*, wherein the thermal regime (including the level of the permafrost table) is preserved, and (2) the *active method*, in which permafrost is completely removed by thawing and excavation. Figure 10.12 shows schematically

FIG. 10.12. Passive method of construction of a fill.

the application of the passive method in the construction of a highway fill with berms on both sides to protect the permafrost from thawing. The earth material of the berm may be placed directly on existing vegetation, or vegetation may be stripped and the active zone of the soil removed and replaced with an insulating blanket, consisting, for instance, of coarse sand and gravel. The active method may be used in this case only if permafrost is thin and the ground after thawing will have satisfactory bearing power to support the fill.

10.15. Runways and Roads in Permafrost Areas. In the construction of runways and roads, both the passive and the active methods are used. The passive method, where practicable, is preferred because atmospheric heat will not penetrate so fast or so deep through a natural vegetal cover as through artificially placed materials, and therefore the permafrost will not thaw in the summer. The insulating capacity of the construction materials should be at least the same as that of the natural cover. Furthermore, in wintertime, cold should not be allowed to penetrate into the ground to such a depth as to affect the flow of ground water and disturb the existing thermal regime. This leads to the necessity of placing a substantial *insulating layer* under the riding surface. Very briefly the passive method in such construction is to push the moss and brush over the proposed site to form a base; this is packed and subgrade placed over it. Under such circumstances, the permafrost table will hold and may even rise. On highways, destruction of the road pavement is inhibited by discontinuing heavy traffic and minimizing general traffic during the critical season, i.e., the fall and early spring.

Presumably, material used as a subgrade should not be susceptible to frost heave.

In the active method of construction, the entire active zone is removed and replaced with porous granular material, such as coarse gravel, and the roadway or runway is placed directly upon this material. This method generally is applicable only in arctic areas where the mean annual temperatures are at or near freezing. Otherwise, on the days during which the temperature rises considerably above the freezing point, the permafrost will thaw owing to the lack of insulating cover, and runway damage is likely to occur. Construction details of both concrete and bituminous pavements in arctic zones can be found in Refs. 20 and 29.

Generally in locating runways and highways in permafrost areas, cuts should be avoided because (1) excavation of permafrost is difficult and thus expensive and (2) if ground-water channels are intercepted, the thermal regime may be disturbed and ice fields may be created. Embankments should be built as far as possible from places where they may produce ice fields, as, for instance, along brooks or close to valley bottoms. Hill slopes and tundra regions are convenient for construction provided there is no danger of tundra mudflows and there are no lobate terraces or lobes on the slopes.

Drainage. As it is everywhere else, the problem of drainage is important in the arctic for both runways and roads and even more important for the latter, since, as a rule, they traverse much rougher terrain. The side ditches should be as far from the crown of the road as practicable; also, narrow, deep ditches are preferred to wide, flat ones. Narrow ditches provide more protection from cold air and are less susceptible to icing. Bridges and culverts should be high enough to clear stream ice, which can build up quite rapidly during the winter and during the spring breakup.

10.16. Dams in Permafrost Areas.[30,31] *Interaction of Dam and Permafrost.* Operating water reservoirs in arctic zones is possible, since only the upper 12 ft of the reservoir water is subject to freezing under the most severe frost conditions. Water under this cover remains unfrozen and available for use. In turn, the relatively warm body of water thaws at least the upper strata of the underlying permafrost. Using thermodynamics, equations may be established showing the relationship between the reservoir depth and the depth to which the permafrost will be thawed. As a rough average, however, the depth of thawing of the permafrost under the reservoir may be estimated at one-fourth the depth of the water in the reservoir. On the other hand, the level of permafrost is raised by a dam as by any other embankment as shown in Fig. 10.10. This ingrowth of permafrost into the body of the dam contributes to the stability of the structure and prevents, at least partially, possible loss of water from the reservoir if the insulation of the upstream face (described hereafter) is not fully efficient. In order to maintain the permafrost table high within the body of the dam through the whole year, the dam is *overcooled* in winter by circulating cold air through aeration ducts or pipes or simply through a coarse gravel layer within the dam. In this way, the permafrost is prevented from thawing during the short arctic summer.

Construction. If it is planned to build an earth dam by the passive method, all locations with considerable ice inclusions should be eliminated from consideration unless the ground is preconsolidated. This preconsolidation involves defrosting of ice inclusions with removal of moisture and grouting of the voids with cement, bitumen, or sodium silicate.

If construction by the active method is planned, vibroflotation* using hot water, steam, or combination of both may be helpful in thawing permafrost to a considerable depth. Also blasting and tamping are sometimes used for this purpose. The active method may be used if the local soil is sandy or gravelly. Soils of silty and clayey character require the use of the passive method.

Types of Dams. Dams in permafrost areas may be constructed either on frozen rock or on frozen soil. Those dams built on frozen rock can be masonry, rock-fill, timber, or earth. Masonry dams are not used on permafrost soil. There may be some disagreement as to the proper choice between a rolled fill or a hydraulic fill in permafrost areas. Since the interior of an earth dam in permafrost should be dry to avoid freezing of the moisture and swelling of the material, some engineers believe that hydraulic fills are not satisfactory. On the other hand, considerable seepage through a rolled fill cannot be permitted on the same grounds. Hence substantial protection of the upstream face against percolating water is required, which is a disadvantage in rolled fill dams.

Similarly if meteoric water happens to penetrate into the body of a rolled fill, it should be promptly removed; this suggests the use of pervious materials for a dam to be built under such circumstances.

The Hess Dam near Fairbanks, Alaska, is a successful combination of a hydraulic fill with a rolled fill.[32] It is 83 ft high, the capacity of the reservoir being 480,000 cu yd. The ground in the reservoir thaws to a depth of about 2 ft. The dam was started as a hydraulic fill in 1939 and reached a height of 56 ft when work was discontinued because of war conditions. During and after the war the dam was fairly well consolidated, and the remainder 27 ft could be placed by the rolled-fill method (Chap. 16). The service of this dam has proved quite satisfactory.

* A "vibroflot" is a huge spud vibrator that stirs a sand mass to which water is gradually added.[31]

Upstream Slope (Face) of a Dam. Figure 10.13 illustrates the protection of the upstream face. The waterproofing blanket *B* should be flexible enough to follow deformations of the ground induced by both temperature and loads. Hence a concrete blanket cannot be used, and bituminous material reinforced with wire is indicated. The waterproofing blanket is protected from the shocks and thrust of ice by blocks *A* of high-strength concrete. The thermoinsulating layer *C* should protect blanket *B* from transmission of cold from the body of the dam into the reservoir. The material for the thermoinsulating layer should be of low heat conductivity, and because of the inaccessibility, it should not require repairs for a considerable time. Light concrete products such

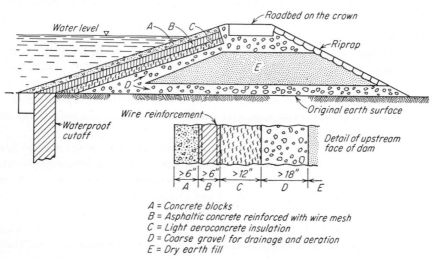

A = Concrete blocks
B = Asphaltic concrete reinforced with wire mesh
C = Light aeroconcrete insulation
D = Coarse gravel for drainage and aeration
E = Dry earth fill

FIG. 10.13. Earth dam in the arctic built by the active method.[30]

as "aerocrete" (*C* in Fig. 10.13) fulfill these requirements. The slope of the upstream face is made flatter than in temperate zones by adding approximately one unit to the coefficient of the slope (e.g., 4:1 instead of 3:1).

Downstream Slope (Face) of a Dam. The downstream face is protected against erosion by meteoric water usually by riprap. Seeding generally is not done in arctic zones because of the low temperatures of the fill.

Measures against Entrance of Water below the Heel of a Dam. The upstream face of a dam may be well insulated, and the permafrost table may be high in the dam, but water still can enter the dam below its heel. Such water could move within the dam both under head and by capillarity (siphon capillarity). To prevent this eventuality, either cutoff walls (Fig. 10.13) or grout curtains can be used (Chap. 15).

Rock, sand, or gravel to be grouted is first defrosted and drained.

Grouting pressures are about 50 per cent higher than in temperate zones. If cutoffs or grout curtains are impracticable, the whole area where water may enter the dam could be "permafrosted" by overcooling it in winter.

Spillways. A concrete spillway for earth dams should be located as far from the dam as possible. In the absence of such a location, the overflow may be discharged through an intake combined with a pipe or tunnel to convey the overflow to the tail water of the dam. Presumably the downstream face up to the elevation of the tail-water level and a short distance higher should be made watertight. Another solution would be to discharge the overflow over the dam by extending the protective blanket (*B* in Fig. 10.13) in the corresponding section over the top of the dam and the downstream face. In all cases of spillway construction, the problems of watertightness and stability of the structure should be given due attention. More details on spillway and generally on dams in permafrost may be found in Ref. 30.

10.17. Foundations of Structures (Other than Embankments).[33] In designing the foundations of a structure in a permafrost area, there are three major considerations: (1) the type and size of the structure, (2) whether it is to be heated or unheated, and (3) the expected permanency of the structure.

1. *Small, Temporary Buildings.* Such buildings can be maintained easily and kept level using shims. They may be built on timber mudsills or on posts set on concrete pads which in their turn are placed on small sand or gravel mats extending about 6 in. above the ground surface. Gravel layers as described will not eliminate settlement or heaving but serve for equalizing movements caused by frost and thaw.

2. *Permanent Structures.* In the design of the foundations of permanent structures, several basic principles as discussed hereafter should be followed. In the first place it should be noted that on large structures, the thermal regime beneath the center of the building is not appreciably affected by edge conditions as in the case of small structures.

Air Space. To prevent degradation of the permafrost, air space is provided between the floor of a building and the ground surface. The floors are then *elevated*, being supported by piling or by sufficiently high posts on pads and gravel mats. In hangars and other buildings with heavily loaded low floors, ducts for circulating winter air in and below the floors are provided; otherwise there will be a progressive increase in depth of thaw under the building. This fact should be considered during the maintenance of the structure. The Army Engineers recommend a 12-in. minimum air space for narrow structures (width less than 20 ft) and a double air space for wider structures.[28b] In existing arctic buildings the air space is often larger (2 to 3 ft). Air circulation in the space left for this purpose may be caused by winds, but for wider structures, induced air circulation may be necessary. Snow should not be permitted to pile against the walls and thus cut off the air circulation. Floors with an air space underneath are cold in winter unless special heating and insulation are provided.

Gravel Mats. Gravel mats may be used, either with or without air space. For larger structures, gravel mats 4 to 5 ft in thickness are not uncommon. When gravel is placed on fine-textured soils such as clays, it should be underlain by a 6-in.-thick clean, well-graded sand. This prevents intrusion of soft soil into the gravel. The top and slopes of the mat should be covered with well-graded materials. The mat should extend beyond the edges of the structure and be sloped to drain precipitation. Gravel mats often are used to support structures, such as oil tanks or airplane hangars, that exert almost uniform pressure on the ground. If the structural loads are not excessive and can be distributed over a large area by proper design, the placement of a thick gravel cover over existing vegetal cover generally provides a satisfactory bearing for the structure.

Thaw-compaction Method.[26,28b] This method is applicable in gravels and sands saturated with ice (see Sec. 10.7, Soil Strengths and Bearing Values). The thawing of the soil may be accomplished by steam and cold water jetting (Sec. 10.20) and preconsolidation by vibration, blasting, and preloading or by a combination of these methods. The surplus of thaw water with suspended fines should be removed from the site immediately after thawing and compaction, for instance by drainage. In this method a foundation of low dry density and low shearing strength is transformed into one of high density and high shearing strength and, hence, a high bearing value. This generally is an expensive method.

3. *Unheated Structures.* An unheated structure placed on the earth's surface stripped of vegetation acts as the fill shown in Fig. 10.10; i.e., it makes the permafrost table rise under the structure. Conversely, a heated structure generates heat downward, causing the permafrost table to degrade. Unheated structures also generate some heat downward; e.g., the concrete of the foundation generates heat during setting. Furthermore, certain types of structures such as steel towers will absorb heat from the sun and tend to transmit it through the tower legs into the foundation.

In the construction of a steel tower in northern Alaska, it was discovered that although the tower was founded on steel piles driven into the permafrost, heat generated by the setting of the concrete in the pedestals at the top of the piles caused water to accumulate around the foundations. It was necessary to drive well points around each footing to remove this moisture and thus prevent heaving once foundation temperatures were stabilized. Additional construction on the pedestals was not permitted until the footings had been firmly frozen in the soil.[34]

In another similar case it was found that the maximum heat flow came to the permafrost during the summer months and was a result of the direct transmission of heat from the steel towers down through the steel piles at the base. The foundation was constructed with 2 ft of compacted dry sand fill above the permafrost, 6 in. of concrete, and an 18-in. overlying insulating blanket, with a ⅛-in.-thick copper sheet under the concrete to dissipate the heat laterally as much as possible. This design was based on the requirement that the heat in each pile should not melt more than 0.05 in. of material around the pile at a rate of 1 ft or less vertically in 24 hr. The piles were designed to withstand excessive settling and high tensional forces caused by wind action on the towers.[35]

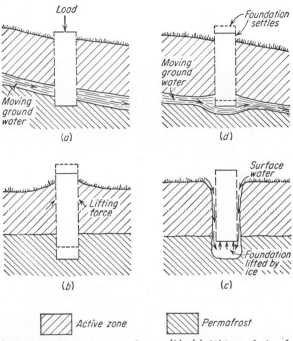

FIG. 10.14. (a) Active zone and permafrost. (b), (c) Lifting of the foundation. (d) Settlement of the foundation.[29]

4. *Heated Structures.* The most satisfactory foundation for a heated structure consists of piles driven into permafrost to a sufficient depth to anchor the structure to the permafrost. Such anchorage prevents possible upward or downward movements that may result from the swelling in the active zone (Fig. 10.14b and c) or from excessive settlement (Fig. 10.14d). The common method of founding a structure on timber piles is shown in Fig. 10.15. If, because of freezing, there is swelling in the fine-grained material of the active zone and hence a tendency toward upward movement, this tendency would be basically opposed by the adfreezing of the pile to the permafrost. The empirical rule of driving piles into the permafrost to a depth equal to twice the depth of the active zone ($2h$ and h in Fig. 10.15) has proved successful in the arctic.

Adfreezing. All ground in contact with a structure which is in a thawed state during construction adheres to the structure as the temperature drops to the freezing point. This form of adhesion, or *adfreezing,* is stronger in sands than in clays. In the design of foundations in temperate zones, adfreezing generally is neglected.

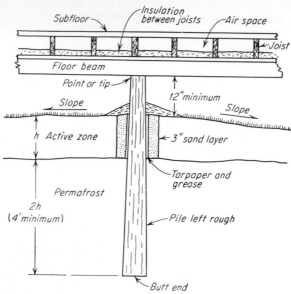

FIG. 10.15. Typical design of a foundation when the active zone may swell in freezing.[29]

Some information on the magnitude of adfreezing forces is given in Table 10.2. This table is based on pile extraction tests performed in 1953 in Fairbanks, Alaska.[28b] The test soil was silt (ML of the Unified Classification), dry density 64 pcf, moisture content 64 per cent. The temperature of the permafrost was 30.5°F. The depth of the pile embedment was 10 ft; the depth of thaw, 3.5 ft. Somewhat higher unit adfreezing forces (up to 4,800 psf) were found at the 17-ft embedment of the piles.

TABLE 10.2. Tangential Adfreeze Strengths

Type of pile	Ultimate tangential adfreeze strength, psf
Timber, 9 in. diameter	3,000
Reinforced concrete, 9 in. square	3,800
Steel pipe, 8-in. diameter	4,200
Steel beam (light), 8-in. depth, 4-in. flange	3,400

Foundations. If there is no danger of heaving in the active zone, spread foundations (Sec. 13.3) under a heated structure can be used. Figure 10.16 shows a reinforced concrete footing isolated from the permafrost by crisscrossed heavy timbers. If the footings have to be founded in the permafrost because of its proximity to the ground surface, they can be built in oversized timber boxes. This permits the placement of sand-gravel insulation on the sides and timber insulation on the base of the footings. Figure 10.17 shows timber footings (pillars) bolted to the timbers resting in permafrost.

The walls of small heated structures, such as small houses, may be placed directly on the earth's surface providing sand or moss is used to protect their lower parts to a height of about 18 in. Air space under the floor with good circulation should be provided. This space usually has

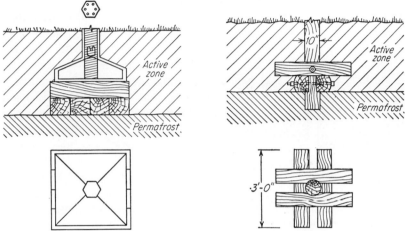

FIG. 10.16. Reinforced concrete footing on permafrost.[36]   FIG. 10.17. Timber footing on permafrost.[36]

shutters around it which can be opened in the summer for cooling and closed in the winter to retain heat in the building. Reinforced concrete slabs placed directly on the ground or gravel, with no basement, are also used.[9]

Refrigeration of Foundations. For some large and costly structures located over permafrost, the foundation materials have been artificially frozen and kept frozen permanently by artificial methods. This method maintains the permafrost table at a constant level and prevents detrimental thawing of the active layer. Thus, a structure (in the case of an earth dam, for example) can be founded directly on the active layer. When this method is used, refrigeration pipes are either buried or laid around the foundation and the refrigerant circulated. Where power is available at a relatively low cost, as in powerhouses, the cost and maintenance of similar installations may prove more economical than some other type of foundation treatment. The refrigerant is circulated during the thawing season, and then only during the daylight hours when the outdoor temperature rises above freezing. It is obvious in this case that the building floor should be sufficiently insulated to prevent thawing of the underlying material.

10.18. Pier and Pile Foundations in Permafrost. As in temperate zones (Sec. 13.4), the structural loads may be transmitted to a reliable layer by piers placed down to and usually into the permafrost. If the active zone is more than about 4 ft thick, piers usually are not considered economical.

Piles.[22,29] The best time to drive piles in the arctic is in the fall as soon as the ground has frozen sufficiently to support equipment. The active zone is excavated, and the permafrost thawed to a depth equal to the length of the piles. This is conveniently done with steam points as explained in Sec. 10.19. The jetted holes should be made just large enough to fit the pile diameters. If the ground has dry layers, the addition of water to the steam speeds the jetting. Usually the piles are driven within 3 or 4 days after the holes are jetted. The jet holes can be left open up to 3 weeks in summer but not more than 1 week in winter.

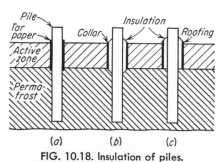

FIG. 10.18. Insulation of piles.

Early arctic builders used timber piles, and these still are in use for small structures. The steel pipe pile is another pile type used in the arctic. For an average structure, the piles are spaced about 6 ft center to center. If their length does not exceed 60 ft, a light, fast drop hammer can be used. When timber piles are used, they generally are 7 to 10 in. in diameter at the top and are driven butt down in order to increase the forces resisting heaving (Fig. 10.15).

Adfreezing in that portion of the pile located within the active zone is detrimental, since the swelling of the active zone tends to move the pile upward. Conversely, adfreezing of the pile to the permafrost is beneficial, since the forces developed in this case tend to keep the pile down. Hence adfreezing of the pile in the active zone should be prevented. One method of prevention is to place a collar on the portion of the pile located in the active zone; thus the ground in the active zone tends to freeze to the collar instead of the pile. If the collar is permitted to shift up and down without tending to pull the pile out, the pile will be safe. This is especially important immediately after the pile has been driven and there has not been enough time for it to adfreeze to the permafrost.

There are several variations of this method. On timber piles the collar may be made simply by wrapping the pile with tar paper (Fig. 10.18a). Sometimes a sand layer is placed around the pile (Fig. 10.15). Such protection may last only 2 or 3 years, but this time interval is sufficient for the pile to be solidly adfrozen to the permafrost and thus resist motion. On steel piles, the collar may consist of a short section of pipe with a heavy coat of grease between the pipe and the pile. Peat or moss soaked in oil may be placed around the pile and held in place by a plastic or metallic cover (Fig. 10.18b), or two loose layers of 1-in. blanket insulation covered with three-ply roofing with top and bottom seals may be used (Fig. 10.18c). A variation of Fig. 10.18b consists of merely planing a timber pile, greasing it, and placing the collar. An efficient method, though based on a different principle, is to weld a steel collar or flange to the lower end of a steel pile with a diameter about $1\frac{1}{2}$ times greater than the pile. The flange acts as an anchor when the pile is adfrozen to the permafrost.

The piles should be driven into the permafrost well in advance of the construction of the superstructure in order to permit the development of proper bearing power through adfreezing. The freeze-in is more rapid on steel than on timber piles. Adfreezing proceeds from the tip of the pile upward. The foot of the pile may adfreeze to the permafrost in about 3 to 6 days, whereas the top may require weeks for that purpose.[22]

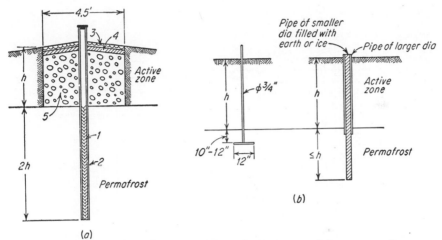

FIG. 10.19. Bench marks in permafrost: (a) permanent, (b) temporary.[36]

It is of interest to note that the pile driving lowers the frost line somewhat. This drop is considerably more for steel piles than for timber piles.

In one case this drop was 1 in. for a timber pile and 4 in. around a 6-in. pipe pile.

10.19. Earthwork in Arctic Conditions. The most favorable season for earthwork is presumably late spring and summer. Before the work starts, moss and turf cover, if any, should be removed. For ease of removal, the cover may first be cut into long strips, which are rolled up like rugs. This removal facilitates drying of the site and contributes to the stability of the fill.

As with any engineering structure it is necessary to establish accurate bench marks.[36] Figure 10.19a shows one type of permanent bench mark in the arctic. The material in the active zone is removed, a perforated pipe (1) is placed into the boring (2), and a mixture of gravel and earth is placed into the hole. Cobblestones (3) and turf layer (4) protect the surface of the gravel fill (5). One of the temporary bench marks shown in Fig. 10.19*b* consists of a ¾-in. rod with a metallic plate 12 by 12 in. at the bottom, the rod being embedded 10 or 12 in. into the permafrost.

Softening the Permafrost. Rock-breaking tools are used to break up the permafrost. The softening usually is done by applying heat. (The primitive thawing method was to set surface fires or place on the surface a riprap layer heated to 350 or 400°F. This riprap method, however, has been abandoned in United States practice.) Flooding the excavation surface with hot or cold water often is used in Alaskan gold mining[37,38] and can be applied to any excavation in permafrost. Cold water is more efficient in this case than hot water. The optimum thawing depth with water is about 2½ ft, after which the softened material is excavated, and the area reflooded if greater thawing depths are necessary.

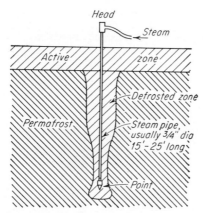

FIG. 10.20. Steam pipe and point for defrosting.

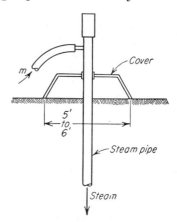

FIG. 10.21. Protective cover for a steam pipe.

Steam jetting is a quick method of thawing and can be used in all types of soils. A typical steam jet is shown in Fig. 10.20. Steam pressures of 30 psi are satisfactory for depths up to 15 ft or so; for greater depths, heavier equipment and higher pressures, such as 50 to 80 psi, should be used. The pipe is first hammered lightly into the frozen ground, the steam aiding in penetration; subsequently heavier hammering and sinking follow. Particularly in pile driving, the steam jet should be about ¾ to 1 in. in diameter and should be about 1 ft shorter than the length of the pile. After the desired depth is reached, the pipe is left in the hole for ½ hr for sandy soils to 3 hr for clays. Figure 10.20 shows only one steam pipe corresponding to the so-called *open method* of steam thawing wherein the steam enters the frozen ground directly through a system of points connected with a boiler. To decrease heat losses, the stem of the steam pipe may be protected by a cover of steel or timber. The cover should prevent steam leakage and, if necessary, is loaded to prevent its being lifted by the steam (Fig. 10.21). In the *closed method* of thawing, the steam circulates through closed pipes or coils and condensed steam is returned in the form of water. Good results are obtained if steam pipes (in the open method) are of two different diameters. The frozen ground is first pierced with small pipes ($\frac{5}{8}$ to $\frac{3}{4}$ in. diameter), then larger pipes are inserted into the same holes. The points of the pipes must have holes at both the tip and the sides to avoid possible plugging.

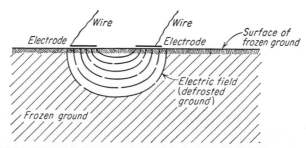

FIG. 10.22. Schematic representation of the electric field of high frequency for thawing frozen ground.

A high percentage of boulders or coarse gravel will impede the sinking of steam points. Also the process of thawing must be carefully supervised, especially in the open method wherein the natural moisture content of the ground is increased by the addition of the condensed steam. Overthawing by steam may convert the frozen ground into a liquid mud very difficult to handle.

High-frequency-current Thawing.[36] Frozen ground is a good conductor of high-frequency current. In using such a current for thawing permafrost, an electrical field is created between two electrodes placed at or close to the surface of the frozen soil; hence a rise in soil temperature occurs (Fig. 10.22). Electrodes (steel rods $\frac{5}{8}$ or $\frac{3}{4}$ in. diameter) may be driven into the frozen ground as shown in Fig. 10.23a or simply placed at its surface (Fig. 10.23b). In the latter case steel strips also may be used.

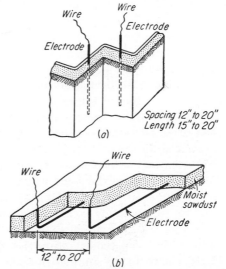

FIG. 10.23. Electroheating of the frozen ground by high-frequency current.[36]

When the electrodes are driven vertically as in Fig. 10.23a, the current is periodically turned off to avoid useless overheating of the already soft upper layer and to permit the heat to penetrate deeper. An alternative consisting of a gradual increase in electrode spacing is less convenient. Horizontal electrodes (Fig. 10.23b) are placed within a layer of sawdust moistened with weak common salt solution. Deep snow should be removed from the frozen ground surface, but a little snow may be left in place. The temperature of the sawdust should not exceed 90°C (194°F). The amount of energy required for thawing clays and silts with a 10 per cent admixture of sand (1) decreases as the spacing of the electrodes increases and (2) increases practically proportionally to the moisture content if the latter ranges from 20 to 35 per cent. For higher percentages the rate of increase diminishes. Larger than 10 per cent sand admixtures require less energy.

Use of Alternating Three-phase Current.[36] Frozen ground is a poor conductor of the alternating current, and when it is used, some good conductor is placed at the ground surface and heated (e.g., a weak solution of common salt or calcium chloride, sawdust, or simply thawed ground). In this way, heat is transferred downward. The amount of energy expended in this method is about 50 per cent higher than in the use of high-frequency current. When the frozen ground is thawed to a depth of 1½ or 2 ft, it is removed, leaving a thin layer of thawed material (4 to 6 in.). Electrodes may be made from any metallic pieces available at the job.

Excavation. One of the simplest methods of excavating permafrost is to do it in the summer and permit the thawing of successive layers by normal daylight temperatures. However, for unstable materials, such as some silts or clays, this method may result in considerable overexcavation because the side slopes collapse and the material flows into the hole. If it is planned to build a structure by the passive method, the temperature of the permafrost can be lowered by exposing it to low temperatures, such as 30°F, for a period of 15 to 25 days. During this period all falling snow has to be continuously removed to prevent it from acting as an insulator. When the structure is to be built by the active method, the permafrost table is lowered 10 to 15 ft below the base of the foundation. If this lowering is done by steam heating, the natural moisture content is increased and the bearing power of the soil decreased. Electroheating contributes to the drying-out of the soil, although insignificantly. Sometimes the bottom of the excavation is stabilized by spreading a layer of gravel, 1½ to 2 in. thick, and compacting the ground. Another method is to drive short piles (about 10 ft long) spaced 5 or 6 ft apart, extract them, and fill the holes with sand.

Excavating permafrost without previous softening can be done by using explosives in the same way as in rock excavation. The resistance to blasting of frozen clay is approximately the same as that of sandstone, whereas frozen sandy soils are like weak limestones in their reaction to explosives. In the summer, blasting should be done without delay after drilling the blast holes; otherwise the holes may fill with fluid mud. Furthermore, in the summer the bore hole must be cleaned and the water removed, unnecessary precautions in the winter.

Removal of the frozen ground from the active zone by the *hydraulic method* is a possibility provided the frozen ground either has been previously thawed by using steam or has been broken by blasting. In permafrost proper, only the thawed crust can be successfully eroded by hydraulic methods unless considerable quantities of water are available. Permafrost with a large ice content is difficult to erode, since water merely slides along the smooth ice surface with little erosive effect.

Excavating and earth-moving equipment as used in temperate zones may also be used in permafrost by adjusting them to local conditions. If the frost zone reaches the permafrost table, the excavation can be done in winter without bracing. In spring, the excavation slopes thaw, starting with the south-facing slopes. This results in considerable snow melt, and hence surface drainage is necessary. On the slopes where intense thawing is expected, protection should be secured by placing canvas or timber mats on the slopes. Bracing, if necessary, can be vertical or horizontal lagging.

10.20. Building-site Selection in Permafrost.[22,26,32] The criteria for the proper selection of building sites can be briefly summarized as follows:

1. The best site should have competent bedrock near the surface, and thus the thickness of the active layer is at a minimum.

2. A good location should have soil that when thawed would still possess adequate bearing properties.

3. Either a permafrost-free area or an area containing continuous permafrost is better than one in which sporadic permafrost islands exist because of the difficulty of properly locating such islands.

4. The site should have good drainage, and there should be no signs of detrimental frost action such as mounds, blisters, or boils.

5. A site at the foot of a slope is undesirable, as it might be affected by ground-water seepage and subsequent icing.

In all cases, the soil survey is of prime importance. Numerous drill holes or test pits may be required for exact determination of foundation conditions; generally more such holes are required than in temperate locations because of the extreme irregularities found in the active layer and below the level of the permafrost table.

REFERENCES

1. Taber, S.: (a) Frost Heaving, *J. Geol.*, vol. 37, 1929; (b) Mechanics of Frost Heaving, *ibid.*, vol. 38, 1930; (c) Freezing and Thawing of Soils as Factors in the Destruction of Road Pavements, *Public Roads*, vol. 11, 1930; (d) Discussion of Frost Heaving, *Highway Research Board Proc.*, vol. 12, 1932.

2. Beskow, G.: "Tjalbildningen och Tjallyftningen," Stockholm, 1935. Translated by J. O. Osterberg and published by Northwestern University.

3. Casagrande, A.: Discussion on Frost Heaving, *Highway Research Board Proc.*, vol. 12, 1932.

4. Johnson, A. W.: Frost Action in Roads and Airfields, *Highway Research Board Spec. Rept.* 1, 1952.

5. Cooling, L. F., and W. H. Ward: Damage to Cold Stores due to Frost Heaving, paper read before the Institute of Refrigeration, London, Dec. 14, 1944.

6. Black, R. F.: Permafrost, "Applied Sedimentation," chap. 14, John Wiley & Sons, Inc., New York, 1950.
7. Muller, S. W.: "Permafrost or Permanently Frozen Ground and Related Engineering Problems," J. W. Edwards, Publisher, Inc., Ann Arbor, Mich., 1947.
8. "Cold Weather Engineering," chaps. 1–5, U.S. Navy Civil Engineering Corps, Bureau of Yards and Docks, Washington, D.C., 1948–1949.
9. Leggett, R. F.: Special Foundation Problems in Canada, *Bldg. Research Congr. Proc.*, div. I, part III, London, 1951.
10. Terzaghi, K.: Permafrost, *J. Boston Soc. Civil Engr.*, vol. 39, 1952.
11. Brian, K.: Cryopedology, *Am. J. Sci.*, vol. 244, 1946.
12. Cederstrom, D. J., P. M. Johnston, and S. Subitzky: Occurrence and Development of Groundwater in Permafrost Regions, *USGS Circ.* 275, 1953.
13. U.S. Engineer Office: A Test Study of Foundation Design for Permafrost Conditions, *Eng. News-Record*, Sept. 18, 1947.
14. Tsytovich, N. A., and M. J. Soumgin: "Basic Mechanics of Frozen Soils," pp. 83–85, Academy Science of the USSR, Moscow, 1937.   In Russian.
15. Siple, P. A.: Ice-blocked Drainage as a Principal Factor in Frost Heave, Slump, and Solifluction, Frost Action in Soils, *Highway Research Board Spec. Rept.* 2, 1952.
16. Conrad, V.: Polygon Nets and Their Physical Development, *Am. J. Sci.*, vol. 244, 1946.
17. Nakaya, U., and J. Sugaya: A Report on Permafrost Surveying (Manchuria, 1943), *Natl. Research Council Can. Technol. Trans.*, TT 382, 1953.
18. Taber, S.: Perennially Frozen Ground in Alaska: Its Origin and History, *Bull. GSA*, vol. 54, 1943.
19. Taber, S.: The Problems of Road Construction and Maintenance in Alaska, *Public Roads*, vol. 24, 1943.
20. Airfield Pavement Design, "Engineering Manual for War Department Construction," chap. 7, part XII, Corps of Engineers, Office of Chief of Engineers, 1946.
21. Ghiglione, A. F.: The Alaska Road Commission, *Western Construction*, vol. 28, October, 1953.
22. Hemstock, R. A.: Permafrost at Norman Wells, NWT, Canada, Typescript for Arctic Institute of North America and U.S. Air Force, February, 1949.
23. Frost, R. E.: Interpretation of Permafrost Features from Airphotos, Frost Action in Soils, *Highway Research Board Spec. Rept.* 2, 1952.
24. Review of Certain Properties and Problems of Frozen Ground Including Permafrost, *Purdue Univ. Eng. Expt. Sta.*, SIPRE Rept. 9, March, 1953.
25. Joesting, H.: Magnetometer and Direct-current Resistivity Studies in Alaska, *Trans. AIME*, vol. 164, 1945.
26. Thompson, S. F.: Construction in Permafrost, *Western Construction*, vol. 28, October, 1953.
27. "Science in Alaska—Selected Papers on the Alaskan Science Conference of Nov. 8–11, 1950 of the National Academy of Science," sections on Public Health and Medicine, National Research Council and Arctic Institute of North America, June, 1952.
28. "Engineering Manual for Military Construction," (*a*) chap. 2, Arctic and Subarctic Construction, Site Selection and Development; (*b*) chap. 4, Arctic and Subarctic Construction, Building Foundations, part XV, Corps of Engrs., U. S. Army, October, 1954.
29. Construction of Runways, Roads, and Buildings on Permanently Frozen Ground, *War Dept. Tech. Bull.* TB 5-255-3, 1947.

30. Lewin, J. D.: Prevention of Seepage and Piping under Dams Built on Permafrost and Related Problems Connected with the Design and Construction of Such Dams, *Proc. 3d Congr. Large Dams*, Stockholm, Sweden, paper R. 66, question 10, 1948.

31. Fruhauf, B.: Wet Vibration Puts Strength in Sand, *Eng. News-Record*, pp. 60–62, June 23, 1949.

32. Huttl, J. B.: Building an Earth Fill Dam in Arctic Placer Territory, *Eng. Mining J.*, vol. 149, no 7, pp. 90–92, July, 1948.

33. Pihlainen, J. A.: Building Foundations on Permafrost, Mackenzie Valley, NWT, *Natl. Research Council Can. Div. Bldg. Research Tech. Rept.* 8, DBR No. 22, June, 1951.

34. Roberts, P.: Arctic Tower Foundations Frozen into Permafrost, *Eng. News-Record*, Feb. 9, 1950, pp. 38–39.

35. Nees, L. A.: Pile Foundations for Large Towers on Permafrost, *Proc. ASCE Sep.* 103, vol. 77, November, 1951.

36. Zhukov, V. F.: "The Earthwork for Foundations in Permafrost Zones," printed in Russian, Moscow-Leningrad, 1946, trans. by Stefannson Library.

37. Wimmler, N. L.: Placer Mining Methods and Cost in Alaska, *USBM Bull.* 259, 1927.

38. Janin, C.: Recent Progress in the Thawing of Frozen Ground in Placer Mining, *USBM Tech. Paper* 309, 1922.

Supplemental Reading

"Permafrost—Studies in Connection with Engineering Projects in Arctic and Sub-arctic Regions," part I: Instructions for Measuring Ground Temperatures, U.S. Air Force Publication, December, 1944.

"Arctic Bibliography," 5 vols., Department of Defense, Superintendent of Documents, 1953–1956.

Carlson, H.: Stability of Foundations on Permanently Frozen Ground, *Proc. 2d Conf. on Soil Mech. and Foundation Eng.*, vol. IV, Rotterdam, paper VI, p. 6, 1948.

Johnston, W. A.: Frozen Ground in the Glaciated Parts of Northern Canada, *Trans. Roy. Soc. Can.*, vol. 24, 1930.

Soumgin, M. I.: Permanently Frozen Soil in the Limits of the USSR, *Acad. Sci. USSR*, Moscow, 1937, and a large number of other publications in Russian on the same subject, many of which are listed in "Arctic Bibliography."

CHAPTER 11

SHORE-LINE ENGINEERING
AND RIVER IMPROVEMENT

Erosion and deposition of solid materials by the action of water are continuous processes along a shore line, be it a coast line of the ocean or of a sea, a lake shore or the banks of a river. Shore engineering is the study of the sequence and the rate of these destructive and building-up processes and the design and use of technical measures to counteract these processes, if necessary. It is necessary to keep the shore in a condition favorable to navigation and to the stability of the adjacent land and structures built on it.

SHORE LINES, WAVES, CURRENTS, AND TIDES

11.1. Shore Lines and Beaches. There are two characteristic types of shore lines: (1) the submergence type produced by the partial *submergence* of the coast such as the irregular, rocky New England shore and (2) the emergence type produced by the partial *emergence* of the ocean floor as seen along the smooth, low, sandy shore line of southern California. A *neutral* shore line shows no effects of submergence or emergence; a shore line of mixed characteristics is a *compound* shore line. The two basic types of shore lines are depicted in Figs. 11.1 and 11.2. The shore or coast immediately adjacent to the water may be a rather flat beach, generally sandy or gravelly (Fig. 11.3); a steep rocky sea *cliff* (Fig. 11.4); or some intermediate landform. A *wave-cut terrace* is a shelf at the base of a cliff covered by shallow water. As water erosion progresses, the terrace width gradually increases and the cliff retrogresses landward.

The Beach Erosion Board (Corps of Engineers, U.S. Army) uses the sand beach terminology shown in Fig. 11.3. Horizontal or almost horizontal parts of a beach are *berms*. The seaward edge of the berm zone divides the beach into a *foreshore* and a *backshore*. Waves continually act upon the foreshore and only occasionally on the backshore.

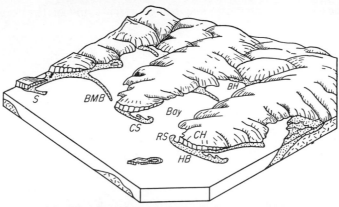

FIG. 11.1. Shore line of submergence. (*From W. H. Emmons et al., "Geology: Principles and Processes,"* 3d ed., McGraw-Hill Book Company, Inc., New York, 1949.)

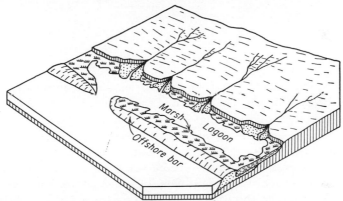

FIG. 11.2. Shore line of emergence. (*From W. H. Emmons et al., "Geology: Principles and Processes,"* 3d ed., McGraw-Hill Book Company, Inc., New York, 1949.)

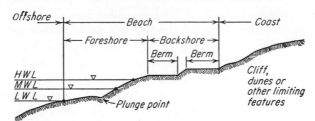

FIG. 11.3. Beach features terminology. (*Adapted after Beach Erosion Board, Corps of Engineers, U.S. Army.*)

Beaches may be classified according to the slope of the foreshore as follows:[2]

Flat (maximum slope 1:120)*
Mild (maximum slope 1:60)
Gentle (maximum slope 1:30)
Moderate (maximum slope 1:15)
Steep (slope over 1:15)

The slope of a beach is related to the grain size of the materials forming the slope. In general, the finer the slope-forming material, the gentler

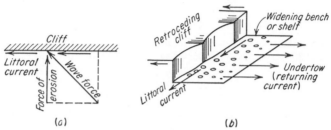

(a)   (b)

FIG. 11.4. Erosion of a cliff.

the slope. The size of the slope-forming material continually changes, however, because of the (1) abrasion of the material that moves back and forth, (2) removal of some material by stormy winter waters, and (3) addition of fresh material eroded from the shore or brought in by gentle waves and currents, mostly in the summertime. A beach may grow seaward or retreat landward. There may be depressions (troughs) in the foreshore or relatively high spots (terraces) protruding permanently or periodically above the water level. The symbols HWL, MWL, and LWL in Fig. 11.3 mean "high water level," "mean water level," and "low water level," respectively.

Additional Nomenclature. A *bar* is a sand deposit parallel to the beach or running at an angle to the shore line. It may or may not be covered by the high water. When a stream, either large or small, flows into another stream, there may be sand deposition in the form of a bar because of the loss of velocity. A *sand ridge* near a shore is practically the same as a bar. *Spits* and *cusps* are sand peninsulas at a shore; a spit is elongated and generally curved, whereas a cusp is roughly triangular. Both spits and cusps may vary in size; both are unstable landforms and may

---

* The symbol 1:120 means that the slope is defined by measuring 120 units horizontally and 1 unit vertically. Slopes of excavations and fills are usually designated in reverse order. For example, 3:1 slope means 3 units measured horizontally and 1 unit vertically. Highway and railroad grades are usually designated in per cent, as 0.8 per cent.

disappear after a violent storm. *Sand keys* are small islands running parallel to the shore.

11.2. Waves. Most waves are generated by the wind. Waves in open water far off shore are *oscillatory waves* (Fig. 11.5). A particle of water in such a wave describes an orbit in a vertical plane; e.g., observation of a floating object shows that most of the movement is up and down in a vertical plane and there is little or no forward motion.

Mathematical theories generally assume a horizontal sea bottom and stable periodic waves. The first known solution of this problem was developed about 1802.[3] This theory stated that the wave surface is trochoidal in form, with a closed orbit ("rotational motion"). According to Stokes:[4] (1) the wave velocity depends upon the wave height as well as upon the wave length and water depth, and (2) orbital motion of water particles is open rather than closed, indicating a certain small translation in the direction of the wave travel (irrotational motion). Experiments have proved the correctness of these assumptions.[5]

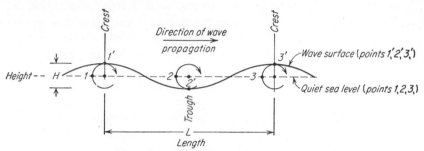

FIG. 11.5. Wave caused by oscillatory movement at each point of water surface.

The height H of waves (Fig. 11.5) is variable, and 20 to 30 ft is not uncommon. The length L increases with the height, but not proportionately. The ratio L/H varies approximately from 15 to 30 with possible deviations. The maximum wave length often is assumed as 400 ft.

Upon reaching shoal (shallow water), the wave starts to drag on the bottom, it increases in height (about 40 per cent in comparison with its deep-water height), and its length decreases. The waves pile up; their fronts become concave and topple forward. These are plunging waves. When the *plunge point* (Fig. 11.3) is reached, the wave breaks up. The position of the plunge point is variable and is located within a certain zone or range (breaker belt). The foamy remnants of broken waves still rushing forward, before they stop and start backwashing, are known as *surf*.

11.3. Currents. The two most common types of current of immediate engineering interest are *undertow* and *longshore* (or *littoral*) currents; huge oceanic currents such as the Gulf Stream are not considered in this book. Whenever the waves attack a shore line, there are returning, or seaward, currents.　Undertow is the returning current *below* the oncoming waves.　When the waves attack a cliff normally (i.e., perpendicularly), the eroded rock fragments are ground back and forth, decrease in size, and may be shifted toward deeper waters.　When, however, a wave attacks a cliff at an angle, the kinetic energy of the wave is only partly spent in impact and erosion of the cliff; the remaining energy contributes to the formation of a longshore (littoral) current (Fig. 11.4).　Such *refraction* also takes place when the waves approach a sandy beach at an angle.　In this case the waves tend to swing parallel to the shore because they are retarded by the shallowing water.[2]　Figure 11.6 shows diagrammatically what occurs in such cases.　The curves without arrows in that figure show how the wave crests tend to curl, whereas the curves with arrows show the wave trajectories.　As the waves break, a littoral current is formed in the direction toward which the waves are sweeping.[2]　A meeting of two longshore currents moving in opposite directions may result in a seaward (rip) current approximately perpendicular to the shore line.

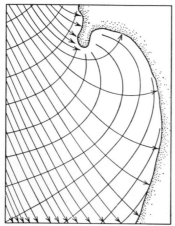

FIG. 11.6.　Wave-refraction diagram of Half Moon Bay, California.　(*After Krumbein.*[2])

Currents connected with the translation of suspended fine materials are known as *density currents* (suspension currents). They often may be detected several miles out from the mouth of a large river flowing into the ocean, as their color (often yellowish) is different from the sea water.　A spectacular example of such currents is at the mouth of the Plata River, in Argentina.　Density currents also occur in dam reservoirs (Sec. 12.9).

11.4. Tides.[6,7]　A tide is a great wave continuously traveling around our planet under the pulling action of the moon and sun.　The tide height is at a maximum or minimum when sun, moon, and earth are in a straight line.　If, in such an instance, the sun and moon are at one side of the earth, their pulls are combined and thus create the highest tides (spring tides).　When the sun is on the opposite side of the earth from the moon, however, the sun's pull decreases that of the moon and the lowest tides (neap tides) occur.　As a lunar day equals 24 hr 50 min, every day the tides occur later than the preceding day.　Usually there are two tides a day, but on some days of the month there is only one tide (diurnal tide).　As the relative positions of the sun and moon at all times are known, tide timetables for each day at any locality can be computed. In the United States such tables are published every year by the USCGS for all our coasts.

The difference between high-tide and low-tide level is the *tidal range*. The maximum tidal range is the difference between the HHW—highest high water level ever observed—and the LLW—the lowest low water level ever observed. The average tidal range is the difference between the MHW—mean high water level—and the MLW—mean low water level.

At New York City, the mean tidal range is about 4.5 ft and the maximum tidal range may be 7 ft. At the Atlantic entrance to the Panama Canal, the tidal range is less than 1 ft, whereas at the Pacific entrance, it averages 12.5 ft. At Cook Inlet near Anchorage, Alaska, the tidal range is one of the largest known—35 ft or more.

Generally tidal ranges are smaller along straight parts of the coast and may be very large at the inlets of funnel-shaped bays and in the bays themselves (as in Cook Inlet). Because of changes in sea level caused by wind, barometric pressure, or other factors, the apparent heights of a tide may be correspondingly increased or decreased. A wind, for example, may increase the height of a tide, sometimes a couple of feet or more.

The oncoming part of a tide is the *flood*, and the retreating part is the *ebb*. These two parts are unequal in duration because the high flood wave travels more quickly than the rather low ebb wave. In harbor and sea-wall design, knowledge of the duration of the ebb is of importance. Water saturates the backfill of a sea wall during the flood tide and descends during the ebb. As the water level behind the wall cannot drop as fast as the ebb water outside, the resultant lag in water level behind the wall contributes to an increase of lateral pressure against the wall. In the case of high tides this lag should be carefully estimated according to the permeability of the material.

Tidal Rivers. The rivers flowing into a sea or the ocean may be tidal or tideless (nontidal), depending on the tidal range. The mouths of rivers flowing into a tidal sea are much larger than those of tideless rivers because a tidal river has to discharge not only the water from its drainage area but also the periodic back-and-forth motion of the tidal flow. A delta does not form at the mouth of a tidal river because there is no abrupt change in the water velocity; instead, the river water gradually mingles with the moving tide water. Some precipitation with resultant sedimentation may occur because of the electrolytic action of sea water.

An *estuary* is that part of a sea where the river meets the tide; this may be the mouth of the river or any bay or lagoon. Usually the estuary waters are more muddy than the river water because the flood and ebb of the tide keep the suspended sediment continually in motion. Near the mouth of a tidal river (point *A*, Fig. 11.7) the tidal range is the same as in the adjacent sea. As the wave moves upstream, its crest also moves until it reaches a maximum elevation at some point C; then the crest declines, and at some point D the tide vanishes. In exceptional cases, if the river channel above point D is large enough, a small tidal wave may be propagated along it (sometimes for a considerable distance as in the Amazon River, where the tidal wave moves upstream about 600 miles). The ebb crest possesses a minimum at point B, which is located close to the upstream edge of the estuary. Thus the tidal range at the estuary and to a certain distance beyond is larger than in the sea itself.

According to Shankland,[6] the low water level at Bristol Channel, England, is 12 ft below sea low water and equivalently high at high water. Generally there is a delay in the flow when the tide turns, i.e., an interval between flood and ebb (point D in Fig. 11.7), termed "slack water." Also, if there is a shallow bar at the mouth

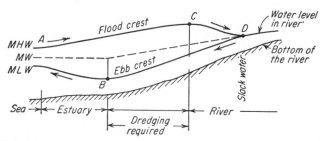

FIG. 11.7. Tide action in a tidal river (exaggerated vertical scale).

of the river, the tide rises more rapidly in the sea than in the river, and the tidal wave may break over the bar to form a *tidal bore.* The positions of points B, C, and D (Fig. 11.7) are not invariable; they should be determined as an average from a series of observations. To quote Shankland, "If the low water channel between B and C be deepened and widened, the point B will be brought higher up the river, and facility given for the point C, the summit of the high water line, to reach higher up. This of itself will push on the point D. Thus dredging between points B and C would permit more tide water to come into the river and thus assist navigation."

DESTRUCTION OF SHORE LINES: LITTORAL PROCESSES

11.5. Destruction of Shore Cliffs and Steep Banks.[10] Steep shore banks made of soil materials are susceptible to failures such as landslides (Sec. 17.3) and to the gradual motion of large earth masses, plastic or otherwise. Clay banks are especially sensitive to wetting, either directly by the waves or by spray. Wetting results in a decrease in cohesion and hence in resistivity of the material to the shearing stresses; this may cause flow or sliding. When wetting is alternated with drying, shrinkage cracks develop in the dry bank, and thus an easy path is provided for the penetration of water into the bank. A common cause of landslides in clayey banks is the erosion of the foot of the slope; in a general case this is known as the "removal of the lateral support" of the slope (Sec.

17.5). When a clayey bank fails, the huge, semiliquid masses of material may engulf one or more protective structures built along that bank and put them out of service. Because of their critical nature the shore clay banks should be disturbed as little as possible and, where practicable, protected by coarse granular material, or riprap.

The more sandy the clay-sand material, the more stable is the bank or cliff, because in this case the drainage is relatively rapid and, hence, the possibilities of high-pore-pressure development are less. In alluvial sand and silt deposits of more or less uniform thickness, the banks of a stream are often vertical. As soon as the erosive action of the stream decreases the stability of the bank to a certain critical value, the bank material falls off in strips. The thickness of these strips is approximately equal to one-half of the height of the bank just before the failure. The failure repeats itself, and the bank retrocedes similarly to what is shown in Fig. 11.4. The phenomenon described often is easily seen in silted dam reservoirs.

Soft-rock Shore Cliffs. The softness of a rock usually is caused by the lack of firm cementing material or the alteration of its mineral constituents into soft compounds. In either case water sorption becomes rapid and erosion intense. Often excessive erosion of cliffs composed of such materials can be temporarily, if not permanently, halted by the erection of concrete walls immediately adjacent to the cliff. The walls have to be of such a height as to prevent water penetration into the soft rock. It must be remembered, however, that destruction of banks and cliffs of soft materials is not entirely without its beneficial effects. The eroded materials are the source of *littoral drift*, the sedimentary material transported by the waves and currents parallel to the shore line. This drift often comes to rest in beach areas where erosion had been making inroads in desirable land. Thus the erosion is temporarily halted, and the beach rebuilt.

Hard-rock Shore Cliffs. Normally, cliffs that are composed of hard rocks, such as granite, provide as good a barrier to erosion as man can build. Even these cliffs, however, are subject to erosion over long periods of time, as exemplified by the numerous caverns and jagged pinnacles often found along granite sea cliffs. Usually there are but few beaches in such areas, and thus every single beach is of great value. The only material to keep these beaches in existence is the littoral drift from erosion of the nearby hard-rock cliffs. Thus the erosion of a hard-rock cliff may be either beneficial or detrimental to the shore shaping according to circumstances.

11.6. Littoral Processes on Sandy Coasts. Ocean or sea sediment transport is basically similar to sediment transport by water over the remainder of the earth's surface. In fact, as far as size, shape, grain density, and particle arrangements are concerned, both types of sediment have identical characteristics. The only actual difference is that sea sediment transportation is controlled largely by waves and tides whereas waves and tides are either absent or only partially operative in rivers. Generally it is assumed that at a depth equal to half the length of a wave (or about 200 ft as a maximum, Sec. 11.2) the wave "feels" the bottom and is retarded by friction and resistance of the bed material to erosion. The part of the ocean where the waves move the bed material generally is called the *littoral berm* or *littoral cone*.[11a]   Figure 11.8 shows the relative position of the littoral berm and the beach with respect to other parts

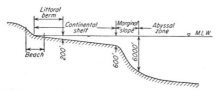

FIG. 11.8. Subdivisions of the ocean floor according to depth.

of the ocean. The relatively shallow portion of the ocean up to about 600 ft in depth (which includes the littoral berm) is the *continental shelf*. The deepest section of the ocean is the *abyssal zone;* it is separated from the continental shelf by an abrupt marginal slope. Littoral currents move sand already eroded from beaches and a little sand coming from the ocean itself. A new supply of sand material constantly is brought to the coast from the upland by streams and may be deposited at the coast or removed by the waves and tides. Sediment can move along the shore and perpendicularly to the shore. Douglas Johnson[1] stated that marine forces attacking a shore would produce, over a certain period of time, a "profile of equilibrium" at which the amount of deposition and removal would balance. This original geologic viewpoint led to the later concept of the *material-energy balance* of deposited and eroded sediment.[8]   At a certain section of the shore ("physiographic unit") the littoral processes are quite independent of those in the adjacent sections. The balance between the rate of supply and that of loss during a certain time determines whether the shore line in this physiographic unit tends to be eroded or built up by the currents. In reality these rates and the acting forces constantly change, and a sandy shore never acquires a steady state or a definite shape. It continues to oscillate back and forth about a certain "profile of equilibrium." The onshore component of a littoral current is offset by gravity (because to move on shore, the material has to move up-slope) and by reflected return currents such as undertow and *rip tides*. The latter are swift local streams in the return flow and are caused by irregularities in the bed of the coast (see rip currents, Sec. 11.3).

Material Sorting. At the seaward limit of the littoral berm the wave influence is very weak, but it increases as the depth decreases. Hence, coarser particles are likely to be moved toward the shore, whereas fines remain at greater depths. Here again the size of particles in equilibrium and the slope of the foreshore of the beach are interrelated. As already stated in Sec. 11.1, generally the finer the sand, the gentler the slope, and vice versa. Theoretically, if the sand is about 0.2 mm in size, the slope is about 1:20, whereas for coarser diameters such as 0.5 mm the slope is about 1:7.[2] The shape and specific gravity of the grains also influence the sorting, and the *sorting by shape* is more pronounced in coarser sands. Abrasion of sand grains changes the size and shape of the sand, but to a minor degree in comparison with the losses attributable to littoral movement. Sand beaches sometimes consist of alternate layers of coarser and finer material (laminae) that range from 1 to 20 mm in thickness. They are most apparent when the coarse laminae are of light-colored material (quartz, feldspar) and the fine ones consist of dark minerals.[12]

Bar Deposition and Removal. Sand may be deposited if the velocity of the current moving it is checked. If the seaward and landward currents moving the sand normally to the shore are balanced within a certain area and during a certain time, an *offshore bar* is formed. Conversely if conditions of equilibrium within the subject area are disturbed and vanish, the bar will be eroded. Thus, during storms or neap tides, sand is shifted seaward to form an offshore bar; during quieter conditions or spring tides, the sand is wholly or partly moved back to the beach. A more or less steady offshore bar creates a semiconfined channel parallel to the shore, disturbing the normal path of return currents. Such a bar also may change the position of the breaking point of the waves to a considerable extent.

Natural Barriers. Any object that disturbs the existing material-energy balance at a given point of the shore line and causes undesirable deposition or erosion is a *barrier* (or littoral barrier). There may be man-made and natural barriers. An abrupt change in shore alignment, such as a prominent headland, may act as a littoral barrier and cause accumulation of the sand material on the updrift side of the barrier. Inlets to tidal bays or estuaries also may act as littoral barriers, especially on the Pacific Coast, where maximum tidal velocities are reached in ebb flow. The littoral material entering the inlet during flood tide is swept seaward by the ebb and deposited offshore in the form of a bar.

Sand beaches can start to erode owing to minor and sometimes unsuspected causes. For example, occasional boulders lying on the sand may cause scour around them and eventually destruction of a very large area. Similarly, dredging just offshore may increase the turbulence of the water and cause erosion of the beach. In fact, the turbulence increases the carrying power of the running water with the consequent removal of sand grains in the form of a suspension. Any object, whether man-made or natural, should be removed from the beach if, in this way, the beach will be protected.

11.7. Currents and Littoral Processes in Lakes. A lake is a more or less large body of water in a steady state which occupies a depression in the earth's surface. The water in such bodies may be fresh (as in many mountain lakes) or salt (Great Salt Lake, Utah). Lakes may be formed in a variety of ways, some of which are (1) glaciation, such as carved the basin for the Great Lakes; (2) diastrophism, which created numerous lakes along the San Andreas fault in California; and (3) damming of natural drainage lines, e.g., by landslides or by lava (Lake Tahoe, California). Geologically speaking, lakes have a short life and can be completely obliterated, often by evaporation, e.g., ancient Lake Lahontan in Nevada. Great Salt Lake in Utah is a remnant of the ancient Lake Bonneville which was partially obliterated by evaporation.

Currents in Lakes. Some of the currents found in lakes are caused by *overturns*, wherein the colder water sinks and the warmer water rises. This process tends to stabilize temperatures throughout the lake and usually occurs in spring and fall. In tropical regions the bottom water of a lake is cooler than the top water. In polar regions the opposite is true, as there are no overturns and the lake water is stagnant. In moderate climates, ice shove in lakes, particularly in small ones, may push and distort the shore deposits into irregular ridges. Timber piles and sometimes masonry (as in dams) may be shorn or moved by such action.

Littoral Processes in Lakes. The shore lines of large lakes have many problems in common with seacoasts. However, tides are practically absent in lakes, the waves commonly are lower, and heavy storms are rare. For example, in Lake Michigan the maximum wave height is about 8 ft, sometimes reaching 10 ft. Cliff erosion and streams from the upland supply sand which may be moved by littoral currents. Natural beaches may be sand or gravel. Where the drift is very lean, or where the shore is protected by sea walls or bulkheads (Lake Michigan front in Chicago), imported sand may be used for construction of artificial beaches. Such work requires serious preliminary studies and geological consultations.

PLANNING AND CONSTRUCTION OF LITTORAL BARRIERS

11.8. Principles of Planning. To prevent or to stop the destruction of a shore line by waves and currents, *protective works* (littoral barriers) have to be planned and built. The action of a barrier may be either favorable or *detrimental* for engineering purposes. Serious attention should be given to the proper location and position of each barrier in order to avoid possible detrimental effects to the shore line as a whole. Each barrier should be provided with an adequate *foundation* capable of resisting vertical and especially lateral forces due to the impact of the destructive agents.

Preliminary investigations of the site should be first concerned with the properties of local rock and soil materials, particularly their *durability*, i.e., their *resistance to abrasion*. Since the impact action of destructive agents is strong and even violent, as in stormy weather, the stability of the local deposits, i.e., their resistance to translation, should be investigated. In fact, some sections of the shore line may be removed or added either suddenly by a storm (Sec. 11.1) or gradually during a relatively long time.

A detailed map of the shore line should be available, and if necessary, additional surveys should be made. To have an idea of what may occur to the sections of the shore line represented on that map, it is necessary to know the geological history of the site: the geological processes that are acting on the shore at the time of investigations and those that have acted on it previously. It should be established whether the shore line is submergent, emergent, or neutral and which factors have been primarily responsible for the actual shaping of the shore line. As an example, along certain portions of the Arctic Coast, glaciation still plays an important role in the shaping of the shore line, whereas in the continental United States, the glaciers have disappeared from the coastal regions. It is important for engineering purposes to know not only the qualitative picture of the processes that may change the shore line but also the *rates* at which these processes may operate or are operating. Some analogies can be drawn from the various studies of shore lines previously made *for engineering purposes* (in the United States mostly by the Army Engineers). It should be noted, however, that each shore line is an individual engineering project requiring individual analysis.

11.9. Summary of Information for Planning Littoral Barriers.[9a] For rational planning of shore protective structures the following information may be needed:

1. The geological history of the site, its stratigraphy, and lithology.
2. Classification of the grain sizes in the given beach or beaches.
3. Petrographic identification of rocks along the shore line.
4. Influence of climatic factors such as rainfall and surface runoff on the cliffs and banks along the shore line.
5. The natural (or even artificial) sources for littoral drift.
6. The dependability of such sources for future supply.
7. The rates of erosion of all shore-line materials in the area or those located at some distance from the area but still serving as a source of littoral drift.

11.10. Artificial Littoral Barriers. Examples of man-made littoral barriers (termed also protective structures, corrective structures, waterfront structures) are jetties, groins, and breakwaters. These structures are of trapezoidal or a similar cross section and of limited transverse dimensions in comparison with their length. Figure 11.11, representing the cross section of a groin, may be considered typical for all structures mentioned. Figures 11.9 and 11.10 give examples of locations of such structures.

One of the purposes of *jetty construction* (Fig. 11.9) is to protect inlets (entrances) to rivers and bays; jetties are also used for the protection of harbor areas. Often two approximately parallel jetties start at each shore of the river and extend some distance out to the ocean or sea into which the river flows. The velocity of flow between the jetties is increased

FIG. 11.9. Jetties and groins at the inlet of a river.

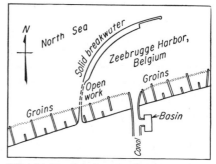

FIG. 11.10. Breakwater (to protect the harbor) and groins (for stability of the shore line).

because of its confinement, and thus, deposition is prevented. By the proper spacing of these two jetties, the channel between them even may be deepened by scour.

A groin (Fig. 11.10) is a littoral barrier starting at and perpendicular or at an angle to the shore line. Sand that is being carried by a longshore current is stopped on the upcurrent (or "updrift") side of the groin and thus aids in the widening of existing beaches. The more acute the angle made by the groin with the shore line, the less is its capacity to form an adequate beach. There are solid groins (Fig. 11.11) and permeable groins (Fig. 11.14); the latter modify the littoral processes as desired but do not obstruct them. The *length* of a groin should be approximately 50 per cent longer than the width of the beach it serves. Groins often are built to reach depths of 6 ft of water; i.e., they are placed in the section of the littoral berm where the most intense drift movement takes place. Groins are usually spaced from one to three lengths of the structure.

On California shores and elsewhere wider spacing is occasionally used, especially if the shore line approximates a long straight line. Large groins at long open beaches and inlets are sometimes referred to as jetties.

Both groins and jetties normal to the shore arrest the littoral drift. At the up side of this kind of structure, accretion of material starts immediately after construction, and in the case of a jetty there are erosion and deficiency of material on the downcurrent side. Erosion on the downcurrent side of groins is less pronounced because of their relatively wide spacing. There is a maximum impounding capacity for jetties and groins, and when it is reached, the preconstruction sediment transport conditions are reestablished. Sediment transport at depths greater than those at the seaward end of a jetty or groin is not affected by the presence of the structure.

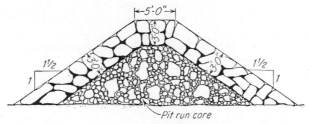

FIG. 11.11. Typical cross section of a rubble-mound water-front structure.[11]

Breakwaters are barriers constructed to break up and disperse the waves of heavy seas, to provide shelter for ships, and generally to contribute to the increase of shipping facilities in a harbor. An offshore breakwater (Fig. 11.10) may or may not be connected with the shore by a bridge. Many breakwaters start directly from the shore.

Another type of littoral barrier used to prevent erosion is a *dredged channel* across the shore line just updrift from the site subject to erosion. Such a channel is dredged periodically, and the dredged material deposited on the downdrift side of the channel. *Artificial nourishment* of the shore, by building beach fills, is used either if it is economically preferable or if artificial barriers fail to defend the shore adequately from erosion. For this purpose, sand is the most desirable material, and its size should satisfy conditions of fill stability in a given environment. Wrong-sized material will be shifted seaward or landward by the currents.[11a]

Sea walls and bulkheads are also protective water-front structures. They tend to protect the shore from wave attack and serve for docking of vessels. Sea walls range from a simple riprap deposit along a stretch of the shore to a regular masonry retaining wall (Fig. 11.18). Bulkheads are vertical walls, either of timber boards or of steel-sheet piling driven into the ground vertically and anchored to the natural ground behind the wall.

11.11. Details of Artificial Littoral Barriers. *Jetties.*[11e] The availability of building materials and construction costs are controlling factors in selecting the jetty type. In *rubble-mound* structures (Fig. 11.11), small stones (the so-called C stones) are placed in the interior of the structure and covered with coarser B and A stones placed in separate layers or mixed. The large A stones on the slopes of a jetty ("armor stones," Fig. 11.12) protect the jetty toe (toe stones) and its top (cap- or topstones). This simple type of jetty is adaptable to any depth, may be placed on any kind of foundation, and absorbs wave impact practically without reflection of water. If damaged, it still operates. The disadvantages of this type are (1) the large amount

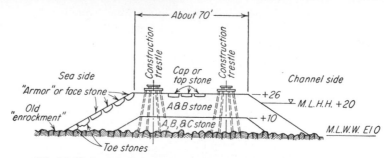

FIG. 11.12. Jetty built on an existing "enrockment" (an example).[11]

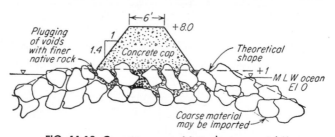

FIG. 11.13. Concrete-capped jetty (an example).[11]

of stone required and (2) the possibility of the wave water passing through the structure, though expensive grading or plugging of voids with finer rock (Fig. 11.13) may be done. European engineers have favored flat side slopes for jetties, whereas American rubble-mound jetties have slopes of 1.1:1 to 1.5:1.

The upper part of a jetty may be built of dimension stones, masonry, or concrete. This part is placed on a rubble-mound base extending from the sea bottom to approximately mean water level (Fig. 11.13). The amount of materials required to build such a structure is less than for the purely rubble-mound type.

Groins.[11f] The longitudinal profile of a groin is designed according to the cross section of the beach it serves. It consists of two horizontal sections, a shore section and a seaward section (Fig. 11.14). The central section slopes according to the beach slope. The seaward horizontal section, generally elevated 1 ft above the mean low water level, is supposed to protect the toe of the slope. It can be omitted, however, if the amount of incoming littoral drift is large, since in this case the toe of the slope will adequately maintain itself. Permeable groins are often simple templets or ribs limiting the space to be filled by the material of the littoral drift.

Breakwaters.[11g] Most breakwaters are of the rubble-mound type (Fig. 11.11). Rubble-mound breakwaters are adaptable to any depth of water, are suitable for nearly all types of foundations from solid rock to soft mud, and can readily be repaired. As in the case of jetties the stone used in rubble-mound breakwaters may be classified as A, B, and C stones (Fig. 11.15). The rough average weight of an A stone is 10 tons.

Vertical-wall breakwaters consist of masonry walls, timber cribs, caissons, or sheet piling, alone or in combination. This type of breakwater is not very much used, but a combined type consisting of a wall placed on top of a rubble-mound structure has many applications in practice.

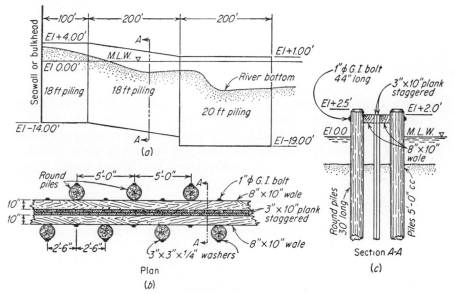

FIG. 11.14. Timber groin.[11]

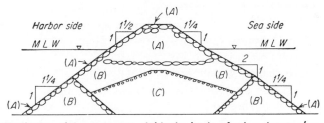

FIG. 11.15. Distribution of the stone material in the body of a breakwater (an example).[11]

HARBORS

11.12. Principles of Harbor Location.   The economic and commercial factors that greatly control the location of civilian harbors are considered in special treatises and thus are not discussed in this book. Military considerations in harbor location also are not discussed.

Harbors may be *natural* or *artificial*, i.e., made by man. Artificial harbors may be constructed either offshore or inside the shore by dredging interior basins and building entrance jetties. An offshore harbor should be protected by a breakwater system; hence their planning should take advantage of natural headlands. A *harbor of refuge* is a temporary haven without elaborate artificial protection where vessels may seek shelter in heavy weather or obtain emergency supplies and repairs.

Shore-line Considerations. The structure of the shore line and particularly the number of river mouths that may be used for the development of harbors is of importance for harbor locations. For example, along the Atlantic Coast of the United States are many natural waterways that may be easily developed into harbors, and hence harbors are constructed at frequent intervals. On the California coast, however, and especially along its southern section, natural facilities are relatively rare. Consequently, in the latter case improved harbors are farther apart and therefore have to be correspondingly larger in size.

Use of the Wind Rose. Wind studies for a proposed location are made, and percentages of time when the wind blows from a given compass direction are determined. These percentages are plotted to scale from one point on the paper along the respective compass directions. In this way a diagram known as the *wind rose* is obtained. (This diagram is analogous to the joint rosette, Sec. 7.10.) The direction of the prevailing gale can be established by a simple glance at the wind rose. In order to minimize the detrimental effect of gales on navigation, the breakwaters

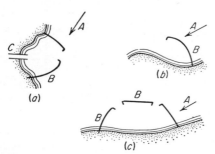

FIG. 11.16. Examples of harbor plans (B = breakwater, C = inner piers).

should be located practically *normal* to the direction of the prevailing winds. The same often is necessary for jetties.

In Fig. 11.16, letter *A* denotes from which direction the worst gale might come. If one imagines the breakwaters as gigantic human arms extended from the harbor to protect the entering ship, it becomes readily apparent that the breakwaters should be *convergent* (sketch *a*). A natural curvature of the shore line may replace a breakwater (sketch *b*). If these "arms" of the breakwaters happen to be far apart, a third breakwater to close the system may be required (sketch *c*). Figure 11.10 also should be referred to in this connection. A great number of possible harbor patterns are given in Ref. 7 (Fig. 99, page 140). More elaborate types of wind rose are used in airport planning. In this case the runways are located *parallel* to the prevailing winds.

Harbor Waves. Generally two kinds of waves should be considered when planning a harbor: (1) those that tend to prevent vessels from entering the harbor and (2) those that may develop in the harbor itself. The *period* of a wave may be short or long. Common ocean waves have short periods; the period of a tidal wave is long (over 12 hr). Common ocean waves are *progressive;* i.e., they move forward until they break.

A tidal wave is also progressive except that at a certain time it changes its direction (point D, Fig. 11.7). If at one point two identical progressive waves travel in opposite directions, resonance may take place. Where the water-particle motions due to each of the waves are in the same direction, they are added together, and thus the motion is larger than that caused by a single wave. Conversely, if the motions are in the opposite directions, the resulting motion may be small or even zero. Such is the

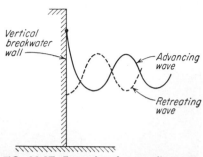

FIG. 11.17. Example of a standing wave (clapotis).

case of waves striking on a vertical breakwater wall (Fig. 11.17). In this and similar cases the wave is not progressive but *standing* (the French term *clapotis* also is used for this concept).

Surge is that wave motion which has a period between that of a common ocean wave and that of the tides; it may be from 1 to 60 min, approximately. In harbors, surge may be caused by the waves entering the harbor from the ocean. The height of surge waves is small (0.4 ft as an average), but the horizontal wave motion may be large. If the surge is due to a progressive wave train, all areas in the harbor experience the same maximum horizontal motion, whereas if the surge is of the standing-wave type, there will be distinct areas of active and quiet water. Unless rigidly restrained, a ship may be expected to move with the water motion caused by surging.

A phenomenon different from surge is *seiching*. This is an oscillation of the water in harbors and lakes and resembles water that is shaking in a bowl. Once started, the motion persists unless brought to rest by frictional forces. Small changes in water level are observed during seiching. This phenomenon sometimes is explained by changes in barometric pressure, although the actual cause is not too well understood. *Seismic sea waves* (*tsunamis*, a Japanese term) are generated by distant submarine earthquakes. Such waves sometimes are incorrectly termed "tidal," perhaps because, similar to genuine tides, they travel a considerable distance around the globe.

11.13. Deterioration of Harbor Structures. In the design and location of harbor structures composed of or founded on piles, the ability of the piling to withstand deterioration should be known. The two major causes of deterioration are *corrosion* and *abrasion*.   Corrosion is caused by the chemical agents in sea water attacking the structure material, by galvanic action if the structure is made of steel, and by chemical or bacterial agents of the soil into which the piling is driven.   Abrasion is due to the erosive effects of wind, water, spray, and water-borne sand. A third cause, which applies only to timber piles, is the attack on the wood by *marine organisms*.   These various agents can work together or separately and sometimes completely destroy a harbor-structure foundation.

FIG. 11.18. Protection for ore-loading dock (Taconite Harbor, Minnesota): (a) breakwaters, (b) temporary causeway to permit construction of breakwater, (c) cellular cofferdams to permit dry construction of sea wall (d).

Steel-sheet Piling.[9b]   Wherever the piling is exposed to continual drenching by sea spray, considerable corrosion will occur, more than that which occurs either above the spray line or below the water level.   Wherever the piling is covered with sand below the water, corrosive effects are practically negligible.   Painting the piling with coal tar or plastic compounds or applying gunite reduces the effect of corrosion.

Steel H Piles.[9d]   The portion of the pile below the water level is subject to corrosion by galvanic action, particularly on the thin edges of the flanges of H piles.   This is due to the electrolysis between the chemicals in the sea water and the metal in the pile; the result is pitting or etching of the steel.   Some designers have allowed for corrosion by increasing the cross section of the pile by 10 per cent.   This has proved uneconomical and often inefficient, since corrosion may attain larger proportions.   It is better to *protect* the pile against corrosion.   An H pile should be protected from about 3 ft above the high water to about 2 ft below the low water.   This can be done, for example, by wrapping the critical zone with two overlapping bent metal plates, fixing them to each other and to the pile, and filling the space between the plates and the piles with concrete.   Instead of two semicircular (in cross section) plates, a metal sleeve may be used.   The plates or the sleeve are mopped with hot asphalt.   Gunite application and painting are also used for corrosion protection.   A particular case of corrosion protection is *cathodic protection*.   The surface of the pile is covered with a coating firmly attached to the pile by cathodic action, this coating being calcareous, plastic, metallic, etc.   Generally cathodic-protection methods are patented (for further information on piles see Sec. 13.6).

Concrete Structures.[9c]   Severe concrete deterioration is caused by alternate cycles of freezing and thawing and is obviously predominant in northern waters.   It is believed that concrete is somewhat affected by the chemical agents found in sea water, though opinions on this subject differ.   For maximum protection, the concrete mix should be made using air-entraining agents and very sound aggregate.

Timber Piles.[9e]   Marine borers, particularly the mollusks, such as Teredo, are the major destructive agents to timber piles.   Members of this family vary in size from $3/8$ in. in diameter and 6 in. in length to larger sizes.   The mollusk, when young, drills a hole and eats up the timber as it grows in it.   The attack usually occurs on that pile section between the water level and the mud line in either salt or fresh water.   Proper pressure treatment of the pile with creosote is the best known protection against these various organisms.

RIVER IMPROVEMENT AND FLOOD CONTROL

11.14. River Improvement for Navigation.[13]   Improvement of a river for navigation requires the creation of a channel deep and wide enough for the passage of commercial vessels.   For example, it can be assumed that as a minimum, a usable inland waterway should be 9 ft deep and 300 ft wide.   Basic methods for building such a channel are (1) dredging, (2) regulation (contraction), and (3) canalization of the river.

If a channel is to be dredged, the bottom of the river should not be lowered so much that the water level is considerably lowered.   If such a case is foreseen, measures will have to be taken to increase the river discharge, which, for example, is feasible in the case of tidal rivers.   Dredging often is used as an auxiliary tool with the regulation or canalization of the river.   If dredging is to be done, a thorough study should be made of the types of the materials in the river bottom and their stability.   On the basis of such study, slopes of a new channel can be safely estimated, and the necessity for blasting, if rock is encountered, can be foreseen.   Sedimentation studies, particularly the determination of the *rates of sedimentation*, are required to estimate the expected life of the channel.   Short life can be expected for a dredged channel adjacent to the mouth of a large stream or that through an alluvial fan or delta in the process of formation, since in all these cases the channel may be filled in with alluvial material.   (See also Chap. 12 and Sec. 16.27.)

River Regulation (Contraction). The sinusoidal shape of many rivers (Fig. 3.5) is due mainly to the variable resistance of the banks to erosion. Hence, the simplest method of river control would be to make this resistance uniform. This type of control could be a bank revetment composed of a subaqueous brush mat, generally of willow branches, covered with riprap protruding above the water surface. Another method utilizes *spur dikes,* or rows of piles spaced about 2 ft apart and tied together. These rows are perpendicular or oblique to the banks and spaced at definite intervals along the river. Drift or sediment carried by the river settles in the intervals between the spur dikes. In this way the banks are naturally raised and stabilized. Longitudinal dikes of a similar material are also used. Sometimes the dikes are built of bundles of willow brush about 12 in. in diameter ("fascines"). More complex systems use masses of riprap extending normal from the bank for a short distance or short piles tied together with heavy wire netting and filled with rock. Many of these or similar methods also are used to control the flood discharges of gullies (Sec. 11.15). In canalization, regulation, or dredging studies, airphotos are particularly useful.

River Canalization. A canalized river is subdivided into several stages, or reaches, by dams located at the downstream end of each stage. Thus, there is an *upper pool* and a *lower pool,* upstream and downstream, respectively, from the dam. A *movable navigable dam* is designed to allow vessels to pass over the sill and through the dam proper. In *fixed dams* at all times and in movable navigable dams at the low-water stages, the vessels approaching the dam have to pass through a *lock.* The lock is provided with two gates, one at the entrance and the other is the exit gate. From the upper pool the vessel enters the lock in which the exit gate is closed, after which the entrance gate also is closed and the exit door opened. The locks are provided with pumps to make traffic possible in either direction.

Dams for canalization of large rivers are mostly earth dams (e.g., the Missouri River). In such rivers adequate rock foundations are deep, and borrow pits for concrete aggregates are quite remote. Thus the construction of a concrete dam on a very large river is often a serious problem, whereas materials for an earth dam may be abundant. Investigations for canalization dams are the same as for any other dam (Chaps. 15 and 16). Settlement in locks may result in a jamming of the gates; thus, investigations (basically similar to those required for spillway chute and dam gate structures, Chap. 15) for locks should be done with great care. It is of interest to note that dams of the future may be designed to pass vessels over the dam by means of ship elevators, thus eliminating the locks (Yangtse Gorge Dam in China, plans in preparation in 1955).

Examples of River Improvement. The following systems of canalized waterways are to be found in the United States and Canada: (1) the connecting waterway between New York and the Great Lakes including the Hudson River and the Erie, Oswego, and Black-Rock-Tonawanda Canals; (2) a continuous channel from the Delaware River nearly to the southern extremity of Florida, parallel and close to but sheltered from the ocean and suitable for barges and non-seagoing vessels (the "Atlantic Intracoastal Waterway"); (3) the continuous inland waterway from Apalachicola Bay, Florida, to southern Texas; (4) the Mississippi River system consisting of four subsystems including the Missouri and Ohio Rivers; (5) the California system consisting of the Sacramento and San Joaquin Rivers, both dredged only; and (6) the Columbia River, which has been improved by dredging. The waterways in the Great Lakes region are improved, including the "Soo Canal" (Sault Sainte Marie) in Michigan along the Canadian border. The Saint Lawrence Seaway, under construction in 1955, will, when completed, permit ocean-going vessels from the Atlantic Coast to dock at ports on the Great Lakes.

11.15. Principles of Flood Control. Flood-control problems generally arise in alluvial plains limited by bluffs. Most of the time the river flows in a well-defined bed in only part of the plain. During heavy discharges after excessive rainfalls, the river bed is inadequate to carry the load, and the river rises over its banks and floods the adjacent plain. The most critical time for such discharges is early spring, when heavy snow melts increase the flow. The problem is of particularly increasing importance in the United States because many major cities are located adjacent to large rivers. As these cities have grown, they have encroached upon the old river channels, and the present-day stream is confined by means of levees or dikes. Continual encroachment results in an everdecreasing channel width and capacity of carrying flood discharges. The flood-control problem is further aggravated by the continually increasing runoff from the drainage area as a result of increasing denudation of the land by cultivation and erosion. Such runoff carries large quantities of silt eroded from the denuded land.

Thus flood control proper is intimately connected to *erosion control.* Some of the measures that are being applied along these lines are reforestation and use of agricultural procedures that permit the retention of as much precipitation as possible in the ground. Three kinds of dams are constructed in this connection: (1) small dams in the gullies to prevent erosion and rapid, high-velocity runoff; (2) larger dams on tributaries to main streams, to act as catch basins for floodwaters in widely distributed areas and thus prevent the entire flood from reaching the main stream (or "the main stem") all at once; and (3) large dams on the main stream at spaced intervals to control siltation and smooth out the flood discharges so an excessive quantity of water does not hit an area all at once.

More information on flood control and corresponding geotechnical investigations may be found in Ref. 14.

REFERENCES

1. Johnson, D. W.: "Shore Processes and Shore Development," John Wiley & Sons, Inc., New York, 1st ed., 1919, last ed., 1938.
2. Krumbein, W. C.: Geological Aspects of Beach Engineering, "Berkey Volume," 1950.
3. Gerstner, Franz: Theorie der Wellen, *Abhandl. Koenigl. Boemischen Ges. der Wiss.*, Prague, 1802.
4. Stokes, G. G.: On the Theory of Oscillatory Waves, *Trans. Cambridge Phil. Soc.*, vol. 8, 1847.
5. Mitchim, C. F.: Oscillatory Waves in Deep Water, *Military Eng.*, March–April, 1940.
6. Shankland, E. C.: "Dredging of Harbors and Rivers," 2d ed., Brown, Son and Ferguson, Ltd., Nautical Publishers, Glasgow, 1949.
7. Minikin, R. R.: "Winds, Waves, and Maritime Structures," Charles Griffin & Co., Ltd., London, 1950.
8. Mason, M. A.: Geology in Shore Control Problems, "Applied Sedimentation," John Wiley & Sons, Inc., New York, 1950.
9. *Proc. 3d Conf. on Coastal Eng.*, Council on Wave Research, University of California, Berkeley, 1953: (a) L. W. Currier, Geology in Shoreline Engineering; (b) A. C. Rayner, Life of Sheet Pile Structures; (c) H. K. Cook, Exposure Research on Concrete in Sea Water; (d) G. L. Wey, Corrosion Studies of Steel Piling in Sea Water; (e) G. E. Knox, Prevention of Deterioration in Waterfront Structures.
10. Minikin, R. R.: "Coast Erosion and Protection," Chapman & Hall, Ltd., London, 1952.
11. *Proc. 1st Conf. on Coastal Eng.*, Council on Wave Research, University of California, Berkeley, 1951: (a) R. O. Eaton, Littoral Processes on Sandy Coasts; (b) W. C. Krumbein, Littoral Processes in Lakes; (c) K. P. Peel, Location of Harbors; (d) V. A. Vanoni and J. H. Carr, Harbor Surging; (e) R. E. Hickson and F. W. Roloff, Design of Jetties; (f) D. F. Norton, Design and Construction of Groins; (g) J. R. Ayers, Seawalls and Breakwaters; (h) K. Kaplan and H. E. Pape, Jr., Design of Breakwaters.
12. Emery, K. O., and R. E. Stevenson: Laminated Beach Sand, *J. Sediment. Petrol.*, vol. 20, 1950.
13. "Engineering Construction—Canalization," 2 vols., The Engineer School, Ft. Belvoir, Va., 1939.
14. "Engineering Construction—Flood Control," The Engineer School, Ft. Belvoir, Va., 1946.

CHAPTER 12

ELEMENTS OF SEDIMENTATION ENGINEERING

The three basic sedimentation processes are erosion, transportation, and deposition of sediments. The term "sediment" is used hereafter to describe soil and rock particles of various sizes and shapes taking part in any stage of the sedimentation processes. Although presently considered a misnomer, the term "silt" is also used rather loosely to express the same concept both in practice and in the technical literature. The term silt thus used should not be confused with the same term designating a certain definite kind of soil (Secs. 3.2 and 4.11). This chapter discusses water as a sedimentation agent in deep-water bodies such as rivers, harbors, and dam reservoirs and in smaller channels such as irrigation canals.

12.1. River Terminology. The area that supplies a river with water and sediment is its *watershed*. The terms "watershed" and "drainage area" are synonymous in this case. The *mouth* of a river is that part where the river joins the ocean, the sea, a lake, or another river. Any stretch or a part of a river along its length upstream from the mouth may be termed a *reach*. It is assumed hereafter that a reach has practically a constant shape and size and constant physical properties along its whole length.* The discharge of a river is the amount of water flow through the cross section of a reach in a unit of time. This discharge is directly related to the geometry of a reach and the frictional resistances to the flow. The latter are due to roughness of the bottom of the river and its banks within the wetted area and losses of energy caused by changes in the shape of the water prism in the stream. The "geometry" of a reach is its longitudinal profile and its cross sections within the wetted range. It should be borne in mind that the longitudinal slope of the water surface in a river and the longitudinal slope of its bed may or may not be identical and that it is the water slope that controls the discharge.

* The stretch of a canal between locks is a reach. Also, an arm of the sea extending deep into the land is a reach.

445

River Bed and Channel. The bottom of a river between its banks is the *bed* of the river. The bed and the banks constitute the *channel* of the river. The channel may be *nonerodible* (rock or rough cobblestone bed) or *erodible* (composed of sand or fines) or both. In the first case an increase in water discharge means a related increase in stage, in other words, rise of the water level. In an erodible channel the increase in water discharge may mean an increase in stage and a change in bed condition. In this chapter, rivers with erodible beds are considered.

TRANSPORTATION AND DEPOSITION OF SEDIMENTS

12.2. Sources of Sediments, Poised River. Sediment moving in a channel, particularly in a stream, is generally termed *load*. Soil and rock particles loosened by weathering and by the force of falling rain are transported to the stream by surface runoff. Streams carrying load mechanically erode soil materials and chip rock particles from the channel walls; this process is known as *corrasion*. The processes described occur in high-velocity stretches where the dynamic effect of running water is high. As soon as water enters a low-velocity reach, deposition of the load will take place. Generally, deposition always occurs *when the velocity of the water carrying the load is checked.* "Degradation," or scour, in stretches of high velocity and "aggradation," or deposition, in the stretches of low velocity lead to the formation of a uniform longitudinal slope along the whole river. This process is termed "gradation of the river."[1] Rivers in such condition are termed *poised* or *graded.* Occasionally the banks may cave into the stream. This temporarily adds to the load carried in suspension, but usually the caved material is shifted toward the downstream banks, especially if the river is poised.

12.3. Mechanics of Sediment Transportation. Water flow may be *laminar** or *turbulent.* Generally, water flow in a river is turbulent. Tributaries from the watershed flow into the river with a certain velocity, the load is not uniformly distributed across cross sections of the reach, and there may be purely mechanical obstacles to the flow. These and similar circumstances contribute to the irregularity of the river flow. The turbulence is accompanied by the formation of *eddies*, both horizontal and vertical, i.e., spots with decreased or no pressure. An eddy may favor a sudden stop of the movement or dropping out of particles to the bottom. Just as an oblique force causes a body on an inclined plane to slide downward, water carrying load is pushed downslope by

---

* In laminar flow, the water moves in parallel, approximately horizontal sheets, or laminae, and all water particles follow parallel straight-line trajectories. This type of flow is rare in nature but sometimes is assumed to occur in sheet runoff, i.e., when heavy precipitation creates sheets of water flowing over normally dry slopes.

tractive force (TF).   This force can be calculated by the following formula:

$$\text{TF} = \gamma_w ds \qquad\qquad (12.1)$$

wherein γ_w is the unit weight of water, 62.4 pcf; d is the depth of the water in feet; and s is a nondimensional number designating the flow gradient.   (A discussion of the influence of various soil sizes on tractive force is given in Sec. 16.27.)

Coarse and fine particles behave differently in a river flow.   A *coarse sediment* particle tends to move obliquely downward under the combined action of tractive force and its weight, since the specific gravity of the sediment is larger than that of water.   Ultimately a coarse sediment particle may reach the bottom of the river.   It may or may not remain there, since the turbulent flow may lift even large rock particles and then drop them again or make them roll and jump along the bed.   This latter movement is termed "saltation."   In the turbulent flow next to the river bottom, water may push a coarse sediment particle in any direction and occasionally not push it at all.   *Fine particles* move along with the river flow, since in this case the tractive force is predominant as compared with the weight, and the probability of their reaching the river bottom is not high.   The *finest sediment particles* are prevented from falling downward by the Brownian movement that is operative in both still and moving water.   The Brownian movement is the continuous bombardment of the suspended soil fines by the water molecules.   As is actually the case, it may be concluded that the river bed generally is composed of particles coarser than those moving through a given reach above the bed.   The presence of a small quantity of finer particles in the voids of the coarser material at the bottom of the river does not affect the validity of this statement.

12.4. Kinds of Sediment in a River.   Immediately above the bed of a river there is a layer where coarser particles move.   The thickness of this bed layer is variable and is controlled mostly by the size of the sediments.   Particles moving in the bed layer constitute the *bed load*. Particles continuously supported by the water and moving outside the bed layer constitute the *suspended load*.   The suspended load may contain particles the size of bed load but generally are finer than the bed load.   These fine particles are known as *wash load*.   The weight of the bed load may be three times as much as that of the suspended load. Wide, shallow streams carry relatively more bed load than deep, narrow streams.   Also, bed load is more important in streams with sandy beds.[12]

Wash load has a different influence from bed load on the stability of the river channel.   The amount of the wash load depends only on the availability of this material in the watershed.   The wash load travels easily through the reach and in most cases has no influence on the stability of the channel, provided the flow conditions do not change abruptly.

It should be recalled that in this chapter, only erodible river beds are considered. In this case, the materials forming the bed and the bed load are essentially the same. A reach in this condition is often termed an *alluvial reach*. Observations and experimental data[2] show that if an alluvial reach is in equilibrium, the channel is stable; i.e., although there is aggradation and degradation of the bed, over all the reach does not change gradient. In other words, an alluvial reach possesses a certain *capacity* for permitting the bed load to pass. The bed load tends to travel until the capacity of the reach is exceeded.[2-5] If the capacity of the reach is exceeded, the excess of traveling bed load is deposited. Conversely, in the case of deficiency of the bed load, the bed would be scoured, the channel widened, or the gradient increased until equilibrium is restored. This also explains why erosion occurs downstream from a dam. When a stream (and its tributaries, if any) reaches the reservoir, the flow velocity is checked and both the wash load and the bed load are deposited (Sec. 12.9) in the reservoir. Thus, the clean water released over the dam tends to scour the material from the channel downstream until a bed layer is formed.[2]

Recapitulation. The amount of wash load traveling in an alluvial reach depends on the availability of the corresponding material in the watershed. The amount of bed load is controlled by the capacity of the channel. The channel stability depends exclusively on the amount of bed load in the bed layer.

12.5. Derivation of Stream-borne Sediment. The availability of sediment in the watershed depends on the factors controlling erosion: soil or rock type, vegetal cover, intensity and duration of precipitation, and soil conditions prior to precipitation.

Soil Erodibility. The degree of erosion and thus the amount of sediment contributed to a stream are directly affected by the type of soil. As can be seen in Fig. 12.1, clayey and coarse-grained sandy soils are less erodible than silty soils under identical flow conditions. Generally the erodibility varies inversely with the size of the particles in the soil; this is true, however, only for soils that are not cemented or tightly packed.[6] Figure 12.1 shows that the erodibility decreases as the grain sizes increase except for very fine-grained soils having the coherence found in clays.[7] This chart further demonstrates that as the velocity of flow decreases, the tendency toward sedimentation increases; similarly, the transporting power of the water decreases as the soil sizes increase. For example, if the stream velocity reached 50 cm/sec it would loosen a soil particle of 0.02 mm from its natural bed. If the velocity dropped to 10 cm/sec, the stream would not loosen (erode) a 0.02-mm particle but would be capable of transporting it.   If the flow were to drop below 0.15 cm/sec, the 0.02-mm particle would be dropped from the stream; i.e., sedimentation would occur.

The density of the soil is important.   For example, a loosely packed (or loose-texture), porous (low-density) soil would permit infiltration of the precipitation and thus inhibit high runoffs.   Conversely, a tight, dense (heavy-texture) soil contributes to a high runoff.   High-velocity runoff is conducive to gully formation and a resultant increase in available sediment.

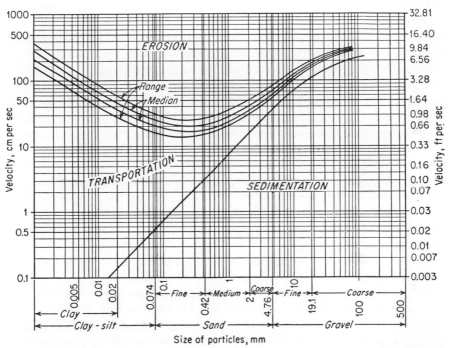

FIG. 12.1. Ratio of particle size to velocity required for erosion, transportation, and deposition. (*After Hjulström.*[7])

Rock Type.   Soft, easily friable (crumbly) rocks may rapidly erode if subjected to intense precipitation.   Some sandstones, siltstones, claystones, shales, and decomposed granites fall in this category.   The particle sizes likely to occur in the stream sediment often can be determined beforehand by a study of the rock type and the method in which its constituents are welded together.   The dip of the formations can affect the erosion characteristics of an area.   Gently dipping rocks are more susceptible to erosion than steeply dipping ones.   The latter, however, contribute to rugged terrain and resulting high-velocity runoff. This may cause erosion of usually erosion-resistant rocks but more particularly can increase the erosion of adjacent soils. The steepness of the topography is not always a sure criterion as to the amount of sediment likely to be obtained from an area. Generally, steep terrain is indicative of highly resistant rocks. Observations have shown that the degree of erosion is about proportional to the slope up to about 20°, reaches a maximum at about 40°, and then decreases in effect.[6]

Vegetation decreases erosive effects, as (1) it tends to increase the cohesiveness of soil and (2) it tends to absorb the impact energy of raindrops. A heavily forested area retains runoff to a great extent because of the usual undergrowth and the action of tree roots in absorbing water and in holding soil in place. If such an area is denuded by a forest fire, small and then large gullies form, and tremendous amounts of soil are carried downhill to the stream. Dense grass and shrubbery tend to inhibit erosion, as they decrease the runoff velocity and act as a protective mat for the underlying soil. It is interesting to note that erosion is increased when the effectiveness of a grass cover is destroyed by the close grazing common to certain animals such as sheep. Any such activities induced by man, including improper agricultural or lumbering practices, may result in "accelerated erosion."[8]

Climate. The amount of humidity and degree of temperature may indirectly affect erodibility of soils. Prolonged low humidity and high or low temperature bake or harden a clayey soil. Thus, the infiltration rate and the surface frictional resistance to runoff are decreased. Although high runoff velocities may occur under such conditions, considerably less sheet erosion occurs. Continually moderate temperatures and high humidity may have a reverse effect, although the usual good vegetal growth under such conditions may modify the expected high erosion.

Short-duration, high-intensity rainfall will, under certain conditions, cause considerable erosion because of the raindrop impact and the high-velocity runoff from a soil that does not have time to absorb the water. Long-duration, high- or low-intensity rainfall also results in considerable erosion, as the soil becomes saturated and the long duration of the rainfall creates a high runoff. Generally, soils are more erodible when dry than when moist.

In summation, watershed sediment results from a high interrelation of the above factors and results from (1) sheet erosion of the land surface by the forces of raindrop impact and surface runoff and (2) formation of gullies eroded by concentrated runoff. Continuous gullying eventually may lead to the formation of canyons. Flood erosion by slowly moving, practically stagnant water may cause banks and slopes in the watershed to collapse, but the sediment load may not increase because of possible redeposition of the material in the watershed.

There are statistical data referring to the sediment productivity of different watersheds. This productivity is expressed in tons per year per square mile of drainage area. Under identical geological and climatological conditions, the larger the watershed, the smaller the unit amount (per square mile of watershed) of sediment coming to a river (or to a basin). This is because the eroded material is deposited in overbank (flood-plain) and dead-water areas. Also, some precipitation water carrying sediment from remote parts of the watershed may never reach a river because of infiltration, evaporation, and drying out.

Some statistics on sediment production follow:

1. Soldier River at Pisgah, Iowa. Drainage area 417 sq miles, annual sediment production 10,384 tons/sq mile.

2. Elm Fork, Trinity River at Dallas, Texas. Drainage area 1,174 sq miles, annual sediment production 1,300 tons/sq mile.

3. Rio Grande at Elephant Butte, New Mexico. Drainage area 26,312 sq miles, sediment production 680 tons/sq mile.

4. San Leandro Creek at Oakland, California. Drainage area 30.3 sq miles, annual sediment production 2,470 tons/sq mile.

5. Santa Anita Creek at Arcadia, California. Drainage area 10.8 sq miles, annual sediment production 6,054 tons/sq mile.

12.6. Silt Survey in Rivers. *Measuring Amount of Suspended Load.*

There are a variety of samplers for measuring sediments, from the very simple to the most elaborate, from portable to permanently installed.[9,10] Suspended load commonly is measured from a sample obtained in a depth-integrating sampler which is moved up and down along a certain vertical line. Occasionally, point-integrating samplers are used; these collect water at one point of the depth where they are held for a period of time, dependent on the capacity of the sampler. The point-integrating sampler requires a certain time to be filled up, whereas in the so-called instantaneous samplers, the sampling operation requires but a minimum of time. (These latter samplers, however, have not been sufficiently adapted to practice.) A homemade depth-integrating device consists of a pint milk bottle equipped with a two-hole rubber stopper through which two $1/4$-in. tubes extend, one for water intake and the other for the air exhaust. The bottle is used in horizontal position with the water-intake tube facing directly into the current. The air-exhaust tube is bent downstream.

There are no established standard rules as to where to measure the suspended load in a river. The USBR samples the Colorado River in Arizona at surface, mid-depth, and bottom in each of three or four verticals located in a cross section of the reach. The area of the cross section is subdivided into three or four parts with equal discharge values, and the verticals where the sediment is measured are at the mid-points of those parts. The practice of the Army Engineers in various Middle Western states is to take daily surface samples at midstream if the stream is narrower than 100 ft and in wider streams in midstream and at one-sixth or one-fourth of the width measured from each shore. Samples of six-tenths of the depth are taken weekly.

Measuring Concentration and Size Distribution of the Suspended Load. The weight of sediment in a sampler is determined and referred to the weight of water carrying it. This is the *concentration* of the suspended load expressed as a number of parts in a million. For instance, a concentration of 40,000 ppm means 4 per cent of the weight of the water carrying it. One method used to determine the concentration is to

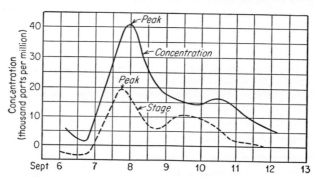

FIG. 12.2. River stage and concentration graphs, Bighorn River at Manderson, Wyoming.[10]

decant or remove the clear water after the sediment settles to the bottom of the sampler, dry the sediment, and then weigh it. The concentration C_s is computed as follows:

$$C_s = \frac{\text{wt. sediment}}{\text{wt. water} + \text{wt. sediment}} \times 1,000,000 \qquad (12.2)$$

Another method, among many, is to pour the suspension through a filter paper fitted in a funnel. The size distribution (gradation) of the sediment is determined by using sieves or a hydrometer or both.

The suspended-load concentration usually increases with an increase in stage of the river (Fig. 12.2). In large streams the peak of the sediment concentration is generally close to the peak of the discharge and on some occasions even precedes it. The total suspended load in tons per day (Q_s) carried by water is computed as follows:

$$Q_s = 0.0027Q_wC_s \qquad (12.3)$$

wherein Q_w is the total discharge of water in cubic feet per second and 0.0027 is a conversion factor. In this equation, it is assumed that the average speed of the suspended sediment is the same as that of the surrounding water. However, next to the bed this assumption becomes misleading.

Measuring Bed Load. For measuring *bed load* three types of samplers are used: (1) box-type samplers or traps in which the velocity is decreased and thus the coarse sediment settles down and the water passes by, (2) pan-type samplers consisting of a bottom and two side walls with baffles that deflect the flow and make the sediment settle, and (3) slot-type samplers with a large hole or slot in a horizontal plate; the coarse sediment passing over the hole falls through and can be measured. In the United States, an entirely successful device for measuring bed load has not yet been found.

12.7. Theoretical Study of Sedimentation in Channels. The so-called "bed-load function" is supposed to give the rates at which flows of any magnitude in a given channel will transport the individual sediment sizes of which the channel bed is composed.[2] Obviously, if an observer finds that at the given reach of a river there is practically no deposition and no scour, the sediment travels "to capacity." If, in addition, he determines that all bed-load material is of the same grain size, and if he is able to measure the discharge of the flow in the reach in cubic feet per second and the amount of transport in tons per day, then he can plot *one point* of the graph of the bed-load function for that reach. Reference 2 is an advanced attempt theoretically to define the shape of the curve representing the bed-load function. This work was done on the basis of experimental evidence, existing theory of turbulent flow, and reasonable speculation. Figure 12.3 is an example of a

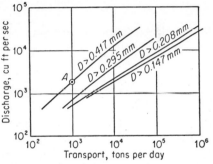

FIG. 12.3. Bed-load function, Big Sand Creek, Mississippi. (*After Einstein.*[2])

family of curves representing the bed-load function for various grain sizes of the bed-load material of a certain reach.

Transportation of suspended sediment by water has been studied experimentally.[11] It was found that a flow can support and, therefore, transport fine sediment more effectively than the theory indicates. Further studies showed that experimental results often do not agree with theoretical computations on the distribution of the concentration of the sediment in a flow. It was also found, among other things, that sediment-laden water flows more rapidly than a comparable amount of clean water.

SEDIMENTATION IN ENGINEERING

12.8. Sedimentation in Reservoirs. When a dam is built in a valley, a reservoir is formed upstream. Sediment coming to the reservoir is largely deposited in it, and sooner or later the reservoir will be silted, i.e., filled up with sediment, and the usefulness of the dam will be lost. The study of reservoir "siltation" is important, as good damsites are scarce and sometimes it is impossible to replace a silted site.

Figure 12.4 shows schematically the process of the gradual silting of a reservoir. First of all, it should be noted that the presence of the dam retards the natural flow; hence the water level in the reservoir is not horizontal, but its surface is slightly curved (back-water curve). When the river flow reaches the reservoir, the flow velocity decreases and coarser particles such as sand are dropped close to the entrance. Thus *fore-set*

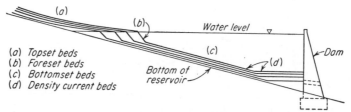

(*a*) Topset beds
(*b*) Foreset beds
(*c*) Bottomset beds
(*d*) Density current beds

FIG. 12.4. Schematic representation of the silting process in a reservoir. For limitations see Sec. 12.8. (*After Lane.*[13])

beds of coarser sediment are formed (letter *b* in Fig. 12.4). The *bottom-set beds* are layers of fine sediment settling slowly across the area of the reservoir (letter *c*). The action of the fore-set beds is twofold: First, they obstruct the flow still coming from upstream and thus contribute to the formation of *top-set beds* (letter *a*); second, as time goes on, the fore-set beds advance into the reservoir to form higher and higher slopes that, because of their height, will not stand so steeply as the slopes formed at the initial stages of the sedimentation. Hence, the high slopes of the fore-set beds have to flatten out and cover a part of the previously laid bottom-set beds. This causes an intermixture of coarser and finer sediment at the bottom of the reservoir.

The location of future silt deposits in a reservoir often can be predicted. For example, the sediment will tend to deposit near the dam if the reservoir water surface is at a low elevation, particularly during floods; if there is a high percentage of clay to fine silt sizes in the sediment; if the reservoir is short and has a steep slope to the original valley floor; if there is little or no vegetation at the head of the reservoir; and if the dam has small outlets at a high elevation. If the converse of the above conditions occurs or there are topographic restrictions between the head of the reservoir and the dam, sediment will tend to deposit in the upper part of the reservoir.[12]

In a wide reservoir, one often can observe a muddy stream loaded heavily with fines and flanked on both sides by clear water. This stream sinks down as a unit and proceeds toward the dam at a low elevation. The behavior of a hypothetical mercury stream entering the reservoir together with water is an exaggerated analogy.   Deep, heavily loaded streams in a reservoir are *density currents* (compare Sec. 11.3).

It is strongly emphasized that Fig. 12.4 is schematic and corresponds to the case of high concentration of sediment and short periods of storage. Under other conditions, density currents are not so regular as shown in Fig. 12.4.   In fact, water reaches maximum density at 4°C (39°F); hence, in colder climates the upper density currents tend to sink down and to cause a turnover (Sec. 11.7) of the warmer bottom water.   In this way the pattern of density currents as shown in Fig. 12.4 is disturbed, and the situation thus created requires a careful investigation of temperature effects on density currents (See Ref. 13, discussion).

Clay and colloidal particles, especially when flocculated, form deposits of low unit weight.   In various reservoirs these have been estimated as 60 to 70 pcf.   During the later stages of sedimentation and close to the bottom, higher densities occur because of compaction.

12.9. Reservoir Life.   The reservoir capacity is expressed in acre-feet (acre-ft = 43,560 cu ft).   The methods of measuring capacity are explained in Sec. 12.10.   The basic factor in estimating the life of a reservoir, i.e., how long it will continue to store a useful amount of water, is the length of time the water stays in the reservoir before being used. Thus, the *useful life of a reservoir* is determined by the rapidity with which sediment accumulates from year to year and is not removed by natural action or man-made devices.   A short retention time does not permit the fines to reach the bottom.   Also, as the storage space becomes less and less, there is an increase in the rate at which fine sediment leaves the reservoir.   Thus, although the sedimentation rate decreases, very little sand-size or larger material will pass by the dam.[13]

The silt which is located at high enough levels in the reservoir to be removable, for instance by dredging, though such an operation generally is difficult and costly, is termed silt in *live storage*.   Silt in *dead storage* cannot be removed.   The reservoir life can be extended by *sluicing operations*.   The outlet gates are opened at appropriate intervals, and the resulting high-velocity flows carry some of the sediment downstream. If these operations can be timed to intercept gravity underflows (density currents), the sediment removal can be very efficient.   It has been found that sluicing of sediment that already has settled generally is ineffective except in the reservoir area immediately upstream from the sluice or outlet intakes.[14]   Some dams are designed specifically to remove silt from irrigation water.   These structures usually have an adjacent *desilting basin* wherein the water velocity is purposely decreased to permit sedimentation at accessible elevations.   The silt then can be removed by dredging, by draglines, or, as is done at the All American Canal desilting basin, by rotary scrapers that deposit the silt in trenches where it is sluiced back to the river.

Trap Efficiency. The percentage of the total inflow of sediment retained in a reservoir usually is termed its *trap efficiency.* This ability of the reservoir to trap silt depends on the ratio between storage capacity and inflow, the age of the reservoir, the shape of the reservoir basin, the type and method of operation of the outlets, the grain size of the sediment, and its behavior under various conditions.

The trap efficiency, or TE, of an existing reservoir can be determined as follows:

$$\text{TE} = \frac{S_i - S_0}{S_i} \tag{12.4}$$

wherein S_i are the acre-feet of silt that enter the reservoir every year and S_0 are the acre-feet of silt that leave the reservoir every year past

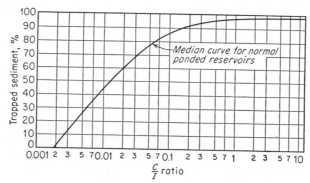

FIG. 12.5. Relation of trap efficiency to capacity-inflow ratio. (*After Brune.*[14])

the dam. As reservoir sediment loads often are given in the literature in tons, the load in acre-feet can be obtained as follows:

$$\text{Tons/acre-ft or } T_{af} = \frac{\text{unit wt. of sediment, pcf, } \times 43,560}{2,000}$$

$$\text{Total acre-ft of sediment} = \frac{\text{total tons of sediment}}{T_{af}} \tag{12.5}$$

If the reservoir was designed merely to pass the major part of an annual flood with little or no storage, the trap efficiency would be very low, as there would be little sediment trapped behind the dam. Conversely, a desilting basin would have a trap efficiency of almost 100 per cent. (Theoretically 100 per cent trap efficiency is not possible; however, actual measurements on some dams have shown 100 per cent trap efficiency.[14])

The trap efficiency of a proposed reservoir can be roughly estimated by dividing the proposed reservoir capacity C in acre-feet by the expected annual inflow (I) of water in acre-feet. This ratio C/I can be compared with special curves such as Fig. 12.5 to obtain an estimated trap efficiency.

In most cases, the trap efficiency is assumed at about 95 per cent or more for proposed reservoirs.   With these data available, an approximation of the useful life of a reservoir R_L can be computed:

$$R_L = \frac{C}{S_i \times \text{TE}} \qquad (12.6)$$

Example of reservoir life computation based on surveys of Lake Corpus Christi, Texas[15]:

Reservoir capacity in 1942.....................................	43,801 acre-ft
Reservoir capacity in 1948.....................................	39,387 acre-ft
Average weight of sediment...................................	35 pcf
Therefore, average annual sediment deposit.....................	1,106 acre-ft
Measurement of annual average sediment carried past dam........	390 acre-ft
Average annual inflow...	738,289 acre-ft

Thus:

$$\frac{C}{I} = \frac{43,801}{738,289} = 0.059$$

According to Fig. 12.5, the trap efficiency would be about 79 per cent.   Actual computations, since the data were available, showed that

$$\text{TE} = \frac{S_i - S_0}{S_i} = \frac{1,106}{1,496} = 0.744 = 74.4\%$$

Therefore, reservoir life remaining in 1942 was

$$R_L = \frac{43,801}{1,106 \times 0.744} = 53.3 \text{ years}$$

This period of life would be applicable only if the trap efficiency remained the same; usually, however, the trap efficiency decreases during the life of the reservoir.

Examples of Reservoir Siltation. The most spectacular example of reservoir sedimentation in the United States is the gradual filling with sediment of Lake Mead.   This is the 115-mile-long reservoir back of Boulder (Hoover) Dam on the Colorado River at the boundary of Arizona and Nevada.   The Colorado River is perennially muddy upstream from the dam because of the enormous area of erodible sedimentary deposits through which it flows.   Boulder Dam began to store water and sediment on Feb. 1, 1935, and in the next 10 years collected over 1,500 million tons of sediment.   Estimates have indicated that the end of the useful life of Lake Mead will occur about the year 2250.[16a]

Another example is the sedimentation of the reservoir of Elephant Butte Dam on the Rio Grande in New Mexico, about 100 miles upstream from the boundary between the United States and Mexico.   This dam was completed in 1916, although impounding of water began a year earlier.   The Rio Grande is perennially muddy to the dam from about 150 miles upstream at the mouth of the Rio Chama, a tributary of the

Rio Grande. About 700 million tons (464,000 acre-ft) of sediment reached Elephant Butte reservoir in the first 32 years of its life. It is estimated that its useful life will end about 2100.[16b]

12.10. Silt Survey in a Reservoir.[17–19] The volume of silt deposited in the reservoir during a certain time interval following its construction may be computed from the comparison of contour-line maps prepared before construction and at the time of the given silt survey. The former map is prepared by usual surveying methods, and the latter from the soundings, or water-depth measurements, at a given water level. The contour-line method in the case of large reservoirs is accomplished by the *range method*. Ranges or directions are first located on the map, making them coincide if possible with tributaries and conspicuous changes in plan of the reservoir, and then marked with permanent reference points on both banks of the reservoir. Measurements are made from a boat moving from one bank of the reservoir toward the transitman located on the other bank. The transitman keeps the boat on the range by using flags or walkie-talkies. Distances from the starting point may be measured with a distance wheel unrolling marked wire. In the case of wider reservoirs, horizontal angles are cut by using two surveying instruments. Depths are measured with a sounding pole or loaded line. The volume of sediment in the reservoir is determined from the field data by plotting a contour map and then planimetering the areas between contour lines. Occasionally, end areas are computed such as in conventional earthwork computations.

Supersonic (or Echo) Sounding. Supersonic equipment determines water depths by utilizing supersonic-pressure waves and precise timing of the wave travel. The two principal devices of this equipment are the projector and the recorder. The projector sends the waves to the bottom of the reservoir and receives the reflected waves. The recorder is electrically connected to the projector and by electronic and mechanical means gives a continuous record of the depth of water through which the waves have been sent and reflected.*

Density of Sediment Deposits. It is essential to determine densities of the sediment deposits at the time of the silt survey. This is a difficult task, since silt deposits consolidate and their density thus increases with time and depth. An average density may be estimated from a study of undisturbed silt samples taken at the time of the survey.

12.11. Sedimentation Survey Report. In order to estimate the useful life of a proposed or existing reservoir, a detailed report should be prepared containing the following data: (1) original reservoir capacity in acre-feet; (2) reservoir capacity existing at time of survey of existing

---

* The depth of water generally is measured in feet or fathoms, 1 fathom being equal to 6 ft.

dam; (3) the percentage of the drainage area covered by each type of bedrock or geologic formation, by each type of soil, and by each type of vegetation and affected by each type of erosion; (4) the relief and type of topography; (5) climatic conditions including temperature and rainfall ranges; and for existing reservoirs, (6) the location of sediment deposits in the reservoir, the volume weights and densities of the sediment, rate of annual sediment accumulation, and acre-feet of sediment per year that passes by the dam.   Reference 20 is an excellent example of the above type of report.

12.12. Sedimentation in Irrigation Canals.[21]   In irrigated areas, water from a dam reservoir or a river is transferred to the fields by main and lateral irrigation canals.   These may be open ditches or closed conduits. The bottom and the slopes of open ditches may be provided with a practically impervious lining such as concrete or asphalt, the lining being designed to prevent or decrease leakage (see Sec. 16.28).   A reach of an open ditch behaves similarly to a river reach as far as sedimentation is concerned.   Silt deposition in canals often is due to the decrease in longitudinal grade and consequent decrease in transporting capacity of the canal.   This effect may be produced, for example, by vegetation that grows in shallow water or on exposed surfaces (as it does under similar conditions on reservoir deposits).   There may be a tendency for erosion if the transporting capacity of the canal increases in passing from one reach to the other; this may be due to an increase in longitudinal grade or narrowing of the cross section.

The silt control in an irrigation canal consists either in completely or partially *desilting* the water coming into the canal (see Sec. 12.9, desilting basins) or in *cleaning* the obstructed channel with removal of water weeds and other plants.   In canals passing through highly pervious materials, reasonable silting is beneficial.   Planned siltation is required in such cases.

12.13. Sedimentation in Harbors.[22,23]   One of the most important problems of harbor engineering is continuously to maintain the *minimum depth* of water required for the convenient movement, anchoring, and berthing of ships.   Since the depth of water in a harbor is decreased by the deposition of silt, the currents bringing the silt to the harbor should be stopped or diverted and the silt given an opportunity to settle down at a place where the silt deposits are not detrimental to navigation. Nevertheless, if silt deposits are being formed within the harbor area, they should be removed.   Generally, this is done by *dredging* and, in relatively rare cases, by eroding and flushing the *shoals*, i.e., shallow places.

Before measures against harbor silting are taken, the following information should be obtained: (1) the directions along which sediment may reach the harbor, whether from the uplands by streams, from the sea by waves, or along the coast by littoral currents; (2) the predominant character of sediment, whether coarse or fine or both; and (3) a rough estimate of expected quantities of sediment.

Hereafter, measures to be taken against harbor silting for different particular cases of sediment or harbor location are discussed.

Coarse Sediment. Coarse sediment such as gravel or even boulders enters a harbor located where a turbulent mountain river flows into a rather placid tidal estuary, e.g., the Washington, D.C., harbor on the Potomac River. (Sand tends to fill many harbors.) Since coarse sediment travels in low waters, near the bottom of the river, it can be stopped or diverted from the harbor by low *training walls.* In this case such walls do not have to protrude above the water level. (Examples: St. Louis, Missouri, on the Mississippi River; Sacramento, California, on the Sacramento River.)

Fine Sediment. Deposition of predominantly or exclusively fine sediment may take place in harbors located at the transition of a slowly moving river into a tidal estuary. Examples are numerous (New York, New York, on the Hudson River; Savannah, Georgia, on the Savannah River; London, England, on the Thames River). In this case sediment may be removed by taking advantage of the tides. A sizable *tidal prism* (i.e., prism of water above the low tide) moves through the harbor located in a tidal estuary. This prism (Fig. 11.7) may be increased. Thus greater velocities are created which flush out the harbor channel. The size of the tidal prism may be increased by improving the hydraulic characteristics of the estuary channel above and below the harbor or by creating a greater tidewater area above the harbor. By proper design of training walls and arrangement of one-way tide gates, the ebb flow may be used for removing the shoal material.

Off-channel Harbors. If the amount of the sediment expected is too large, *off-channel* river and tidal harbors may be constructed. In this case the harbor is built in a special basin connected with the river or the estuary by an entrance channel. The latter should be kept clear of sediment. This can be done, for instance, by a proper arrangement of training walls. The entrance channel may be provided with a lock or a gate to isolate the basin from the river or the estuary during periods of turbid water. Examples of off-channel river harbors are Freeport, Texas, and Greenville, Mississippi; Texas City and Houston, Texas, have off-channel tidal harbors.

Shore-line Harbors. In this special kind of harbor the danger of shoaling by littoral currents should be foreseen during the design of the harbor. Properly located breakwaters and jetties will prevent such shoaling. A detached (island) harbor constructed completely separate from the shore is protected from shoaling, as then the littoral drift is permitted normal movement along the shore. Improvements to tributary flow and to the tributary watershed to decrease the amount of sediment may be helpful.[23]

REFERENCES

1. Mackin, J. H.: Concept of a Graded River, *Bull. GSA* 59, 1948.
2. Einstein, H. A.: The Bed-load Function for Sediment Transportation in Open Channel Flow, *USDA Tech. Bull.* 1026, 1950.
3. Einstein, H. A.: Determination of Rates of Bed-load Movement, *Proc. Fed. Inter-agency Sedimentation Conf.*, U.S. Bureau of Reclamation, Denver, 1948.
4. Einstein, H. A.: Estimating Quantities of Sediment Supplied By Streams to a Coast, *Proc. 1st Conf. on Coastal Eng.*, Long Beach, Calif., published by the Engineering Foundation, 1948.
5. Einstein, H. A., and J. W. Johnson: The Laws of Sediment Transportation, "Applied Sedimentation," John Wiley & Sons, Inc., New York, 1950.
6. Linsley, R. K., M. A. Kohler, and J. L. H. Paulhus: "Applied Hydrology," McGraw-Hill Book Company, Inc., New York, 1949.
7. Hjulström, F.: Studies of the Morphological Activity of Rivers as Illustrated by River Fyris, *Uppsala Univ. Geol. Inst. Bull.* XXV, 1935.
8. Brown, C. B.: Effects of Soil Conservation, "Applied Sedimentation," John Wiley & Sons, Inc., New York, 1950.
9. A Study of Methods Used in Measurement and Analysis of Sediment Loads in Streams, a series of 9 reports by 7 organizations including Iowa Institute of Hydraulic Research, University of Iowa, published by St. Paul Engineering District suboffice, Iowa City, Iowa, 1940–1948.
10. Benedict, P. C.: Determination of the Suspended Sediment Discharge of Streams, *Proc. Fed. Inter-agency Sedimentation Conf.*, U.S. Bureau of Reclamation, Denver, 1948.
11. Vanoni, V. A.: Transportation of Suspended Sediment by Water, *ASCE Trans.*, vol. 111, 1946; also Developments of the Mechanics of Sedimentation Transportation, *Proc. Fed. Inter-agency Sediment Conf.*, U.S. Bureau of Reclamation, Denver, 1948.
12. Maddock, T., Jr., and W. M. Borland: Sedimentation Studies for the Planning of Reservoirs by the Bureau of Reclamation, *Proc. 4th Congr. on Large Dams*, R. 41, question 14, New Delhi, India, 1951.
13. Lane, E. W.: Sediment Engineering as a Quantitative Science, with discussion by P. D. Trask and A. S. Fry, *Proc. Fed. Inter-agency Sediment Conf.*, U.S. Bureau of Reclamation, Denver, 1948; also Some Aspects of Reservoir Sedimentation, *Irrigation Power J.*, vol. X, nos. 2 and 3, April–July, 1953.
14. Brune, G. M.: Trap Efficiency of Reservoirs, *AGU Trans.*, vol. 34, no. 3, June, 1953.
15. Brown, C. B., V. H. Jones, and R. E. Rogers: Report on Sedimentation of Lake Corpus Christi and the Water Supply of Corpus Christi, Texas, *USDA SCS Rept.* SCS-TP-74, December, 1948.
16a. Thomas, H. E.: First Fourteen Years of Lake Mead, *USGS Inform. Circ.* 346, 1954.
16b. Seavy, L. M.: Sedimentation Surveys of Elephant Butte Reservoir, Hot Springs, N. M., *USBR Hydrol. Div. Rept.*, February, 1949.

17. Fry, A. S.: Sedimentation in Reservoirs, *Proc. Fed. Inter-agency Sedimentation Conf.*, U.S. Bureau of Reclamation, Denver, 1948.
18. Eakin, H. M. (rev. by C. B. Brown): Silting of Reservoirs, *USDA Tech. Bull.* 524, 1939.
19. Corfitzen, W. E.: Silt Problem, *Civil Eng.*, vol. 12, 1942.
20. Seavy, L. M.: Sedimentation Survey of Guernsey Reservoir, Wyoming–Nebraska, *USBR Hydrol. Div. Rept.*, November, 1948.
21. Golzé, A. R.: Problems of Irrigation Canals, *Proc. Fed. Inter-agency Sedimentation Conf.*, U.S. Bureau of Reclamation, Denver, 1948.
22. Caldwell, J. M.: Sedimentation in Harbors, *Proc. Fed. Inter-agency Sedimentation Conf.*, U.S. Bureau of Reclamation, Denver, 1948.
23. Shankland, E. C.: "Dredging of Harbors and Rivers," Brown, Son and Ferguson, Ltd., Nautical Publishers, Glasgow, 1949.

The following is an excellent quick review of available literature applicable to sedimentation engineering:

24. "Annotated Bibliography on Sedimentation," Subcommittee on Sedimentation, USDA SCS Federal Inter-Agency River Basin Commission, Sedimentation Bulletin 2, 1950.

CHAPTER 13

BUILDINGS: SITE
EXPLORATION AND FOUNDATIONS

This chapter contains primarily the principles of correlation of the type of foundations for a building with the geological factors of the site and its neighborhood. The term "exploratory program" as used hereafter involves all geotechnical operations to be performed on the site and in the laboratory in order to obtain data required for an intelligent design of the foundations of the building and a rational execution of this design. For those interested in the details of foundation engineering, a list of references is given at the end of the chapter.

13.1. General Considerations. The exploratory program for the foundation of a building essentially depends on two factors: (1) the *weight* of the building and other *forces* acting on it and (2) the *service* of the building, or the purpose for which it is being built. Generally for lighter structures the depth to which the investigations are carried is limited, whereas for heavier structures it is commonly necessary to explore the entire depth of soil covering the rock and even penetrate into the rock. The drilling work should be limited to a strictly necessary number of holes, which should not be excessively deep. To accomplish this work, general knowledge of local geological conditions is required, combined with experience in handling building projects. The influence of the regional geology may be striking; e.g., in some cities in the glacial zones, the sequence of strata is well known, and the problem consists only in the determination of their thickness, which can be done by rudimentary methods such as wash borings. Careful examination of the behavior of the neighboring buildings sometimes eliminates part of the geotechnical investigations. The fact that in some cases field investigations are not essential or are very simple does not mean that in really involved cases a detailed investigation can be unduly shortened. The same principles govern the laboratory soil and rock testing. These tests have to be considered as an integral part of the exploratory program and merit every attention, especially if there is possibility of excessive settlement of the building (consolidation test) or threat of a shear failure (shear test).

As to the *service* of the building, the correct outline of the exploratory program taxes the ingenuity of the geotechnical investigator and draws on his experience with buildings. The organization of an exploratory program is discussed herein by means of simplified examples for four categories of buildings, namely, (1) residential buildings including housing projects, (2) commercial buildings, (3) industrial buildings, and (4) power and pumping plants. In actual practice, however, there are many more categories and sometimes even individual buildings that do not fall into any category, such as monuments or palaces.

Besides the soil and rock exploration proper, all information should be obtained concerning the *earthwork to be done* in connection with the building. Such earthwork consists of the excavations for the foundation and grading of the site, which sometimes involve high fills and deep excavation. The topography of the site before and after construction is not the same, and the final elevation (or elevations) of the ground after completion of the building is known as the *finished grade*.

The magnitude and, especially, the cost of the exploratory program depend on the importance and the cost of the building. For example, the owner of a small residence perhaps cannot spend a few hundred dollars for explorations, even if they are needed. However, a concern planning the construction of a million-dollar commercial or industrial building can easily spend a sum of $10,000 for exploration, provided, of course, that the expenditure is justified.

13.2. Structural Loads. The weight of the building itself including its appurtenances is the *dead load* (DL). The weight of the loads applied to the building intermittently is the *live load* (LL). In the case of a school, the weight of the students is the live load, and so is the weight of the stored merchandise in a warehouse. In the majority of the buildings, the dead load is larger than the live load, but in some cases (e.g., one-story warehouses) the opposite is true.

Except in such cases and for rough computations only, the live load is sometimes estimated at 50 per cent of the dead load. The live load generally is variable and cannot be accurately computed; an assumption with a certain degree of probability is made according to the service of the building.

Generally the dead and the live loads are transmitted to the foundations *vertically*. Such is the case when all the loads are taken up by the walls and the columns and then passed to the footings. Roofs over large halls, such as auditoriums or mess halls in large military quarters, are sometimes designed in the form of *arches*. The arch action is schematically shown in Fig. 13.1a. The arch produces a vertical pressure on its

abutments and tends to push them apart, thus producing a *thrust*.   If
the abutments are tied together with a steel or reinforced concrete tie
(Fig. 13.1*a*), the thrust may be balanced.   The roof of a hall may be
arranged in the form of a *rigid frame* (Fig. 13.1*b*).   In this case some
horizontal action also may be ex-
pected at the *hinges A* and *B*.

Besides the dead and live loads,
there are lateral forces acting on the
building.   These   are   mostly   wind
and earthquake forces.   The latter
are discussed in Chap. 18.   Wind
forces are intermittent and tend to
push a building sideways.   The force
is transmitted to the foundations,
where they must be resisted by the
shearing strength of the material in
which   the   foundations   are   built,
and the passive resistance of the soil
mass supporting the building.

Other   forces   acting   on   some
buildings are *vibrations*.   These are
particularly evident in power and
pumping plants or in industrial buildings   containing   large   machines.

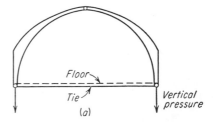

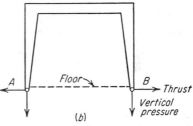

FIG. 13.1. (a) Three-hinge arch.   (b) Rigid frame.

For vertical and lateral forces, the *building code* of the city where the
building is to be located should be consulted.   Typical floor and wind
loads may be found in Ref. 1 (pages 551 to 555).

TYPES OF BUILDING FOUNDATIONS

13.3. Spread Footings.   The term "foundation" may designate either
the lower part of a structure in contact with earth or rock (e.g., a pile
foundation) or the upper part of the soil or rock mass in contact with
the structure (e.g., a rock foundation).   The term "footing" is sometimes used for any kind of foundation, but strictly speaking, this term
refers to *spread footings* only.   The terms "spread foundations" and
"spread footings" are synonymous.   As the term itself indicates, a
spread footing takes up the weight of a part of the building and spreads
it over a larger area in order to decrease the *unit load*.   (The latter is
expressed in tons per square foot or in pounds per square foot.)   The
larger the area of the footing in contact with earth, the smaller is the unit
load transmitted to each square foot of the subsurface material on which
the footing rests.   There are three basic types of spread footings: (1)

individual footings, generally supporting a column (either an outside wall column or an interior column); (2) *continuous* or wall footings all along the outside and inside walls (but not partitions); and (3) the *mat* or raft foundation in the form of a slab under the entire building or a part thereof.

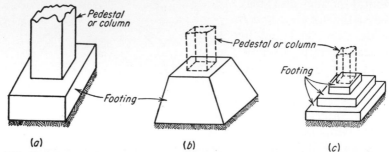

FIG. 13.2. Types of spread footings: (a) shallow slab, (b) battered, (c) stepped.

Individual footings may be any one of three varieties: (1) shallow slab type (Fig. 13.2a), (2) battered type (Fig. 13.2b), and (3) stepped

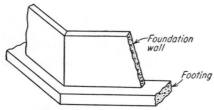

FIG. 13.3. Continuous footing.

type (Fig. 13.2c). Individual footings sometimes accommodate two or, in exceptional cases, more than two columns. In a cantilever footing when a column of an outside wall is just at the property line and its footing cannot be extended beyond that line, the wall column and an inside column are interconnected by a strap beam and both columns work together.

The *continuous footing* (Fig. 13.3), or the foundation wall, is merely a wall with a widened base resting on the subsurface material. This type is used preferably for light buildings such as residences, one-story school buildings, or small, one-story warehouses. A variation of this footing is called "pier and grade beam" construction. In this type the foundation consists of regular footings (piers) supporting wall columns placed at correct design intervals; the piers are connected with beams or simply with *curtain* walls generally thinner than the beams. A curtain wall has no foundations and extends only 4 to 6 in. below the finished grade. Curtain walls support no structural load other than their own weight.

Sometimes the roof of a building is supported by the outside wall at certain intervals rather than continuously. At these intervals *pilasters,* or local widenings of the wall and its foundation, are placed (Fig. 13.4a). Pilasters may be located on either the interior or exterior of a wall. They extend to the roof and are placed on the local widenings of the continuous wall footings. This arrangement is convenient for one-story

buildings with heavy roofs.   For larger structures with roofs in the form
of an arch of a rigid frame, *buttresses* (Fig. 13.4*b*) are used to oppose the
possible thrust from the roof.   The buttress is wider at the bottom than
at the top and usually extends from the base of the roof beam to the
footing level.   Thus, the roof load is transmitted through the buttress
and spread over a larger area at the base of the buttress.   This design

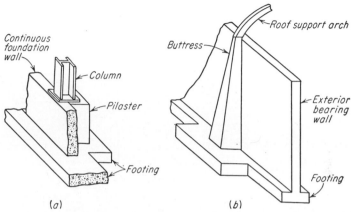

FIG. 13.4. (*a*) Pilaster.   (*b*) Buttress.   (For pilaster see also Fig. 13.5.)

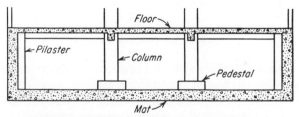

FIG. 13.5. Mat foundation.   (After Dunham.[8])

may be used where weak soil conditions would tend to cause the footings
to rotate or where there is a threat of overturning a high pilaster or
wall column.

The *mat*, or *raft* (Fig. 13.5), foundation is a continuous, reinforced-
concrete slab that covers the entire area of the structure.   It generally
is used where the soil is of a very low bearing capacity or where it can
economically be incorporated as a utility portion of the structure, such
as the basement floor.   The building loads are spread over a very large
area, and thus the actual per-square-foot load on the underlying soil is
low.   In exploring for this type of foundation, it is important to determine carefully all conditions underneath the proposed mat.   For example,
if a relatively hard spot should occur under one point of the mat, it
would tend to act as a fulcrum and cause a failure of the mat.

Ground-water conditions are critical, as the mat is very susceptible to hydrostatic uplift owing to its relatively light load. The maximum possible elevation of the ground water should be carefully determined for the design of the slab. In a particular case (called a "floating foundation"), when the mat is placed at the bottom of an excavation and the weight of the excavated earth material equals the weight of the mat, the walls, partitions, and columns placed in the excavation, the pressure at the bottom of the excavation will be the same as before construction. The hydraulic uplift tending to move the slab upward is

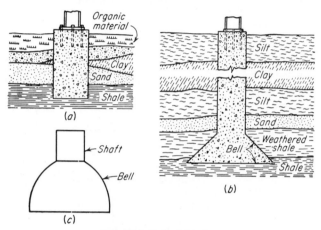

FIG. 13.6. Drilled-in pier.

helpful in some cases. When the excavation for the mat is made, there will be some elastic rebound, or upward motion, of the bottom of the excavation; subsequently, placing of the mat will cause some settlement, or downward movement, of this raised excavation bottom.

High buildings on shallow mats undergo the danger of movement by heavy lateral forces. This movement is resisted by the friction of the mat against the underlying material and by the passive resistance of the material into which the mat has been built. It should be remembered in this connection that the passive resistance of a cohesive soil equals at least its compressive strength (Sec. 4.23), and performance of the corresponding test may prove necessary.

13.4. Caissons or Piers. A drilled-in pier (termed also "concrete pier" or even "caisson") is a variation of an individual footing. Such piers vary in diameter from 6 in. to 6 ft and more, the most common sizes being 24 and 36 in. (Fig. 13.6a). The hole for the pier is usually drilled with special machines, and after the excavation is made, the hole is filled with concrete and sometimes reinforced with vertical steel bars. A pier, as described, supports the load by the resistance of the soil at its base and by friction of the sides of the concrete against soil (termed "peripheral friction"). The blowing wind tends to push the building, together with the piers, and the latter are pushed against the soil. In the case of cohesive soils, as already stated, this push is resisted by the compressive strength of the soil. If it is necessary to spread the load carried by the pier over a wider area than the actual drilled diameter of the pier, the bottom of the hole is "underreamed" or "belled" (Fig. 13.6b and c); i.e., the lower end is enlarged into a cone or bell-like design by a special device, or the bell is carved by hand labor into a step-shaped form. The depth of a concrete pier is controlled *either* by the resistance of the base and peripheral friction to the load *or*, if the upper strata are not reliable, by the necessity of carrying the load to a reliable but deep soil or rock material. In the former case the length of a concrete pier may be 5 or 6 ft or more and, in the latter case, very great, depending on the depth of the reliable material.

The Woolworth Building, New York City, is founded on piers which reach solid rock (Manhattan schist) located 115 ft below the curb level. The piers supporting one of the heaviest loaded columns (loaded over 9 million pounds) is 18 ft 9 in. in diameter. The unit load on the rock is about 17 tons/sq ft. For information on pneumatic caissons that sometimes are used for building foundations, consult Sec. 14.6.

13.5. Bearing Value and Load Tests. The unit load at the base of a spread footing should not be larger than the safe bearing power of the supporting soil. The *safe bearing power* is the ultimate bearing power divided by a safety factor of 2 or 3. The *ultimate bearing power* of a soil is the unit load under which a failure of the supporting material occurs. It is well known that failure may appear as a squeezing of soil from underneath the footing (shear failure) or by excessive settlement detrimental to the structure.

The safe bearing values are given in the building codes. Sometimes a building code suggests performing load tests and recommends procedures to be followed. A load test usually consists of loading a certain area with the design unit load (i.e., the load in pounds per square foot with which the foundation will be loaded) and afterward with 150 per cent thereof and observing settlements. The latter should not be larger than some prescribed values (e.g., $\frac{3}{4}$ in.). The simplest load test is to load a timber post 12 by 12 or 17 by 17 in. (i.e., 1 sq ft or 2 sq ft) in cross section. A platform is placed on top of the post and loaded by increments of about 200 lb with pig iron, stone, water, etc. Protective measures against tipping of the post and against unfavorable weather such as rain should be taken. No loading tests on frozen or swelling soils should be performed. The bottom of the test post preferably should be placed in an excavation at the design level of the base of the future

footing.   Figure 13.7 shows a more elaborate test in which a hydraulic
jack acts against a loaded steel beam and strain gauges measure the
settlement.   A different type of load test is discussed in Sec. 2.9.

13.6. Pile Foundations.   There are timber, concrete, and steel piles
and various combinations thereof.   Piles are driven by means of a
hammer.   The pile and the hammer are placed between a pair of vertical

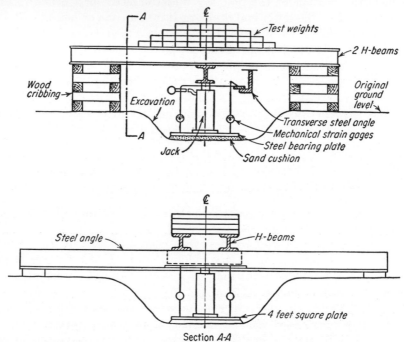

FIG. 13.7. Load test.   (*Eng. News-Record,* Apr. 16, 1953, p. 51.)

guides, or *leads* (Fig. 13.8), which are carried by a frame (pile driver) and
may be driven either vertically or in an inclined position (battered piles).
A drop hammer is simply a heavy weight lifted up and dropped by manual force.   Such hammers are not in general use in the United States.
In a steam hammer, a heavy ram is lifted up by the steam power and
dropped on the butt of the pile.   Steam hammers may weigh from $1\frac{1}{2}$
to 7 tons and more; the hammer strikes the pile 35 or more blows per
minute.

Wood or timber piles have an average diameter of about 10 or 12 in.
They are wider at the *butt,* i.e., the top of the pile on which the hammer
falls, and have a blunt or sharp point at the tip which may be protected
by a steel shoe.   A 60-ft depth is practically a limit for a timber-pile
foundation because of the difficulty in obtaining timber piles of greater
lengths.   Shorter piles may be spliced together, which, however, decreases

their bearing capacity. Creosote treatment protects timber against fungi, termites, and marine borers if driven in the sea (see Sec. 11.13).

Concrete piles may be *precast*, i.e., fabricated before driving, and *cast in place*, i.e., made during the driving process or immediately thereafter. Precast piles are generally octagonal or square in cross section and are reinforced with steel bars. For cast-in-place piles, a thin steel shell with a steel core (mandrel) inside is driven into the ground, after which the mandrel is removed and the shell filled with concrete. In other types there is a heavy drive pipe driven with the mandrel and gradually

FIG. 13.8. Driving H-beam piles.

removed as the concrete is poured in. A brief but complete review of the types of cast-in-place concrete piles is given in Ref. 15 (pages 176 to 177).

Steel piles are of two major types: concrete-filled steel pipes (tubes) and steel H piles. Open-end heavy steel tubes up to 30 in. in diameter are driven through soft deposits to hardpan or rock; the material inside the tube is washed out with a water or compressed air jet; then the tube is redriven to final position and filled with concrete. This type of pile is often used in New York City. The H piles (Fig. 13.8), so called because of the similarity of their cross section to this letter, have all three components of the letter H equal (10, 12, and 14 in. wide). Usually the smaller (10-in.) type is used for buildings. The H piles, like the steel tubes, are used to carry the weight of the building to deeper, more reliable strata.

A more detailed review of existing pile types may be found in Ref. 15. Attention should be called, however, to the following two pile types: (1) the pile formed by

FIG. 13.9. Mixed-in-place piles.   Man's right hand indicates piles.   (*Intrusion-Prepakt, Inc.*)

vibroflotation methods[20] and (2) the intrusion grout mixed-in-place pile. In the former type, a vibrating device is worked into the ground with aid of a high-pressure jet of water. After the device has penetrated to the predetermined depth, it is vibrated at an extremely high frequency rate and gradually pulled to the surface while sand is poured into the hole. This pile has had considerable success in sandy soils, as it leaves a compact column of sand which is capable of bearing considerable load. The intrusion grout mixed-in-place pile (Fig. 13.9) is formed by forcing a grout mixture through a hollow shaft. An augerlike head on the bottom of the shaft rotates and drills into the ground, and at the same time grout is forced down the shaft and mixed with the soil. After the head is retracted from the soil and the "soil-grout" mixture allowed to set, a pile with compressive strengths as high as 4,600 psi may result, depending on the soil type.

If the soil through which the pile is driven acts as the main support for the pile owing to the adhesive and frictional forces on the sides of the pile, it is said to be a *friction pile* (Fig. 13.10a). However, if the pile is driven to a firm nonplastic material at its tip, it is then known as a *point-bearing* pile (Fig. 13.10b). Usually, every pile has some combination of both friction and point-bearing forces acting upon it. However, the frictional forces acting on a point-bearing pile often are not considered in design.

If a pile must penetrate through dense layers of sand and gravel, this material is loosened by a water jet. Obviously jetting decreases the friction of the pile against the surrounding material and therefore must be used with caution.

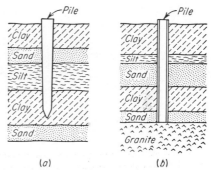

FIG. 13.10. (a) Friction pile. (b) Point-bearing pile.

Generally as the pile is driven, more and more blows are required to produce 1 in. of penetration. About 10 blows per inch is considered the refusal. The *set* in such a case is $\frac{1}{10}$ in. The value of the set is used to check the bearing power of a pile in pile-driving formulas, such as the *Engineering News* formula published in all American civil engineering textbooks and handbooks concerned with this field.

The case of clays requires particular consideration. Any pile driven into clay remolds it; the remolded or disturbed cylindrical body on an average is about three times wider than the pile itself. Thus, *as an average*, a 12-in. timber pile remolds the clay in the form of a cylinder 36 in. in diameter or to a distance of 12 in. from the outside surface of the pile. If the clay is of a low sensitivity to remolding (Sec. 4.6), the disturbed zone will be compacted in a manner similar to sand. The compressive strength of the material around the pile may even be increased in this connection. If the clay is of high sensitivity, however, the displaced material will be squeezed out to the surface and cannot be compacted. In the clays of medium sensitivity, the clay around the pile itself may be somewhat disturbed and there may be some heave around the pile. If there is a heavy overburden above a clay stratum, heave at the surface is unlikely. It should be noted that in all clay deposits, remolding next to the ground surface is more pronounced than at a depth and sometimes the piles themselves are lifted up. Some redriving may be helpful in such cases.

The well-known phenomenon of "setup" of a pile takes place in thixotropic clays in the following manner: The pile keeps moving down under the blows of a hammer without showing any sign of refusal, but after a period of rest (sometimes one day or less), the pile will withstand a reasonably high static loading.

13.7. Negative Friction (Drag). A building founded on end-bearing piles driven through a highly compressible material (soft clay, organic silt, etc.) to a relatively reliable layer may suddenly start settling and cracking. This usually occurs if an additional fill is built around the building. The weight of that part of the fill located between but at a certain distance from the piles is supported by the reliable layer on which

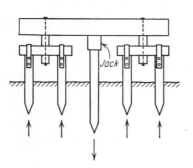

the pile points rest, but the soil directly surrounding the pile hangs onto it, so to speak, and increases the load carried by the pile. This weight cannot be larger than the weight of the remolded material, however (Sec. 4.6). For a submerged condition and a 12-in. timber pile, the drag, or additional load on the pile, may be estimated at 300 lb/lin ft of the submerged length. This value obviously varies with local conditions. The phenomenon described, scientifically termed

FIG. 13.11. A pile load test.

negative friction, is most likely to develop in the shore-line sections of a country.

13.8. Pile Load Tests. An essential feature of the geotechnical investigation for a pile foundation is the pile load test. This should not be done until at least 30 days after driving to permit the pile-soil adjustment to be completed. Figure 13.11 schematically shows a simple pile load test using a calibrated jack with auxiliary tension piles on both sides of the central test pile. The design load, i.e., the load to be carried by the pile (or, as sometimes required, 150 per cent of the design load), should show no appreciable settlement during 24 hr. (On impervious, saturated materials, however, it usually is advisable to maintain the maximum load at least 10 days to permit pore pressure relief in the material under the load.) For more details consult the building code of the city where the building is located and Ref. 18. Groups of piles (clusters) may be tested by placing a loaded platform on their tops. All other conditions being identical, the *bearing capacity of one pile* in the group is *less* than that of an individual pile (not in a cluster) because of pile interference.

SELECTION OF FOUNDATION TYPE

13.9. Sources of Preliminary Geological Data for the Foundation Design. Prior to the start of the exploratory program, as much geological information as possible should be obtained from local sources. Many cities in the United States and Europe have compiled boring records for the whole city or a part thereof, and usually these are available to the public. Such records, either in published form or as only partially analyzed material, may be found in local libraries, city engineering or building inspectors' offices, or local universities. These compilations may provide a good description of foundation materials in the city or a large part of the city, but although of help in the initial stage of exploration, they generally will not provide sufficient details to eliminate the necessity of geotechnical studies for a *proposed* building. Another data source is the publications mentioned in Chap. 7, particularly USGS bulletins. These will give a general geologic description from which it may be possible to *deduce* the nature of the materials likely to be found at the given site. Other sources are personal discussions with local utility companies, architects, contractors, and residents. *Any verbal information should be regarded as only general* until confirmed by actual written records. Previous knowledge of the materials likely to be found at the building site not only permits the development of the most economical exploratory program but is vitally important for the preliminary selection of the type of foundation, as explained in the next section.

13.10. Preliminary Selection of the Foundation Type. With preliminary geological data in hand (Sec. 13.9), the type of foundation to be used in a given case may be intelligently guessed. In this case, the objective of the exploratory program would consist of finding out if a particular type fits the situation. The following examples show various simplified geological conditions and corresponding foundation types.

Case 1 (Fig. 13.12). The geological conditions at the site are a 40-ft layer of compact sand and gravel overlying firm sandstone and water table about 30 ft below the surface. What type of foundation should be used? *Solution.* The angle of internal friction of this sand and gravel material is probably high, and a shear failure (squeezing out of the sand) is not likely. Settlement in similar soils is known to be limited—not over 2 in. under large structures. Spread footings are indicated. The base of such footings should be below the frost line (4 ft below surface in New York City) and 2-ft minimum in frostless regions.

If sand is compact only at a certain depth, say 10 ft, there may be some settlement at the edges of the foundation because of squeezing out of loose sand. This may occur if the building is not rigid, e.g., a newly built brick apartment house. If the water table is high, e.g., close to the surface, but the material is compact, there may be more settlement than in the case of a low water table but a shear failure is unlikely.

Case 2 (Fig. 13.12). The depth of rock and rock itself are the same as in Case 1, but the overlying material is stiff, nonfissured clay. The foundation type is spread footings on the clay.   Driving piles into the given good-bearing material is the same as driving nails into a mirror in order to reinforce it.   The older generations of engineers were very suspicious of clay and, if hesitant in the choice of the foundation type, drove piles.   The modern viewpoint is: Do not drive unnecessary and harmful piles, and if there is any question on whether or not to use piles, *do not use them!*

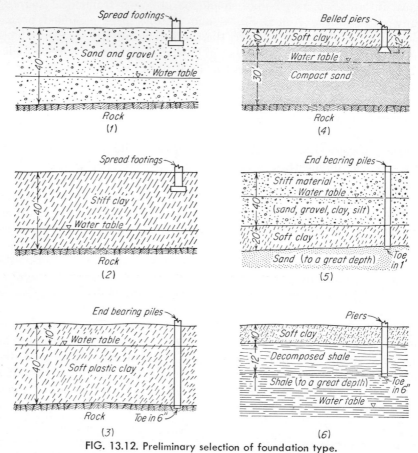

FIG. 13.12. Preliminary selection of foundation type.

Case 3 (Fig. 13.12).   The conditions are the same as Case 2, but the material overlying rock is soft clay.   A correct solution in this case is to drive end-bearing piles to rock.   The question now arises: What kind of piles should be used?   Timber piles are always preferred to other types if they are available.   Their great advantage is the ease of increasing and decreasing their length.   The butts of timber piles should always be under water, however, since fluctuations of the water table cause deteriorations ("dry rot") in the timber.   Hence, if in this case timber piles are to be used, they should be composite piles with the upper part made of concrete, for example. To avoid complications, it may be better in this case to use entirely concrete piles provided they are cheaper than the composite wood-concrete pile.   Obviously, each timber pile is cheaper than a concrete pile, but the number of timber piles required is larger than the number of concrete piles because of the comparatively limited bearing capacity of timber.   If concrete piles are chosen, precast piles could be used only if the depth of the rock floor is well investigated, since their lengths should be ordered and it is very inconvenient and expensive to cut them.   Furthermore, the cutoff stumps are useless.   Apparently cast-in-place concrete piles would best fit the given situation.   Steel piles are too heavy for the given condition.   Generally they are used if it is necessary to drive through resisting strata to a deep, high-bearing-value material.   It should be noted that H piles may be cut and welded again very easily. Note also that for small projects, timber piles exclusively are used.

   Case 4 (Fig. 13.12).   The rock is 40 ft deep, the overburden consists of compact sand, but the upper 10 ft is soft clay.   A solution is to use concrete piers a couple of feet longer than the thickness of the upper soft layer.   Since the piers rest on sand and not on a firmer material, it is convenient to bell them out.   The position of the water table permits sinking of the piers, since, generally, pier sinking in water is undesirable. If the water table were higher, e.g., close to the ground surface, short timber piles to sand would be indicated provided their heads are kept under water.

   Case 5 (Fig. 13.12).   The upper soil deposit is about 40 ft thick, but it rests on a 20-ft-thick layer of soft clay underlain by rock.   It is useless to try spread footings or friction piles in this case because the soft clay will consolidate and there will be considerable settlement of the planned structure together with the friction piles, if they are used.   A solution would be to drive end-bearing piles to sand.   Timber or concrete piles in this case would be effective.   If the clay deposit were thicker, for instance, 40 ft thick, H piles would be indicated.

   Case 6 (Fig. 13.12).   A shale deposit is covered with a sandy clay mantle 10 ft thick, and the material of the upper 12 ft of shale is decomposed.   A solution would be to sink concrete piers to sound shale.   This case calls for more careful investigation of both the decomposed and sound shale, however.   The piers perhaps could be shorter, or the decomposed material could be removed by rippers (rooters) as stated in Sec. 16.3, and the space used for the construction of a basement.

The examples discussed show that in some (but not in all) cases a good idea of the type of foundation may be formed from the preliminary data (Sec. 13.9).   In some cases, necessary simplifications and modifications of the exploratory program become clear from the preliminary data; e.g., in Case 1, no geotechnical investigations are needed unless the owner of the project insists on performing them, and in Case 3, all that is needed are wash borings to explore the topography of the rock floor to determine the length of the piles to be driven.   These examples show that the position of the ground-water table is a vital factor which influences the selection of the type of foundation and that this item should be given serious attention in the exploratory program.   [It is also convenient when gathering preliminary data (Sec. 13.9) to explore the wells in the neighborhood of the site, particularly in regard to water-level fluctuations.]

FOUNDATION PROBLEMS AND EXPLORATORY PROGRAMS

Some foundation problems are common to all kinds of buildings and are discussed first.   These are excavation problems, unstable foundation material problems, and ground-water problems.   Afterward, problems characteristic of different types of buildings will be considered.

13.11. Foundation Excavation.   In a geological analysis of proposed excavations both for buildings and for major structures such as bridges and dams, primary consideration should be given to (1) the type of side slopes for the material in the excavation walls, (2) the probable difficulty in excavating, (3) the stability of the excavation floor, and (4) ground-water conditions.   The following comments refer to all kinds of foundation excavation.

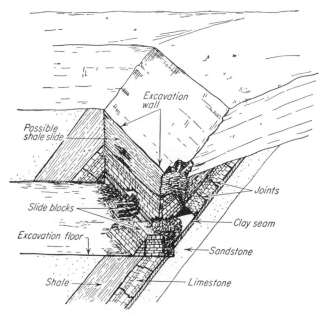

FIG. 13.13. Possible slides in rock excavation.   (*Drawing by J. Vitaliano.*)

If the *excavation is in hard rock* not affected by air-slaking or water action, the walls usually will be stable at steep angles.   However, the possible presence of fault seams which include gouge in connection with an unfavorable dip of the rock beds may cause difficulties.   In fact, any large slabs of rock overlying gouge and generally clay may become unstable when the slab is fully exposed and its toe support removed (Fig. 13.13).   The dip of the beds with relation to the excavation should be paid due attention.   For example, if a series of alternating sandstone and shale strata has a steep dip toward the excavation, ground water or the infiltration of heavy precipitation may lubricate the shale and cause the overlying sandstone slabs to slide into the excavation.   This problem is aggravated wherever such strata are cut by the excavation into isolated slabs.   Generally, in sound igneous and metamorphic rocks (except for some very platy schists), little trouble is had with stability, which is not the case with sedimentary rocks, particularly shales and claystones. The latter and some siltstones may air-slake when exposed and thus continually ravel into the excavation.   This raveling may progress so rapidly as to form unsound materials on the floor and walls of the excavation. Also, sound beds may be undercut by the raveling of underlying softer beds.

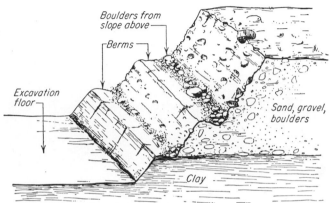

FIG. 13.14. Excavation in unconsolidated materials.   (*Drawing by J. Vitaliano.*)

Where it is necessary to place concrete directly on the rock under such circumstances, two methods can be used to ensure a firm bearing surface: (1) Approximately 6 to 12 in. of unexcavated material is left above the final excavation grade; then immediately prior to the concrete pouring, this undisturbed layer is rapidly removed and concrete immediately placed, which minimizes air-slaking of the rock; or (2) the foundation is excavated to final grade and then immediately coated with either asphalt emulsion or gunite.

When *soil materials* are being *excavated*, the stability of the banks is a major concern.   Sands and gravels generally are fairly stable on about 1:1 slopes.   However, these materials are subject to raveling and undercutting by runoff from rainstorms.   Furthermore, heavy vibrations (such as construction machinery and blasting) tend to cause continual raveling.   Silt and clay slopes of considerable height require accurate soil mechanics analyses before the side slopes can be safely excavated (Chap. 4).   Occasionally in glacial and alluvial materials, large boulders are present (Fig. 13.14); erosion may undermine such boulders and thus create a hazard to workmen and equipment in the cuts below.   To decrease such hazards (as well as to increase stability), the banks may be intersected at critical intervals with berms.   A berm (Fig. 13.14) is an approximately level bench cut into the face of a side slope to (1) collect material rolling from above, (2) control surface drainage, or (3) flatten the over-all slope of a high bank to secure better stability.

Various methods are used to stabilize excavation walls: (1) flattening the slopes below an angle critical for the given material at the given height (Chap. 4); (2) shoring (Fig. 13.15), which, in narrow excavations, may consist of simple wood beams and planks to cross-brace one wall against the other or, in larger excavations, steel or timber beams which brace one wall against the floor of the cut; and (3) well points (Fig. 5.25) to draw down the water surrounding the excavation. Since buildings

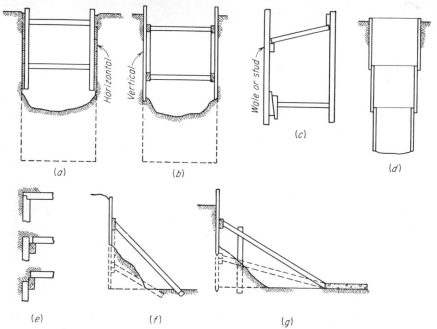

FIG. 13.15. Some methods of shoring excavations. (*From Dunham.*[8])

are usually located in relatively dry places, such arrangements as coffer-dams to protect the working area from water (see Chap. 14) are seldom used for building-foundation excavations. Also under exceptional circumstances, chemical stabilization may be used on the slopes of building excavations. This consists primarily of the injection of chemicals, mostly patented, which tend to solidify the soil. Stabilization by electroosmosis (Sec. 5.31) also has been occasionally used.

Backfilling of the Excavation. A structure is somewhat smaller than the excavation, since space is needed for concrete forms, bracing and shoring, and, sometimes, operational maneuvering. Upon the completion of the structure, the empty spaces are backfilled. Trenches for utility lines in residences and for sewers in housing projects are also backfilled. The carelessness with which this work is often done should not be permitted. The trench should be gradually backfilled and compacted either by a jet of water or by tamping with portable tampers similar to the devices used for breaking pavement. Tamping results in higher densities of the material, though jetting is generally less expensive.[24] Well-tamped mixtures of silt and clay around foundation walls (with a small addition of bentonite to increase watertightness) give satisfactory results. To take care of the possible settlement of the material, a small mound about 4 in. or more high (which gradually disappears) may be left on top of the trench. Large trenches are compacted with rollers.[25]

Before excavation for the foundations of an important structure, a detailed description of the type of foundation materials and of the surface and subsurface water conditions should be prepared. Photographs are desirable and in some cases compulsory. These data are required both for the design and for an adequate formulation of contracts.

The person in charge of the excavation should clearly realize his legal obligations for supporting adjacent properties. A discussion of the legal side of this work is presented in Chap. 19.

13.12. Foundations on Unstable Ground. The two basic types of unstable foundation materials are (1) those sensitive to water, namely, expandable (or expansive) soil and rocks, and (2) those subject to rapid settlement when saturated, predominantly loess (Sec. 3.15). Expandable materials are montmorillonite clays, soils containing anhydrous sodium sulfate, and some shales. Of the clay-type expandable materials, a United States Southwestern adobe is well known (Sec. 4.10). The best solution (but often prohibitive because of high cost) in this case is to remove the expandable material completely before construction and replace it with compacted, nonexpandable material. In some cases, only the upper strata of the expandable material should be removed, e.g., in San Francisco Bay area, where adobe is abundant, only the upper 3 ft of it swells. A good precaution is, of course, to avoid building on an expandable material when it is in a desiccated state. Watering of such dry material prior to construction also has been advised.

Difficulties arise when the foundation material starts to swell under a finished building. There may be two cases: Either the whole building is lifted bodily, or the walls remain in place but the floors start to buckle up and crack. The bodily lifting of the building is generally differential, the center being lifted more than the walls. The latter crack, usually quite severely. The lifting of the floors alone generally occurs gradually as moisture penetrates inside the building, either in the form of capillary water or in the form of vapor. To prevent or at least decrease the infiltration of water from outside under the footings, drain tiles at the footing level are placed all around the building (Fig. 13.16). The tile is covered with sand and gravel which grades from a coarse material immediately above the pipe to a fine material at the surface of the 18-in.-thick covering commonly used.   In spite of such an arrangement, however, water may be lifted by capillarity from below the foundation, sometimes to heights as great as 100 ft.[27,28]   Sometimes the rising and cracking of the floor slabs in large halls are attributed to the heating.   According

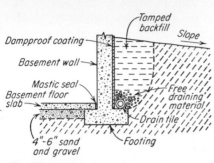

FIG. 13.16. Water-protective construction for a basement.

to this explanation the heat may cause drying of the soil under the slab and thus temporarily increase its bearing value, but simultaneously the capillary action is stimulated, and more water than before may flow to the floors.   There also is a thermoosmosis theory (with pros and cons) which hypothesizes that in hot, arid areas moisture from outside the building moves toward the inside.[29-31]   All in all, there is no inexpensive method of preventing building failures caused by the expansive action of the underlying soils.[32-34]

It should be noted that the presence of sulfates in considerable quantities in soils is detrimental to concrete, and in such cases special cements are required.

As is well known (Sec. 3.15), loess shows rapid and considerable settlement if saturated.   The point is to prevent excessive water from reaching the foundation.   This can be done by drainage or by placing an impervious pavement around the building and sloping the ground to drain away from the building.   However, these are palliative measures only.   Soils containing soluble salts also belong in the category of the materials in which sudden unexpected settlement may be critical, since soluble salts may be leached out by fluctuating ground water and leave cavities or hollows.

Soils sensitive to water, both expandable and settling when wet, may be detected in the consolidation test.   In both graphs of Fig. 13.17 curves AB indicate testing of the sample at natural moisture content.   At point B water is added to the sample.   The downbreak BC in Fig. 13.17a indicates settling soil such as loess, whereas the upturn BC in Fig. 13.17b indicates swelling.   In the latter case the loading is continued until point D of the diagram is at the level of point B, which gives an idea (but not a correct measurement) of the expansion pressure of the material (about 8,000 psf in this case).

13.13. Ground-water Problems in Foundation Engineering.   Deep ground-water tables are generally of no concern in building foundations unless subterranean garages or explosives storage are to be constructed. Conversely, a high water table should be paid considerable attention if the structure or a vital part of it such as a basement will be underground. The underground part of the structure should be thoroughly water-proofed in such cases, and measures should be taken against the possible damage of the structure by the hydrostatic uplift which may lift floor slabs and crack walls. Ground-water fluctuations should be studied with care, since these data are important to the owner of the building, to the designer, and to the excavation contractor. If, for example, it is established that the ground water will not rise above the basement floor, the costly waterproofing of the basement walls can be eliminated.

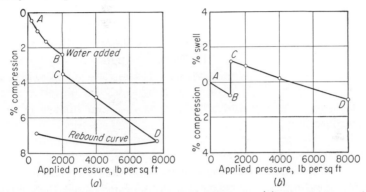

FIG. 13.17. Compression diagrams of water-sensitive soils: (a) soil settlement when wet (loess), (b) expandable soil. (*After report by Woodward, Clyde & Associates, Denver, Colorado.*)

To secure ground-water data, observation pipes should be left in some of the bore holes. Readings on the water table then should be taken at regular intervals as the program progresses and during all seasons of the year after the program is completed and before construction starts. These pipes can be 1- or 1¼-in.-diameter, galvanized iron, open at the bottom. A steel tape with a weight attached is lowered into the hole until the water level is reached. All readings taken in holes drilled with water should be used with caution unless the holes have been bailed dry or almost dry after drilling.

An examination of the rainfall records correlated with the records of the ground-water fluctuations in the given region may be of help, particularly in legal cases (Chap. 19).

13.14. Residential Buildings. Geotechnical investigations for residences are seldom necessary, particularly if the foundation conditions are evident from comparison between the building site and those of neighboring residences, including a study of the crack patterns in the latter, if any. Sometimes, local building inspectors require the performance of geotechnical investigations by competent firms only. In some cases an examination of the building site by competent persons before purchasing it is advisable, particularly in hilly regions where slides may be expected. Difficulties in residences generally arise *after construction* and are due mostly to water conditions or to settlement of footings because of poor work.

Exploratory Program. Since residences are light structures, their foundations are shallow. Usually these are spread footings or small-diameter piers. In a general case three bore holes should be drilled, two about 5 ft below the eventual base of the footings and one about 10 ft deeper to explore for soft deeper strata. On some occasions more than three bore holes may be needed; on others, a few hand auger holes will suffice. If rock is encountered in the exploratory holes, this fact should be brought to the owner's attention. Possible excavation in rock may considerably increase the cost of the building and should be known before the contract with the builder is signed. If the rock is exposed (and not subject to attack by frost action), footings may be placed directly on rock with some cleaning and with benching of the surface if the rock slopes considerably.

If a shallow water table is found, waterproofing or dampproofing of the basement should be recommended. An allowance for the increase in the amount of ground water after construction should be made, especially if a group of residences is considered. Such an increase may be due to lawn irrigation, breaks in the utility line, leakage of septic tanks, etc.

13.15. Commercial Buildings. These buildings are characterized by a heavy concentration of loads, generally transmitted to the foundation by columns. Many of them have deep basements, a fact that calls for a careful ground-water investigation.

Exploration Program. As in the majority of buildings, the extent of the exploratory program depends on the cost of the structure, which generally is high but varies within a wide range, and, of course, on the expected stratification of subsurface materials. In many cases the exploratory program depends on the availability of good access to the site to be explored. Often new buildings are constructed where the old buildings are still standing at the time the bore holes are to be drilled. Thus, holes can be drilled only around the exterior of the buildings unless special equipment is used to drill from within the building.

The number of holes depends upon the relationship between the expected variability of the foundation materials and the areal extent of the structure. The ideal program would be to have one bore hole at every column location, but ordinarily, sufficient information about stratification may be obtained from the bore holes drilled on the building corners and at the locations of the interior columns which will carry the heaviest loads (Fig. 13.18a). If not, additional holes for exact correlation of the strata should be drilled. The depth of the holes depends upon the expected building loads, the depth of the basement, and the materials encountered. A rough criterion much in use is to carry the holes to a depth below the proposed final footing grade equal to 1½ times the proposed width of footing. This depth should be increased, however, if geological studies indicate critical bearing materials at greater depths but still within an appreciable influence of the building-load pressures.

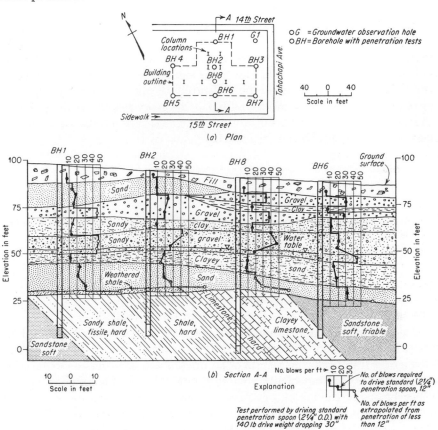

FIG. 13.18. Exploratory program for multistory building.

The standard penetration test usually is suitable for exploring soil materials in the field, providing correlative values have been established by laboratory tests on some of the samples (Sec. 6.11). Generally, for an important structure, at least one large-diameter hole (4 to 6 in.) should be drilled to obtain satisfactory laboratory test samples. Large-diameter drive samples also can be used. In some cases thin-walled drive samples are used instead of the standard penetration test. If the local soil types are fairly well known and easily identified, power augers may be used to explore the site, in which case, the various soil layers are identified from the auger cuttings. This method requires considerable experience, however, as a certain time is needed to bring the deep samples to the surface and it may be difficult to identify the exact depth from which they have come. If hard, cemented material or rock is encountered, or whenever the number of blows per foot in the penetration test exceeds 60, it is advisable to use a rotary drill with a core barrel for sampling in order to prevent excessive damage to the penetration spoon and considerable disturbance to the samples.

Careful logs of all bore holes should be prepared. Geologic cross sections across characteristic sections of the building should be drawn. On these sections the logs of the holes should be plotted, the results of the penetration tests should be indicated (if used), the ground-water level in each hole shown, and finally, if laboratory tests have been made, the density, moisture contents, and visual classification results also should be plotted adjacent to the critical logs.

Figure 13.18b is a typical geologic cross section showing such data. The excavation for the structure would extend to elevation 75, which means that the layer of competent gravel was too thin to support the very heavy column loads. The sandy clay layers from about elevation 65 to 70 and from 45 to 50 also were considered (on the basis of unconfined compression tests) as being incapable of supporting the building loads. Furthermore, the irregularity (lenticularity) of the compressible strata indicated that differential settlement could occur. As a result of these considerations, it was decided to drive piles to the bedrock. In this case, the results from the penetration tests were used only to correlate the various types of strata; the actual bearing powers of the different soils were computed from results of unconfined compression tests performed on samples obtained from two large-diameter bore holes.

The geologic nature of the subsurface materials may be important in the interpretation of the laboratory tests. If, for example, the rock is incipiently fractured, the results of an unconfined compression or triaxial shear test may be but of little value though the sample appears sound. Whether or not the laboratory results from a bore hole can be applied to a relatively large area will, in many cases, depend on the geological identification of the materials. For example, in alluvial soils and in glacial till the strata are likely to be erratic, and thus laboratory test results may apply only to the very small area adjacent to the bore hole from which the samples were taken. Conversely, if the samples have been obtained from a loess deposit or from an evenly stratified sand or sandstone, a few laboratory-test values may well apply to the entire building area.

13.16. Industrial Buildings. This category is intended to include warehouses, large garages, manufacturing plants, and arenas such as auditoriums and coliseums. The structures to be discussed are of considerable height, though of only one or two stories; they may have heavy roof and heavy wall loads. In manufacturing plants, floor loads may be very heavy and individual footings for heavy machines may be required. Any of the footing types discussed in Sec. 13.3 may be used as support. In manufacturing plants, besides the usual live and dead loads, it may be necessary to consider the vibration effects, unless the vibrations are damped by specially designed foundations. Particularly sensitive to vibrations are relatively loose sands and gravels, and compaction of these materials by vibrations is responsible for settlement of footings placed on them. Besides machinery in the plant, vibrations may be caused by the nearby movement of railroad rolling stock or, in exceptionally critical soils, by the wheels of heavy trucks pounding on adjacent highways. Cases of settlement by vibration on machinery footings on pile foundations have been reported; this settlement apparently is the result of repetitional, very small penetrations of the piles into relatively soft rock.

A minor foundation problem in industrial buildings is the influence of saturated wastes dumped close to the building. This can cause a decrease in the bearing capacity of the foundation materials or, if the wastes contain deleterious chemicals, the deterioration of the concrete of the footings. Thus, the waste disposal is an item to be seriously considered in the foundation design in such cases.

Exploratory Programs. Because of the large area generally covered by industrial buildings, soil conditions may materially change across the building area. Therefore sufficient bore holes should be drilled to locate possible critical changes in subsurface materials, particularly where concentrated loads are to be imposed on the soil. It is advisable to have a bore hole at each footing location unless the stratification follows a regular pattern. The depth of the holes and the types of drilling are as discussed in Sec. 13.13. High ground water is of concern, since it may cause uplift on floor slabs which are placed directly on the soil in these buildings, cause buoyancy on the footings, or interfere with excavation.

13.17. Power Plants and Pumping Stations. In the design of the buildings of this category, two features should be given special attention. These are (1) intense influence of vibrations and (2) sensitivity to settlements. Continual economic operation of pumps, turbines, and generators is possible only if the settlement of the foundation is reduced to an absolute minimum. Even a slight excess of settlement above that minimum can throw the turbines and generators out of alignment and cause excessive bearing wear as a result. Furthermore, excessive settlements can break the penstocks at their points of connections with a hydroelectric power plant (for penstock, see Fig. 13.19).

There are two major types of power plants: (1) the hydroelectric plant,

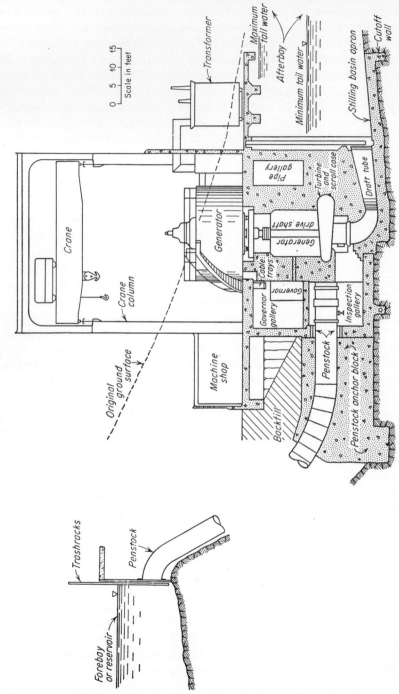

FIG. 13.19. Transverse section of power plant (through center line of unit). (Adapted from USBR Polehill Powerplant, Colorado.)

wherein water power drives the turbines, and (2) the plant wherein coal, gas, or oil is used to generate steam to drive the turbines or the turbines are direct-connected to a gasoline- or diesel-driven motor. Figure 13.19 shows a typical cross section of a hydroelectric plant and identifies the essential components which are of interest to the foundation engineer. A plant using steam or motor power would differ only in that the features related to water shown in Fig. 13.19 would be omitted. The ever-present water around a hydroelectric plant may drastically change the character of the original subsurface materials on which the plant has been built. This water could seep from the reservoir feeding the plant or be a result of back seepage from the afterbay (Fig. 13.19).

If the source of energy of a power plant is fissionable materials, massive structures are required for radioactive protection. Suitable radioactive waste disposal may require very deep excavations.

In a plant where the turbines are steam- or motor-driven, water is generally used to cool them. Natural lakes, artificial reservoirs, or wells supply that water. If wells are used, a study should be made of the suction effects upon the foundation materials and possible changes of the natural hydrostatic conditions underneath the plant foundations. If a plant is operated by coal, there may be large ash dumps near by. The percolation of rain or snow through these ashes can form sulfurous acid and sulfates that finally reach the foundations and may attack the concrete or leach certain salts from the soils under the footings and thus contribute to their settlement.

Vibrations.[35] The effect of vibrations on the foundations and walls of a power or pumping plant is twofold: (1) There may be induced settlement of the footings, especially those built on soils sensitive to shaking such as sands and gravels, and (2) the walls may be damaged if the vibrations caused by the machinery are out of phase with the natural vibrations of the walls (compare also Sec. 18.5). The possible damage to the walls may be prevented, however, by keying them to the surrounding rock with steel rods and thus transmitting the vibrations to the rock. In such cases the rock has to be suitable for the placement of such rods and should resist disintegration from continuous vibration in the rod. This is a geotechnical problem of concern.

Landslides, Snowslides, Rockfalls. In the case of powerhouses placed at the foot of a hill or on a steep hillside, the uphill rock should be examined for fissures, cracks, closely spaced joints, and presence of cliffs. The landslide problems are discussed in Chap. 17, and for rockfalls particularly, Sec. 17.9 should be consulted. In critical cases the change of location of the powerhouse should be recommended.

In regions with long winters and heavy snowfalls, snowslides are a serious problem.   If there is abundant snow, any slope over 22° can be subject to snowslides provided that the temperature and wind conditions favor them.   Vegetation on the slope has little influence on the occurrence of snowslides, contrary to common opinion.   Protecting a powerhouse from snowslides (avalanches) is difficult.   Sometimes heavy masonry walls ("deflector" walls) are constructed on the hillside in the possible snowslide path in order to deflect the snow away from the plant.   Such barriers, however, may become filled with snow and cease to act as a deflector unless the accumulated snow is periodically removed—a difficult task.   In hydroelectric plants, the construction of surface pen-stocks may contribute to snowslides, since the clearing for the penstock may become a funnel for the snow.   Where snowslides are exceedingly large and thus cannot be effectively controlled, it may be desirable to consider the possibility of placing the powerhouse underground as has been done in Switzerland.

Exploration Program.   Because of the aforementioned factor of sensitivity to foundation settlement, the subsurface investigation for a power-house should be very detailed and thorough.   As a general rule, holes are drilled at the approximate location of the plant corners.   Additional holes should be bored at the center of each turbine location and at the location of any heavy bearing wall within the plant (Fig. 13.20).   These holes then will have to be supplemented by such additional borings as may be required by the subsurface geology.   All faults, shear zones, weathered zones, closely spaced joints, changes in rock type, and formation contacts should be thoroughly studied by boring and surface mapping.

All bore holes, as a general rule, should go to a depth equivalent to $1\frac{1}{2}$ times the width of the power plant.   This depth should be measured from the proposed final grade of the power plant (i.e., the elevation at the bottom of the excavation for the plant).   In rock with apparently adequate bearing characteristics, the bore holes need to penetrate only 25 ft *providing the geologist is certain that there are no softer materials below this depth.*   When the final grade is estimated, it should be noted that the foundations of hydroelectric plants generally are 30 or 40 ft or more below the original ground surface.

As far as the type of drilling equipment is concerned, standard penetration tests may be used for soil materials and NX core barrels (rotary drills) for harder materials.   Large-diameter (NX) core is necessary to study joint patterns and fracturing properly.   As the loads imposed by a powerhouse may be exceedingly heavy, shear tests on rock samples are desirable.   It is preferable to place the plant on one kind of rock only and thus use only one value of the modulus of elasticity for settlement computations.   If two kinds of rock have to support the plant, the determination of two moduli of elasticity becomes necessary and there will be a possibility of differential settlement.

Auxiliary problems that require solution during the geotechnical explorations are (1) the determination of the slope stability of the plant excavation, (2) the determination of the stability and erosion characteristics of the tailrace channel, and (3) exploration for the foundations

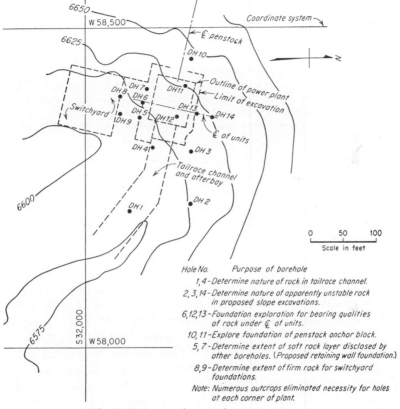

Hole No.   Purpose of borehole

1,4 – Determine nature of rock in tailrace channel.

2,3,14 – Determine nature of apparently unstable rock in proposed slope excavations.

6,12,13 – Foundation exploration for bearing qualities of rock under ℄ of units.

10,11 – Explore foundation of penstock anchor block.

5,7 – Determine extent of soft rock layer disclosed by other boreholes. (Proposed retaining wall foundation.)

8,9 – Determine extent of firm rock for switchyard foundations.

Note: Numerous outcrops eliminated necessity for holes at each corner of plant.

FIG. 13.20. Power plant exploratory program.

of high stacks (chimneys) in steam plants. The main purpose of the studies in the tailrace channel is to determine if lining should be used, especially in fractured rocks subject to plucking action by water (particularly basalt). The stack foundations may be eccentrically loaded because of wind or earthquake action, and generally they impose high pressure on the supporting materials. Other structures usually adjacent to power plants that require exploration of their foundations are switchyards and transformers. The weight of such equipment may be relatively light, but if rigid electrical connections (bus bars) are used, slight settlements or expansions will cause breakage of the bus bars.

Underground Plant Construction.[36] Many of the tunneling problems encountered in underground plant construction are similar to those found in transportation tunnels but may be of considerably greater magnitude. Underground chambers may be used for manufacturing plants, hydroelectric plants, or large storage rooms.

During World War II, the Germans constructed aircraft factories and other war manufacturing industries underground. Some of these were built in existing caves, but others were hewed out of solid rock (chalk in many cases). Underground power plants have been constructed in many European countries, in Canada, and in the United States. The plants in Sweden, placed underground purely for hydraulic considerations, also proved to be more economical than surface plants. Of secondary importance in design was their safety in air attacks in the event of war. The Swiss located some of their plants underground in order to secure maximum protection against snowslides; at the same time their plants are almost bombproof. One of the earliest underground power plants in the United States is located near Snoqualmie, Washington.

An outstanding feature of the investigations for underground plants is the increase in magnitude of what formerly would be regarded as minor geological problems. A fault zone which can be crossed with only minor difficulty in a small tunnel presents a serious problem if encountered in the large chamber required for an underground plant. In fact, in such a case it may be necessary to design an extraordinarily heavy roof for the plant, which might result in so high a cost as to call for the relocation of the chamber. If at all possible, the main chamber is located in rock of sufficient competency to preclude the use of heavy concrete roof supports or a heavy continuous arch. An evaluation of rock stresses is necessary to determine the roof design (Sec. 2.5). The presence of slabby or raveling rock will have to be foreseen in order that in such cases concrete side walls can be designed. This problem of raveling rock also is increased owing to the vibrations from the plant machinery.

Explorations for underground plants should include at least one exploratory drift driven to the approximate location of the plant chamber. From the end of the drift it is possible to drill holes to explore all sides of the, as yet, unexcavated chamber. Detailed surface mapping of joints and faults is very necessary because of their influence on the plant construction. Determination of the possibility of water flows in the plant excavation should be made.

13.18. Building Foundations on Fills. Shortage of dwellings in the cities and shortage of desirable sites on which to build them have led to the appearance of housing projects (and often their schools and shopping centers) built on sites that previously were considered undesirable for building purposes. These "sites" are (1) old fills, including city dumps; (2) reclaimed lands at the ocean and bay shores and swamps; and (3) uneven or rugged topography.   The general approach to these areas is to construct new fills using proper procedures (Chap. 16) and to place the residences on them.   Warehouses, schools, and multistory buildings also may be founded on fills.

Old fills generally have a hard crust that has a certain bearing power. An old fill which is capable of supporting a one-story building often shows a considerable increase of settlement if this bearing capacity is only slightly exceeded.   If the crust is impervious, which is often the case, it prevents the underlying material from being drained and consolidated. To build on such a fill, it may be desirable to remove the crust and some of the underlying material to a total depth of at least 3 ft and then place a new fill.   Observations have shown that the settlements of the old material on which a new fill is placed generally are small.   If a new fill is built over an old one, the following information should be obtained: (1) material and age of the old fill, (2) is it consolidated or not, and (3) material underlying the fill and, if this is clay or a similar material, whether or not it is consolidated.   It is not always possible to drill through an existing fill because of the variety of materials.   In such cases either a shaft is made or bore holes are sunk outside the old fill if the underlying strata are continuous.   Generally field exploration should provide samples to test for shear failure and possible settlement of the fill.

Areas *reclaimed from water bodies* are either (1) shallow-water areas permanently or periodically flooded, such as tidal flats, or (2) deep-water areas filled practically to the top with soft material, such as organic silt. In both cases the buildings are supported on a well-compacted, rigid fill in which the footings are constructed as in natural ground.   In the former case, however, the shallow-water area is first filled above the high tide hydraulically, i.e., with dense soil suspension which is permitted to consolidate somewhat in order to be able to carry the rigid fill.   The new fills should be several feet in thickness, starting from a few feet in frostless areas; in the frost zones they should be thick enough to permit placing the base of the footings below the frost line.   If piles are used to support a building, they should be driven through the fill to firm bearing.

Since a compacted fill is very rigid in comparison with the underlying soft material, especially if it is made using very heavy equipment, it causes not only vertical compression (consolidation) but also lateral flow and "packing" of the soft, usually loose material.   This is apparently one of the reasons why retarded settlement of such fills is sometimes observed.   A serious problem in the case of fills on reclaimed lands is their dishlike settlement, and if the areas involved are large, this differential settlement should be considered in the design and placement of utilities, such as sewers, to avoid breakage of conduits or detrimental flow gradients.   Presumably the surface drainage also would be affected by this kind of settlement.

In *graded sites*, fills and cuts generally alternate.   The column and wall loads of heavy buildings may be carried to the virgin soil under the fill or incorporated in the fill[34] if the latter is thick enough and the bulk of its settlement is completed.   Rugged topography is entirely remade by building deep cuts and high fills reaching to heights of 70 ft or more (e.g., Westlake housing project near San Francisco, California).   Critical geotechnical problems in this case are concerned with the stability of fills which exert heavy pressures on the underlying ground and often with the stability of high cuts at the periphery of the area.   Ground-water flow may drop considerably because of deep excavations and thus constitute an additional problem.

EXAMPLES OF BUILDING FOUNDATIONS

13.19. Building Foundations in Glacial Zones.   The *New York City downtown section*[37] is located on Manhattan Island, a rock massif built of Manhattan schist or gneiss surrounded by ancient canyons up to 300 ft deep that are filled with glacial drift deposits (gravel, sand, clay, and, of most concern, the characteristic organic silt).   The Greater New York area is encircled on the south by a terminal moraine.   The rock floor in the downtown section is deep and covered with a thick layer of drift but gradually rises in a northern direction and finally outcrops.   The representative structure in the deep drift zone is the Woolworth Building founded on deep piers (Sec. 13.4).   The Empire State Building, which has 85 stories and is 1,248 ft tall, is built in the shallow drift region, and its columns are placed on large concrete blocks just resting at the surface of rock.   Other Manhattan Island buildings are supported by piles, large open tubes filled with concrete (Sec. 13.6), and numerous caissons.

The *Chicago downtown section*[38] (Loop area) at one time was under the waters of a large glacial lake, the present remnant of which is Lake Michigan.   The corresponding soil profile is shown in Fig. 13.21.   The essential part of the soil deposit consists of blue (or rather bluish-gray) clay underlain by glacial hardpan (Sec. 3.11) and rock (Niagaran limestone).   Similar (but not identical) soil deposits are found in Boston, Detroit, and Winnipeg, Canada.   The area of the last city is a portion of the bottom of the extinct glacial Lake Agassiz (named after the geologist who first explained the phenomenon of glaciation).   Clay deposits in some glacial regions are covered with a crust about 5 ft thick, formed apparently by desiccation during exposure.   In Chicago particularly, this crust is capable of supporting lighter structures but settles under the weight of heavy structures such as the tower of the Chicago Auditorium (about 1898).   The corresponding time-settlement curve is shown in Fig. 13.22.   Belled concrete piers (Sec. 13.4) originated in Chicago and therefore are often called "Chicago caissons."

The thick deposit of the *Boston blue clay* has been used for foundation purposes in two different ways.   The foundations of the New England Mutual Life Insurance Company Building[39] have been designed as a reinforced concrete box placed on the crust of the blue clay deposit (floating foundation, Sec. 13.3).   Another monumental building in the same locality, the John Hancock Mutual Life Insurance Building,[40] is supported by H piles driven through clay to the underlying hardpan and rock.

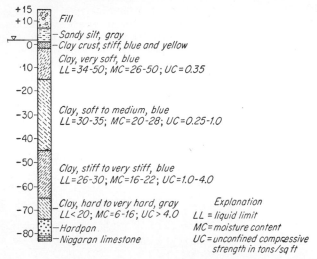

+15
+10 — Fill
— Sandy silt, gray
0 — Clay crust, stiff, blue and yellow
Clay, very soft, blue
·10 — LL=34-50; MC=26-50; UC=0.35
-20
-30 — Clay, soft to medium, blue
LL=30-35; MC=20-28; UC=0.25-1.0
-40
-50 — Clay, stiff to very stiff, blue
LL=26-30; MC=16-22; UC=1.0-4.0
-60
— Clay, hard to very hard, gray   Explanation
-70 — LL<20; MC=6-16; UC>4.0   LL = liquid limit
                                                  MC= moisture content
-80 — Hardpan   UC= unconfined compressive
— Niagaran limestone   strength in tons/sq ft

FIG. 13.21. Typical soil profile, Loop area, Chicago, Illinois. (After Peck.[38])

13.20. New Orleans Foundations. The attention of engineers was invited to New Orleans foundations about 1939 in connection with the considerable settlement of the then new Charity Hospital. The new Veterans Administration Hospital[41] built in 1951 stands across the street from the Charity Hospital (Figs. 13.23 and 13.24). The Pleistocene deposit is clay covered with a crust and its alluvial overburden consists of intermittent sand and clay strata. The foundations of the Veterans Administration Hospital represent a combination of floating and pile foundations. The timber piles used reached deep sand layers (marked S_3 and S_4 in Fig. 13.23). Use of a water jet in the pile driving caused heavy pore pressures in the soil strata, which were measured with piezometers during construction (Fig. 13.23).

Finer sediments are carried farther by a stream than the coarser ones. Hence the alluvial foundation soils in the Upper Mississippi Valley (St. Louis, Cairo, etc.) are coarser and more appropriate for spread foundations than those in the Mississippi Delta region.

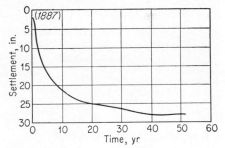

FIG. 13.22. Time-settlement curve for tower, Chicago Auditorium.

13.21. Foundations in California. The city of San Francisco[42,43] is mostly underlain by the materials of the Franciscan formation (somewhat metamorphosed sandstones and shales with intrusions of igneous rocks). The downtown section is located over or very close to the ancient canyons now filled with eroded materials, and a part of it is built on land reclaimed from San Francisco Bay. Piles and caissons support the tall buildings.

The Greater Los Angeles area[44] is defined by the watersheds of the three streams discharging into the Pacific Ocean: Santa Ana, San Gabriel, and Los Angeles Rivers. The physiographic properties of the area are variable, and so are the properties of the foundation materials. Sand deposits vary from compact to loose and often are spotty. Clays vary from black adobe to kaolinitic types. Decomposed granite, locally known as "DG," if well-compacted, is an excellent foundation material. Except for some water occasionally trapped in faults, the water table in the city is low, which explains the wide use of belled caissons. In many places in the area, practically vertical slopes can stand unsupported to amazing heights, and, as a rule, borings are done without casings. The city has to protect itself from the debris coming from the erodible slopes of the adjacent ranges (mostly the San Gabriel Mountains). A series of dams with periodically cleaned reservoirs serve to detain the moving debris.

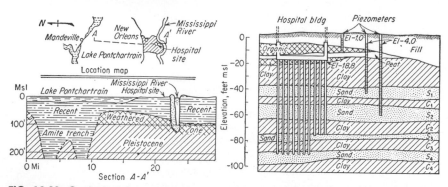

FIG. 13.23. Geological formations at the site of the Charity Hospital and the Veterans Administration Hospital, New Orleans, Louisiana. (*From Eng. News-Record, Sept. 13, 1951.*)

FIG. 13.24. Charity Hospital (upper right) and Veterans Administration Hospital (lower left), New Orleans, Louisiana. (*Courtesy of Favrot, Reed, Mathes, and Bergman, Architects.*)

Both California areas described are subject to landslides and earthquakes.   Expansible soils, particularly adobe, are not uncommon.

13.22. Foundations in Colorado.   Some critical foundation conditions have been observed in the state.   Many of the rock formations contain a high percentage of gypsum, which occurs as lenses up to several inches thick and is partially soluble in water; such occurrences have been noted in the Pueblo and Denver areas.   A few areas such as Rangely in the northwest corner, Grand Junction on the west, and Denver in the east-central part of the state are plagued with unusual clays.   These

FIG. 13.25. Subsidence of Mexico City illustrated by protrusion of well casing originally flush with surface in 1923 and projecting 5.5 m in 1956.   (*Photo by Ing. J. Marsal.*)

clays have uncommonly high expansion characteristics in some cases and severe settlements in others.   The expansive clay in Denver has been known to raise a concrete pavement as much as 6 in., whereas within a matter of a few hundred feet, loessial deposits occur that are subject to severe settlements when moistened by rain or irrigation.[45]   Numerous cases of serious structural distress have been noted in houses founded on loess.   This distress is primarily due to severe foundation settlement caused by excessive lawn watering and improper surface drainage.

Another peculiarity has been observed in the bedrock (the Denver formation) underlying the city of Denver.   This bedrock is generally composed of weakly cemented sandstones, siltstones, and some very hard shales.   Exploratory drilling for the foundations of multistory buildings has disclosed the presence of very soft clays,

12 in. or more thick, within this hard shale. Drilled-in piers with belled bottoms are commonly used for the foundations of heavier structures, and occasionally cast-in-place piles are used. Small-diameter drilled-in piers reaching to the underlying bedrock have been used to support house foundations that otherwise would rest on loess.

13.23. Foundations in Mexico City. Mexico City[46] is built on top of a heterogeneous deposit of clay (mostly montmorillonitic, with PI values up to 215), sand, pumice, and organic admixtures which fill most of the territory that was occupied by a huge ancient lake. Lake Texcoco (with salt water) is one of the remnants of that lake. The deposit was formed by volcanic ashes loosely deposited by wind and waters flowing into the lake. The material of the deposit is very compressible, especially in its upper portion (average voids ratio 6.93), which is separated from the lower portion by a discontinuous crust of cemented, calcareous sand about 10 ft thick (*capa dura*). The piles supporting the buildings generally reach the crust. The deposit is up to 165 ft thick. The city settles as a unit but not uniformly. The total settlement in the period 1898–1950 was from 10 to 17 ft at different points in the city.[47,48]

REFERENCES

1. Urquhart, L. C. (ed.): "Civil Engineering Handbook," 3d ed., McGraw-Hill Book Company, Inc., New York, 1950.
2. Terzaghi, K.: "The Influence of Modern Soil Studies on the Design and Construction of Foundations," Div. I, part III, pp. 139–145, Building Research Congress, London, 1951.
3. Harding, H. J. B., and R. Glossop: "The Influence of Modern Soil Studies on the Construction of Foundations," Div. I, part III, pp. 146–155, Building Research Congress, Div. I, London, England, 1951.
4. Judd, W. R.: Exploration Principles for Major Engineering Works, *Proc. ASCE Sep.* 550, November, 1954.
5. Sowers, G. F.: Modern Procedures for Underground Investigations, *Proc. ASCE Sep.* 435, May, 1954.
6. Andersen, P.: "Substructure Analysis and Design," Irwin-Farnham Publishing Co., Chicago, 1948.
7. Cummings, A. E.: Lectures on Foundation Engineering, *Univ. Illinois Eng. Expt. Sta. Circ.* 60, December, 1949.
8. Dunham, C. W.: "Foundations of Structures," McGraw-Hill Book Company, Inc., New York, 1950.
9. Hool, G. A., and W. S. Kinne: "Foundations, Abutments, and Footings," 2d ed., McGraw-Hill Book Company, Inc., New York, 1943.
10. Jacoby, H. S., and R. P. Davis: "Foundations of Bridges and Buildings," 3d ed., McGraw-Hill Book Company, Inc., New York, 1941.
11. Minikin, R. R.: "Structural Foundations," Crosby Lockwood & Son Ltd., London, 1948.
12. Peck, R. P., W. E. Hanson, and T. H. Thornburn: "Foundation Engineering," John Wiley & Sons, Inc., New York, 1953. Part B is an excellent textbook on foundations for beginners.
13. Prentis, E. A., and Lazarus White: "Underpinning," Columbia University Press, New York, 1950.
14 Soil Mechanics and Earth Structures, *U.S. Navy Bureau of Yards and Docks* TP-Pw-18, Oct. 1, 1953.

Piles and Pile Foundations

15. Chellis, R. D.: "Pile Foundations," McGraw-Hill Book Company, Inc., New York, 1951.
16. Cummings, A. E.: Pile Foundations, *Proc. Purdue Conf. on Soil Mech. and Its Appl.*, September, 1940.
17. Cummings, A. E.: Stability of Foundation Piles against Buckling under Axial Load, *Highway Research Board Proc.*, vol. 38, part II, 1938.
18. Dunham, J. W.: Pile Foundations for Buildings, *Proc. ASCE Sep.* 385, January, 1954.
19. Minikin, R. R.: "Piling for Foundations," Crosby Lockwood & Son Ltd., London, 1948.
20. D'Appolonia, E., C. E. Miller, Jr., and T. M. Ware: Sand Compaction by Vibro-flotation, *Proc. ASCE Sep.* 200, July, 1953.
21. Cummings, A. E., G. O. Kerkhoff, and R. B. Peck: Effect of Driving Piles into Soft Clay, *ASCE Trans.*, vol. 115, 1950.
22. Housel, W. S., and J. R. Burkey: Investigation to Determine the Driving Characteristics of Piles in Soft Clay, *Proc. 2d Conf. on Soil Mech. and Foundation Eng.*, vol. V, Rotterdam, paper VII, p. 21, 1948; also Housel's discussions of Ref. 15.
23. Florentin, J., and G. L'Hériteau: About an Observed Case of Negative Friction on Piles, *Proc. 2d Conf. on Soil Mech. and Foundation Eng.*, vol. V, Rotterdam, paper VII, p. 23, 1948; and *Travaux*, vol. 32, 1948.

Foundation Problems and Exploratory Programs

24. Coffman, B. S., and C. H. Bryant: Backfilling Trenches: Jetting Versus Tamping, *Highway Research Board Abstr.*, vol. 24, May, 1954.
25. Finn, F. N.: Soil Mechanics in Trench Backfilling, *Gas*, vol. 28, 1952.
26. Dawson, R. F.: Test Tile Footings on Expansive Soils, *Arch. Record*, vol. 108, November, 1950.
27. Reiner, M.: Stripfootings in Shrinkable Clay Soils, *In the Field of Bldg. Bull.* 1, Hebrew Inst. of Technology, Haifa.
28. Neumann, I. H.: Classification of Cracks in Buildings, *In the Field of Bldg. Bull.* 3, Hebrew Inst. of Technology, Haifa.
29. Jennings, J. E.: Foundations for Buildings in the Orange Free State Goldfields, *J. S. Africa Inst. Engrs.*, vol. 49, 1950.
30. Jennings, J. E.: The Heaving of Buildings on Desiccated Clay, *Proc. 3d Conf. on Soil Mech. and Foundation Eng.*, vol. I, Zurich, 1953.
31. Du Bose, L. A.: A Full Scale Investigation of the Thermo-osmotic Hypothesis, *Proc. 3d Conf. on Soil Mech. and Foundation Eng.*, vol. I, Zurich, 1953.
32. Mielenz, R. C., and C. J. Okeson: Foundation Displacements along the Malheur River Siphon as Affected by Swelling Shales, *Econ. Geol.*, vol. 41, May, 1946.
33. Wooltorton, F. L. D.: Movements in the Desiccated Alkaline Soils of Burma, *ASCE Trans.*, vol. 116, 1951.
34. Dalrymple, G. B.: Fill Utilization for Building Foundations, *Proc. ASCE Sep.* 417, February, 1954.
35. Judd, S.: Vibrations in Hydroelectric Plants, a paper given at the ASCE convention, Denver, Colo., June, 1952, preprinted by the U.S. Bureau of Reclamation.
36. Jaeger, Charles: Present Trends in the Design of Pressure Tunnels and Shafts for Underground Hydroelectric Power Stations, *Inst. Civil Engrs.* (*London*) *Paper* 5978, Nov. 16, 1954.

Examples of Building Foundations

37. "Rock Line Map of New York," an unpublished 1935 WPA project under the direction of Dr. C. P. Berkey.
38. Peck, R. B.: History of Building Foundations in Chicago, *Univ. Illinois Bull.* 29, 1948.
39. Casagrande, A., and R. E. Fadum: Application of Soil Mechanics in Designing Building Foundations, *ASCE Trans.*, vol. 109, 1944.
40. Casagrande, A.: The Pile Foundation for the New John Hancock Building in Boston, *J. Boston Soc. Civil Engrs.*, vol. 34, 1947.
41. Huesmann, H. A.: Foundation Problems in the New Orleans Area, a paper given at the ASCE Convention, New Orleans, March, 1952; also *Eng. News-Record*, vol. 147, 1951.
42. Trask, P. D., and J. W. Rolston: Engineering Geology of San Francisco Bay, California, *Bull. GSA*, vol. 62, 1951.
43. Lee, C. H.: Building Foundations in San Francisco, *Proc. ASCE Sep.* 325, 1953.
44. Personal communications by Prof. F. J. Converse and Mr. Paul Baumann.
45. Judd, W. R. (ed.): "Boring Data and Its Engineering Applications in Denver," Hotchkiss Map Co., Denver, Colo., 1954.
46. Marsal, R. J., F. Hiriart, and R. Sandoval: "Hundimiento de la Ciudad de México," Ingenieros Civiles Asociados, Mexico City, 1951; also "Arcillas del valle de México," *ibid.*
47. Zeevaert, L.: Compressibilidad de la arcilla volcanica de la Ciudad de México, *Revista Mexicana de Ingenieria y Arquitectura*, vol. 30, 1952.
48. Zeevaert, L.: Characteristics of the Unconsolidated Sedimentary Deposits in the Valley of Mexico, *Actes du IV Congrès International du Quaternaire*, Rome-Pisa, 1953.

CHAPTER 14

BRIDGES AND PAVEMENTS

Many route features, e.g., earthwork and special structures such as tunnels, are discussed in other chapters. In this chapter foundations and site explorations for bridges are considered, with the understanding that the crossing of the valley where a bridge is located is an integral part of the bridge location. Pavements are discussed briefly with emphasis on the moisture under pavements, a geotechnical factor of great importance to the stability of pavements but as yet insufficiently clarified.

BRIDGES AND BRIDGE SUPPORTS

14.1. Classification of Bridges. A bridge consists of a superstructure and substructure. The weight of the superstructure and the loads imposed on it are taken by the *supports* of the bridge and transmitted to the foundation; thus, every force acting on the superstructure ultimately reaches the foundation. To design the foundation and to perform the necessary geotechnical investigations intelligently, it is necessary to know not only the magnitude of the forces transmitted to the foundation but also the *manner* in which the forces are transmitted. From this point of view, there are three categories of bridges: (1) the vertical loads acting on the superstructure are transmitted vertically to the foundations (Fig. 14.1*a* and *b*), (2) besides the vertical forces transmitted to the foundations by the supports, the horizontal *thrust H* pushes the supports outward (Fig. 14.1*c* and *d*), (3) the vertical forces are transmitted to the foundations vertically, but for stability the superstructure has to be anchored to rock or a large concrete mass, and there are forces that tend to pull out the anchorage (Fig. 14.1*e*).

Figure 14.1*a* represents a *simple beam*, or girder, on two supports. The superstructure may be steel, reinforced concrete, or timber. It rests on or is fixed to the *abutments*. Abutment refers to a terminal support of the bridge. Obviously, a bridge has two abutments, and the traffic has to pass over both abutments in order to enter on the bridge and leave it. Abutments generally are made of concrete, plain or reinforced, although in some bridges, other materials are used, e.g., steel or, in old bridges, rough rubble masonry (sometimes without mortar). The concrete abutments sometimes are faced with dimension stones (Sec. 8.1). Generally, dimension-stone facing may be applied to any exposed concrete surface of a bridge.

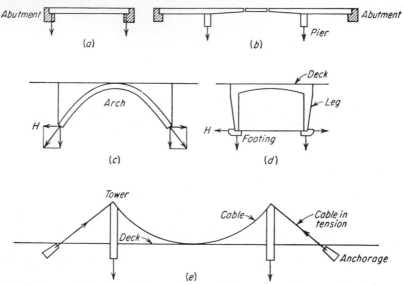

FIG. 14.1. Types of bridges: (a) simple beam, (b) cantilever, (c) arch, (d) rigid-frame, (e) suspension.

The bridge in Fig. 14.1a has one *span* only supported by the abutments; this is a very common type of small bridge for highways. If the bridge consists of several spans of equal or variable length, the intermediate supports (between abutments) are *piers*. The term "pier" is used in engineering loosely; besides the concrete piers supporting buildings (Sec. 13.4), it also means a harbor structure for landing, loading, and unloading ships. A *multispan* bridge may consist of a number of mutually independent girders supported at both ends on the piers and abutments, or a long girder may cover several spans ("continuous" girder). In all cases of girders, the girder must be allowed to move a little for temperature expansion. For this purpose the girder should be fixed firmly on an abutment or pier and placed on *rockers* or rollers on other supports.

Figure 14.1b represents a *cantilever* bridge with two piers and two abutments. The weight of the superstructure is essentially carried by the piers, and the structure may be so balanced that the load on the abutments is negligible, if any. The girders have protruding arms (cantilevers) and carry a relatively small, simple beam at the center of the bridge. The term "cantilever" generally is applied to a structure or a part of the structure, horizontal or nearly so, fixed firmly at one end and unsupported at the other end.

The arch and the rigid frame (Fig. 14.1c and d) are also discussed in Sec. 13.2. The bridges of these systems produce a vertical pressure and tend to exert a horizontal thrust on their supports. Arch bridges may be steel, concrete, or timber. A number of old masonry arches still exist. As in the case of buildings, an arch bridge may be provided with a tie that takes up the horizontal thrust caused by the arch instead of the abutments doing so. A rigid-frame bridge may be steel or of reinforced concrete and is commonly of one or two spans.

A suspension bridge (Fig. 14.1e) consists of two cables, generally spun of strong wire, which rest on saddles firmly fixed at the top of the steel towers (some old bridges have masonry towers). The traffic deck is suspended on the cables. The loaded cables have the tendency to pull the towers inward, and to oppose this tendency, the cables are anchored either in natural rock or in a massive block of concrete that holds down the ends of the cables. In some suspension bridges, a stiffening truss is added to the structure to prevent undue deflection or oscillation of the cables. The longest span in a suspension bridge is the 4,200-ft span of the Golden Gate Bridge in San Francisco, California.

14.2. Abutments and Piers of a Bridge. The abutment connects the bridge to the roadway. This may be an embankment of variable height or merely the ground surface with perhaps a little grading. Both cases are illustrated by Figs. 14.2 and 14.3 (case of embankment) and Fig. 14.7 (case of natural ground surface). There are a number of intermediate cases.

If the bridge access is an embankment, the abutment has to hold it back to prevent the earth from moving into and obstructing the waterway between the bridge supports. The abutment has to offer a *seat* for the superstructure and at the same time be a retaining wall for the embankment. This is done by designing winged abutments. In Fig. 14.2a and b, a *straight-wing* abutment is shown. Under item a the embankment is shown, and under b the abutment.

The plan of footing for a *beveled-wing* abutment is shown at the bottom of Fig. 14.2 under item c. The straight-wing abutments are somewhat weaker than the beveled-wing type; the wings of the latter reinforce the straight retaining wall. The slight difference in simplicity and cost of construction, however, favors the straight-wing type.

When the land is inexpensive, U-shaped abutments can be used (Fig. 14.3). In this type only the central part of the embankment is contained between the wings, and the slopes are permitted to fall outside. If the substructure is narrow, this represents a strong structure, rather

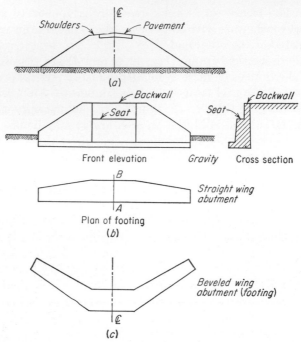

FIG. 14.2. (a) Embankment in connection with the abutment.   (b) Straight-wing abutment.
(c) Beveled-wing abutment.

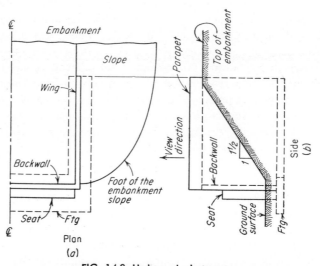

FIG. 14.3. U-shaped abutment.

economical on account of savings in the concrete work.   In Fig. 14.3
only half of the plan of the abutment is shown (item a); the side view
(item b) shows how the abutment looks if sighted in the direction of the
arrow.   All portions covered with earth are shown as dotted lines in
Fig. 14.3.

Besides these simple abutment types, there are a number of other
arrangements serving the same purpose.   Some of them are shown in
Figs. 14.4 to 14.6.   Figure 14.6 shows a flanking-span abutment in which
the wings are omitted and the front wall, instead of being solid, may be
provided with openings.   The superstructure rests on the front wall,

FIG. 14.4. Precast-concrete slope protection; no abutment.

which is connected by short girders to a secondary wall placed on the
embankment.   The slopes of the embankment are permitted to fall
free.   Figure 14.7 shows that arch bridges are very suitable for spanning
waterways located between two rocky shores.   It also shows that the
bases of the two abutments of an arch may be located at two different
levels.   If one or both abutments of an arch should be located in soil
materials, large massive concrete abutments would be required to prevent
sliding, with an increase in cost of the structure.

Sometimes the local topography and the presence of sound bedrock
clearly indicate the abutment location.   In a general case, however,
the proper emplacement of an abutment is a problem requiring considerable experience and judgment.   Such are the cases of swampy low

FIG. 14.5. Mass concrete abutments.

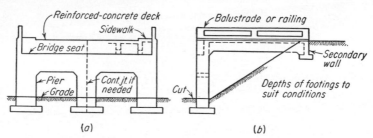

FIG. 14.6. Flanking-span abutment.   (*From Dunham.*[4])

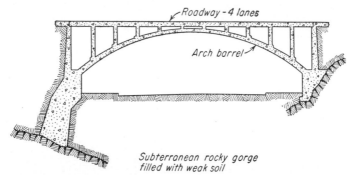

FIG. 14.7. Abutments of a reinforced concrete arch at different elevations. (*From Dunham.*[4])

shores extending a large distance from the bridge or meandering streams that may change the location of their channels. In questionable cases it is advisable to find the proper location by comparing the results of geotechnical investigations at several possible locations. In some cases of bridge construction, a channel is excavated and the bridge constructed in the dry, after which water of the stream is directed to the new channel. In the simplest type of abutment the superstructure is placed on the surface of the embankment without any concrete, but a steel or concrete contact plate is used for better pressure distribution. This can be done in frostless zones, e.g., California, provided the embankment is built of nonexpandable, properly compacted materials and proper drainage measures are taken.

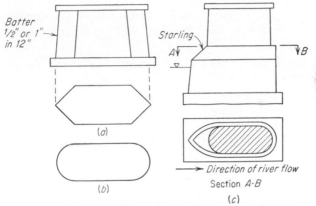

FIG. 14.8. Types of piers: (a), (b) solid shafts, (c) solid shaft with starling (schematic sketches).

Piers. These intermediate bridge supports are built mostly of concrete, often with granite facing. Occasionally steel is used or even timber in bridges formed by piles protruding over the high-water level. As a general rule, the larger the stream, the higher the piers and the deeper the foundations. There are many exceptions, however. In wide, shallow rivers, the piers are generally low and foundations rather shallow; the bridges of the Pennsylvania Railroad over the Susquehanna River near Harrisburg consist of a great number of small span arches. Highway piers are long perpendicular to the general direction of the bridge; e.g., for a four-lane bridge, a pier may be up to 50 to 60 ft long. Railroad bridges generally are much narrower, their width depending on the number of tracks they have to carry. Small bridges generally have no piers but in rare cases may have one or two. Figure 14.8a and b represents a solid concrete or masonry shaft for a medium-size bridge. It has triangular (or rounded) ends directed against the current, though in many cases the piers have symmetrical ends. The pier is *battered* on all sides, though in modern bridge piers the downstream end is often vertical. The pier may be provided with a *starling* (Fig. 14.8c), most of which should be located below the high-water level. The function of the starling is to regulate the passage of water and, particularly, to serve as an ice breaker in the spring. Ice lumps tend to ascend the starling and break under their own weight. A few types of hollow piers for medium-size bridges are shown in Fig. 14.9. In a long structure with a con-

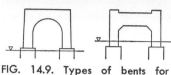

FIG. 14.9. Types of bents for medium-size bridges.

siderable number of spans, piers similar to those shown in Fig. 14.9 are generally called *bents*. Piers for larger bridges generally are hollow and somewhat similar in shape to the diminutive bents shown in Fig. 14.9. They consist of combinations of high vertical shafts, straight, stepped, or circular (cylindrical) with portals, and other architectural features (see page 535 of Ref. 4).

14.3. Bridge Foundations: Some Design Features. In the design of bridge foundations, the *settlement* of the bridge and its *stability* should be considered. For settlement computations, the only important factor is the dead load (DL) of the bridge. The dead load is found by applying unit weights to the computations of volumes. Since a part of the bridge is permanently submerged, the influence of the buoyancy should not be overlooked.

The total vertical load on the foundation of a bridge consists of the dead load plus the vertical live load (LL), e.g., the weight of rolling stock such as trains passing over the bridge. If the bridge support stands on a spread foundation, the total vertical load (in pounds) divided by the area of contact of the foundation (in square feet) gives the *soil pressure* in pounds per square foot on the base of the support. The soil pressure should not exceed the value of the unconfined compression strength of the underlying materials.

Stability of the bridge is affected by *lateral forces* and *scour* (a type of water erosion). The lateral forces, of which *wind* is the most important, essentially cause an overloading of the foundation on the lee side. This overloading is generally of short duration only and, practically, does not affect the settlement of the bridge, but in exceptional, very rare cases, it may cause the support to tip. Some state highway departments have tables of ultimate bearing powers that should not be exceeded for different soils; a small safety factor is usually incorporated. If the laboratory tests of the supporting soil materials are available, the maximum shearing stress acting on the supporting soil material should not exceed their shearing strength, using some small safety factor.

Besides the wind, lateral forces acting on the bridge and transmitted to its foundations are pressure of the running water, wave action if any, ice and drift pressure, and shocks from passing vessels if the piers are unprotected by special fenders of piles driven around the pier. In railroad bridges, the longitudinal forces, mostly due to braking of the trains, may be of importance. For modern bridge supports on deep foundations the lateral forces are of little consequence, except they may be critical for weak timber piers. For earthquake action on the bridges, see Chap. 18.

Scour. When the bridge supports and often a portion of the access embankments are placed in a waterway, it becomes narrower. Thus scour is caused because the water velocity increases and the bed deepens until some state of equilibrium is reached (Chap. 12). Scour also may be induced by rectifying a meandering stream in soft alluvial deposits. Such a rectification is combined with an increase of the flow gradient (because of the shortening of the distance) and hence in velocity. Scour may be observed also in streams under natural conditions without any bridges. In all cases, scour is intensified during the high-water periods.

Terzaghi's[8] tentative empirical rule is that for each foot of rise of the high water above the ordinary water level, the scour is likely to assume values of 3 or 4 ft. This scour may be permanent, or more likely, the stream will be fully or partly refilled. Studies of scour phenomena in the Rio Grande Valley, New Mexico, have shown that the material excavated in the narrow sections of the river is redeposited in its wider sections and the bottom of the river as a whole does not lower.[10] According to some observations, in the Southwestern[9] United States scour extends to twice floodwater depth in the river.

There is no efficient method to prevent scour. One of the palliative methods is placing riprap around a pier; another one is to drive piles under a pier to a depth greater than required by stability of the pier. Piles thus driven, which will be exposed between the periods of high water, should be protected against dry rot, for instance by impregnation.

Though the terms "scour" and "erosion" are practically synonymous in geotechnics, there is a little refinement in their usage, e.g., scour of the bottom of the channel and erosion of the banks of the channel.

Abutment Foundations. There is a similarity between the geotechnical investigations for a building and for a bridge abutment. In the latter case the influence of water in scouring or in softening and swamping the soil should be duly considered. Abutments are built mostly on spread foundations, though pile foundations also have been used. Abutments of large bridges are large structures in their own right.

Pier Foundations. The piers of a bridge are more subject to the action of lateral forces and scour than the abutments and often are founded on deeper foundations. There are *water piers*, permanently or periodically standing in water, and *land piers*, or viaduct-type piers. (The term "viaduct" is usually applied to large bridges with no stream or just with small brooks under them.) The piers of the viaduct type should be founded below the frost line in the same way as building footings, and if the highest ground-water level is above the base of the footing, the resultant decrease of the bearing power of the foundation should be taken into account by the designer when proportioning the pier. Water piers which stand in water only periodically should be founded below the frost line or at a level which ensures a good seat of the pier and safety from erosion during the high-water period, whichever of those two levels is lower. In any case the bottom of the pier footing even for a small bridge should be at least 3 ft or so below the finished grade or the bottom of the stream.

Data on live loads are published for highway bridges by the AASHO (American Association of Highway Officials) and for railroad bridges by the AREA (American Railroad Engineering Association). Data on wind pressure may be found in the "Engineering Handbook" (Ref. 1, Chap. 13).

14.4. Investigations for a Medium-sized Bridge.

On a new route there are many culverts and small one-span bridges. As a rule, no geotechnical investigations are made for these structures. The theory is that the cost of investigations would probably exceed the cost of the repair to possible damage caused by the lack of investigation. Occasionally one bore hole per structure is drilled. However, investigations for every medium-sized bridge, one that spans a stream up to 200 ft or so wide, should be done. These are discussed hereafter.

Preliminary Data. Before the exploratory program starts, complete geological data concerning the given region should be obtained. This is more difficult than for city buildings, since the printed geological information in this case is generally scarcer. Publications of the USGS should be fully explored (folios, water papers, etc.). Equally important is to obtain firsthand information on the service and behavior of adjacent bridges that span the same stream under similar geological situations. Personal interviews with the engineers in charge of these structures are advisable. *Direct measurement* of the waterways of these bridges is important. The bridges should be examined for evidence of settlement or cracking. Local farmers may provide data about the general regime of the stream. All information obtained should be critically examined and recorded in writing.

Environment Studies. Many principles and details of geotechnical investigations for bridges are identical with those for buildings (Chap. 13) and will not be repeated here. The construction and, afterward, the service of the bridge are more influenced by the *environment* than is a building. A stream has to be harnessed, so to speak, and to do so successfully, it is necessary to study the site where the structure is to be built, to collect all possible information on the behavior of the stream at its different stages, and to determine the possible damage it can do to the planned structure in order to be able to avert that damage. In the case of a large stream, the corresponding hydrologic and hydrographic studies may be carried on by special consultants, but for a middle-sized stream these studies should be done by the same person who is in charge of the geotechnical investigations.

The following outline can be used as an approximate guide for the hydrologic and hydrographic studies for a bridge. Such studies are of extreme importance in all cases, but especially if the bridge has to be built abroad in a country with a limited amount of information along these lines.

1. *Basin of the stream:* area; length; elevation; character, including vegetation, soil, and steepness; drainage, including tributaries and existing dams and reservoirs.

2. *Stages of the stream:* elevations and seasons of high high water, ordinary high water, low water; velocities at different stages.

3. *Estimated discharge:* at different stages, including the combined action of snow thawing and rainfall which very often gives the maximum possible discharge; frequency of floods.

4. *Stream bed:* character of bed material upstream and downstream of the proposed bridge site; obstructions to the passage of water and natural tendencies to scour or silting around these obstructions; meandering tendencies and whether or not channel regulations are needed; if so, what kind of channel regulations should be used such as deepening, widening, construction of levees (Sec. 16.1); or bank protection.

5. *Drift* or *debris* carried by the stream: character, amount, and size; approximate vertical clearance of the superstructure needed for the passage of the drift or debris; spans that may be required for that purpose.

Preliminary Selection of the Foundation Type. A middle-sized bridge is usually founded on either spread footings or pile foundations. Generally from the study of local geology and the results of the first few borings, one may guess rather correctly what kind of foundations should be used. The final selection of the type can be made, however, only when the exploratory program is completed.

Exploratory Program. Because of the great variety of foundation types and especially environment conditions, the exploratory program for a bridge, even a middle-sized one, cannot be standardized. General features of such a program are discussed hereafter, however.

The borings are sunk usually at the center line of the planned bridge or at the opposite ends of two neighboring piers. Borings should be numbered and also give the highway station and the distance (to the right or to the left) from the center line of the highway.

As a rule, there should be at least one boring under each bridge abutment and each pier. In a geologically uniform locality, when the stratification does not change considerably from pier to pier, the number of borings can be less. In the case of a low bridge with a considerable number of bents, it is advisable to sink borings at every other pier or even less frequently. If a change in geological conditions is disclosed, then the intervals between holes should be decreased.

If spread footings are planned and the results of the boring are favorable to this preliminary assumption, it is advisable, nevertheless, to investigate the nature of the sediments at least 10 ft below the base of the proposed footings. Coarse sand if compacted represents an excellent foundation for bridge piers. Compacted fine sand may also be good for foundation purposes if confined and thus prevented from flowing from underneath a pier. Even if all piers are planned to be on spread footings, it is advisable to drill one deep hole to rock. However, if rock or some soft layer is not present to a depth of 80 to 100 ft, discontinue the drilling. Softer layers beyond this depth usually are not likely to cause dangerous settlement of the bridge, since the unit soil pressure at that depth is already very limited.

It is advisable to drill the deep hole as early as possible in the course of investigations because if a soft layer is disclosed at a rather shallow depth, it may be necessary to change the foundation design and use *pile foundations*. Piles used to support medium-sized bridges are generally timber or concrete. Steel H piles (preferably 14 in. in size) are commonly used for large bridges and very seldom for medium-sized bridges.

Where foundations are to be situated below the water table or where high soil permeability is likely to exert a major influence on the construction operations, some pumping tests should be made. These tests are usually conducted in drill holes or in special wells. Also to be remembered here is the significance of nonreturning wash water in the borings (Sec. 6.25).

The equipment to be used in the field should give enough 2-in. samples for simple identification tests, such as moisture content, gradation, and dry density. It is desirable to drill some 3- or 4-in. bore holes and use the Shelby tube for extracting samples for unconfined compression and consolidation tests. Triaxial tests are seldom used for medium-size bridges. It is advisable to perform simple tests in trailer laboratories properly equipped for the purpose. Facilities for digging and bracing of the pits should be available without delay if drilling is not feasible because of a high percentage of large rock fragments in the foundation material.

Pile Foundations. In all cases of pile foundations, piles driven to rock or other reliable material (e.g., a thick sand and gravel deposit) are preferable to friction piles. Hence if pile foundations are considered, a preliminary search should be made for rock or some other reliable layer. The most convenient method to use is wash borings. In a few of them (sometimes one is sufficient) the standard penetration test should be done at intervals of 10 ft or less according to local conditions. If rock is found at a depth of 40 or 50 ft and the whole soil deposit is soft, this is a clear-cut case of pile foundations. In this case wash borings merely supplement the explorations (compare also Case 3 of Fig. 13.12a). In some cases, e.g., in some glacial zones, the presence of a clay layer may be anticipated from a very elementary knowledge of local geology. In such instances one or two trial borings should be made initially in order to disclose how deep and thick the clay layer is and whether or not it is topped by a crust. If the layer is *deep* and either thin or topped by a substantial crust or both, there is a possibility of founding the bridge on spread foundations. In all cases, however, when a soft layer is located below the bottom of a spread foundation, computations of settlement should be made and sampling should be done to provide samples for consolidation tests.

If at a certain depth, say 25 or 30 ft, a thick gravel layer (e.g., 20-ft thick) is located, this layer may serve as a support for the piles carrying the piers. A brief geological investigation as to the origin, extent, and supporting power of this gravel slab is required. In no case should piers be placed too close to the edges of a gravel slab.

If friction piles have to be used, some idea should be formed on the value of the unit friction (in pounds per square foot) on the surface of the pile. Friction values for different materials given in textbooks and handbooks should be used with great care, if at all. The total frictional resistance of a pile of a given material, of given dimensions, and driven into the given soil may be determined by a pulling test. The total frictional resistance of such a pile is measured by the force required to pull the pile out minus the weight of the pile. A safety factor, e.g., 2 or more, should be applied. Since the maximum unit friction of the pile directly depends on the shearing strength of the material into which the pile is driven, good samples for unconfined compression tests should be taken in clayey materials. In sand-clay mixtures some other shear test should be used (e.g., three-ring shear test) and samples taken accordingly.

A small detail about pile driving should be added. Piles driven to a steeply dipping rock often slide down under the hammer blow, thus producing a false impression of penetration. The dip of the rock, if required, may be found in this case from a few wash borings. Loading tests are used oftener in the investigations for bridges than for buildings. The results of a loading test should be interpreted in conjunction with the examination of the samples taken close to the test location. The load tests are also discussed in Secs. 13.5 and 13.8.

The load test at a proposed pier location in an overhead structure on U.S. Highway 99 near Visalia, California, was conducted in large-diameter holes drilled with a power-driven bucket auger. The load was applied on plates with a surface area of 2 sq ft by means of a power-operated hydraulic-pressure cell. Constant pressure was maintained by use of a nitrogen-loaded gas accumulator and a mercury pressure switch. Twelve-inch expanding anchors fixed in drill holes furnished load reaction.[11]

COFFERDAMS AND CAISSONS

14.5. Cofferdams.[12] If an engineering structure, such as a bridge pier, has to be built in an area covered with water, e.g., in the middle of a river, the area where the work has to be done is surrounded by a

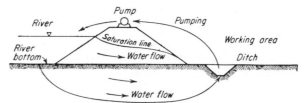

FIG. 14.10. A simple water cofferdam (sketch).

cofferdam. A cofferdam is a wall made of earth materials, of steel or timber sheet piling, or of a combination of various materials. Since under actual working conditions it is impossible to build a perfectly impervious cofferdam, there is always some *seepage* through the cofferdam, and the water has to be pumped out of the working area. Seepage differs from *leakage*, as the latter is water that works out a passage in the cofferdam (or some other earth structure) and abnormally increases the natural seepage discharge. Cofferdams also are used to protect a working area against a large influx of subsurface water ("land cofferdams" used mostly for building excavations).

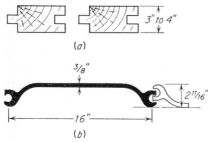

FIG. 14.11. Sheet piling: (a) timber, (b) steel (section MP 112, U.S. Steel Co.).

A simple type of water cofferdam is shown in Fig. 14.10. Water seeps both through and under an earth embankment built in the river. An essential part of the cofferdam is a *ditch* dug within the working area and parallel to the embankment. This ditch acts as a center of attraction for the flow lines of the seeping water. The pumps standing on the embankment throw the water back to the river (Fig. 14.10).

Sheet piling used for the protection of the working areas may be of timber for water depths up to 10 ft (Fig. 14.11a) or of steel of a great variety of cross sections (Fig. 14.11*b*). The illustrations show the cross sections of the sheet piles as seen from above (in plan) when driven.

A *one-wall* sheet-piling cofferdam may be constructed by driving a sheet-piling wall and placing earth on both sides of it to form an embankment similar to that shown in Fig. 14.10. Another type of one-wall sheet-piling cofferdam is shown in Fig. 14.12. The steel sheet piling is

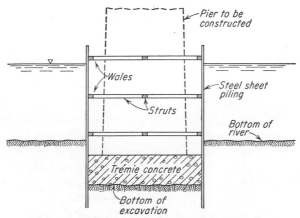

FIG. 14.12. One-wall sheet piling cofferdam (sketch).

driven to a depth below the base of the proposed pier until it is well embedded in the surrounding soil, and the material inside the cofferdam is excavated, usually by dredging. Gradual unwatering of the cofferdam by heavy pumping permits the placement of the bracing (wales and struts, ordinarily of heavy timbers) to oppose hydrostatic and lateral earth pressures. A thick layer of concrete is placed in the water (the so-called tremie concrete), the cofferdam is dewatered, and then the pier is built. Pumping is required during the whole construction period because of the leakage through the fissures

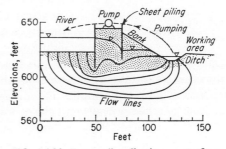

FIG. 14.13. Two-wall cofferdam; note flow lines.

between the sheet piles. If the pier has to be built on piles, these usually are driven from a barge and tremie concrete is placed on their heads.

Two-wall sheet-piling cofferdams are generally used to protect larger working areas (Fig. 14.13). The space between the two rows of sheet piling is filled with compacted earth material, and an earth bank inside the working area is added for stability. A ditch (as in Fig. 14.10) is an essential part of the arrangement. Note the flow lines inside and under the cofferdam.

Cellular cofferdams, also used for protection of larger areas, generally consist of huge vertical cylinders (30 to 40 ft and more in diameter) placed close to one another and filled with gravel and sand. A variation of cellular cofferdams are diaphragm cofferdams. In this case the cells are rectangular instead of circular.

14.6. Caissons. The term "caisson" literally means "box" (French) and hence is loosely applied to concrete piers under buildings (Sec. 13.4).

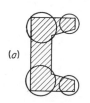

(*a*)

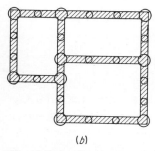

(*b*)

FIG. 14.14. Circular open caissons.

Whereas a cofferdam is removed after the structure is completed, a caisson remains in place (with a few exceptions) and forms an integral part of the structure. During the construction period the caisson functions as a cofferdam. A *box caisson* is a watertight timber or reinforced-concrete box having a bottom but no top. Its use is convenient when there is no excavation and the bottom of the river is more or less level; a box caisson also may rest on top of piles. The caisson is constructed on shore and floated to the site. This type of caisson may be used in the construction of medium-sized bridges.

An *open caisson* is a box without top or bottom, made of timber, metal, or concrete. In the United States both metal and concrete open caissons are used for bridges and buildings. If the river to be bridged is deep, the caisson is built on shore and floated to the site; in shallow water it may be constructed on barges at the site. In the latter case, as the caisson is sunk, the barges are dragged from underneath it. An open caisson has heavy walls and sharp wedgelike edges which allow it to sink with the aid of additional temporary loads and jets of water while the inside material is dredged out. Thus the sinking of an open caisson proceeds at atmospheric pressure, and theoretically there is no limit to the depths of sinking. Practically, there are many difficulties, since the excavation is done under water and cannot be properly controlled; this often results in tipping and wedging of the caisson. The flow of the adjacent ground into the caisson constitutes another difficulty, especially in plastic soils. Open caissons may be rectangular or cylindrical. The latter may be used when the structure is too large in plan to be supported by one rectangular caisson. Such, for instance, would be the case of a large abutment (Fig. 14.14*a*) or a building (Fig. 14.14*b*). In the latter case the large caissons are placed at the corners and wall intersections and are completed by placing smaller cylinders under the walls. The walls should be self-bearing between the caissons. Small steel-cylinder caissons may be driven with a drop or steam hammer or by other means such as jacking. In such a case there is very little difference, if any, between open caissons and the concrete-filled steel pipes discussed in Sec. 13.6. Figure 14.15 shows the process of sinking an open caisson.

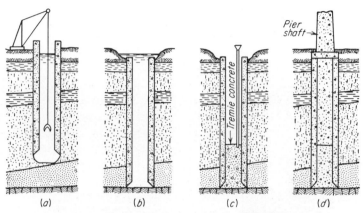

FIG. 14.15. Sinking of an open caisson. (After Dunham.[4])

When it is not practicable to excavate wet ground in the open, *pneumatic caissons* are used. A pneumatic caisson consists basically of a working chamber and tubular shafts (generally two) provided with air locks. One of the shafts has a materials lock which is used for removing muck from the working chamber; the other has a man lock which permits the labor force to travel in and out. Figure 14.16 shows a view of the caisson with a materials lock and chute for disposing of excavated material either directly to the water or to a barge. Compressed air is forced into the working chamber to balance the water and soil pressure and thus permit excavating in the dry. The walls of the working chamber have sharp edges, which facilitate the sinking of the caisson under its gradually increasing weight. Placing of concrete starts at the roof of the working chamber and is carried out in such a way that as the caisson sinks, the work is always protruding above the water level approximately the same distance. When the caisson reaches the design elevation, the sections of shaft are pulled out and the working chamber and the shafts are filled with concrete.

Pneumatic caissons are made of metal, concrete, and timber. Shafts and locks are always steel. The depth of sinking is limited to about 110 ft below the water level. At this depth a man has to work under an air pressure of 50 psi, which is about maximum for feasible work.

Figure 14.17 shows the way a pneumatic caisson is entered through the man lock. The figure represents schematically the man lock and the upper portion of the tubular shaft leading to the working chamber with a ladder. The man enters through door *a*, door *b* leading to the working chamber is closed, and there is atmospheric pressure

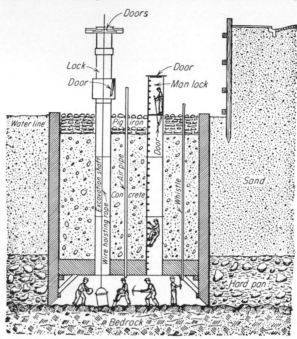

FIG. 14.16. Pneumatic caisson. (*After Hool and Kinne.*[6])

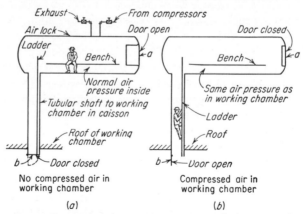

FIG. 14.17. "Locking-in" operation for entering a pneumatic caisson. (*From Dunham.*[4])

in the man lock. The man waits until a lock tender closes door *a*, turns on the air pressure, and lets the pressure in the lock build up until it equals that in the working chamber. The door *b* is then opened, and the man descends to the working chamber using the ladder. When he leaves, the procedure is reversed.

Compressed-air work is strictly regulated by law. All persons desiring to enter a pneumatic caisson have to undergo a physical examination. There is a caisson disease (the bends, a kind of severe rheumatism) generally caused by too rapid a decompression which traps part of the compressed air in the man's system.

14.7. Exploratory Program for a Pneumatic Caisson. The considerations presented hereafter apply to both open and pneumatic caissons. The geotechnical investigations should indicate the depth to which the caisson has to be sunk in order to ensure the stability and the safety of the bridge pier or other structure it will support. At least one bore hole,

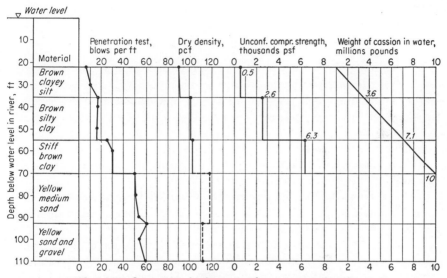

FIG. 14.18. Geotechnical information for a pneumatic caisson.

two for a deep caisson, should be drilled to rock. The drilling may be discontinued at a depth of 110 or 120 ft if no rock is found. Shallow borings should be avoided because they will not disclose soft deep layers. Figure 14.18 shows the most essential field operations in this case. The penetration test should be done every 10 ft as the drilling progresses and also at all changes of the earth material. When the drilling passes through clay, at least one sample per layer should be taken for consolidation tests (although the tests will not necessarily be done), dry density, and unconfined compressive strength. If sand samples cannot be taken, moisture content should be determined from disturbed samples. In a saturated sand a low moisture content indicates dense material; assuming a specific gravity (e.g., 2.65) it is possible to estimate the dry density theoretically (Sec. 4.8).

According to the diagrams of Fig. 14.18, the caisson can be stopped in the stiff brown clay layer if there is no danger of considerable settlement of this layer under load; otherwise the caisson should be carried to the underlying sand.

In the case in Fig. 14.18, the necessary computations could be made as follows: The weight of the caisson at the beginning of the operation and at some final position should be estimated. From a rough straight-line assumption (Fig. 14.18) the weight of the caisson at various positions may be found. It is reduced by friction, the maximum unit friction being about one-half of the unconfined compressive strength in each clay layer. A reduction of this strength by about 25 per cent would be wise in this case. The soil pressure from this computed reduced load should be less than the unconfined compressive strength of the material on which the caisson is placed. A consolidation test on a sample of the stiff brown clay would be helpful.

RIVER AND VALLEY CROSSINGS

Determining the waterway for a proposed bridge is a somewhat complicated hydrological and hydraulic problem and will not be discussed here. Obtaining data for the design of approaches to the bridge or crossings of the valley where the bridge has to be built is a purely geotechnical problem, however. Working abroad, and particularly in the so-called "undeveloped" countries, is given special attention in the following discussion. Extensive use of airphotos in the study of all kinds of crossings is recommended.

14.8. A New Bridge Emplacement: General Considerations. Geotechnical investigations for a new crossing and a new bridge should be planned according to the following principles:

1. The bridge and its approaches should be considered and designed *as a unit*, not only from the point of view of continuity of traffic but mostly from a point of view of statics. The instability of approaches may affect the stability of the bridge itself.

2. The hydrography and hydrology of the region must be considered, as they control the area of the waterway, often the subdivision of the bridge into spans, and practically always the regulation of the river (Chap. 11). Some of the hints given in Sec. 14.4 for this kind of investigation are fully applicable to a large bridge with the difference, however, that all data should be more complete and be *firsthand* information. Particular attention should be paid to the origin, discharge, and periods of high water, since there may be quite unexpected phenomena in this respect. As an example, in an investigation for the bridge over a flood plain in an arid region abroad, it was disclosed that the high water arriving in the middle of the hot summer season came from the melting of snows covering a quite remote mountain chain.

3. Planning of the geotechnical investigations is controlled by the same *economic factor* that governs the design and construction of the proposed bridge. As an example, in the United States, where steel and concrete

are relatively inexpensive, the approaches to a high bridge across a flood plain often are designed as viaducts. In such a case, the latter consist of a series of small spans, gradually sloping down on both sides of the bridge. Only at the ends of the approaches are there short and low earth embankments. In countries where steel and concrete are expensive and the manual labor cheap, large access embankments are designed and constructed by manual labor with the occasional assistance of horses and mules. In these two cases, geotechnical investigations are different in character. In the former case the attention of the exploratory party is directed mostly toward data for the foundation design. In the latter case the location of suitable borrow pits, testing of borrow materials, and planning of the earthwork are of major importance.

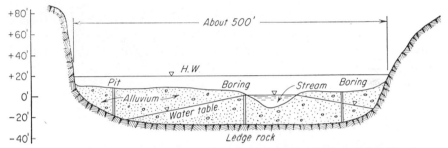

FIG. 14.19. Preliminary site investigations for a bridge at the middle reaches of a stream (sketch).

A correct dimensioning of the waterway of a bridge with respect to the possible scour is another example of an economic problem to be solved by the geotechnical investigations. If the waterway is too small, any savings on the superstructure may be lost by the expense of preventing scour and repairing the damage it caused. The geotechnical studies should give material for financial estimates required in this case.

14.9. Crossings at Different Reaches of a Stream. An example of a stream originating in the mountains and flowing into an alluvial valley (or on an alluvial fan) will be considered. If the stream is crossed at an upper reach or at the beginning of the middle one, a V-shaped narrow valley has to be spanned. In such a case there is only a little alluvium in the stream, the latter is not wide, and thus a one-span bridge with abutments on rock can be used.

At its middle reaches, the stream gradually approaches the characteristics of a mature stage. The cross section of the valley is wider and is troughlike (Fig. 14.19). There are boulders, gravel, and some coarse sand on the shores. At the edges of the valley rudimentary terraces may have developed (although not shown in Fig. 14.19). If the freshet (high water) is abundant and the water level high (as in the figure), usually a high, one-span bridge is needed. The immediate objective of the subsurface investigations is to disclose the configuration of the rock floor and the suitability of the rock as a foundation material. If there are no quarries in the vicinity, as may be the case of a valley crossing in an undeveloped country, the valley itself may be converted into a quarry. The quality of the rock material in the cuts on both shores of the valley should be investigated. This rock material, together with alluvium in the valley, may be used for concrete and, in any case, for building the approach embankments to the bridge.

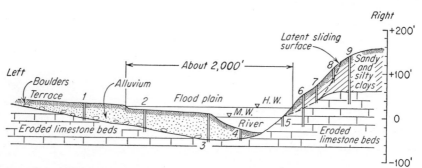

FIG. 14.20. Site investigations for a bridge in an alluvial valley (sketch).

In the cases similar to that shown in Fig. 14.19, the transverse gradient of the water table is probably from the stream rather than toward the stream (inversed infiltration). The water table may even be completely absent as in the case of the pit shown to the left in Fig. 14.19.

Crossing a Lower Stream Reach. An example of such a crossing with a relatively shallow rock floor is given in Fig. 14.20, which represents a modified geologic section of a Volga River crossing.[13] Assume that the valley has been formed by erosion of original limestone beds and is filled with alluvium, sand, gravel, and boulders (left shore). The right shore is high and formed by sandy and silty clays developed on eroded and weathered limestone. Also assume that the following set of structures has to be planned: (1) left-shore approaches to the bridge in the form of embankment on the terraces, (2) a large bridge with a part of supports in the flood plain (viaduct or land piers) and another part in the stream itself, and (3) a deep cut through clay on the right shore to connect the bridge with the adjacent country. Accordingly, the geotechnical work consists also of three parts, the simplest one being the study of the terraces on the left shore. Here, simple auger holes probably will do (boring 1, Fig. 14.20). A preliminary examination of the right-shore slope should be made to spot any possible sliding or fissuring. If there is some evidence of danger, investigations are required that in some cases may be even more complicated or at least more detailed than the investigation for the bridge itself. Some of the borings to be sunk in this connection should be deep enough to reach into the rock and investigate its stability against sliding (for instance, borings 6 and 9); some others, as borings 7 and 8, may be shallow to spot only the latent sliding surfaces. Emphasis should be laid on the study of ground-water conditions (direction and discharge). If the slopes of the cut will be in limestone, samples of the latter should be tested for solubility.

Large rivers in their lower reaches are very wide, and the general procedure for locating borings, applicable to the case of Fig. 14.20 (and to any other similar case), would be as follows: When the position of the longitudinal axis of the bridge is established and the abutments located, the topography of the river bottom along the bridge axis should be obtained. In a wide river there may be channels or depressions which are unfavorable for the emplacement of piers, and conversely subaqueous islands, ridges, or outcrops may be suitable for this purpose. The transverse section of the river along the proposed bridge axis is prepared by measuring the depth of the river in the way described in Chap. 12 (particularly Sec. 12.10). In this connection, it is very desirable to sink several wash borings in order to obtain a preliminary idea of the geology of the river bottom. This topographic and geologic information may influence the plans of the engineering staff as to the subdivision of the bridge into spans and the type of bridge foundations. In some cases this preliminary information is sufficient, e.g., in the case of a valley filled with a very thick alluvium. In the majority of cases, however, large-diameter borings should be made at the site of each pier and undisturbed samples taken. Drilling in the deep section of the river is generally done from a barge; in shallow sections various arrangements, e.g., that of Fig. 6.22, may be used.

In large navigable streams, the location of the piers may be controlled by the large vertical and horizontal clearances that are required. A large span covering the deepest part of the stream channel, with a considerable vertical clearance (e.g., 30 ft) above the high-water level, may be required for bridges covering navigable streams near their deltas. This increases the height of the piers and, of more importance, increases the length of the bridge. In this case a bridge proper degenerates into viaducts on both shores of the stream. The abutments are thus moved far away from the stream, often to an elevated dry locality with a simplification of construction and no worry about proper location. A required vertical clearance also demands exceedingly accurate settlement computations of the bridge supports in the navigable channel, since an excessive settlement may unduly decrease the clearance.

Crossing a River Close to Its Delta. A navigable river in this case is always spanned by a high bridge with land piers on both sides. Under certain geological conditions, e.g., when the river flows over a huge alluvial fan and is higher than the adjacent locality, the bridge would be extremely high; thus attempts should be made to locate the railroad or the highway in a tunnel under the river. Since a bridge crossing a river at its delta is close to the shore line, the possible slow movement of huge sand masses toward and along the shore line should be considered (Chap. 12). This precaution is especially important if a bridge has to be founded on friction piles. If the latter are too short, they may move downstream and deform and twist the superstructure. Such cases, though rare, have happened.[14]

14.10. Crossing a Valley Underlain by a Thick, Soft Deposit. In this case the bridge supports are always on deep foundations, e.g., pneumatic caissons.

Case of Sandy or Sand-gravelly Thick Alluvium. The procedure recommended for the relatively thin alluvial deposits (Fig. 14.20) has to be modified if the alluvial deposit is several hundred feet or so thick. The pier foundations have to be held in place mostly by friction and therefore should be deep in order to develop the necessary frictional resistance. H piles are not used as friction piles, so the probable solution would be a deep pneumatic caisson. The use of an open caisson (Sec. 14.6) is too risky if the alluvium consists of loose sand that may flow into the caisson. Exploratory borings and penetration tests should show if at a relatively shallow depth there is some firmer layer on which the caisson could stand, thus developing some vertical support in addition to the skin friction. The reliable layer in question may be dense sand, sand and gravel, or sometimes a thick, stiff clay layer, in which case it is necessary to sink the caisson several feet into the clay. Abutments or land piers in this case are generally built on friction piles (e.g., concrete or timber). Typical of this case are bridges in the lower Mississippi River Valley.

Case of an Organic Silt Deposit in a River. Deposits of organic silt (mud) often obstruct the lower reaches of large rivers in glaciated areas (Potomac River in Maryland, Thames River in Connecticut, and others). The usual geological conditions in such cases are 20 to 30 ft of water at the middle of the river and 80 to 100 ft of soft material underlain by bearing strata (often successive layers of dense sand, glacial hardpan, and rock). Geotechnical studies are as simple as in the case of alluvial deposits. It is necessary only to wash through soft material and, if it seems necessary, use the standard penetration test in the underlying material.

Case of a Valley Underlain by a Continuous Clay Layer. This is a case analogous to those described in Sec. 13.19, concerning building foundations in glacial zones. Floating foundations, however, cannot be recommended for bridges of a certain magnitude. For such structures the soft clay layer should be penetrated by heavy (size 14) H piles founded on rock or other reliable material. If the competence of the material is doubtful, penetration tests should be performed.

The H piles may be 80 to 90 ft long, and if the clay is not very soft, an attempt should be made to found the bridge at a shallower depth, perhaps 50 or 60 ft, using caissons.

If a clay layer supports both the bridge abutment and the adjacent embankment, the latter should be constructed first to preload or prestress the clay layer. Otherwise, the abutment will settle twice, namely, after its construction and during and after the construction of the embankment. Settlement due to the embankment load may be more dangerous than that after construction and may cause a failure of the abutment.[15] Building the embankment first is generally inconvenient for the contractor, since it obstructs the work front. However, it is the duty of the geotechnical party in charge to report the possibility of double settlement.

14.11. Minor Cases of Crossings. *Case of a Meandering Stream.* An extensive use of airphotos is perhaps more essential in this case than in other types of crossings. The airphotos should be used to prepare a map of the river for about ½ mile along its course (or more if necessary); then the direction of the channel to be excavated should be established. If there are no facilities for airphotos, the map should be prepared on the basis of a land survey. The cross section of the channel should be determined by simple hydraulic formulas. If the cross section of the channel thus computed appears to be too large, a smaller, but lined, channel should be used. If a clay borrow pit of an adequate capacity can be found in the neighborhood, clay may be used for both the lining and the earth dikes that should be built at all intersections of the new channels with the existing river channels. These dikes prevent the water from entering the new channel prematurely. The usual geotechnical studies for the bridge supports should be performed. For canal linings see Sec. 16.28.

Case of a Dry Valley with Periodic Shallow Flood. In this case probably no bridge is needed, and a low paved embankment with some culverts for the passage of the water would satisfy the requirements of predominantly horse-drawn traffic the whole year around. During the highest stage of freshets when the embankment is flooded, the automobile traffic has to be discontinued.

Case of a River Crossing in a Jungle Area. The problem is to trace an access to the selected bridge location across 12 to 15 miles of jungle area. A method used for many years is to start a big fire at night next to the bridge location. The fire is observed with a transit from an elevated point on the planned road outside the jungle. The transit is left in place overnight with the screws tightened. Next morning the layoff and clearing of the right-of-way work start.

CASE HISTORIES OF LARGE BRIDGE FOUNDATION EXPLORATIONS

Hereafter the basic foundation conditions and foundation exploration for three large bridges are discussed. The bridges are situated on the

East Coast, on the West Coast, and in the Mississippi River Valley, respectively.

14.12. George Washington Bridge. This bridge was built between 1927 and 1931 to span the Hudson River and connect New York City

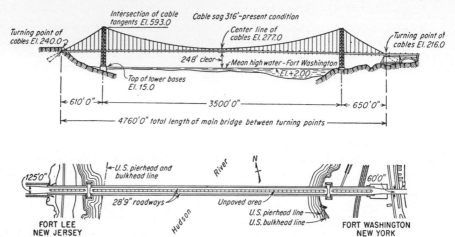

FIG. 14.21. Elevation and general plan, George Washington Bridge. (*ASCE Trans., vol., 97, p. 101, 1933.*)

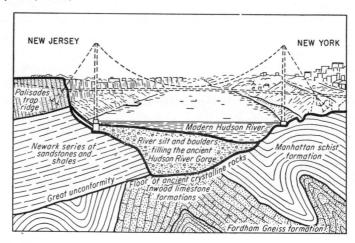

FIG. 14.22. Geological cross section at the site of George Washington Bridge. (*ASCE Trans., vol. 97, 1933.*)

with New Jersey cities. It is a suspension bridge with two towers and two cable anchorages.[16] The central span is 3,500 ft between the towers with a clearance of 248 ft above high water (Fig. 14.21). The geological section at the site (Fig. 14.22) indicates a river gorge filled with silt and boulders, perhaps over 400 ft deep, extending down to the ancient rock floor. The general arrangement of the structure is shown by dotted lines in the figure.

Whereas all the ledges exposed on the New York shore were of sound and strong Manhattan schist, a fault in the rock about 200 ft east of the site was disclosed. There was, however, no indication of any objectionable rock features under the site. In the rock excavation for the New York tower, a clay-filled seam was found, and this required that the rock be removed to a greater depth than planned. It was not possible to place the New York anchorage in the rock; therefore it is a mass of concrete, approximately 200 by 300 ft in plan, rising to 85 ft above the adjacent street level (Riverside Drive).

On the New Jersey shore the ancient rock floor is overlain by sandstones and shales of the Newark formation. The slightly sloped Newark formation is overlain by the high Palisade trap ridge. The term "trap" in this case designates a compact mass of hard diabase. The anchorage tunnels are driven about 250 ft into this traprock. Any shifting of the tower toward the river meant economy in the cost of superstructure but an increase in cost of the tower foundation. A set of 30 borings was sunk for clarification of this problem, and from various possible emplacements the one offering the hardest and most compact rock structure with no seams or boulders was chosen. The New Jersey tower was finally founded on a reliable sandstone in the Newark formation. This rock had a compressive strength of 12,000 to 15,000 psi; this gave a safety factor of about 30.

14.13. San Francisco–Oakland Bay Bridge. San Francisco Bay, separating the city of San Francisco from the Oakland-Berkeley area, was formed by a slow sinking (in late geologic times) of a portion of the California coast.[17] The bay is filled with sediments (bay mud, previously mentioned, and fine- to medium-grained sands). The bedrock in the vicinity of existing and proposed bridges over the bay consists of the Franciscan formation (feldspathic sandstone, graywacke, siltstone, shale, interbedded chert, and shale and various types of rock altered to greenstone and serpentine). The formation is broken by numerous small inactive faults.[17]

The San Francisco–Oakland Bridge (Fig. 14.23) consists essentially of the San Francisco (West Bay) portion and the Oakland (East Bay) portion. They are separated by the rocky Yerba Buena Island. The two portions are different geologically. As the rock floor in the San Francisco portion is shallow, this portion is spanned by two suspension spans with piers placed on rock. The cables are anchored in the rocks of Yerba Buena Island and in a concrete massif on the San Francisco shore. The island is crossed by a tunnel combined with a viaduct. The rock floor in the Oakland portion is very deep, and the superstructure

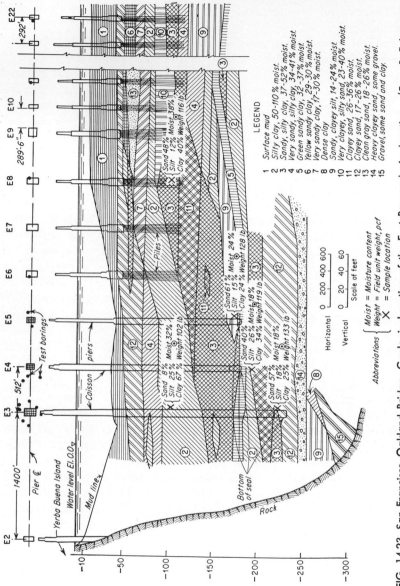

LEGEND

1   Surface mud
2   Silty clay, 50–110 % moist.
3   Sandy, silty clay, 37–52% moist.
4   Very sandy, silty clay, 34–41% moist.
5   Green sandy clay, 32–37% moist.
6   Yellow sandy clay, 29–30% moist.
7   Very sandy clay, 17–30% moist.
8   Dense clay
9   Sandy, clayey silt, 14–24% moist.
10 Very clayey, silty sand, 23–40% moist.
11 Clayey sand, 26–36% moist.
12 Clayey sand, 17–26% moist.
13 Clean gray sand, 18–26% moist.
14 Heavy clayey sand, some gravel.
15 Gravel, some sand and clay.

Abbreviations
Moist = Moisture content
Weight = Field unit weight, pcf
X = Sample location

Scale of feet
Horizontal 0   200 400 600
Vertical   0   20   40   60

FIG. 14.23. San Francisco–Oakland Bridge: Geological section of the East Bay and pier location. (From R. F. Leggett, "Geology and Engineering," McGraw-Hill Book Company, Inc., New York, 1939.)

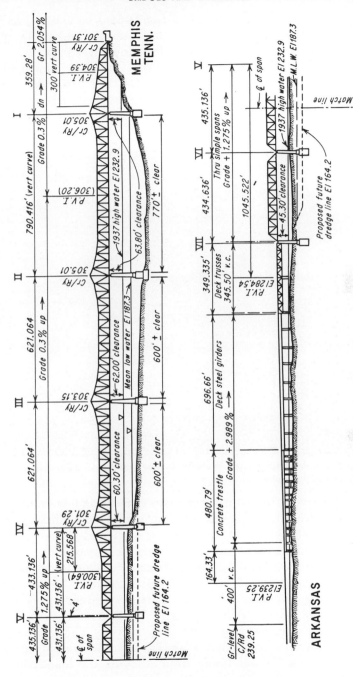

FIG. 14.24. Memphis-Arkansas Bridge: Schematic elevation.

consists of a 1,400-ft cantilever and 19 small spans on caissons and pile clusters.   Because of the unusual depth of caisson sinking, a special type of caisson (domed cofferdams with false tops) was used.[18,19]

Original systematic investigations, started in 1930, compared five different crossings.[20]   Final investigations (1932) obtained data on the location of the rock floor in the West Bay and load tests on rock.[21] Detailed soil investigations were made in the East Bay.   In the latter investigation, soil samples were taken and analyzed in a field laboratory

FIG. 14.25. Memphis-Arkansas Bridge: Weaving a willow mattress.   (Courtesy of Modjeski and Masters, Consulting Engineers.)

installed on a barge.   Undisturbed samples were 2 and 5 in. in diameter. Loading tests were run for pile foundations and on rock under caissons. The seismic risk also was analyzed and was not found alarming.   More details may be found in the consulting geologists' reports.[22,23]

14.14. Memphis-Arkansas Highway Bridge.[24]   This bridge connects Tennessee and Arkansas and was built in 1945 to 1949.   Three middle spans of the bridge cover the river at its mean low-water stage (elevation 187), whereas the whole structure is required to pass high water (elevation 233) such as occurred in 1937 (Fig. 14.24).   Though the whole flood plain of the Mississippi River at that location is sandy or sand-gravelly in character, it was found from studying the foundation conditions of existing bridges and checking soil investigations that the caissons of the new bridge could be sunk to a clay deposit.   This deposit

FIG. 14.26. Memphis-Arkansas Bridge: Moving floating caisson for Pier II. (*Courtesy of Modjeski and Masters, Consulting Engineers.*)

FIG. 14.27. Memphis-Arkansas Bridge: Building a pier. (*Courtesy of Modjeski and Masters, Consulting Engineers.*)

had a bearing power of 10,000 psf and was less than 100 ft from the surface. Existing bridges and the new bridge on the Memphis shore are founded on a 40-ft-thick tough clay overlying a 10-ft sand layer on top of the 10,000-psf clay. Since the sand layer is exposed to running water, it is protected against erosion by willow mattresses and riprap (Fig. 14.25). Similar mattresses and riprap were also provided for two river piers. Other construction features of the bridge are shown on Figs. 14.26 through 14.28.

FIG. 14.28. Memphis-Arkansas Bridge: Erection of span III–IV. (*Courtesy of Modjeski and Masters, Consulting Engineers.*)

Twenty-four-hour load tests on the soil were made to a pressure of 10 tons/sq ft in the working chambers of the caissons, piers I to IV. An 18-in.-square concrete bearing block embedded 1 ft in the soil at the foundation elevation was loaded by jacking against the roof of the working chamber. Settlement was measured relative to steel angle frames driven into the soil well away from the test blocks. Settlement of test blocks at design pressure in piers I and III was on the order of ⅛ in. and reached a ⅜-in. value under a 10-ton/sq ft pressure.

PAVEMENT FOUNDATIONS

14.15. Rigid and Flexible Pavements. Pavements for highways, airport runways, parking lots, etc., that are properly designed (1) distribute wheel loads over a sufficient area of the underlying earth materials so that the induced stresses do not exceed the bearing value of the soil and (2) provide a durable, weatherproof surface that has adequate stability and cohesiveness.

Pavements are constructed of layers of different processed materials placed on the natural, generally treated ground surface. This ground surface is the *subgrade*, or the foundation of the pavement. According to modern concepts, *all layers* of different materials placed on the subgrade constitute the *pavement*. The top of the uppermost layer of a pavement is the *surface course* (or wearing or riding surface).

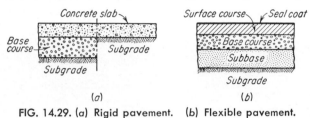

FIG. 14.29. (a) Rigid pavement. (b) Flexible pavement.

There are *rigid pavements* (such as concrete slabs) and *flexible pavements* (such as various asphalt and other bituminous surfaces). The old-fashioned water-bound macadam made of rolled and compacted crushed stone, using water only as a binder, is also a flexible pavement. These two types of pavements are schematically shown in Fig. 14.29. A concrete pavement (Fig. 14.29a) consists of a rigid concrete slab, 6 in. (or more) thick on highways and considerably thicker on the airport runways and taxiways. The slab is placed either directly on the subgrade or on a base course. The surface course of a flexible pavement (Fig. 14.29b) may be a layer of asphalt concrete (AC) from 2 to 6 in. in thickness, or it may be a quite thin bituminous dust palliative. The base course may consist of crushed stone, gravel, or soil mixed with binders such as asphalt or portland cement (stabilized soil). The subbase is usually composed of pit run material or "selected" soil from an adjacent road excavation or a borrow pit. Whereas a rigid pavement is a structure supported by the subgrade, a flexible pavement is considered only as a protective blanket of the subgrade which carries the loads directly. As time passes, the subgrade densifies under the action of the loads and vibrations. Deep, soft layers consolidate very little under the action of intermittent loads; a sizable consolidation can be produced only by large steady loads. Furthermore, total settlements computed according to the theory of consolidation cannot be fully realized in this case because of the limited life span of a pavement. The situation is quite different if an airport is built on a fill insufficiently compacted or on a soft, settling foundation, e.g., a deep swamp. In this case a heavy settlement of the fill and deterioration of the pavement are to be expected. Building airports on swamps or reclaimed lands is not an unusual condition, however, since desirable locations in the neighborhood of the cities are usually already taken for other purposes. Proper drainage and compaction are necessary in such cases.

14.16. Cross Sections and Joints of a Pavement. Figure 14.30a is an example of a highway cross section, and Fig. 14.30b is a cross section of a runway. As a matter of terminology, a *runway* is the pavement and its accessories where planes land and take off; a *taxiway* is used for service movements of the planes within the airport; the *apron* is the place where the engines of the planes are warmed up before the flight starts.

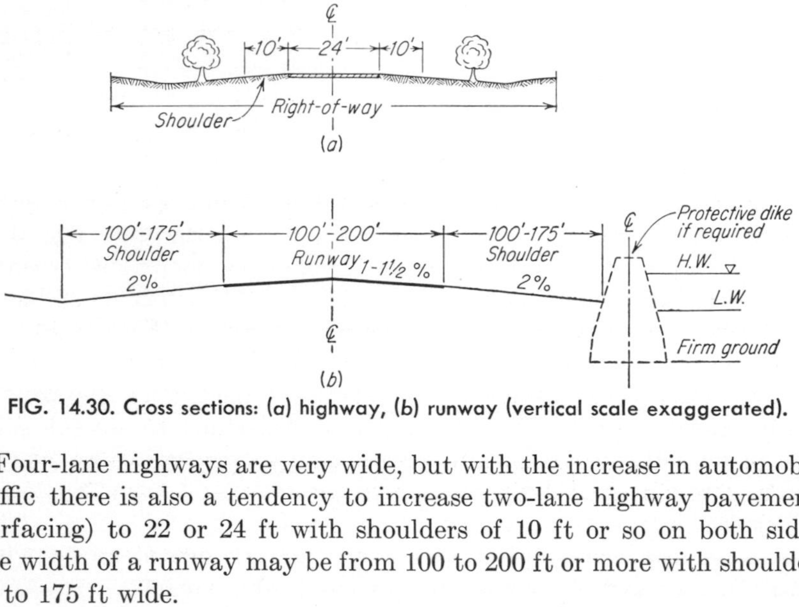

FIG. 14.30. Cross sections: (a) highway, (b) runway (vertical scale exaggerated).

Four-lane highways are very wide, but with the increase in automobile traffic there is also a tendency to increase two-lane highway pavements (surfacing) to 22 or 24 ft with shoulders of 10 ft or so on both sides. The width of a runway may be from 100 to 200 ft or more with shoulders up to 175 ft wide.

Under the action of temperature and moisture variations, concrete pavements tend to expand, contract, and warp. To prevent high stressing and cracking of the slabs, joints are arranged. There are transverse and longitudinal joints. Transverse joints are of two different types, namely, expansion and contraction joints. Construction joints are transverse joints placed at the end of a day's construction or in the case of some other interruption. All joints generally are provided with dowels or tie rods. The objectives of the latter are different for different types of joints, though all have to ensure the integrity of the pavement and a proper "load transfer" from one slab to the adjacent one.

14.17. Geotechnical Investigations for Pavements. *Field Work for Highways.* Highway personnel are often familiar with soil conditions in the locality where they work. Sometimes there are only four or five soil varieties in a region or even a state, and their properties and behavior are well known from years of practice. The investigations are reduced to making auger holes along the center line of the highway, widely spaced, and drilled to about 5 or 6 ft in depth. The soils then are inspected visually, without any laboratory work. More general is the case when soil samples are taken and laboratory tests are made. In contrast to the geotechnical investigations for buildings and bridges, soil samples for pavement investigations are mostly surface samples, taken only from shallow depths. Also the tests for which the samples are taken are different. Compressibility and shearing strength are rarely studied; pavement samples usually are tested only for mechanical analysis, consistency limits, moisture content, dry density, and expandability (in regions having expansive soils). Some additional tests are described at the end of this section.

It is convenient to use pedological maps in this work (Sec. 7.11). If the soil type is more or less the same for long distances and the locality is flat, the samples need be taken only every 1,000 ft or so. Obviously, if pedological maps or the samples indicate changes of soil types at shorter intervals, sample spacing should be reduced accordingly. Change in soil type may mean a design change, i.e., increase or decrease in thickness of the pavement, addition or omission of the subbase, etc.

Field Work for an Airport. For airport work, the layout plan and the grading plan (showing finished grades) should be consulted. Preliminary locations of the proposed bore holes should be plotted on this layout. Then a field examination should ascertain if all the proposed borings are necessary. Topography, including relief and drainage conditions, often has a considerable influence in the arrangement of the borings in plan.

As the boring operations progress, the field data should be gradually examined and, if necessary, additional bore holes requested or some of the proposed holes omitted. The borings should give enough information for tracing an accurate geological section of the upper layers of the area (at least to a depth of 15 ft) *along any direction* within the area. Old fills and old roads crossing the area should be spotted. Old fill may be a source of detrimental settlement, and old roads, because of their hard ridges, may prevent settlement of the new fill. Either problem can eventually produce a corrugation of the pavement. Ground water should be given attention in airport studies, and places where the water table tends to approach the ground surface should be spotted.

Figure 14.31 is an example of preliminary bore hole planning for a small airport with 100-ft-wide runways and 40-ft-wide taxiways. The length of the runways in this example is about 4,000 ft. In large military

airfields and modern civilian airports, this length may attain 10,000 ft
and more. As in the case of highways, the bore holes are drilled at the
center line of a narrow runway (Fig. 14.30); in wider runways, two bore
holes on each transverse section may be needed. The bore holes on the
runways (Fig. 14.31) are spaced 500 ft apart and on the taxiways even
more. On aprons and under the hangars, a more detailed investigation
of the subgrade should be planned, as there are vibrations from warm-
ing-up engines on the apron and heavy static loads, e.g., from columns
in the hangars.

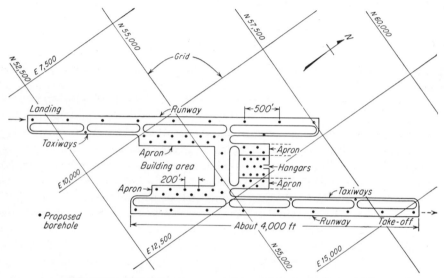

FIG. 14.31. Exploratory program for a small airport (note grid).

Investigations in the building area (Fig. 14.31)' are performed as is
usually done for building foundations (Chap. 13). In the modern airport, the building area is not crowded but provision is made for the
possible expansion of runways without demolishing the major buildings.
Hence it often is advisable to investigate the whole building area at once,
rather than only the areas under the proposed buildings, in order to have
geotechnical records for future developments.

Airport on the Shore Line of a Sea Bay. New runways and taxiways are to be constructed on the adjacent tidal lands (tidal flats). Geotechnical studies will be made
for a runway 10,000 ft long located approximately parallel to the shore line. The
bottom of the bay under the longitudinal axis of the proposed runway is assumed to
be practically horizontal. Assume its elevation at −1, the datum plane being the
level of the mean low tide. The elevation of the runway crest should be at the level
of the high tide and assumed at +6. As the runway has to be protected from high
tide by a dike (Fig. 14.30b, right) with a freeboard of about 3 ft, its top elevation will
be +9. Therefore the total height of the dike would be 9 + 1, or 10 ft. The dike

should encircle the runway and other extension arrangements and be continued upshore to meet elevation +9 at two points to form a closed space.   Assume that the bottom of the bay is covered with 15 ft of organic silt (or "mud") underlain by firmer clay, practically incompressible in comparison with the mud.   Assuming that the high tide and low tide each act half the time, the consolidation of the mud should be estimated.

Another basic problem in this case is the stability of the mud deposit against squeezing out, as controlled by its shearing strength.   The latter should be determined by the triaxial test in the laboratory and by the vane test in the field.   Both tests should check.   Shelby tube samples should be obtained for consolidation and shear tests by making four or five large-diameter borings (e.g., 4 in.) about 30 ft (or more if necessary) into the underlying clay.   A number of shallow borings only through the mud should be made for checking the tests.   The runway and the dike have to be built as hydraulic fills, and a good, preferably sandy borrow should be sought for in the same bay.   The dike should be based on firm clay; hence dredging to firm clay and disposal of the waste are two other problems to be solved in this example.

14.18. Special Field and Laboratory Tests for Pavements.   *Modulus of Subgrade Reaction.*   The supporting value of the subgrade soil is conventionally expressed by the modulus of subgrade reaction K, which is the ratio of unit load to the subgrade deflection.   Thus if a unit load of 10 psi caused a deflection of the subgrade of 0.05 in., the value of K is $10 \div 0.05 = 200$ psi.   This value corresponds to a satisfactory subgrade. Smaller K values indicate a rather weak subgrade.   By artificial means, such as compaction or stabilization with portland cement or bitumen, the value of K may be considerably increased.   The K value is determined in the field by means of a plate bearing test.   The plate usually is 30 in. in diameter, and the load is applied by jacking from a special heavy truck-trailer.

To design for a saturated condition of the subgrade, if required, undisturbed subgrade soil samples in natural and saturated condition are placed into two consolidation devices and compressed by identical unit loads.   The field K value is then decreased proportionally to the ratio of the two settlements observed.

California Bearing Ratio Test.   California bearing ratio (CBR) is used for expressing the properties of materials such as crusher run, pit run, gravel, etc., by a single number.   The larger the CBR, the better the material.   The test is described in Sec. 4.18.

A more complete description of the two tests in this section may be found in the publications of the U.S. Armed Forces,[26,27] also in Ref. 28 and in soil mechanics textbooks.   For the design of flexible pavements in California, the Hveem stabilometer is used.[28,29]

Expansive Soils.   Building highways on water-sensitive expansive soils often is a major road problem.   Expansive materials in the subgrade may be identified by the usual methods explained in Sec. 4.10, and the expansion or swell pressure should be measured.   If local soils are well known, simple sieve analysis and quick Atterberg tests sometimes permit identification of expansive materials without further testing.[25e]   On many occasions adverse soils covered with a bituminous blanket and a layer of imported, nonexpansive borrow material do not develop appreciable lifting.[25a]   On 27 California projects the thickness of such layers was from 4 to 18 in. but thicker layers may sometimes be needed.   In more severe cases stabilization of the subgrade is required, achieved, among other methods, by admixtures to the scarified and pulverized in-place native soil.   In Texas and other localities 1 to 3 per cent (by weight) of hydrated lime is used, with a maximum of perhaps 4 or 5 per cent.[25b,c]   Quicklime, an anhydrous lime product, and in recent years fly ash, a waste material from coal burning, also are used.   (Fly ash mixed with lime forms synthetic, roman cement.)   In California portland cement, 1 to $2\frac{1}{2}$ per cent by weight, is substituted for lime for economic reasons.[25d]

Damage to concrete pavements caused by expansive, nonstabilized subgrades consists in warping of the slabs (up to 0.8 ft, as reported in a Colorado publication[25e]), with longitudinal and transverse cracking. The variable moisture and temperature differential between the top and the bottom of the concrete slab causes the ends of slabs to rise off the subgrade in the night and to contact the subgrade in the daytime.   In addition to this cyclic movement, surface water may penetrate to the subgrade through open cracks and if the material of the subgrade is expansive, causes it to expand.   If the lifting is uniform throughout the length of the slab, the smoothness of the slab is not affected; but if the accretion of water is localized in the vicinity of cracks and joints, a long-lasting, *curled condition* results.   It may be relieved if the moisture content along the whole slab becomes uniform, but this is a slight probability.[25a]

In regions with prolonged dry periods special attention should be paid to using sufficient water in construction to prevent cracking of the drying surface and subsequent swelling during the rains.   The treated soil is compacted at its optimum moisture content (Sec. 16.11) or even at its liquid limit.

14.19. Moisture under Pavements.   Presence of water under pavements, i.e., by definition in the subgrade, has been observed in a large number of localities, from humid to semiarid regions with a negligible amount of precipitation.   For the measurement of moisture content in place, see Sec. 4.2.

Origin of Moisture under Pavements.   Water may enter the subgrade from above, below, or the sides.   Meteoric water, such as rainfall or melting snow, may flow down through joints and cracks of a concrete pavement.   In all kinds of pavements, the meteoric water may wet the shoulders and, from there, penetrate to the subgrade by horizontal capillarity. Water comes to the subgrade from the water table in the form of an upward flow, either of capillary moisture if the water table is close to the ground surface or of vapor if the water table is deep. In the frost areas additional water is lifted to the surface in winter (Sec. 10.3), freezes there, and melts and drains back in the spring. The water vapor from the water table may condense on the cold surface of the concrete slab and moisten the subgrade. This effect is considerably moderated, however, by subsequent evaporation. Moisture may be brought to the pavement from remote water bodies by air currents, often in a most unexpected way.

Several miles of pavement in the driest part of the desert, about 3 miles west of Salton Sea in California, failed so rapidly that it could not be repaired and had to be abandoned. A test hole some 20 or 30 ft in depth, put down a few feet from the edge of the pavement, disclosed not even a trace of moisture in the soil; at the same time, moisture accumulated under the pavement and reduced its supporting power to that of a thin liquid.[30]

Degree of Saturation and Plastic Limit of the Subgrade. In various observations throughout the United States made by the states and the U.S. Armed Forces,[31-34] the moisture content of subgrade materials was determined as usual in per cent by dry weight of the soil. This moisture content was then expressed in terms of the degree of saturation of the subgrade material (Sec. 4.7). Since in very dense materials a high degree of saturation does not always mean a flooded condition, the moisture content also was related to other soil characteristics, namely, the plastic limit and the standard optimum moisture content (Sec. 16.11). It should be recalled that clays offer the best engineering service at their plastic limit and that the supporting power of a clay subgrade is decreased when the moisture content exceeds the plastic limit.

Concrete and Flexible Pavements. Except in a few cases, the degree of saturation of the subgrade is higher under concrete pavements than under flexible ones. One of the reasons for this is the presence of joints and cracks in the concrete which, if improperly sealed, offer an easy path for meteoric water into the subgrade. Apparently there are other reasons that are not clearly understood.

Rainfall Variations. Under certain conditions, the degree of saturation of the subgrade materials may reach a value of 90 or even 100 per cent in regions with abundant precipitation. Such high values have been observed in arid and semiarid sections only exceptionally.

Seasonal Variations. The amount of moisture under pavements fluctuates during the year. In North Carolina studies,[33] high moisture contents were observed in the granular-type subgrades and in the bases from January through May. In the silt-clay subgrades, the high degrees of saturation were found from March to September, the highest value being in March (93.5 per cent as compared with an 88.7 per cent average for the year).   An analysis of the data from Illinois and Texas indicated the highest moisture in March and the lowest in October (see Ref. 33, discussion).   This information, though incomplete, indicates that spring precipitation strongly influences the moisture distribution under pavements.   In the frost areas and in frost-sensitive materials, the ice formed from water lifted up from deeper strata during the winter (Sec. 10.2) melts in the spring.   Added to the precipitation, this may cause a complete soaking of the subgrade and, in severe cases, a failure of the pavement.

Soil Variations.   In all cases, the maximum moisture accumulation was observed in silt and clay subgrades and the minimum in sandy subgrades.   The moisture accumulation in other soil varieties is between these two extremes.   The fine-grained soils such as clays exhibit a marked tendency to attain moisture contents in excess of their plastic limits.

Position of the Water Table.   A shallow water table, e.g., about 4 ft or less from the ground surface, contributes to the full saturation of clays or silts.

Moisture Content under Pavements and CBR.   The laboratory specimens for CBR design tests are soaked for 4 days in water prior to testing.   This requirement may be too severe for (1) sandy and gravelly subgrades everywhere and for other subgrades located in the low-rainfall zones and (2) subgrades in which sorption of moisture is accompanied by high densification from traffic.[34]

14.20. Pumping and Rutting.   If free water accumulates under a concrete pavement, the deflections of the slab under the action of heavy axle loads may agitate the accumulated free water and erode the subgrade material, particularly if the latter contains a considerable percentage of fines.   A thin suspension is then formed under the pavement which is ejected (pumped) through joints and cracks and along the edges of the pavement under traffic, especially if a heavy axle crosses a transverse joint at a limited speed.[35]   "Free water" in this case may be defined as water above the subgrade which is not able to enter the voids of the already fully saturated subgrade material.   This is predominantly the infiltrating meteoric water and water formed in the upper layers of the subgrade during the spring breakups (Sec. 14.18).

Soil Variations.   Pumping may be chiefly expected at the places and times of major moisture accumulation under pavements (Sec. 14.18). Hence as far as the soils constituting the subgrade are concerned, fine-grained plastic soils should be primarily considered as conducive to pumping.

It has been reported from various localities[35] that pumping did not occur on pavements placed on subgrade soils containing 55 per cent or more of the fraction retained on sieve No. 200 or on sand and gravel (Sec. 4.3).   From other sections it has been reported that pumping was confined largely, but not wholly, to fine-grained soils. It also has been observed that as the amount and the weight of the traffic increase, pumping spreads to less and less plastic soils.

Where so-called "hardpan" or "claypan" is being formed in subsurface horizons (Sec. 3.11), the upper strata may be naturally waterlogged at certain seasons. Particularly this is the case of the "old drift" deposited by earlier glaciations and exposed in southern Ohio, Indiana, Illinois, Iowa, Nebraska, and some parts of Kansas and Missouri. Addition of just a small amount of free water may cause pumping in these and similar cases.

Rutting. Excessive amounts of moisture soften the flexible pavements, which in this case are easily cut (rutted) by the passage of heavily loaded axles. The same phenomenon is observed on the shoulders of all pavements, both concrete and flexible.

Corrective Measures. In all cases of pumping and rutting, excessive moisture should be removed by drainage, using open or covered ditches (subdrains). If there are pervious deposits underlying the upper impervious layers, free water may be removed by vertical drains. In all cases of covered drains, the possibility of clogging by the muddy suspension should be considered.

When a new concrete pavement is designed, the chances of moisture accumulation and eventual pumping should be studied, and good geotechnical advice in this case is invaluable. Heavy pumping of existing pavements is often relieved by *mudjacking;* i.e., hot bituminous material or a slurry is forced under the slab after the muddy suspension has been removed by compressed air. Jacking up of the slab may accompany the operation.

REFERENCES

1. Steinman, D. B., and S. R. Watson: "Bridges and Their Builders," G. P. Putnam's Sons, New York, 1941.
2. Johnstone, Taylor F.: "Modern Bridge Construction," 2d ed., The Technical Press, Ltd., Kingston Hill, Surrey, England, 1951.
3. Smith, H. S.: "The World's Great Bridges," 2d ed., Phoenix House, Ltd., London, 1953.
4. Dunham, C. W.: "Foundations of Structures," McGraw-Hill Book Company, Inc., New York, 1950.
5. Jacoby, H. S., and R. P. Davis: "Foundations of Bridges and Buildings," 3d ed., McGraw-Hill Book Company, Inc., New York, 1941.
6. Hool, S. B., and W. S. Kinne: "Foundations, Abutments and Footings," 2d ed., McGraw-Hill Book Company, Inc., New York, 1943.
7. Feld, Jacob: Abutments of Small Highway Bridges, *Highway Research Board Proc.*, vols. 23, 24, 25, and 26, 1943–1946.
8. Terzaghi, K.: Failure of Bridge Piers due to Scour, *Proc. 1st Intern. Conf. on Soil Mech. and Foundation Eng.*, vol. II, Harvard, 1936.
9. Stewart, R. W.: Safe Foundation Depths for Bridges to Protect from Scour, *Civil Eng.*, vol. 9, 1939.

10. Lane, E. W., and W. M. Borland: Riverbed Scour During Floods, *Proc. ASCE Sep. 254*, August, 1953.

11. Harned, C. H.: Foundations for Highway Bridges and Separation Structures on Unconsolidated Sediment, "Applied Sedimentation," John Wiley & Sons, Inc., New York, 1950.

12. White, L., and E. A. Prentis: "Cofferdams," 2d ed., Columbia University Press, New York, 1950.

13. Viktorov, A. M., *et al.*: "Geology and Soils," Dorizdat, Moscow, 1947. Printed in Russian.

14. Nord, C. L., *et al.*: A Case of Settlement of a Bridge Pier, *Proc. 1st Conf. on Soil Mech. and Foundation Eng.*, vol. I, Harvard, 1936.

15. Engel, H. J.: Over 1,000 Feet of Continuity, *Eng. News-Record*, May 5, 1938.

16. George Washington Bridge issue, *ASCE Trans.*, vol. 97, 2d part, 1933.

17. Trask, P. D., and J. W. Rolston: Engineering Geology of San Francisco Bay, California, *Bull. GSA*, vol. 62, 1951.

18. Purcell, C. H., C. E. Andrew, and G. B. Woodruff: Deep Open Caissons for Bay Bridge, *Eng. News-Record*, vol. 113, 1934.

19. Proctor, C. S.: Foundation Design for Trans-Bay Bridge, *Civil Eng.*, vol. 4, 1934.

20. Report of the Hoover-Young San Francisco Bay Bridge Commission to the President of the United States and the Governor of the state of California, August, 1930.

21. Stocks, A. J.: Final Report, Foundation Investigations, San Francisco–Oakland Bridge, Department of Public Works, State of California (in the California state archives), 1932.

22. Lawson, A. C.: Geological Report on Foundations, San Francisco–Oakland Bridge (in the California state archives), 1932 and 1938.

23. Buwalda, J. P.: Geologic and Seismic Conditions Affecting San Francisco–Oakland Bridge (in the California state archives), 1938.

24. Modjeski and Masters, Engineers: Memphis and Arkansas Highway Bridge over the Mississippi River, Final Report and Supplement, submitted to Arkansas State Highway Commission and Department of Highways and Public Works, State of Tennessee, 1950.

25a. Various information from F. N. Hveem, Materials and Research Engineer, California Division of Highways.

25b. Lime Stabilization of Roads, *Bull. 323, Nat. Lime Assn.*, Washington, D.C., 1954. Contains bibliography.

25c. McDowell, Chester, and W. H. Moore: Improvement of Highway Subgrades and Flexible Bases by the Use of Hydrated Lime, *Proc. 2d Conf. on Soil Mech. and Foundation Eng.*, vol. V, Rotterdam, 1948.

25d. Zube, E.: Experimental Use of Lime for Treatment of Highway Base Courses, *Tech. Bull. 181, Am. Road Builders Assn.*, Washington, D.C., 1952.

25e. ASCE Committee on Denver Subsoils: "Borehole Data and Engineering Applications in Denver," Hotchkiss Mapping Co., Denver, Colo., 1954.

26. "Engineering Manual," Corps of Engineers, War Department, 1946.

27. Soil Mechanics and Earth Structures, *U.S. Navy Bureau of Yards and Docks*, TP-Pw-18, Oct. 1, 1953.

28. Horonjeff, R., and J. H. Jones: "The Design of Flexible and Rigid Pavements," Institute of Transportation and Traffic Engineering, University of California, 1950.

29. Hveem, F. N., and R. M. Carmany: The Factors Underlying the Rational Design of Pavements, *Highway Research Board Proc.*, vol. 28, 1948.

30. Personal communication from Mr. J. M. Lackay, District Engineer of the Asphalt Institute, Los Angeles, Calif.; also see *ASCE Trans.*, vol. 115, p. 574, 1950.

31. Kersten, M. S.: Progress Report of Special Project on Structural Design of Non-rigid Pavements, *Highway Research Board Proc.*, vol. 24, 1944, and vol. 25, 1945.

32. Palmer, L. A., and J. B. Thompson: Pavement Evaluation by Loading Tests at Naval and Marine Corps Air Stations, *Highway Research Board Proc.*, vol. 27, 1947.

33. Hicks, L. D.: Observations of Moisture Contents and Densities of Soil Type Bases and Their Subgrades, *Highway Research Board Proc.*, vol. 28, 1948. Discussion by M. S. Kersten.

34. Redus, J. F., and C. R. Foster: Moisture Conditions under Flexible Airfield Pavements, *Highway Research Board Spec. Rept.* 2, 1952.

35. Final Report of Project Committee 1, as Related to the Pumping Action of Slabs, *Highway Research Board Proc.*, vol. 28, 1948.

CHAPTER 15

MASONRY DAMS: GEOTECHNICAL STUDIES

Three features distinguish dams from other engineering structures: (1) unusual accumulation of large masses of building materials and water on a limited area of the earth's surface and, hence, exceedingly heavy pressures on the foundation; (2) destructive influence of the water in the reservoir on the foundation and on the structure itself which may cause leakage, erosion, or even failure of the structure; and (3) emplacement always in a valley. Consequently, dams depend on the environmental conditions, particularly on the geology of the site, more than other engineering structures, and in the case of large dams, adequate geologic studies are always made. Even in small dams, however, it sometimes is more economical to retain a geologist than to suffer the consequences of a dam failure due to unforeseen conditions.

The proper construction and maintenance of these types of structures are vital, as failures may and have caused hundreds of deaths and millions of dollars in major property damages. Thus, an additional distinguishing feature of dams is that unlike most other engineering structures, the failure of a dam may result in *severe* loss of life and property.

TERMINOLOGY AND DEFINITIONS

15.1. Classification of Dams. Dams are constructed for water storage for community and industrial use, for irrigation and flood control, for the development of hydroelectric power, for river canalization, and for silt or debris control. A dam that serves more than one such purpose is called a *multipurpose dam*. A *diversion dam* is one which serves primarily for diverting water from the river. Dams also are classified according to the material from which they are constructed. Thus there are masonry, earth, and rock-fill dams. (The last two are discussed in Chap. 16.) Dams are rarely built of steel or timber. Timber generally is used only if the dam is temporary, and steel occasionally is used as the upstream face of a rock-fill dam. At the present time, most large masonry dams in the United States are built of concrete (some foreign dams still are being built of large blocks of cut stone). In this chapter, the terms "masonry dam" and "concrete dam" are used synonymously.

The type of dam to be used in a given locality often is a matter for considerable study. *Safety* is the first consideration, then cost, which involves both the initial expenditure and the annual maintenance. Safety considerations require that the foundation and abutments be adequate for the type of dam selected.   Cost studies must, among other factors, take into account the construction treatment of the foundations, the suitability of the topography, and the availability of construction materials.

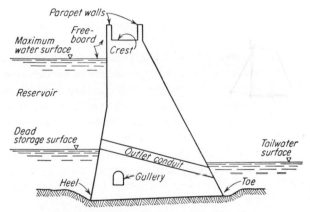

FIG. 15.1. Schematic cross section of a dam.

15.2. Terminology.   Some of the more common terms used to describe certain portions of a dam (Fig. 15.1) are as follows:

Abutments.   Either the sloping sides of the valley upon which the dam is built or the actual part of the dam that rests on this portion of the valley. The geologist generally uses the word "abutment" in the former sense, whereas the engineer commonly uses the latter definition, by analogy with bridge abutments.

River or Channel Section.   The center portion of the dam that directly overlies the river channel or that portion of the valley that is so situated.

Heel of the Dam.   The upstream portion of the dam where it contacts the bearing surface (i.e., the ground or rock foundation).

Toe of the Dam.   The downstream portion of the dam where it contacts the bearing surface.

Crest.   The top of the dam.   If walls are placed along the top of the dam to afford safety to a road or walkway, these walls commonly are called parapet walls.

Freeboard.   The distance between the highest level of water in the reservoir and the top of the dam.

Axis of the Dam. An arbitrary imaginary line drawn either along the exact center of the plan of the crest or along the contact between the upstream part of the crest with the upstream face of the dam. Actually this line is merely a reference line and practically can be established at almost any position on the dam.

Dam Cross Section. Usually drawn on a vertical plane that is perpendicular (normal) to the dam axis.

Galleries. Formed openings within the dam. They may run either transversely or longitudinally and may be level or have a sloping grade. They provide means for draining water seeping through the face or the foundation, act as openings to drill grout and drainage holes, and provide access to equipment within the dam and for observing its performance.

Dead-storage Water Surface. The elevation of the reservoir below which water stays permanently in the reservoir and cannot be withdrawn. Also includes the *silt storage*, which is that portion of the reservoir basin reserved for storing any silt which may enter and be deposited. The silt storage space should be properly considered in all computations for determining the actual amount of water to be stored. If, for instance, a reservoir is to hold 100,000 acre-ft* of water and the expected silt accumulations are 10,000 acre-ft, the dam must be high enough to store 110,000 acre-ft.

Tail Water. Water at the downstream base of the dam resulting from backup of water discharged through the spillway, outlet works, or powerhouse. The term sometimes is used to refer to the extreme upper end of the reservoir, i.e., the "reservoir tail water."

Minimum Water Surface. The lowest elevation to which the reservoir can be lowered and water still withdrawn by means of the outlet works (see Sec. 15.7) or powerhouse penstocks (see Sec. 15.7).

Maximum Water Surface. The highest elevation at which water can be stored in the reservoir without overtopping the dam or being released through the spillway.

TYPES AND APPURTENANCES OF DAMS

15.3. Gravity Dams. A gravity dam is constructed of concrete or solid masonry. Its axis may be a straight line (Fig. 15.2), slightly curved upstream (Fig. 15.3), or a combination of curves and straight lines to take the best advantage of the topographic conditions. Its cross section usually is roughly trapezoidal, approaching a triangle (Fig. 15.4). Although sound rock is desirable for the foundations of gravity dams, they have been built successfully on badly fractured, variable rock and

* One acre-ft is the volume of water covering the area of 1 acre to a depth of 1 ft.

FIG. 15.2. Straight gravity dam with controlled overflow spillway. Note outlet works on both upper left and right abutments and in channel section. Friant Dam, California. (*USBR photograph.*)

FIG. 15.3. Curved gravity dam. Note penstocks to powerhouse at upper left center. Shasta Dam, California. (*USBR photograph.*)

even on river fill. If the topography presents a wide canyon with gentle slopes, a gravity-type dam may prove the most economical as well as the most feasible type of dam.

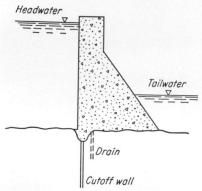

FIG. 15.4. Schematic cross section of a gravity dam.

15.4. Buttress Dams. A buttress dam is composed principally of (1) a sloping upstream deck of reinforced concrete that takes up the water load and (2) buttresses (vertical walls) with their axis normal to the plane of the slab that support the slab and transmit the water load to the foundation (Fig. 15.5). The buttresses may be single-wall (Fig. 15.6a), hollow, or double-wall (Fig. 15.6c). There are several types of buttress dams: (1) the flat-slab type wherein the concrete deck slab spans the dis-

FIG. 15.5. Slab and buttress dam, Stony Gorge Dam. (*USBR photograph.*)

tance between adjacent buttresses (Fig. 15.6a), (2) the multiple-arch type (Fig. 15.6b) wherein each unit of the water-supporting member is an

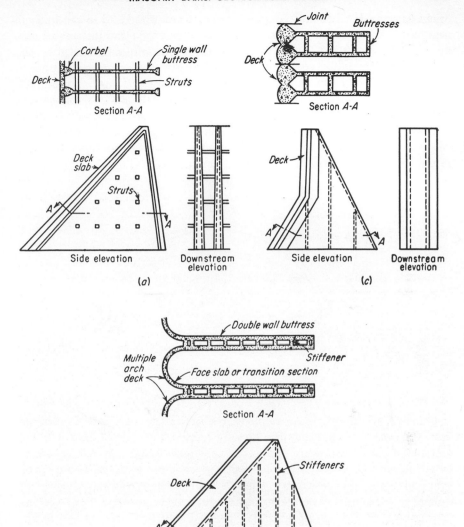

FIG. 15.6. Types of buttress dams.[4]

inclined arch barrel supported by the buttresses, and (3) the massive-head type (Fig. 15.6c) in which the water-supporting member is merely an enlargement of the upstream head of the buttresses. Various modifications of these types have been built.

The buttress dam may prove more economical than other types because of the comparatively small volume of concrete required and rather limited foundation excavation yardage.   The space between the buttresses can be economically used for locating outlet works and powerhouses.   In localities where labor costs are high, the buttress dam may be uneconomical because of the considerable amount of concrete forming required.

In the case of a site located in a wide canyon with gently sloping walls and where conditions require an outlet works through the dam, the buttress-type dam may be the most economical type.   If under similar site conditions an overflow spillway (Sec. 15.6) is required, the buttress dam might prove more economical than an earth or rockfill dam.

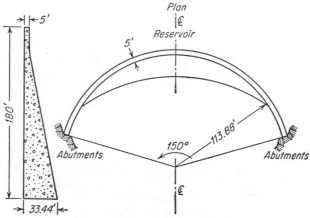

FIG. 15.7. Example of a constant-radius arch dam.   (After Justin, Creager, and Hinds.[2])

Usually the buttress dam requires good foundations.   The buttresses are rather narrow and act like heavily loaded walls, thus exerting tremendous unit pressures on the underlying foundation.   Since the space between the buttresses is not loaded, the possibility exists that in poor materials, the buttress may "punch" into the ground, causing an upheaval of material between buttresses.   In a gravity dam, on the other hand, such a shear failure is not so likely to occur, as the entire base of the dam confines the underlying beds.

15.5. Arch Dams.   An *arch dam* is composed of a single concrete wall curved in plan with its convex face upstream, and a part of the water load is transmitted to the adjacent rock abutments by the arch action. In all curved dams, part of the water load is imposed directly upon the foundation by being transmitted through the mass of the dam and part is transmitted to the abutments by the arch action.   If this distribution of load is approximately equal, then the dam is considered as a gravity-arch or arch-gravity dam.   Where a considerable proportion of the load is transmitted by arch action to the abutments, the dam may be referred to as a thin-arch dam (Fig. 15.7).

In order that the arch action be really effective, the abutments must be as immovable as possible.   As explained in Sec. 15.17, the abutment rock should satisfy certain rigorous requirements and the arch itself be well keyed into the abutments.   Existing arch dams may have a series of rock steps in the abutments, which fact may cause stress concentration and cracking in the concrete.   Modern dam designers believe, therefore, that effective keying action may be accomplished merely by the thrust of the arch and by cantilever action of the dam mass.

15.6. Spillways.   This term is applied to a dam appurtenance which permits water to pass over or around the dam.   The *spillway* is a concrete structure that conveys floodwater from the valley upstream from the dam to the valley downstream without damaging the dam or reservoir walls or eroding the foundation or toe of the dam.   Usually the spillway starts to operate when water in the reservoir rises over the maximum water surface.   In the simplest type of spillway, the water is permitted to flow over the crest of the dam (overflow dams and weirs).   The crest may even be submerged (Fig. 15.8), or the spillway may

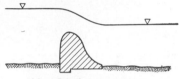

FIG. 15.8. Submerged dam.

be designed as an opening in the crest of the dam, provided with abutments and piers similar to a bridge.

Crest gates of different types permit opening and closing the spillway opening completely or partially.   Such gates may be automatic, in which case by means of a float system, the gates automatically open when the water reaches a certain elevation, or they may be manually raised and closed.   In small dams, the spillway sometimes is closed by *stop logs* or *flashboards*, which are merely large planks that are manually dropped into slots in concrete piers across the spillway crest.   An *uncontrolled* spillway is one in which gates are not used, and in a *controlled* spillway gates are used.   A controlled overflow spillway is shown in Fig. 15.9.

Types of Spillways.   A spillway may be designed in the form of an open channel of rectangular or trapezoidal cross section that permits direct flow of the water from the reservoir to the downstream valley.   A spillway of this type usually is called a *normal* spillway or *normal control* spillway and may consist of a short channel followed by a steep chute made of concrete or paved (Fig. 15.10).   The term *side-channel* spillway designates structures wherein the water flow passes over a weir that is perpendicular or at an acute angle to the dam axis and is carried past the dam by an open channel or tunnel running practically normal to the dam axis (Fig. 15.11).   In a *shaft* spillway, more often termed *morning-glory* or *glory-hole* spillway (Figs. 15.12 and 15.13), the excess of water in the reservoir drops vertically or obliquely into a funnel and is conducted downstream more or less horizontally in a concrete pipe, generally under the body of the dam.   The lip of the funnel is at the maximum water surface in the reservoir.   The *ski-jump spillway* has a projecting lip on the face of the dam (used on concrete dams only).   This lip causes the water to "jump" into the air and land a safe distance downstream from

FIG. 15.9. Controlled overflow spillway, Olympus Dam, Colorado.   (*USBR photograph.*)

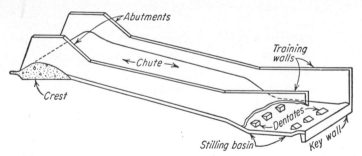

FIG. 15.10. Chute or normal spillway.   (Compare with Fig. 15.14.)

the toe of the dam, thus protecting the toe against erosion and also dissipating the water energy in the air.

Parts of a Normal Spillway.   These are shown in Fig. 15.14, a normal spillway under construction.   The *crest* of the spillway is its highest part over which water flows.   The spillway *chute* is the downstream portion of the spillway over which water flows from the crest to the stilling basin (see below).   The concrete walls on either side of the spillway chute are referred to as *training walls*.   The training walls upstream from the crest are sometimes referred to as the *intake training walls*.   The *stilling basin* is a deep basin located at the bottom of the chute and is designed to reduce the turbulence of the spillway flow (and its velocity) before it enters the river channel below the dam.   This basin sometimes is referred to as the spillway "bucket."   The turbulence and energy of the spilling water sometimes are further dissipated by means of large, toothlike projections of concrete referred to as "dentates" or "energy dissipators."

FIG. 15.11. Side-channel spillway, Taylor Park Dam, Colorado.   (*USBR photograph.*)

Emergency Spillways.   This type of spillway is designed for extreme flood conditions when the normal spillway capacity might be exceeded. For an emergency spillway, a natural topographic saddle, if any, in the reservoir rim may be used, or an excavation, sometimes with a low weir across its crest, is made in the rim.   None of the other normal spillway appurtenances (chute, walls, etc.) are constructed.   Considerable damage to the foundation of an emergency spillway can be expected, and if the foundation rock is not resistant enough, it will be deeply scoured and the reservoir partially or fully emptied.   Considerable damage to the downstream valley may thus be caused, as in the case of a regular dam failure.

FIG. 15.12. Glory-hole spillway, Hungry Horse Dam, Montana. *(USBR photograph.)*

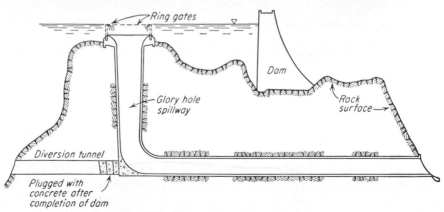

FIG. 15.13. Glory-hole spillway cross section.

15.7. Outlet Works and Penstocks. There are primarily two types of outlets: (1) the *tunnel* type wherein the water is carried in a lined or unlined tunnel through the abutments of the dam and (2) the conduit type wherein the water is carried in a pipe through or under the dam. The outlet works are used in irrigation dams to divert some of the reservoir water into irrigation canals, municipal water supply distribution systems, and hydroelectric plants for power generation and for other uses.

FIG. 15.14. Chute spillway under construction, Cachuma Dam, California. (Compare with Fig. 15.10.) (*USBR photograph.*)

Penstocks. These are steel pipes of large diameters crossing the body of the dam for the purpose of conducting the reservoir water to the hydroelectric generators of a powerhouse (see Fig. 15.3). The latter may be located either just at the toe of the dam or a certain distance downstream, according to topographic or economic considerations.

PROBLEMS AND FAILURES

15.8. Forces Acting on a Masonry Dam.[2-4] Forces acting on a masonry dam during and after construction may be static or dynamic. *Vertical static forces* acting downward are the weight of the concrete and steel reinforcement and of the bridges, gates, and other superimposed features and the weight of the water and sediment (silt) that settles from the water and is deposited over the sloping faces of the dam (*AB* and *CD*, Figs. 15.15 and 15.16). A large part of the dam is submerged, and thus the buoyancy force, or uplift equal to the weight of the displaced water, acts upward and decreases the sum of the static forces acting downward. Uplift is discussed in more detail in Sec. 15.12.

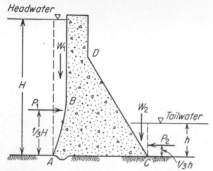

FIG. 15.15. Water loads on a gravity dam (W_1 and W_2 are weights of water; P_1 and P_2 are lateral water pressures).

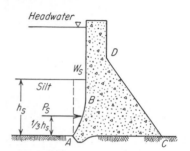

FIG. 15.16. Silt loads on a gravity dam (W_s is weight of silt; P_s is lateral pressure exerted by silt).

Horizontal forces acting on the dam are primarily due to the lateral pressure of water and silt deposited in the reservoir behind the dam and (in winter) occasionally due to the expansion of ice cover in the reservoir during temperature rises.[5] Because of the plastic flow of ice pressing against a dam, the strain is relieved and the ice pressure is smaller than one might expect it to be from the shearing strength of ice, though damage caused by ice pressure has been recorded.

Pore pressure (Sec. 4.16) is a variable static force acting in all directions, which often decreases vertical forces with detrimental effect. It is operative both in the body of the dam and in its foundation. It is recommended that the reader becomes familiar with Ref. 6 dealing with this subject.

Temperature changes may cause additional stresses in masonry; the temperature stresses may be particularly detrimental in arch dams and therefore are considered in design. The most important *dynamic forces* acting on a dam are wave action, overflow of water, shocks, and seismicity, which is discussed in Chap. 18.

15.9. Sliding Failures of Masonry Dams. Horizontal forces tend to push the dam downstream and, if excessive, may cause the dam to slide. Numerous cases of masonry dam failures are given in Ref. 7a and b; however, a few characteristic sliding failures are discussed here.

Bouzey Dam, France. This dam, 72 ft high, failed in 1895 after 15 years of service, when the upper part of the dam slid on the lower part which remained in place.[8]

According to modern standards, this was a poorly constructed gravity dam built of rubble masonry placed on sandstone. However, a considerable number of better constructed gravity dams that followed it, failed.

Austin, Texas, Dam. This dam, 66 ft high, was built on the Colorado River of Texas in 1892. The foundation was limestone, porous, and soluble. The strata were jointed and faulted; during construction a fault 75 ft wide was disclosed. A break in the dam occurred in 1893 and was repaired. It appears that the toe of the dam was gradually undermined until the dam slid downstream on its limestone foundation[9] on Apr. 7, 1900, after a prolonged heavy rain. Figure 15.17a shows in plan the initial and final position of the broken part of the dam. Solid lines between B' and D' denote the parts that remained standing after failure.

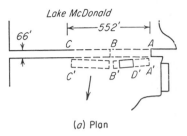

(a) Plan

Austin, Pennsylvania, Dam. This dam was 50 ft high and built in 1909–1910. It started to show failure signs even in January, 1910, in the form of slipping and cracking. The failure occurred finally by sliding *within* the sandstone foundation. The sandstone, though hard, consisted of thin, loosely adherent strata with inclusions of shale and gravel.[10,11] The dam broke into seven large segments on Sept. 30, 1911.

St. Francis Dam,[11-15] near Saugus, California, was built on San Francisquito Creek. This was a solid gravity dam, curved on a radius of 500 ft, the crest being 16 ft thick, the base 175 ft thick, and the height of the dam 205 ft. Storage of water began on Mar. 1, 1926, and the failure occurred on Mar. 12, 1928. Leakage through the

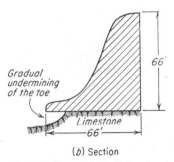

(b) Section

FIG. 15.17. Failure of the Austin, Texas, Dam. (ABC = original position; A'B'C' = position after failure.)

dam, especially through the foundation, preceded the failure. The bottom and one slope of the canyon at the damsite were underlain by laminated mica schist, whereas the opposite slope was underlain by a reddish conglomerate. The contact between the two rock types was along a fault. The dam was placed astride the fault, partly on schist, partly on conglomerate. The latter is very hard when dry but in water disintegrates into loose sand and small rock fragments. Failure occurred near the fault and basically was due to the softening and disintegration of the conglomerate by percolating water. Figure 15.18 shows how the dam was broken. A water flow up to 125 ft in depth rushed down the valley below the failed dam, causing the death of more than 236 persons and several million dollars' property damage. Concrete blocks from the dam were carried as far as 3,000 ft downstream by the water. Examination of the strewn concrete blocks showed that the dam itself was built properly of good materials.

Lake Gleno Dam, Italy,[16] a combination of multiple arch and gravity, near Bergamo, failed Dec. 1, 1923, after heavy rains. The dam was 143 ft high and 863 ft long. The central part that failed had a gravity base of stone masonry surmounted by multiple arches of reinforced concrete. The rock foundation dipped downstream, and this fact, together with some construction deficiencies, caused sliding. About 600 persons lost their lives because of this failure.

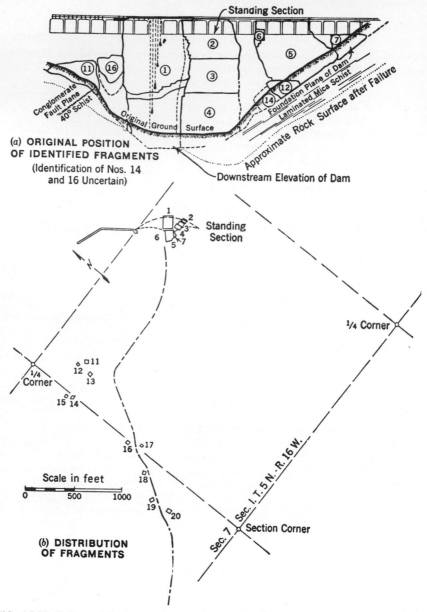

FIG. 15.18. Failure of St. Francis Dam in California. (From *ASCE Trans.*, vol. 94, 1929.)

15.10. Discussion of the Sliding Problem. In Fig. 15.19, symbols W and P stand for the sums of all vertical and horizontal forces, respectively. Force P tends to displace the dam horizontally and to make it slide. The resultant of forces W and P (dotted line and designation R in Fig. 15.19) intersects the base of the foundation at a distance e from the vertical force W, the value of e being the *eccentricity* of applied forces. The trapezoidal pressure diagram underneath the base AC of the dam indicates that when the reservoir is full, the toe C of the dam is overloaded and the heel A relieved.

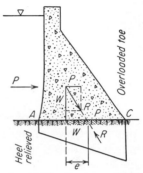

Since the forces W and P undergo changes both during and after construction, the designer seeks the most unfavorable combination when P is at a maximum and W at a minimum. The smaller the ratio P/W, the more stable is the dam. In any event the value of this ratio should be less than the value of the coefficient of friction f of the dam on its foundation. Engineers assume that the value of the coefficient of friction for masonry on masonry (and approximately for masonry on "good" rock) is between 0.6 and 0.7.[2]

FIG. 15.19. Pressure distribution at the base of a masonry dam (case of filled reservoir).

According to the state of the rock (e.g., in seamy rock) this value of f may have to be decreased. For dams on gravel or sand foundations the value of f would be 0.5 or 0.4.

A study of dam failures (Sec. 15.9) shows that sliding occurs usually in the masonry dams of the gravity type. The sliding (or shearing) surface may be located within the masonry as it was in the Bouzey Dam; at the contact of the dam with its foundation as in the Austin, Texas, Dam; or within the rock foundation as in the Austin, Pennsylvania, Dam. At the present time, concrete dams are built of excellent materials and correct building procedures are used. A failure of the masonry in shear, as at Bouzey Dam, is highly improbable. Sliding tendency at the base of the dam, i.e., at its contact surface with the rock foundation is, however, pronounced in the case of slippery rocks such as some shales and tuffs. These rock types may contain montmorillonite, which greatly limits the value of the coefficient of friction f between masonry and rock. To prevent sliding at the base of the dam, proper bond between the dam and the rock should be ensured. The rock surface should be roughened, and keying the structure some distance into the rock may be advisable. This can be done by building a key wall or by providing a cutoff wall at the heel of the dam (Fig. 15.4).

A cutoff wall is constructed primarily for the purpose of decreasing the flow of water under the base of the dam but may simultaneously serve the purpose of keying the dam into the underlying rock. This may be either a sheet-piling row or a concrete wall penetrating deeper into the rock than a key wall. A key wall, on the other hand, is constructed primarily for the purpose of reducing the sliding factor of the foundation and may only incidentally serve as a water cutoff wall. A key wall is built by excavating a trench at the heel of the dam and pouring concrete in it simultaneously with the concrete work of the dam. Key walls are often built under other types of structures, such as power plants, spillways, or simple retaining walls, if these structures are subject to sliding. In some cases the key wall may be located at the downstream end of the structure (Fig. 15.10). Key walls are not essential in earth dams, where the heavy weight of the structure provides a component efficiently opposing sliding.

Another method of reducing the sliding factor is to give a downward slope to the base of the dam in the upstream direction of the valley rather than in the downstream direction, which probably contributed to the failure of the Gleno Dam. In all cases it is the duty of the engineering geologist to study carefully the rock under the dam and appraise the possibility of sliding on or within the foundation. Special attention should be given to stratified rocks, such as shales, schists, siltstones, claystones, and some slates and clayey sandstones. Basalts that are likely to contain interflow zones of ash also may be critical in this regard.

Some idea of the value of the coefficient of friction of masonry on the foundation rock may be obtained from a sliding test. Two samples of concrete and rock are pressed together by a certain known load, and the force required to make them slide with respect to each other is measured. In some cases, a soil shear machine may be used for this test, although a field test is preferable.

At Harlan County Dam, constructed in Nebraska by the Corps of Engineers, a field sliding test was performed to determine the friction characteristics of the bentonite layers in the chalky shale foundation. The design required that some portions of the massive concrete spillway rest on these slick layers. A trench was excavated into bedrock until a block of shale with bentonite layers was exposed. A concrete cap was placed on the shale in one case and on a bentonite layer in another case. The blocks were loaded vertically, and horizontal thrusts were applied to the blocks with the use of hydraulic jacks. The coefficients of friction of the concrete cap on bentonite, the concrete cap on shale, and the shale on bentonite were determined.

15.11. Sliding Problems: Action of Water in Stratified Rocks. Most of the dam failures recorded (Sec. 15.9) took place after heavy rains. Storm water is a serious contributing factor to dam sliding failures (as it also is in many landslides, Chap. 17). The action of water in stratified rock in dam foundations is threefold:

1. Water lubricates dry rock surfaces and decreases the dry coefficient of friction between them (except, apparently, with quartz). In moist

rocks this lubricating effect is produced, however, even before a storm, since thin moisture films are always present in moist foundation rock, and there is no additional effect from the storm water.

2. Water moving between strata not only dissolves the cementing material but erodes it by a purely mechanical action, thus increasing the volume of openings and forming caverns.

3. If the interstices between the strata are saturated, i.e., completely filled with water, the latter stands under hydrostatic pressure directly proportional to the hydraulic head, or vertical distance to the water level in the reservoir.   As this level rises during and after a rainstorm, the pore pressure increases accordingly.   The increasing pore pressure tends to lift the overlying rock strata and the dam itself, thus decreasing the shearing strength of the rock, i.e., the resistance to sliding inside the rock, though the value of the coefficient of friction at the base of the dam may not be altered.   In brief, during and after a rainstorm, a sliding failure may develop within a stratified rock foundation through a *detrimental increase of pore pressure* accompanied by *solution and erosion of cementing material.* Critical strata should be located and removed or solidified, or their detrimental

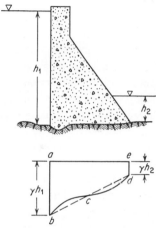

FIG. 15.20. Uplift: simple case.

effects decreased by a key wall, and access of storm water to the dam foundation should be prevented as far as possible.

15.12. Uplift Problem.[17-19]   Uplift is a force acting against the weight of the dam and caused by the hydraulic head of water at the upstream and downstream faces of the dam (h_1 and h_2, respectively, in Fig. 15.20). As a matter of terminology, though the uplift is measured in terms of stress, it should be distinguished from the "pore pressure" (or more accurately "pore water pressure") in the earth and rock material under the base of the dam and in the body of the dam itself.   In the simplest case, it is assumed that the difference in hydraulic heads h_1 and h_2 is dissipated uniformly (linearly) between the heel and the toe of the dam.   The actual distribution of the uplift force at the base of the dam has been measured in many instances and has not proved to be exactly linear.   Examples may be found in Ref. 19 together with descriptions of measuring devices. Various views on the distribution of the uplift force and the area on which it acts may be found in the discussion of Ref. 19.

Many share Terzaghi's view[17] that the uplift acts on the whole area of the base (Fig. 15.20) independently of the porosity of the underlying rock or soil material. In such a case, the total uplift force would be approximately represented by the trapezoid *abde*, in which the ordinate *ab* and *de* equal γh_1 and γh_2, γ being the unit weight of water. Some designers[19] assume, however, that the uplift acts over two-thirds of the base area.

15.13. Settlement and Rebound Problems. Under the action of its weight and other vertical forces (or vertical components of forces) imposed upon it, the dam settles. Filling the reservoir causes an additional settlement of the dam. In the case of extremely large reservoirs, the weight of the reservoir water and silt also may cause the entire region surrounding the reservoir to subside, e.g., in the Hoover Dam area in Nevada. It should be noted that any settlement consists of an elastic (reversible) part and a so-called plastic (irreversible) part. The latter constitutes a small, progressive sinking of the dam into the underlying material and is absent in the relatively old dams in which this small amount of sinking is already completed.

The settlement problem is simple if the foundation rock is sound and strong, such as only slightly fractured and jointed granite. In dealing with weaker rock types, especially clayey in character, such as soft shales, claystones, or siltstones, the question of deformation under heavy load becomes critical. Some idea of the total initial settlement of a dam, i.e., combined elastic and plastic deformation, may be obtained from a compression test on the given rock and from the comparison of the results of the compression tests with analogous tests on rock foundation materials of other dams. Field loading tests on relatively large areas may be helpful in these cases. The settlement of a dam may be increased by the presence or formation of cavities under the existing structure, for instance by removal of salt and gypsum beds by solution.

When a part of the load acting on the rock under the dam is removed, the dam will move slightly upward. This upward action is called *elastic rebound* or simply "rebound" (Sec. 2.11). The value of the rebound depends on the modulus of elasticity of the rock. The larger the modulus of elasticity, the smaller the rebound. In the case of dams, the rebound problem may be serious if, during construction, a thick layer of unreliable rock material is removed to prevent excessive settlement of, or leakage under, a dam. In this way the pressure on the underlying rock is relieved. The rebound process in rocks generally requires considerable time for its completion but may start after the dam is built if the expansive pressure developed by the rock foundation is larger than the weight of the structure. If the excavated rock has been exposed for a certain time, it may swell, and a structure built on that foundation will settle more than expected. Elastic rebound often occurs in many types of shale, and deep spillway cuts in such materials may cause buckling of the spillway slabs. Nonuniform rebound may be responsible for **differ-**

ential settlement of the whole structure. Rebound on a smaller scale occurs after each emptying of the reservoir. When an empty reservoir is refilled, the vertical reaction W (Fig. 15.19) moves in the downstream direction for a distance equal to the eccentricity e and becomes the oblique reaction R. The heel of the dam becomes relieved, and the toe overloaded. There is rebound at the heel and settlement at the toe.

Of particular note in studying settlement and rebound are the difficulties encountered when the dam is placed on varying rock types. If a portion of the foundation were to overlie a hard sandstone and another portion a soft shale, differential settlement and rebound could be expected. The internal stresses thus imposed on the structure could be disastrous if not accounted for by the design. If the dam cannot be located on one type of rock only, the designers should be aware of the possible variances in the moduli of elasticity (compare also Sec. 15.17).

15.14. Reservoir Problems: Seepage and Leakage. When water is held back by a membrane, such as a masonry dam, there is a permanent movement of water from the reservoir under and around the membrane and at the rims of the reservoir. This normal phenomenon is *seepage*, which is to be distinguished from *leakage*. The latter is the abnormally large escape of water from the reservoir through the fissures and openings in the rock. In appraising the leakage possibilities, attention should be paid to the dam itself and its reservoir, since leakage in the reservoir may be more serious than under or around the dam. In the study of leakage possibilities, attention should be given to the pattern of joints, seams, and vertical and oblique channels; their frequency and degree of openness; together with general topography and geology of the locality.

It is not uncommon to find "buried" channels in the reservoir rim. Such channels are deeply eroded surfaces in the bedrock due to ancient streams that were subsequently backfilled with what may be highly impervious material. In glacial terrain, such channels should always be searched for, particularly as surface topography may give little or no clue to their subsurface location. It is very common to find such buried channels in the Great Plains of the Middle West in the United States. Their presence usually is due to the vagaries of ancient streams caused by glacial outwash waters. In some glaciated terrain, buried channels are formed by advances and retreats of the glacier, and present stream channels may bear little or no relation to such glaciation effects. Buried channels also can be formed as a result of ancient stream piracy (Sec. 3.6) when, for geologic or topographic reasons, the flow of water in two channels was diverted into one channel and the dried up channel was backfilled by soil materials brought by occasional floods and surface erosion, leaving little or no topographic evidence at the surface. If the bedrock surface in a buried channel filled with pervious materials is below the maximum water surface of the reservoir, measures against possible leakage may be required.

Leakage often may be due to the solubility of the dam foundation or reservoir material, such as limestone, gypsum, or salt rock.   These rocks are especially noted for their solubility under long-time water action, particularly if the water contains carbon dioxide.   If this solvent action occurs previous to the construction of the reservoir (and this is usually the general case), it results in the formation of caverns, underground channels, and extensive interconnected fissures in the rock.   Very often, the presence of the underground open channels is not readily apparent on the surface and may even be undetected during drilling at the damsite. Furthermore, the openings may have been sealed with clay or similar impervious but soluble material.   Once the reservoir is filled, however, the clay materials will be gradually washed out, and as a result the reservoir will not hold water.   Various methods used to seal such openings are rarely economical unless the defective rocks are under the body of the dam (and not in the reservoir) and have a reasonable thickness or are confined by impervious rocks on both sides of the reservoir basin.   The most efficient methods involve grouting (see Sec. 15.19).

The nonsoluble rocks such as granite or sandstones usually do not present serious leakage problems unless severely fissured.   Igneous or metamorphic rocks that are badly broken, intensely jointed, or severely faulted should, however, be closely investigated as to their leakage characteristics.   Such investigations would include pressure testing in drill holes to determine water losses and the excavation of large-diameter shafts or drifts in order to observe the fissure characteristics of the rock directly. Fortunately such rocks usually are amenable to grouting without great difficulties.   Some of the rocks in this category may include sandstones that have a highly pervious structure such as the St. Peter and Dakota sandstones.   Grouting in these rock types may not be entirely successful, as the grout will not permeate the tiny pore spaces (although water under high head will).   Some relief can be obtained, however, by constructing deep cutoff walls under the dam and by placing some dependence on the silting action in the reservoir to seal the openings.

A rock classification of great concern when encountered in a potential reservoir is basalt, including both the light- and dark-colored varieties (or lava flows).   Because of the rapid cooling of these rocks during their formation, there are extensive cracks and fracture patterns in them, which are rarely sealed with residual materials.   Natural tunnels in basalts may be miles long and up to 30 ft or more in diameter.   One of the greatest difficulties in analyzing the leakage properties of a prospective reservoir in basalt is that the water pressure tests in drill holes (as described hereafter) may be completely misleading.   One drill hole may prove very tight when subjected to high water pressures, whereas another hole only a few feet away may show very high water losses. The absence of defects at the surface is not a proof of tightness of the rock. Because of this uncertainty, grouting of such materials may become difficult and require a great number of grouting holes.

15.15. Water Pressure Tests in Drill Holes. During the drilling period for a prospective dam, it is advisable to run pressure tests in the drill holes. The objective of these tests is to obtain at least a qualitative estimate of possible water losses in the reservoir (Sec. 5.17) and a concept of the amount and pattern of grouting that will be required. A rule of thumb for the amount of pressure to be used in the tests is as follows: The maximum amount of pressure in pounds per square inch applied at each test should be equivalent to 0.4 times the expected height of the dam. Thus, in the case of a 200-ft dam, the pressures should be raised up to 80 psi. It is advisable to start such tests with low pressures and build up to the maximum pressure by increments, decreasing it afterward by decrements until the original low pressure is reached.

Several precautions are necessary in these tests: (1) The packers, i.e.,

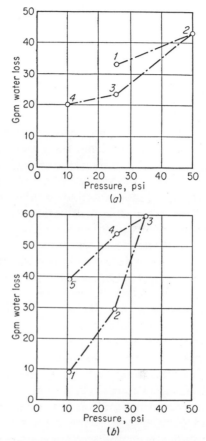

FIG. 15.21. Water pressure test curves.

devices covering the top of the tested section, should be tight in order to prevent leakage of the water under pressure; (2) lifting of rock beds by water pressure should be prevented by carefully observing the pressures and reactions of adjacent outcrops (erratic test results generally are indicators of such lifting); and (3) the discrepancy between the surface gauge readings and the actual pressure should be negligible, this discrepancy being caused by high friction losses in the drill pipe.

Figure 15.21a is an example of actual pressure testing in soft limestone, which was known to contain some clay-filled fissures and some open fissures. The first pressure applied was 25 psi at which 33 gpm (gallons per minute) was lost, after which the maximum pressure of 50 psi was used. Then the pressures were reduced to 25 and finally to 10 psi, with losses of 24 and 20 gpm, respectively. It appears from the experimental curve that either the clay was forced back into the open fissures and thus finally blocked them (resulting in a lower water loss than at the initial stage of the test) or a "water block" occurred; i.e., the fissures were finally filled with water, thus preventing the further admission of water. In Fig. 15.21b, the reverse situation is shown. This curve seems to indicate that clay-filled fissures were gradually flushed out, resulting in very high losses under the final application of the original low pressure. If the fissures were open but interconnected and solution or flushing did not take place, then points 1 and 5 in Fig. 15.21b would coincide.

15.16. Reservoir Problems: Ground Water. If the water from the main stream on which the dam is to be built easily escapes into the ground and joins the ground-water flow, there may be serious losses of water from the reservoir by seepage. Such is the case when the ground water is flowing from the main stream (i.e., the water table slopes down and away from the water surface in the stream). Also, if a ground-water divide is found to exist under the rim of the reservoir, and if its elevation is lower than the water surface of the filled reservoir, the reservoir is likely to be subject to serious leakage. Other ground-water occurrences should be studied also, e.g., the presence of springs in or immediately downstream from the reservoir or the appearance of artesian water in drill holes. It is sometimes believed that the mere existence of a natural lake in the potential reservoir site indicates that the reservoir basin will be tight. This is not always so, however. The increase of hydraulic head caused by the water in the new reservoir may open leakage paths that under the natural regime did not exist; also siltation of the natural lake over a period of years may have effectively sealed its bottom, whereas effective siltation in the new reservoir may require considerable time, if, indeed, it can be depended upon at all as a seepage barrier.

15.17. Abutment Problems. The foundation material should be resistant to disintegration and erosion. This obvious statement cannot be overemphasized as far as the dam abutments are concerned. The slaking test is compulsory for every case of suspicious materials in the abutments, particularly clayey rocks and conglomerates. Abutments built on such materials, if dried out during long periods of exposure to air, as in arid and semiarid regions, and subsequently saturated by the reservoir, may endanger the stability of a dam. The failure of St. Francis Dam (Sec. 15.9) was attributed by some to this cause. To obviate the danger, stripping to sound rock where moisture never dries out completely is recommended. Exposed surfaces thus obtained should be protected until concrete is poured. This protection can be achieved by such methods as spraying with gunite or use of asphalt emulsion or by leaving a protective layer of rock in place (6 to 12 in. thick) that is stripped immediately before pouring concrete.

Instability of the valley slopes where the abutments are to be constructed may constitute another difficulty.   Such is the case of rocks of poor shearing strength, as, for example, some shales.   If the abutments are built by pouring concrete on steps carved in rock, it is recommended that the steps be of reasonable height as determined by full-sized experiments and that the concrete be poured from lower to higher elevations and as soon as possible after exposure to ensure proper bond between concrete and fresh rock.   Sometimes the dip of the beds or the joints is such as to cause the entire abutment slab to be unstable when excavated, especially if undercut even slightly.   Such rocks should be removed or, as is sometimes done, anchored to the underlying stable rock.   A suitable anchor can be obtained by drilling holes through the loose slabs into the underlying strata and inserting rock bolts (Sec. 9.6) or some other form of steel anchor into the holes.   At a dam near Ketchikan, Alaska, such loose slabs were underpinned with steel H beams, which were left in place when the main mass of dam concrete was poured.[20]   Another type of abutment slope instability results from pressure relief because of excavation (Sec. 9.4).   In this case, the overlying or adjacent rock is loosened and may slide into the excavation.   Usually the sliding movement is slow enough to permit the concrete to be poured before a major slide occurs, and thus further loosening of the rock is stopped.   This phenomenon sometimes has been known to occur after the dam is constructed and cause slight movements in the abutments immediately downstream from the dam.

Arch-dam Abutments.   The abutment problems in arch dams are of particular significance.   The rock that has to take the thrust of the arch should be strong enough to resist the pressure without being crushed and should be free from deep weathering, alterations, faulting, and critical stratifications.   It should be able to withstand the action of shearing stresses, which inevitably develop in the rock in conjunction with pressure.   Rocks intersected by joints and faults may be amply strong as far as pressure is concerned but may yield easily under shear.   In this connection, the presence of joints and fissures approximately parallel to the direction of the thrust may be dangerous, especially if the fissures "daylight" (i.e., are visible) on the downstream side of the abutment.*   In these cases it is the duty of the geologist to determine whether the fissure system will develop toward the exposed surface of the rock or toward the main massif of the hillside; in the latter case, the abutments will not be subject to movement.

Figure 15.22 represents a critical abutment condition caused by the strike of the abutment rock being nearly parallel to the thrust T and the

---

* The verb "daylight" as used in engineering means to uncover a structure or surface from debris or soil accidentally or temporarily covering it, e.g., after an earthquake or flood.

dip being nearly vertical. A block (crosshatched in Fig. 15.22a) is then compressed by a force T_1, nearly equal to thrust T. The smaller the modulus of elasticity of the rock, the greater is the yield of the abutment rock (schematically shown by the black strip in Fig. 15.22a). In addition to compression, the block in question tends to expand laterally, and the larger Poisson's ratio of the rock, the larger is its lateral expansion tendency (refer to Fig. 2.8). If there is a clay-filled seam next to the block in question, it will be compressed laterally by the expanding block, and the yield of the abutment will be greater (Fig. 15.22b). Should the clay-filled seam located within the block be considered, the greater will

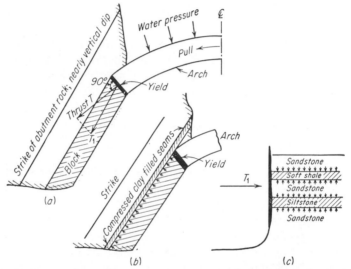

FIG. 15.22. Arch dam: critical abutment conditions (sketch).

be the yield of the block. The greater the yield of the abutment, the stronger is the pull of the concrete material from the center of the arch (Fig. 15.22a). This pull may finally cause fissures owing to the tensile stresses. An analogous situation takes place when the bedding strata are nearly horizontal and softer materials are confined between harder beds (Fig. 15.22c).

Unfortunately the values of the modulus of elasticity and Poisson's ratio as determined in the laboratory on small rock samples do not reflect the influence of the rock defects in the field and thus are mere approximations. To be on the safe side some designers use half of the laboratory values of the modulus of elasticity.

15.18. Channel-section Problems. The common problem in the channel section of the site is to determine the depth of unsuitable material to be removed. This usually is done by exploratory drilling sometimes combined with geophysical methods. The latter have to be used with some caution, however, if it is suspected that the channel fill contains numerous closely packed, large boulders or narrow, buried channels, gorges, or potholes. The existence of potholes or narrow gorges may be undetected even by drill holes.

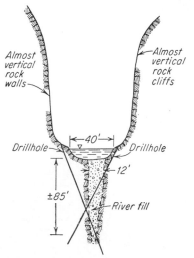

At Hoover Dam, deep potholes that were not disclosed by very extensive drilling were found in the river bed during foundation excavation. Conversely, in the exploration for a dam near Sitka, Alaska (Fig. 15.23), the drilling fortuitously located directly under the damsite a pothole that was only 12 ft wide and at least 85 ft in depth.

FIG. 15.23. Damsite near Sitka, Alaska.

Deep buried gorges can be expected to occur in rugged, mountainous terrain of fairly young geologic age or in steep-walled canyons with modified topography of very ancient geologic age. Also any region that has undergone extensive glaciation in past epochs may have concealed buried gorges. On the other hand, in broad plains country such as the Middle Western United States, *broad* buried channels are common.

A river valley should always be suspected of concealing a fault. Usually when a river tends to carve its way through a mountain range, it often finds that the soft material in a fault zone provides the easiest path. If the rock type on one abutment of the proposed dam is considerably different from the rock on the other abutment (Fig. 15.24a), it is possible that a fault at the center of the valley is responsible for this. Similar situations can occur, however, because of igneous intrusions into metamorphic or sedimentary beds. The zone of contact of such intrusions should be examined by drilling. Very often, this contact zone contains highly altered, soft material unsuitable for dam foundations or is highly pervious. Abrupt changes in the dip (Fig. 15.24b) of strata from one side of the valley to the other side may indicate faulting. It is impossible to list all such geological anomalies likely to occur, but any unusual changes in the dip or strike or in the rock type from one abutment of the proposed dam to the other should be carefully investigated. (The reader should consult Refs. 21 to 23 for further examples of geological anomalies.) In order to locate faults or contact zones under a river bed, the drilling of oblique (angle) holes may prove the most effective method (Fig. 15.24). In narrow stream valleys this will save the trouble of drill-

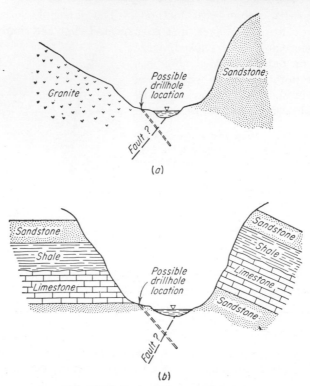

FIG. 15.24. Fault evidence at damsite.

ing from the water surface on barges or bridges.   For wide rivers, drilling from barges or other special devices may have to be resorted to, however, as in bridge investigations (Chap. 14).

FOUNDATION WORK: GROUTING[4,22,24–27]

15.19. Preparation of the Foundation.   If it is economically feasible, all material under the base of a proposed dam should be removed that can cause excessive settlement or leakage.   Otherwise, the dam design should be modified to account for such material.   Sometimes it is necessary to carry this removal to considerable depths in isolated areas of the foundation.   Such localized excavation is referred to as *dental work*.   The general over-all removal of unsound material is termed *stripping*, whereas the removal of loose masses of rocks on the abutments is termed *scaling*. In connection with this work, the engineering geologist retained for the project has to determine the expected depth of weathered or unsound rock or overburden that must be removed, estimate (at least approximately) in advance of construction the amount of dental work that may be required, and assist the engineering staff in estimating the amount of scaling to be done.

This processing of the surface of the rock excavation on which the dam concrete is to be poured is not final, however.   Additional treatment sometimes is extended to the abutments and the reservoir rims.   Geologic conditions that require special treatment are faults, joints, and

FIG. 15.25. Exfoliation cracks in dam abutment.

cracks; seams of altered rock that are too soft for the expected dam loads; solution channels and caverns in limestone and other soluble rocks; flow tunnels and large cavities in basalts (or lavas); and soft layers or inclusions such as are often found in shales and siltstones.

One unusual geological situation was noted during the construction of a dam near Steamboat Springs, Colorado (Fig. 15.25).   When the estimated stripping of weathered rock had been accomplished, a large crack was noted in the rock on one abutment. The crack was about 6 to 10 in. thick and of unknown length.   Apparently the crack was due to exfoliation affecting a considerable depth (over 10 ft) of the granite bedrock.   Corrective measures included removal of as much of the cracked rock as possible, and the remainder of the cracks were pumped full of concrete or grouted (Sec. 15.20).

The unsuitable material may be removed by *tunneling* or by sinking *vertical shafts* to the poor material, after which the latter may be excavated by mining methods.   All openings thus created are filled with concrete.   If the leakage channels are too large or too many, the leakage zone may be intersected by a subterranean concrete *cutoff wall*. The most common remedial treatment is by grouting.

15.20. Grouting.   Grouting is a method of injecting suitable mixtures of cement and water or other admixtures into a foundation for the purpose of (1) creating a seepage barrier to reduce uplift and (2) consolidating the rock sufficiently to make it act as a monolith under the structure. The geologic character of the foundation is of prime importance in establishing the need for grouting a foundation and the method to use.   The grout mixture may be injected through holes drilled specifically for this purpose from the foundation surface, tunnels may be constructed under the foundation or in the abutments primarily for the purpose of grouting these areas more effectively, the grout holes may be drilled from galleries within the dam after construction, or the grout may be injected into pipes left projecting through the main body of the dam for this purpose. Hereafter, only the foundation grouting will be discussed.

15.21. Grouting Materials.   The most common type of grout is a mixture of portland cement and water (also called neat portland cement grout); occasionally a cement-sand-water combination is used.   The distance which cement grout can be forced through minute cracks depends upon its viscosity or thickness.   The latter varies according to the size of the cracks; commonly some experimentation at the start of operations is required to determine the proper mixture.   The wider the cracks, the more viscous a mix may be used.   Instead of using cement grout, it may prove more effective under certain conditions to grout with a mixture of cement-rock flour, cement-hydrated lime, cement-calcium chloride, cement-diatomaceous earth, lumnite cement (for quick setting), cement-clay, or cement-bentonite.   Some of these substances may be used without cement.   Various combinations of them also are used.   Asphaltic emulsions and other bituminous compounds have been effectively used for grouting.   In some cases, for example where large water flows are encountered, hay or straw may be forced down a large hole and saturated with an asphaltic emulsion.   The straw retards the grout sufficiently to permit it to set before the water carries it away.   Various chemicals are also used for grouting with varying degrees of success, as, for example, the calcium chloride–sodium silicate process.   The choice of a grouting material obviously depends on its cost, availability, and efficiency under given conditions.

15.22. Grouting Equipment.   Grouting *pumps* are used to inject grout, but if the grout material is abrasive, *pneumatic injectors* may be used. An injector of this type consists of a tank filled with grout, which is cautiously squeezed out into the ground opening by compressed air.[24] A *grouting unit* often consists of a mixer and one or more pumps.   Pressure gauges are installed at the grout holes in order to maintain proper line pressures. In addition, careful notes are taken of the amount of grout placed in each hole in order to determine the "grout take." Excessive quantities of grout required in one particular hole and difficulty in maintaining adequate pressure are signs that the grouting procedure is not efficient and should be modified.

FIG. 15.26. Drilling grout holes.

Grout Holes. The grout is forced into vertical or oblique (angle) holes, the diameter of which may be from $2\frac{1}{2}$ to 5 in. These holes are drilled with ordinary diamond-drilling equipment (Fig. 15.26), and as cores are rarely taken, a "plugged" bit is used. For shallow holes, wagon drills may be used (these are jackhammers mounted on wheels and can handle long lengths of drill steel). If such air-type hammers are used, the holes should be flushed with water, after drilling, to prevent the rock particles from sealing the seams which are to be grouted. The spacing and depth of the holes are determined by the persons in charge and should be based primarily on the geology and the amount of hydrostatic head to which the foundation will be subjected.

15.23. Low-pressure Grouting. There are primarily two types of grout holes: low-pressure or shallow grouting and high-pressure or deep grouting. The terms "low" and "high" pressure are relative only, as may be seen from the ensuing discussion.

Low-pressure grouting precedes the high-pressure grouting and usually is done before any concrete is placed in the body of the dam. Its primary purpose is to consolidate the rock and to seal all major crevices and openings. Another purpose is to create a strong monolithic slab at the heel area of the dam to resist the high pressures that will develop in this area during the second stage of the grouting program. These high pressures will tend to heave or even destroy the rock at the heel area of the dam foundation. Low-pressure grouting should be extended throughout the whole foundation if the state of the rock requires it.

The holes in the low-pressure-grouting stage are from 20 to 50 ft deep and may be spaced 20 ft apart. In the dam shown in Fig. 15.27 there are two rows of shallow holes (labeled *B* and *C*, although more rows may be used). The grouting starts on holes spaced 40 to 80 ft apart, and if they accept the amount of grout required to fill the seams, intermediate holes may be omitted. The holes are drilled vertically unless the best contact with the seams can be obtained by using oblique holes. The pressures should be insufficient to heave the overlying rock or destroy the friction and cohesion keeping the strata together, a damage that cannot be compensated for by grouting. To observe heaving, accurate surveying methods should be used. In fact, the heaving may be quite imperceptible until a large block of rock comes loose and slides into the excavation (as sometimes occurs when an abutment area is being grouted).

15.24. High-pressure Grouting. The objective of high-pressure grouting is to create a deep curtain at the heel of the dam which will prevent leakage from the reservoir and will prevent uplift of the dam by the water under pressure caused by the high level of the reservoir water. This type of grouting is done after the low-pressure grouting is completed and some structural concrete is placed. The weight of the concrete together with the strong slab at the heel of the dam (as already mentioned) will hold down the effects of the high grouting pressures. The drilling is done from galleries within the dam or from pipes projecting through the heel of the dam, and therefore drilling must be done before storage of water in the reservoir starts. Usually the holes are placed in a single row on about 5-ft centers (*A* holes in Fig. 15.27*a*).   The holes are either vertical or oblique; their inclination basically is controlled by the dip of prominent joints or seams in the foundation rock.   Usually oblique holes make 10 to 15° angles with the vertical.   Their depth usually is from 20 per

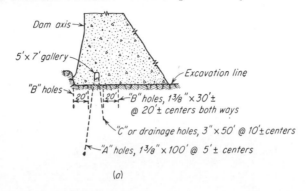

(*a*)

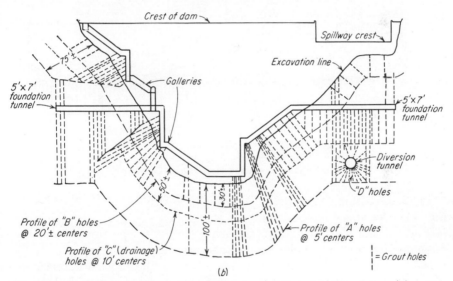

(*b*)

FIG. 15.27. Grout pattern at Kortes Dam, Wyoming: (*a*) Typical section of dam, (*b*) profile showing grout holes.   (USBR.)

cent of the hydrostatic head in hard, dense rock to 70 per cent in poor rock.

An additional line of grout holes sometimes is drilled (usually on 5-ft centers) from the upstream face of the dam to intersect the row of the high-pressure holes.   This is done when the foundation galleries are at some distance from the heel of the dam. The purpose of this arrangement is to reduce the uplift between the heel and the gallery.   These supplementary holes usually are drilled prior to the drilling of the high-pressure holes.

Figures 15.28 and 15.29 represent the grouting patterns at Norris Dam (Tennessee Valley Authority). The limestone and dolomite foundation of the dam, though strong enough to support the structure, constituted serious leakage hazards because of numerous cavities. Hence the grouting was done on the whole area of the foundation. The complex grid system (Fig. 15.29) consisted of elementary patterns repeating themselves.[22]

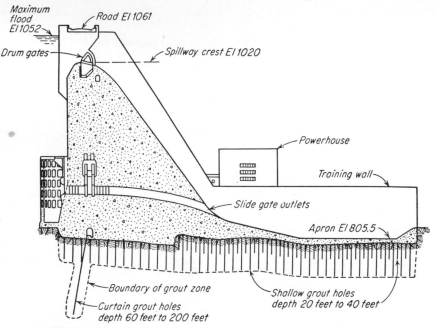

FIG. 15.28. Foundation grout holes, Norris Dam, Tennessee. (TVA.[22])

15.25. Grouting Methods. There are basically two methods of grouting: (1) the stage method and (2) the packer method. In the *stage method* the holes are drilled to the seam closest to the surface and the seam is grouted. The holes proper then are cleaned of grout, mostly by washing, and drilled down to the next seam. The process is repeated, using increasing grouting pressures, until the planned grouting depth is reached. In the *packer method* the holes are drilled to the full planned depth. A zone of a certain thickness then is grouted, and a seal, or packer, is inserted into the hole to a level corresponding to the top of the grouted zone. The overlying zone, for a certain thickness, then is grouted, using decreased pressure. The process is repeated until the uppermost seam is grouted.

Flushing or Washing Grout Holes. Clay seams, if any, should be thoroughly flushed out with water prior to grouting. Clay cannot be grouted and if permitted to remain in the seam will contribute to the instability of the foundation.   In shallow holes, the flushing water is kept flush to the ground surface, whereas other holes are capped or left open.   All seams thus are washed simultaneously, and dirty water emerges from the uncapped holes.   During the washing, the caps are switched around to change the path of the wash water.   When deep holes are washed, the lowest seam is washed first by forcing water under pressure to a part of the hole depth.   Should a seam be washed individually, as in the case of grout curtains, it is isolated by special tools.[22]

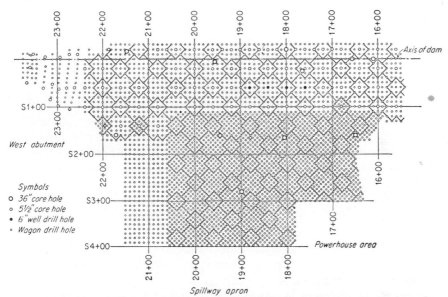

FIG. 15.29. Layout of consolidation grouting, Norris Dam, Tennessee.   (TVA.[22])

Grout Losses.   Even at normal, i.e., nonexcessive, pressures, there may be grout losses because of leakage.   The latter usually is controlled by calking the surface seams with wood wedges and oakum or by grout caps and gunite blanket as described in Sec. 16.19.   If the grouting pressure is excessive, the grout may move upward in the rock adjacent to the hole until it erupts on the surface at some distance from the hole.   It is not unusual for grout leaks to appear over 50 ft from the hole, and in some cases grout has been observed at the surface several hundreds of feet away from the grouted hole.[28]

15.26. Foundation Drainage.   Water seeping under the dam and collecting at its base is still under the hydrostatic head from the reservoir (less friction loss).   It exerts uplift on the base of the dam and should be removed.   This usually is done by drilling vertical holes immediately downstream from the main grout curtain or from the cutoff wall, if any (Sec. 15.5 and Fig. 15.4).

The diameter of drain holes varies from $1\frac{7}{8}$ to 3 in.; in any case the holes should be large enough to prevent their clogging from rock or concrete particles eroded by the seeping water.   (If the water contains a high percentage of chemical salts, the drains are likely to be clogged by the deposition of these salts.)   The holes may be spaced on about 5-ft centers, and their depth varies from 20 to 40 per cent of the hydrostatic head.   If the whole foundation is grouted, the drainage holes should be drilled through the grouted zone.   Drilling is done from the lowest galleries of the dam or from its upstream face.   The drainage holes are drilled after grouting is completed.   The numerical data and procedures explained herein may vary according to the local geologic conditions and effectiveness of the grouting.

An effective foundation-drainage system sometimes eliminates the necessity for grouting.   If the cracks in the rock are too fine to permit effective lateral penetration of the grout but still will allow seepage water to penetrate the foundation rock, the dam and foundation drains should be designed to accommodate for such conditions.   Usually it is desirable to grout one or two holes experimentally before starting the complete program.   After these few holes are grouted, cores should be obtained in holes drilled near the grout holes.   If close examination of the cores fails to disclose any grout, it may prove too expensive to continue with a truly effective grouting program, owing to the close spacing required to obtain a continuous grout curtain under the foundation.

SPILLWAY EXPLORATIONS

15.27. Geological Factors in Spillway Design.   If the spillway is of considerable size and is not an integral part of the dam, the spillway structure may resemble a small concrete dam, may cost as much as the main dam, and may require almost identical investigations.   In the investigation primary consideration should be given to the foundations of the crest, the chute, and the training walls.   These must be investigated for sliding and for total and differential settlement.   The chute foundations must be studied for the same factors and also for permeability and frost-action damage (as some chute slabs have been heaved out of place by frost action).   The foundation of the crest must be thoroughly explored for the competency of the bearing materials, for sliding factors, and for frost heave.

Erodibility of Foundation Materials.   The erodibility of the materials in the channel immediately downstream from the chute and the stilling basin must be investigated.   In some spillways where there is good rock at the bottom of the chute, a concrete stilling basin may be eliminated. One of the major factors in studying erodibility is to determine if the rock will be subject to quarrying or "plucking," i.e., removal by the velocity or turbulence of the spillway flow.   Badly fractured rock and some blocky materials such as basalts are likely to be badly eroded by spilling water. High water velocities are likely to occur near the intake training walls, and if the foundation material is susceptible to erosion, the walls may be damaged.

Vibrations. Turbulence in spilling water causes vibrations that affect all parts of the spillway. Vibrations in the stilling basin may cause consolidation of the underlying material with resultant settlement and damage of the concrete in the basin. Such settlement is particularly liable to occur if the basin foundation is made of sands and gravels of low density. Hence it is necessary to have proper compaction in such materials when they are used for foundations subject to vibrations. Morning-glory spillways are noted for very high vibratory effects, and their foundations demand special attention in this respect.

Effects of High Hydraulic Head. Water under different parts of the spillway is still under a high hydrostatic head from the reservoir. In this connection, seepage and piping (erosion from the bottom of the chute in an upward direction) under the chute and the crest and uplift of the stilling basin are possible. The downstream lip of the spilling basin may be damaged by such water under head and especially by spilling water.

For those spillway structures built as an integral part of the dam, only a few additional topics should be investigated beyond those investigated for the dam proper. The only foundation problems that may cause difficulties in such spillways are the possibility of erosion at the downstream end of the chute and the vibratory effects on that portion of the dam.

15.28. Site and Type of an Isolated Spillway. The topography and geology at a damsite play an important role in selecting the proper site for a spillway and in determining the type of spillway to be used. A natural topographic saddle near or adjacent to the dam is an ideal location for a normal spillway. To be economical, such a location requires that good foundation rock be available near or at the surface. A considerable depth to firm rock may necessitate a different type of spillway, such as the glory hole, or a spillway that is merely an overflow section on the dam. Glory-hole types are convenient for those locations where the canyon has steep walls, where the firm rock is quite deep, and particularly where the diversion tunnel can be utilized to form a part of the spillway and carry the discharge (Fig. 15.13). (A diversion tunnel is used to carry the waters of the main stream during construction of the dam.)

Whether or not the spillway should be provided with gates for regulating the passage of water may depend on the likelihood of differential settlement of the crest. Such an occurrence could cause the gates to jam and thus put the spillway out of service.

GEOLOGICAL INVESTIGATIONS FOR A MASONRY DAM

15.29. Site-selection Criteria. From the analysis of the dam problems, failures, and hazards as discussed in this chapter, it may be concluded that a damsite should satisfy the following requirements:

1. The rock should be sound and resistant to the expected static and dynamic forces, including earthquakes.

2. The valley slopes should be stable when the reservoir is full; this requirement also applies to the abutments.

3. The dam foundation should be safe from sliding, especially in the case of gravity dams.

4. The rock of the foundation should be approximately of one geologic classification to avoid variations in the value of the modulus of elasticity.

5. The foundation and the reservoir walls should be watertight. Remedial measures should be established for each prospective (alternate) site and compared. It also should be borne in mind that in the case of flood-control dams considerable water losses from the reservoir are permissible provided the stability of the dam is not affected.

6. The rocks at the site should be resistant to solution, erosion, decomposition, and other detrimental effects of wetting and drying, freezing and thawing.

7. The reservoir drainage area, including rocks and overburden, should be resistant to erosion and therefore not likely to contribute such heavy silt loads to the reservoir that its useful life span is severely decreased (Chap. 11).

8. In the case of an arch dam, the topography and structural features of the rock in the abutments should be favorable for the proper accommodation of the thrusts and stability of the arch.

9. Geologic and topographic conditions should permit a favorable location of (a) spillway and diversion tunnel and (b) powerhouse and outlet works, if any.

10. The location of construction materials, mainly concrete aggregate, should be within an economically justified distance from the project.

For each prospective site, it should be determined if existing railroads, highways, and canals will have to be relocated and if access roads have to be built. Attention should be given to the erosion characteristics of the tailrace channel (i.e., the river channel immediately below the dam which is subject to erosion by water released by the spillway and other appurtenances). Also for each prospective site the amount of stripping, scaling, and dental work to be done should be estimated.

The final selection of the damsite is based on the comparative analysis of all data discussed in this section, the criteria for the comparison being cost and availability.

15.30. Reconnaissance. Before field investigations start, an exhaustive study should be made of all geological literature pertinent to the reservoir site. Publications of the USGS and Soil Conservation Service (particularly those concerned with soil characteristics controlling siltation) and state and municipal geological reports should be consulted. Information can be obtained from oil and mining companies, state highway departments, and local residents. Topographic and geologic maps thus obtained generally are of a small scale, however, and rarely cover the selected damsite in detail. Additional information can be obtained from personal reconnaissance trips to the site and from airphotos. The latter may be specially flown, or sources indicated in Sec. 7.14 consulted. If the reconnaissance gives positive indications as to the suitability of the site, the explorations are pursued; otherwise, another site is sought. Starting with this initial period, close contact between the geologists and the design engineers is maintained through the whole period of explorations, and special field parties consisting of a geologist and an engineer (or several more) are formed.

15.31. Preliminary Investigations. At the end of this stage, a general picture of the regional geology should be clear to the investigators. For this purpose a sufficiently large area on both sides of the proposed project site should be covered by investigations. The geological map of the area should be prepared and should cover all geological features significant to the project. For the details of mapping, see Chap. 7. Simultaneously, sources of construction materials are located, briefly explored, and plotted on the map. For a masonry dam, these usually are concrete aggregate and the materials for the cofferdam (Chap. 14) if the latter has to be used. The cofferdam may require both rock-fill and impervious-type materials. Samples of rocks and construction materials are collected for future reference and, in some cases, for laboratory tests. Possibilities of reservoir leakage should be clarified, at least in general terms, and plans of further leakage exploration made. All cultural features such as cities, quarries, mines, and historical or archeological sites that will be flooded by the reservoir are located and, where required, appraised. In the case of archeological sites, local or national museum authorities should be informed, as they may want to do some additional excavations.

Vertical and horizontal survey control is established. This is required for the preparation of the topographic map of the damsite (on a large scale) and the reservoir basin (on a smaller scale). A grid is established to which the bore holes will be referred. Although detailed drilling is not done at this stage, the geologist with the aid of a hand auger and shovel may make a few tentative explorations of the subsurface conditions. All data collected at this stage should be assembled into a report.

This report is used as a guide to further explorations by both direct and indirect methods (Chap. 6).

The geological report should primarily give a general outline of the regional geology and the damsite geology and indicate apparent geological problems at the damsite. It should be illustrated with a surface (areal) geologic map and airphotos showing the proposed emplacements of the various parts of the dam, views looking upstream and downstream at the dam, and geological features to which special attention should be paid such as faults or bedding in the outcrops.   A brief description of the construction-material sources should be given.   The report should be preceded by a brief introduction describing geographic location of the project, topography, climate, and accessibility (roads, railroads, etc.).   The report also should conclude with definite recommendations as to what and where additional exploration should be performed.

15.32. Detailed Explorations.

Usually by the time this stage is reached, rough designs of the dam have been prepared, and on the basis of these designs an exploratory program is outlined.   As this program is being materialized in the field, its results are continuously reported to the designing forces.   Accordingly, the design may be changed, and conversely, changes in the design may result in modification of the exploratory program.   This interdependence of the design and the field geotechnical investigations during the exploration stage should be especially emphasized as an efficient way to obtain a rationally planned and constructed dam.

The basic field problem during this stage is to determine in detail the character of the overburden and the bedrock on the abutments and in the river section of the dam.   Potential reservoir leakage zones are thoroughly investigated.   Ground-water observation wells are established, and pumping tests are done if required.

During this stage serious consideration is given to exact location of the dam axis and to the type of dam, as influenced by geological conditions. If a large fault crosses the damsite, it may be avoided by shifting the dam upstream.   The sliding danger on an abutment can be remedied by shifting the axis.   An extensive cavernous limestone zone may place in doubt the feasibility of the entire project.   The rock conditions may be more favorable for a gravity or earth dam instead of an arch dam.   If there is a paucity of concrete aggregate but ample quantities of embankment materials, it may prove more economical to construct an earth instead of a concrete dam.   Conversely, some sites, although adequate for an earth dam, can be used only for a concrete dam because little or no embankment material is available within an economically feasible haul.

When this stage of exploration is completed, another geological report is prepared.   The best interpretation of the geological conditions as based on the subsurface work and the results of the studies of construction material sources (with recommendations of additional sources) should be given.

The report should discuss in full the suitability of the site for the planned structure and suggest modifications if necessary.

Drilling. Most of the detailed geological information during the exploration stage of the investigations is obtained from drilling. The initial bore hole locations generally are based on the following criteria:

One drill hole is made on each abutment, about halfway between the top and the bottom of the abutment (unless rock is exposed for the full surface of the abutment, and even then holes are drilled if it is suspected that soft zones interlayer firm rock), and one or more drill holes in the river section to determine the depth of river fill (or overburden) (Fig. 15.30). From the results of these holes, additional holes can be located.

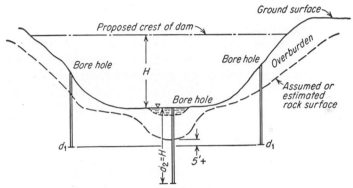

FIG. 15.30. Reconnaissance drilling program.

Definite criteria on the location of these holes cannot be given, however, since these criteria depend entirely on the type of structure, the geological situation, and the designer's requirements in this matter. Sometimes holes are drilled at the intersections of a grid system with square sides, say from 200 to 1,000 ft, as has been done on some large dams. Availability of funds is a very important factor in determining the amount of drilling operations for dams as for all structures in general.

The depth of holes depends on local geological conditions, such as the presence of large boulders, soft zones or strata, faults and joints within the influence of the dam load, and thickness of the valley fill. In some special cases, e.g., when exploring for spillway and powerhouse foundations, it is necessary to know the nature of the rock *below* the finished foundation grade, and therefore the holes must be drilled deep enough to obtain this information. Ordinarily all drill holes should penetrate the overburden and all unsound or weathered rock until sound rock is encountered. If it is found, however, that unusual weathering or alteration conditions have resulted in extremely deep weathered rock, the depth of holes should be agreed upon by the designer, who may change his design accordingly. It may be that the weathered or altered material is pervious; in such a case water from the reservoir may be lost via this material (i.e., under the dam).   To estimate these losses it is necessary to know the depth (thickness) of the ground-water stream that will flow above the rock, and for this purpose the elevation of the sound rock should be known.

The presence of boulders above the rock formation at the damsite may be misleading if they are of the same composition as the local solid rock (and they commonly are).   Therefore when rock is hit by drilling, it is recommended to drill about 15 ft into rock (for lighter structures this additional drilling may be 10 ft or even less as explained in Sec. 6.24). If the boulders are closely nested with little or no fine material filling the interstices, it will be necessary to closely examine drill cores for rounded or weathered surfaces to determine if the hole is in sound rock or boulders.

Some geologists advise that abutment holes should be drilled to approximately the elevation of the river bottom (d_1 in Fig. 15.30) and that the depth of the river-bottom holes approximates the height of the dam (d_2 in Fig. 15.30).   For very high dams this arrangement may be unnecessarily expensive, however, and the geologist will have to be guided by judgment and the designer's wishes in determining depth.   If the spillway is to be located in an area away from the main dam, its foundations should also be drilled.

In addition to bore holes, pits may be sunk.   Exploratory drifts (i.e., tunnels with approximately horizontal floors and a small cross section) may be driven into the abutments if the dam is of major size and/or unusual geologic problems can be solved by such drifts.   The construction-materials sources are investigated by shallow auger holes or test pits, and samples extracted for laboratory tests.

Geophysical Explorations for Dams.   As an example of the use of geophysical methods for dams, the following description of a seismic investigation at Cedar Bluff Dam on the Smoky Hill River in Kansas is presented (Fig. 15.31*a*).   The exploratory stage drilling had located a possible buried channel on the left abutment which appeared to be about 1,000 ft wide and about 60 ft deep to the shale bedrock and filled with pervious sands and volcanic ash (Fig. 15.31*b*).   It soon was recognized that the channel might cross the axis of the dam, then turn sharply and discharge near the toe of the proposed dam.   This could create a serious leakage condition due to the short percolation path.

A bedrock contour map was established by seismic exploration (Fig. 15.31*c*) wherein 20 seismic exploration lines with some 90 bedrock depth points were obtained.   The previously drilled core holes served as a control for checking the seismic data.   In fact, in this case where the seismic depth points were near drill holes, the calculated depth to shale checked the depth obtained by drilling within 5 per cent in most cases. As a result of the geophysical studies it was decided that the buried channel did not have an outlet near the toe of the dam and thus was not a potential hazard.   The time spent and the cost of this program proved to be considerably less than the detailed drilling program that would have been required to obtain similar data.   The

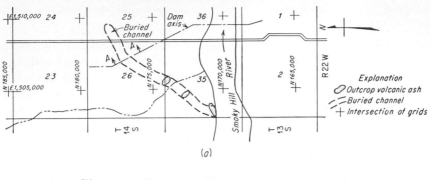

(a)

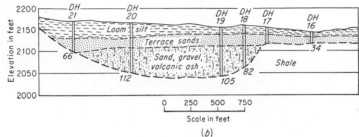

(b)

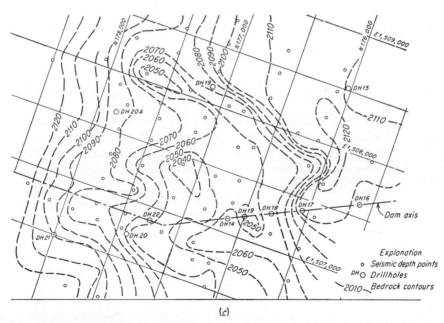

(c)

FIG. 15.31. Geophysical investigations for a buried channel, Cedar Bluffs damsite, Kansas. (USBR.)

completed map is comparable to a surface topographic map, and sufficient depth points were procured to permit contouring at 10-ft intervals.

15.33. Preconstruction Stage.   At this stage, the plans of the structure are well advanced, and in comparing the design features with the available geological information, the engineering staff may ask for some additional geologic evidence as to the capacity of the rock to carry some heavily stressed parts of the dam.   In this connection some additional holes may be drilled.

If a final report is required, it should contain primarily all information obtained during the main exploratory stage (logs of holes and pits, description of drifts if any). Geological sections based on the exploration data should be prepared, particularly some perpendicular to the dam axis, along the dam axis, and parallel to the dam axis, e.g., along the cutoff trench site and possibly along the heel and the toe of the dam. Cross sections of the material sources also are given.   All information should be plotted on topographic maps.   The report should contain descriptive paragraphs as to the geology of the damsite, reservoir, and appurtenant structures.   Recommendations as to construction procedures may be needed.

The various reports prepared at different stages of the investigations are, in reality, the evolutionary stages of a single final report which portrays the geologic evidence as it is gradually obtained and analyzed.

The description of exploration procedures as given in this chapter refers to large dam projects designed and built by major engineering organizations.   In the case of smaller dams, owned by a municipality or a private concern or person, a geologist may be retained for investigations.   In this case generally the time is too short and funds are too limited, so the geologist has to confine himself to the study of the most characteristic geologic features that may control the behavior of the planned structure. He generally submits a single report approximately corresponding to the report at the end of the main exploration stage as described above.   As previously discussed, even in these cases, the geologist also keeps close contact with the designer of the dam.

15.34. Construction Stage.   Subsurface exploration made for obtaining data for design purposes cannot give sufficient information to solve various specific problems which arise during construction.   Therefore, drilling is not ended with the end of the geotechnical investigations for a dam.

To determine the definite depth of excavations, drill holes may be needed as close as 50 ft center to center.   If the depth of the excavation is incorrectly given to the contractor and the excavation has to be deepened, this automatically involves widening the excavation and an increase in yardage.   Additional drilling also may be necessary at the sections where grouting has proved to be ineffective.

Relocation of structures, particularly highways and railroads, requiring construction of high fills or large bridges is an investigation in itself (Chap. 14).   Other problems that arise may be connected with the drainage of swamps, sloughs, and sinks adjacent to the reservoir due to the necessity for malaria control.[22]

The ground-water observation wells or holes established during the explorations should be maintained in good order.   Readings of the water level in these holes should be done periodically (e.g., once a month) for at least one year after the reservoir is filled or until the postconstruction ground-water regime is established.   Homes and entire towns near the reservoir depending upon ground water for their water supplies may find that their wells are drying up, especially downstream from the dam. The ground-water observations may be useful in determining beforehand if such situations are likely to develop and thus aid in taking preventive or remedial measures.   In the case of claims concerned with ground-water damage to farm crops or basements, the ground-water-level records often may be used to dispute such claims, especially if the damage has been produced by causes other than the water in the reservoir.

At the end of his participation in a dam project, the geologist prepares a map of the foundation excavation for the dam.   The map should indicate the location of all major joints and fracture systems in the foundation, the location of any fault zones or other structural defects that cross the foundation area, a description of the rock and soil types in the foundation and the walls of any cutoff trenches, the location of any springs if encountered, the strikes and dips of all formations in the foundation, and any other geological features that could conceivably affect the future performance of the structure.   This map can prove very valuable in the case of possible difficulties with the dam, such as excessive seepage, differential settlement, or local instability.

15.35. Example of Explorations for Gravity Dam.   Explorations for a gravity dam described hereafter refer to Kortes Dam constructed by the USBR on the North Platte River, about 60 miles southwest of Casper, Wyoming.   The dam is about 225 ft high above the level of the foundation rock.   The foundation rock is remarkably consistent, being mostly granite in varying stages of alteration with numerous joints and fracture systems.   A diabase dike occurs on the left abutment.   The primary problems to be solved were (1) the severity and openness of the joints, (2) the location of any altered zones that would require dental work, (3) the depth of several large talus-filled draws on the left abutment, (4) the possibility of faults, and (5) the amount of scaling of loose rocks required for protection of the power plant and for satisfactory dam abutment foundations.

Figure 15.32 illustrates the location of holes and drifts used in solving these problems as plotted on the geologic map prepared.   The purposes of the various holes are as follows:

Drill Holes 1, 2, *and* 3.   To locate depth of bedrock in river section.

Drill Holes 5, 6, 7, 8, *and* 9.   To determine if a fault or deeply altered zones underlay the channel.

Drill Holes 10, 11, 12, *and* 13.   To explore the severity of the joint systems known to be present in the abutments and also to determine the depth of stripping to be required.   It should be noted that these holes were drilled at angles that would intersect the greatest number of joints, as the joints generally had very steep dips.

Drifts 1 *and* 2.   On the left and right abutments near river grade to provide a more

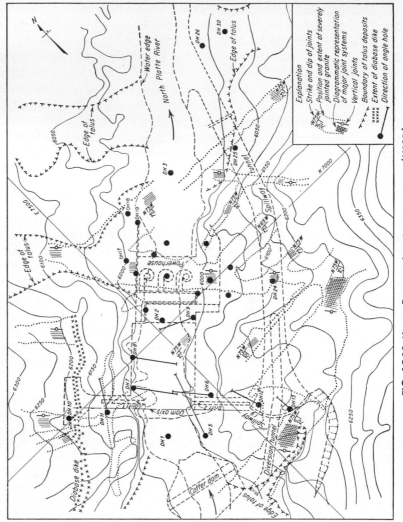

FIG. 15.32. Kortes Dam exploration program.   (USBR.)

complete picture of the joint systems at a depth.   From these drifts it was possible
to determine the openness of the joints, the condition of the rock with regard to
fracturing, and the nature of some of the altered zones.

Drill Holes 17, 18, 19, and 30.   To determine the depth of talus in the draws.*

Drill Holes 24, 25, and 26.   To determine the condition of the rock at the invert
grade of the diversion-spillway tunnel.

All remaining holes were drilled to explore in detail the powerhouse foundation.

Pressure tests were run in most of the drill holes in order to determine the openness
of seams and thus the type of grout curtains required.   The final grout patterns used
are illustrated in Fig. 15.27.   The severity of the joint system is disclosed by the
number of grout holes required.   The profiles shown in the figure illustrate the depth
to which each series of holes (A, B, and C) were drilled.   The holes actually shown on
the figure are diagrammatic only and do not represent the actual number of holes
drilled and grouted.

The above program is not to be regarded as completely typical of a concrete dam
exploratory program.   If large faults had been found, considerable additional drilling
might have been required to determine their extent; if a deep, buried gorge in the
river was disclosed, a number of additional holes would have been necessary to outline the gorge fully; and any similar geologic anomalies would have had to be explored
in detail.   On the other hand, if the rock had been comparatively unbroken, angle
holes might not have been required and fewer exploratory holes would have been
necessary.

REFERENCES

1. Hinds, J.: Side Channel Spillways, *ASCE Trans.*, vol. 89, 1926.
2. Creager, W. P., J. D. Justin, and J. Hinds: "Engineering for Dams," 3 vols.,
   John Wiley & Sons, Inc., New York, 1945.
3. Houk, I. E., and K. B. Keener: Basic Design Assumptions, A Symposium on
   Masonry Dams, *ASCE Trans.*, vol. 106, 1941.
4. "Treatise on Dams," particularly chaps. 4, 9, 10, 11, 12, and 13, U.S. Bureau of
   Reclamation, October, 1950.
5. Rose, E.: Thrust Exerted by Expanding Ice Sheet, *ASCE Trans.*, vol. 112, 1947.
6. Harza, L. F.: The Significance of Pore Pressures in Hydraulic Structures, *ASCE
   Trans.*, vol. 114, 1949.
7a. Jurgenson, L.: The Record of 100 Dam Failures, *J. Elec.*, vol. 44, 1920.
7b. Sutherland, R. A.: Dam Building Reaches a Climax, *Eng. News-Record*, vol. **117,**
   1936.
8. Dumas, A.: Rupture de la digue de Bouzey, *Génie civil*, vol. 31, 1897.
9. Taylor, T. V.: The Failure of the Austin Dam, *Eng. News-Record*, vol. 43, **1900,**
   also *USGS Water Supply Paper* 40, 1900.
10. Halton, T. C.: The Austin Dam and Its Failure, *Eng. News-Record*, vol. 68, 1912.
11. Freeman, J. R.: Some Thoughts Suggested by the Austin Dam Failure, *Eng.
    News-Record*, vol. 66, 1911.
12. Causes Leading to the Failure of the St. Francis Dam, Committee Report for
    the State, California State Printing Office, Sacramento, 1928.
13. Hill, L. C., H. W. Dennis, and F. H. Fowler: Essential Facts Concerning the
    Failure of the St. Francis Dam, Report of the ASCE Committee, *ASCE Trans.*,
    vol. 94, 1929.

* Holes 17, 18, and 19 are at the upper center portion of Fig. 15.32, slightly south
of the 6,000 contour.

14. Ransome, F. L.: Geology of the St. Francis Dam Site, *Econ. Geol.*, vol. 23, 1928.
15. Longwell, C. R.: Lessons from the St. Francis Dam, *Science*, vol. 68, 1928.
16. Stella, A.: Il disastro del Gleno e la natura geologica della regione, *Rass. min., metallurgica chim.*, vol. 61, 1924.
17. Terzaghi, K.: Simple Tests Determine Hydrostatic Uplift, *Eng. News-Record*, vol. 116, 1936.
18. Leliavsky, A., and M. A. Selim: Uplift Pressure in and beneath Dams, A Symposium, *ASCE Trans.*, vol. 112, 1947.
19. Keener, K. B.: Uplift Pressures in Concrete Dams, *ASCE Trans.*, vol. 119, 1945; also *Proc. ASCE Sep.* 25, 1945.
20. Shannon, W. D., and W. L. Shannon: Connell Dam Provides Water Supply for Alaska's First Pulp Mill, *Civil Eng.*, vol. 24, no. 6, p. 37, June, 1954.
21. Burwell, E. B., Jr., and B. C. Moneymaker: Geology in Dam Construction, "Berkey Volume," 1950.
22. Geology and Foundation Treatment, *TVA Tech. Rept.* 22, 1949.
23. Crosby, I. B.: Geological Problems of Dams, *ASCE Trans.*, vol. 106, 1941.
24. Minear, V. L.: Notes on the Theory and Practice of Foundation Grouting, *Proc. ACI*, vol. 43, 1947 (with discussion).
25. Simonds, A. W., *et al.*: Treatment of Foundations for Large Dams by Grouting Methods, *ASCE Trans.*, vol. 118, 1953.
26. Thompson, T. F.: Foundation Treatment for Earth Dams on Rock, *ASCE Reprint* 88, San Francisco Convention, 1953.
27. Lane, E. W.: Security from Underseepage: Masonry Dams on Earth Foundations, *ASCE Trans.*, vol. 11, 1935.
28. "The Foundations of the Portuguese Dams," Sondagens Rodio, Ltd., Lisbon, Portugal.

Additional General References

"Low Dams, A Manual of Design for Small Water Storage Projects," National Resources Committee, Washington, D.C., 1938.

Rajaraman, S.: Geology for Dams, *J. Inst. Engrs. India*, vol. 29, 1948.

Lugeon, Maurice: "Barrages et géologie," Dunod, Paris, 1933.   A pioneer geologic book on dams.

Leggett, R. F.: "Geology and Engineering," McGraw-Hill Book Company, Inc., New York, 1939.

U.S. Bureau of Reclamation: "Dams and Control Works," 3d ed., Government Printing Office, Washington, D.C., 1954.

Wantland, D., and W. Judd: Influence of Geotechnical Factors on Arch Dam Design, *Proc. XX Intern. Geol. Cong.*, Mexico City, 1956.

CHAPTER 16

EARTHWORK

"Earthwork" is the extraction of materials from the earth crust and their utilization in the building of earth dams, highway and railroad embankments, airfields, and fills for supporting buildings and parking lots. Earthwork also comprises the digging of cuts and excavations, including canals.

16.1. General Terminology. An *embankment* is a body built of "borrowed" earth material. As a rule, the length of an embankment is considerably greater than its width or height. There are two characteristically different classes of embankments: (1) highway and railroad embankments and (2) earth dams crossing a valley. Intermediate classes are the levees and dikes parallel to water streams for the purpose of flood control or river regulation and embankments (or berms) enclosing water reservoirs.

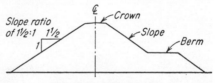

FIG. 16.1. Embankment.

[Strictly speaking, the term "berm" should be applied only to widened portions of an embankment or to benches in a cut (Sec. 16.3).]

A fill is an artificially built earth body which has a height relatively small in comparison with its other dimensions. Grading for a new housing project may, for instance, consist of fills occasionally alternated with excavations. The terms "embankment" and "fill" often are interchanged, and so are the terms "cut" and "excavation." A cut usually refers to an excavation for a "way" (highway or railroad) or a canal. A way or canal may be designed in fill or in cut, or part in each, as on sidehills. Cuts and sidehill sections are provided with *ditches*, triangular or trapezoidal in section. If earth material from the ditch was used in the construction, it is termed a *borrow ditch*. A long, narrow excavation is a *trench*.

The steepness of a slope is expressed by the ratio of the *horizontal* base to the *vertical* height (Fig. 16.1). A usual slope ratio in *small* earth embankments is $1\frac{1}{2}:1$; in shallow earth cuts, $1\frac{1}{4}:1$. A high slope may include berms or benches. For berms, see Sec. 16.3.

16.2. Earth-moving Equipment.[1-6] The material for a fill or an embankment is excavated in a *borrow pit* or an adjacent excavation and *hauled* to the construction. Equipment for moving earth ("earth movers") may be towed by tractors or self-propelled (Fig. 16.2). Rubber-tired tractors operate satisfactorily on relatively dry loams, clays, sand, and gravel. On wet, plastic soils, tractors with treads usually are

FIG. 16.2. Compacting and placing earth dam: (A) sheep's-foot roller, (B) bulldozer, (C) earth mover "euc," (D) sprinkler. (*USBR photograph.*)

required to push or pull the earth mover. Such treaded or tracked vehicles are termed bulldozers, dozers, or caterpillars ("cats"). *Scrapers* are used for cutting earth material, transporting it, and depositing it at the proper places. The scraper may be a small one such as is used in shallow highway work or a very large one used for earth dams. The latter type often is termed "can" and is connected to a dozer or rubber-tired tractor to form a "carryall" unit, used efficiently in huge earthwork operations. For shorter hauls, bulldozers are used. They have wide blades on the front and can push soft earth. *Blade graders* ("motor patrols") do similar work on highways.

FIG. 16.3A. Power shovel loading dumpster. (*USBR photograph.*)
FIG. 16.3B. Dragline bucket. (*USBR photograph.*)
FIG. 16.3C. "Orange peel" placing riprap. (*USBR photograph.*)

For performing excavation in a concentrated area, power shovels or draglines may be used. *Power shovels* (Fig. 16.3a), which may be steam-, gasoline-, or diesel-operated, are used primarily to excavate against a steep slope and load the excavated material into hauling equipment. The digging bucket of a power shovel is rigidly fixed to a long steel beam ("boom"), and the material is excavated by pushing the toothed bucket against the slope. The bucket of a *dragline* (Fig. 16.3b) is attached to long cables, and the digging is accomplished by pulling the bucket over

FIG. 16.4. Euclid loader filling earth movers. (*USBR photograph.*)

the earth toward the machine. The dragline bucket can be replaced by an "orange peel" (Fig. 16.3c), which is a cylindrical bucket containing large controllable steel leaves or movable teeth on the bottom. This equipment often is used to excavate loose rock and deep but narrow holes (such as water wells) or to place riprap on earth dams. A *loader* is an excavating device used in large borrow pits. The machine moves along the face of the earth slope, cutting the dirt and placing it on a conveyor belt which dumps it into trucks following the loader (Fig. 16.4). For very large embankments, a conveyor belt sometimes has been used to transport the earth from the borrow pit to the dam. Excavation under water (dredging) is done by *dredges*. These excavate the material either mechanically by dragline methods or hydraulically by sucking or pumping earth in suspension. The latter type has an augerlike device which rapidly rotates, cutting the soft material and pushing it back into a pipe. It then is pumped to the surface.

When it is necessary to transport large quantities of soil rapidly over relatively long "hauls," specially built trucks with extremely large capacities (20 cu yd or more) are used. These often are termed "eucs" (abbreviation of Euclid, in this case the name of the manufacturing firm) (Fig. 16.2). A "dumpster" is another variety of a large-capacity truck (Fig. 16.3a). The dirt-carrying part of a euc may be a trailer attached to a two- or four-wheeled rubber-tired tractor. The dirt is dumped on the fill by opening gates on the bottom of the truck bed, tipping the bed sideways, or raising one end.

CUTS AND EXCAVATIONS

16.3. Rock Cuts. The shearing strength of the rock in a cut controls the method of excavation. Explosives are used in hard rocks, such as some granites or strongly cemented sedimentary rocks. In highly fractured or fissured rock, only a limited amount (if any) of explosive may be required. Decomposed shales and other materials with weak shearing strength in natural state may be loosened by a ripper or rooter. The latter is a toothed, rakelike device attached to a dozer and often used also for extracting roots and stumps. The loosened material is removed by a scraper, dozer, or shovel.

Slopes in Rock Cuts. The critical (i.e., the maximum possible) height of a rock slope is roughly and, in some cases, accurately proportional to its shearing strength. Rocks that require stronger means for breaking may stand up steeper than the weaker ones. Practically vertical and even overhanging slopes may be designed in nonfissured granites or basalts. In hard shales and sandstones, slopes of $\frac{1}{2}$:1 and $\frac{1}{4}$:1 are generally safe. Increasing fissility and decreasing shearing strength of rocks mean flatter slopes. Slopes flatter than 1:1 in rock, even in weak shales, are rare. All such rules of thumb are modified by the strike and dip of the rock formations with relation to the excavation, and the degree and direction of major joint or fracture systems. Observations on natural rock slopes and slopes in old quarries or road cuts are sometimes instructive for choosing excavation slopes.

High rock slopes should be subdivided into sections with gradual flattening toward the top (compare Sec. 13.11). The sections are separated from one another by benches (berms) that may be practically horizontal in the direction perpendicular to the axis of the cut but must have longitudinal slopes to avoid accumulation of water. The objective of the berms is threefold: (1) to lessen the rock load acting on the lower part of the slope, (2) to prevent water from the top of the cut and from parts of the slope from coming down to the grade, and (3) to prevent rock fragments and debris from falling to the grade. The berm width depends on the depth of the cut, character of the rock, and the topography above the highest point of the cut. In the general case of a slope 40 ft high in slightly decomposed shale, a berm 3 or 4 ft wide may suffice.

In higher slopes, wider berms are required. The Clear Ridge Cut near Everett on the Pennsylvania Turnpike, in sandy shales and sandstones, is 153 ft high and has two benches, one 30 ft and the other 85 ft above grade. The lower bench is 23 ft wide. The slope from grade to the lower bench is ½:1, that between the benches is ¾:1, and it is 1:1 near the crest.[7]

The stability of the cut will be increased if the strike of dipping rock is perpendicular to the axis of the cut. If the strike is parallel to the axis, the strata dip toward one slope only, and if there is danger of sliding, a *nonsymmetrical* cut with one slope flatter than the other should be designed (Fig. 16.5). Designers usually prefer symmetrical cuts, but maintenance practice shows that the increased cost due to slides and traffic interruptions make nonsymmetrical design preferable in the long run.

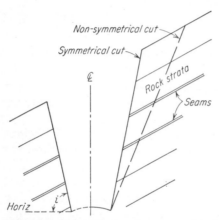

FIG. 16.5. Symmetrical and nonsymmetrical cut.

If the weaker strata in a cut have to support the stronger ones, the cut stability should be carefully studied. The slopes then may be designed by the procedure given in Sec. 16.4. Two difficulties may arise: the uncertainty of the value of the shearing strength of the rock and the applicability of the method to homogeneous materials only. The investigating geologist should consider observable features such as faults, fissures, joints, and bedding planes.

Rock Haul from Cut to Fill. If a fill is made from an adjacent cut, the average ratio between the volume of rock in the cut before excavation and that in the finished embankment should be determined. Preliminary estimates sometimes are based on the rough approximation that 1 cu yd of excavated rock makes about 1.4 cu yd of rock fill. A definite ratio should be finally established for each case, however. Filling large experimental boxes or trucks of known volume with excavated rock and averaging the results of several experiments may give some indication of the correct ratio. The geologist should give his opinion on this ratio and also know whether or not the rock is subject to detrimental alteration upon exposure because of chemical (or rather "geochemical") changes.

16.4. Earth Cuts. The most common slope ratios in cohesive earth slopes are 1:1; $1\frac{1}{4}$:1; $1\frac{1}{2}$:1; and 2:1, though others, generally divisible by $\frac{1}{4}$, are used. The height of the cut slopes (or depth of cut) is generally shown on the longitudinal or transverse profiles of highways, railroads, runways, and canals prepared during the surveys preceding construction. The heights of slopes for building sites and housing projects in hilly regions are indicated on the grading plans. Theoretically, the slope stability in cuts is determined from the steepness of the slope if the height of the latter and the shearing strength of the slope material are known. In low cuts, not over 25 or 30 ft, the steepness of the slope is usually determined by comparing the proposed slope with existing slopes safely standing in similar materials. For higher slopes stability is determined on the basis of laboratory tests and the various methods of stability analysis (Sec. 17.2); in addition to a mathematical analysis, a *careful geotechnical appraisal* of the situation should be made. The main purpose of the investigation is to determine the variability of the shearing resistance of the material at various points of the cut and also seasonally. Fissured and expansive clays are particularly critical. If there are thin fissures close to the foot of a high slope to be built in such clay, excavation relieves the stress. This permits the heavily compressed clay to rebound, which may open existing fissures. The latter then may become filled with water. The subsequent softening of the material may cause a failure of the slope. On the basis of the geotechnical appraisal and available engineering experience, the correct angle of excavation for the slope is chosen. As in rock cuts, high slopes in cohesive soils are flattened toward the top with possible inclusion of berms.

Century-old experience in road construction has proved that a proper mixture of sand and clay furnishes an excellent road material; i.e., it possesses good shearing resistance in a compacted state. Hence, a slope in sandy clay or clayey sand should stand safely at a slope of $1\frac{1}{2}$:1 (that of sand) and possibly somewhat steeper.

If there is a rigid boundary at a certain depth under the ground surface, such as a rock surface or cemented sand, the slope is reinforced by the presence of that boundary. Therefore the slope can be made steeper than in the case of a deep homogeneous soil mass. However, if the rigid boundary occurs in the face of the cut slope and dips toward the open excavation, the over-all stability of the slope may be sharply reduced, particularly if considerable or even moderate downslope water drainage occurs (Sec. 16.3).

Cut Slopes in Sand and Loess. The $1\frac{1}{2}$:1 slope used in excavation in dry sand corresponds to an angle of internal friction of about 34°. This is the average angle made with the horizon by the lee slope of a dune.

Moist sand can stand on steeper slopes until it dries out. Loess slopes may stand practically vertical[8,9] as explained in Sec. 3.15.

The humic acid formed by decay of vegetation on top of a sand deposit penetrates the sand mass and makes it absorb much water during rain. This results in a decrease in shearing strength and possible failure.[10] Formation of a crust on an exposed sand slope due to the cementing action of silicic acid generally is favorable to the stability of the slope provided this crust does not handicap the flow of water.[10]

Volume of Earth Materials. The volume of earth material in its natural condition often is designated as *in-place yards*. After removal, an apparent increase in volume takes place, and these are the loose yards used to express the carrying capacity of earth movers. Material after compaction would be *compacted yards*. If 1 cu yd is the in-place volume of a given earth material, the loose and compacted yards would be as follows:*

Type of material	No. of yards	
	Loose	Compacted
Sandy-clayey loam or common earth..........	1.25	0.90
Clay..	1.43	0.90
Sand..	1.11	0.95

16.5. Sidehill Sections. A sidehill cross section of a highway or railroad in a mountainous region is shown in Fig. 16.6. In the case of the haul from cut to fill, the volumes should be balanced in the longitudinal direction, whereas in the case of sidehill locations, the volumes should be balanced in both the transverse and the longitudinal directions. This is not a rigid rule, however, as in some cases an excess of excavated material and consequent increase in width and stability of the fill may eliminate the need for a retaining wall. Some of the waste material from the excavation placed at the foot of a retaining wall contributes to its stability (Fig. 16.6). If, however, a retaining wall has been built without preliminary estimate of the volume of material to be excavated, it may be finally completely covered with excavated material.

If a natural rock slope, very high and steep, has been exposed for a long time geologically, the rock may be locally deteriorated and fissured. Sometimes there is no exterior evidence of such damage, and geological examination of a similar slope is difficult. Therefore, the proper choice

* The above terminology and numerical data are mean values used by the Corps of Engineers, Use of Road and Construction Equipment, *War Dept. Tech. Manual* TM 5-252, pp. 45–46, 1945.

of slope of a hillside rock section (Fig. 16.6) is a very uncertain task.   In
high igneous rock slopes, the dip of natural fissures may be a guide.
(The Russians utilized such knowledge in the reconstruction of a railroad around a portion of Lake
Baikal in Siberia, located high in
the mountains.)

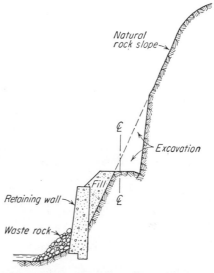

Benches used on high rock hillside slopes considerably increase
the yardage of rock excavation.   A
slight shifting of the alignment of a
hillside section, uphill or downhill,
may considerably change the yardage and often change the geological
conditions.

In designing sidehill cross sections, the engineer and geologist
should be sure that there is no
threat to the stability of the planned
section from the mountain or high
natural slopes *above* (e.g., natural
rock slope in Fig. 16.6).   Such
threats may be water torrents,
slides, isolated falling rocks, creep

**FIG.   16.6.   Sidehill   construction   in   a
mountainous region.**

of the soil mantle covering rock massifs, mudflows, snow avalanches,
etc.   In extreme cases, galleries or concrete or steel roofs (timber may be
used for temporary installations), arched or sloping, protect the way and
ensure the continuity of the traffic.

Sidehill sections often are used in valleys when a railroad or highway
or a part thereof is conveyed from a lower to a higher elevation to take
advantage of the valley gradient.   If, because of geological or other considerations, the alignment has to be shifted from one side of the valley
to the other, the valley has to be crossed by a bridge or a high fill with
large culverts.   Proper location of such bridges or fills may be a matter
of serious geological concern (Sec. 14.8).

EMBANKMENT FOUNDATIONS (PRIMARILY OF RAILROADS
AND HIGHWAYS)

16.6. General Criteria.   All modern embankments are (or at least
should be) built of properly selected materials with proper compaction.
If this is done, defects will be absent in the interior of the embankment
and any deformations, including settlement or readjustment of the material, generally will occur during or shortly after construction.   The

embankment and its foundation represent a unit and act together. In discussing deformations of embankments in this chapter, "consolidation" refers to the readjustment and subsequent compression of the soil in the embankment proper; "settlement" is similar readjustment in the foundation of the embankment.

The best embankment may fail because of a defective foundation. A defective foundation possesses or may possess insufficient resistance to shearing and other stresses acting on the embankment and its foundation. Three major types of defective embankment foundations are as follows: (1) Foundation material is or may become soft to a considerable depth, (2) soft material extends only to a limited depth and is underlain by harder strata, and (3) a foundation consists of alternate strata of hard and soft materials. In the regions with expansive soils as California, Texas, or Oklahoma a foundation failure generally may occur under only a very low embankment (or in a cut), since with a high embankment the expansion pressure is balanced by the weight of the embankment (see Refs. 12 and 14, and also Secs. 4.10 and 13.12 of this book).

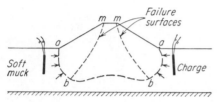

FIG. 16.7. Blasting to sink a fill.

A thorough exploration of the foundation, including boring, sampling, testing, and study of local topography and ground-water conditions, should precede the design of a large embankment.

16.7. Deep, Soft Foundations. These are primarily peat, swamps, and deep, soft clay deposits. The most effective corrective method is to remove the undesirable material completely and replace it with selected material. Where the cost of such an operation is too high, the following methods may be used. *Peat* is improved by excavating its upper part and displacing the rest by constructing a sinking embankment which will squeeze out the underlying soft material. Sinking of the embankment may be accelerated by *jetting* it, i.e., filling its pores with water to increase its unit weight. *Blasting* can be used advantageously in peat excavation and displacement methods.[9] Figure 16.7 illustrates the sinking of a fill by using explosives at the toes of the slopes. In this case, hydrostatic pressure on sides *ab* of the fill is relieved by blasting. This causes an instantaneous increase in the weight of the wedges separated from the body of the embankment proper by shearing surfaces *mb*. If the embankment crosses water, blasting in the water may cause longitudinal fissures along *mm* in Fig. 16.7.

Construction on a swamp covered with a thick mat of decayed vegetation (Sec. 3.22) requires that the mat be first sunk to the bottom of the swamp. This often is done by building a low fill on the mat and skill-fully cutting the mat so the fill and the mat sink together to form the base of the final permanent embankment. The embankment will settle, however, before it attains a state of definite stability. Compaction generally is not used in such cases.

An embankment built on top of a soft clay deposit gradually settles because it causes the clay to consolidate. As consolidation proceeds, however, the bearing capacity of the clay gradually increases and the

FIG. 16.8. Ponding loess foundation of Trenton Dam, Nebraska. (*USBR photograph.*)

settlement of the embankment may stop. Soft clay deposits are often covered with a hard crust. This crust may safely carry a fill several feet high, but if the weight of the fill breaks the crust, the fill will sink and may even disappear with formation of "mud waves" or bulging on the sides. Thus, careful investigation is required where high embankments are to be built on soft clay, even if the clay is covered with later, firmer deposits. Very often in such cases, the embankment slopes are flattened sufficiently to reduce the per-square-foot load on the underlying clay to a value less than that which will cause destructive settlement.

An apparently reliable foundation may become soft because of moistening, swelling, or chemical deterioration; therefore, caution should be exercised in the design. Loess and loess derivatives are especially sensitive to water (Sec. 3.16); these soils, when completely saturated, slump or settle considerably.

Loess foundations of earth dams in Nebraska[11] were consolidated by boring holes along the up- and downstream slopes of the dam and injecting thick slurry through 2-in. pipes. Initial weight of loess averaged 75 to 87 pcf, and increased by 20 per cent after treatment. The slurry was prepared from local materials. Similar work was done on irrigation canals to stop leakage. Other attempts have been made to "preconsolidate" loess foundations by flooding them prior to construction (Fig. 16.8). Incomplete saturation practiced in this case caused immediate but only partial settlement when the embankment load was placed on loess. Complete preconsolidation was not feasible for economic reasons because of the great depth to the water table.

Alluvial deposits of considerable thickness generally are satisfactory for embankment foundations. For embankments containing reservoirs,

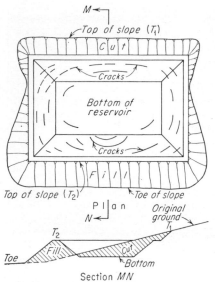

FIG. 16.9. Cracks in lining of water reservoir.

however, alluvium deposited on sloping ground by turbulent streams, e.g., some alluvial fans, may be unsatisfactory. Failures on such foundations have occurred in arid and semiarid regions, particularly in reservoirs built half in excavation and half in embankment (Fig. 16.9). In such cases, the soil in the excavation and under the embankment is very dry; water from rainstorms runs over it practically without moistening it. The foundation materials may contain sand, gravel, or even boulders with a very fine sand binder that is susceptible to water action. Because of the character of their deposition, such materials usually are found in a state of limit equilibrium (i.e., with a safety factor close to 1) and arched in some places. A considerable increase in moisture content easily destroys this equilibrium and breaks the arching. Cracks then are formed in the reservoir lining and may extend beyond the limit of the reservoir slopes (Fig. 16.9). A similar crack pattern also would develop around an excavation in loess if filled with water.

Reservoirs built on sloping ground in any locality may be damaged by a heavy influx of subsurface water even before lining is placed. As in any other excavation under such circumstances, drainage is indicated.

16.8. Shallow, Soft Foundations. Such foundations often are found where a highway or railroad passes close to the edge of some water basin (pond, lake, etc.).   Particularly, many highways close to the ocean shore and Gulf of Mexico are in this condition. As a rule in such cases, the swampy clay or silt layer is only several feet thick and there is a ground-water flow under the embankment toward the water basin; drainage therefore is advisable.   The subsurface flow should be intercepted and conducted to the basin through the embankment by one or a few culverts. A somewhat similar situation occurs when the planned embankment has to cross a large, shallow depression in a flat country.

If the proposed embankment is high and, hence, heavy, it may be advisable to remove the soft part of the shallow foundation partially or completely.   The embankment then is founded on underlying firm ground, provided the latter is not steeply sloping and is above the ground water. If the embankment is low, the solution may consist in removing a few feet of the soft foundation and replacing it with compacted coarse sand

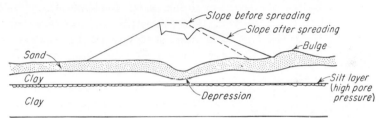

FIG. 16.10. Embankment spreading.   (*After Terzaghi.*)

or, better yet, gravel and rock.   A thick crust covering the soft material and capable of supporting a low embankment or fill by partial arching should not be removed.

16.9. Foundations of Alternate Hard and Soft Strata.   Failures of such foundations often occur because of a deficient (or completely lacking) soil investigation.   Such is, for instance, the case of a soft layer sandwiched between two harder layers and overlooked in the investigation.   There may be a settlement of the embankment and cracking of the pavement or, exceptionally, sliding of the upper layer along a slippery underlying material.

An especially grave situation develops if cohesionless sand or silt is enclosed between two impervious (e.g., clay) layers and water *under pressure* (e.g., from an adjacent high slope or hill) is driven into that pervious material, thus creating a high pore pressure in it.   The embankment instability then is similar to that of a masonry dam on a stratified rock foundation in which high pore pressure has been built up (Sec. 15.8). The pressure in the sand or silt tends to lift the embankment where its weight is at a minimum, i.e., close to the toes of the slope (Fig. 16.10). Since there is no normal pressure on the sand layer under the toes of the embankment, the shearing strength of the sand at those points is zero.

The sand layer fails by wrinkling and spreading, and the embankment above fails in a manner similar to the fall of a person on a rug that has slid and wrinkled under the victim's foot. The spreading embankment failure is one-sided, i.e., in the direction offering the least resistance to sliding. Glacial regions with varved clay deposits formed by intermittent clay and sand layers are favorable for the occurrence of such spreading failures.

16.10. Settlement of Embankments. The decrease in height of a properly compacted embankment by consolidation is insignificant; thus, the detrimental settlement of the embankment, if any, is controlled by the compressibility or yielding of its foundations. Often the settlement of an embankment is rather inconsequential, as, for instance, in the case of the settling of railroad embankments. These can be easily repaired by periodically tipping new material on top of them from ballast trains. Long-time settlement of earth dams is compensated for by the *camber*, i.e., a slight longitudinal hump placed on the embankment during construction to make the dam a little higher at its center than at its abutments. Highway engineers often postpone for 2 or 3 years the placement of the final pavement on a fill that is expected to settle considerably. Where housing projects are to be constructed on fills, especially on reclaimed lands, the amount of the settlement of the fill is particularly critical (also see Sec. 13.18). The fill settlement may be compensated by adding to the fill sufficient earth to equal approximately the value of the expected settlement. This difficult problem requires experience and judgment, since the settlements of actual fills are often in disagreement with computations by the theory of consolidation.

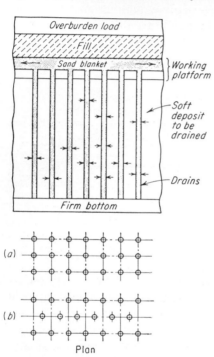

FIG. 16.11. Vertical sand drains.

The settlement of an embankment or a fill cannot be avoided, but it can be *accelerated*. If the fill is to support structures, this permits earlier use of the fill for construction purposes. One method of such acceleration consists of the use of vertical sand drains ("California wells") as described in Refs. 13 to 16.

The sand drains are usually from 14 to 20 in. in diameter and spaced from six to nine diameters center to center, on either a square or triangular pattern (Fig. 16.11, plan *a* and *b*).   On a very soft deposit, a shallow fill is first put down to form a *working platform*, which is covered with a blanket of clean, pervious sand 3 ft or more thick. The drains are installed by driving a casing with a valve at its lower end; sand is poured into the casing as the latter is withdrawn.   When the drains are completed, the proposed fill is placed with a temporary *overburden load* on top (usually earth piled several feet high).   This temporary load squeezes out the water from the soft deposit into the vertical drains and then into the horizontal sand blanket, which acts as a drain.   At the end of the process, the overburden load is removed.   In this method, advantage is taken both of the horizontal permeability of the soil, which generally is greater than the vertical, and of short paths of percolation.   Most sand drain installations have been successful.   Accelerated consolidation on a different principle as used in Sweden is discussed in Ref. **17**.

EMBANKMENT CONTROL

16.11. Compaction.   The shearing and compressive strength of an earth material can be increased by *soil stabilization*.   This process sometimes involves admixtures of portland cement and bitumens or chemicals.   *Compaction* is the artificial densification of an earth mass and is a particular kind of soil stabilization.   Usually it is the simplest and often the most efficient and economical stabilizing process for both cohesive and free-draining materials.   In many cases, compaction is the principal process in making new land to support structures on marginal areas reclaimed from water, on swampy land, or simply on uneven or sloping ground that is leveled by importing earth materials that are placed and densified.   The density of the newly compacted fill land, *if properly built*, usually equals or exceeds the density of the natural ground.   Structures may be built on such new ground as though they were on natural ground, and a duly compacted (sometimes called "engineered") fill may even serve as a bridge abutment as is the case of some California bridges. Knowledge of the basic principles of compaction thus is becoming compulsory for the engineer and the geologist.   Proper compaction ensures stability, minimum consolidation, and an acceptable percolation rate of the fill.

Compaction Equipment.   Large areas generally are compacted by *tamping rollers*, i.e., cylindrical drums several feet in diameter on which projecting feet are mounted.   The latter often are shaped like a sheep's foot, hence the term *sheep's-foot roller* (Fig. 16.2).   Feet of other shapes are also used in tamping rollers.   A three-drum sheep's-foot roller can compact about 150 cu yd/hr with 12 passes of the roller over a 6-in.-thick layer of earth.   Tamping rollers compact the material to the full length of their feet and often deeper.   Rubber-tired (pneumatic) rolling equipment and smooth (flat) steel-tired rollers also are used, the latter chiefly on highways (Fig. 16.12).   Another compaction device combines vibration with a heavy, rubber-tired roller (the Cedarapids compactor).

Small areas (e.g., backfill of a footing excavation, Sec. 13.11) can be compacted by a simple hand tamper consisting of a small, flat steel plate on the end of a shovellike handle or, better, by hand-held power-actuated *tampers*.   Such tampers have a cylindrical-shaped foot on the end of a steel shaft that is driven up and down by air pressure or eccentric cams. A variation of this tamper has three such tampers placed in a wheelbarrow-like holder and operated by one man.   The three feet pound the earth at alternate intervals, and thus the machine can be "walked" (Triplex tamper).

FIG. 16.12. Flat roller compacting fill.   Note upraised scarifier blades on tractor used for harrowing fill.   (*USBR photograph.*)

Embankment Density.   The earth to be compacted is spread in layers or *lifts*, generally 6 in. or less in thickness (also see Sec. 16.14).   For a given soil, there is a certain *maximum dry density* (maximum unit weight in dry condition at 100°C) which can be obtained by a given *compactive effort*.   The latter depends upon the type of compaction equipment and the way it is used (e.g., the number of passes it makes), and the final dry density depends also on the moisture content of the material being compacted.   Clayey and silty soils cannot be compacted to dry densities of over 100 or 110 pcf, even with the highest available compactive effort. Sandy soils and sand-gravel mixtures can be compacted to 120- to 125-pcf dry densities.   Actual ("wet") densities obtained in the field are above these maximum values.   With a rock content below one-third by weight, very satisfactory compaction can be obtained.   Fill containing over two-thirds rock by weight cannot be effectively compacted by usual methods. It is good practice to eliminate rocks having any dimension greater than 6 in. from the usual earth fill.   Rocks larger than this prevent the equipment from exerting complete compactive effort on the underlying fill, as the rock tends to bridge over the underlying material.   The term "rock" generally used in connection with embankment compaction is *purely conventional* and means pebbles or rock fragments retained on a No. 4 sieve (4.76-mm openings). Fines, or the *binder*, are particles passing a No. 200 sieve (0.074-mm openings). Sand containing small admixtures of clay or fines gains in stability because of an increase in cohesion during compaction.

Compaction Test. This test is used to determine the so-called *optimum moisture content* at which the given soil has to be compacted in order to attain the *maximum dry density*. The laboratory compactive effort is

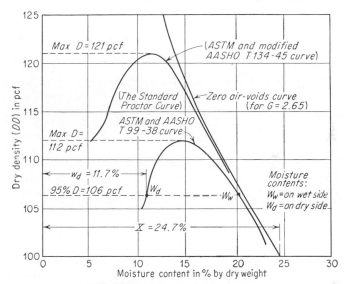

FIG. 16.13. Moisture-density curves.

standardized, and it is assumed that it duplicates the field compaction effort, and vice versa. The material is compacted at various moisture contents, using a *fresh* portion of soil (i.e., not as yet used in the test) for each moisture content applied. As the moisture content increases, dry density D increases to a certain maximum but then decreases. Plotting dry densities against moisture contents gives a *moisture-density curve* corresponding to a given compactive effort. In Fig. 16.13, two such curves are represented. (Note the ASTM and AASHO designations for the respective compactive efforts, printed next to the curves.)

A greater compactive effort (upper moisture-density curve in Fig. 16.13) results in greater density at smaller optimum moisture content. The material compacted on the fill under construction is considered acceptable if its density is 90 or 95 per cent of the maximum laboratory density. This percentage is established by specifications for each project. A horizontal line below the point of optimum moisture content of a moisture-density curve intersects the latter at two points, one on the "dry side" and the other on the "wet side" of the optimum (symbols W_d and W_w in Fig. 16.13). Only in particular cases is the choice of the correct moisture content obvious. Very hot weather during rolling suggests that the fill be worked on the wet side, whereas such materials as decomposed granite may give successful results when rolled on the dry side regardless of the temperature. Perfectly correct values of maximum density and optimum moisture content can be determined only by setting up field test sections, which is sometimes done on large projects.[18]

The "zero-voids curve" in Fig. 16.13 indicates the moisture contents for various dry densities if the material is fully saturated, i.e., all its pores are filled with water. It means that in a regularly compacted earth material, there is *some air* in the pores.

In the standard or Proctor* compaction test (lower curve in Fig. 16.13), the soil is compacted in a steel cylinder having a volume of $\frac{1}{30}$ cu ft, using 25 blows of a 5.5-lb hammer dropped freely from a height of 12 in. on each of three soil layers of equal thickness. In the modified Proctor test (upper moisture-density curve in Fig. 16.13) a 10-lb hammer is dropped from a height of 18 in. on each of five soil layers of equal thickness in the standard cylinder. The wet unit weight of the compacted material in pounds per cubic foot is obtained by multiplying the weight of the compacted material in the cylinder by 30. If the moisture content of the compacted material is W as expressed as a fraction of one unit (e.g., 8 per cent = 0.08), the dry density would be obtained by dividing the wet unit weight by $(1 + W)$.

An allowance of 2 per cent (plus or minus) in the optimum moisture content generally is granted.[19] As a horizontal line intersects the moisture-density curve at two points, a dry density can be obtained by compacting the material on the *wet* or on the *dry side* of the laboratory optimum moisture content. In dry side compaction, added roller passes may increase the density.[20] However, the shearing strength rather than the density is the true criterion for appraising the quality of the compacted material. This is because the shearing strength really controls the resistance of the material to the shear stresses tending to deform it. Figure 16.14 shows the results of a test in which the maximum shearing strength, determined as one-half of the unconfined compressive strength, corresponded to about 92 per cent (and not 100 per cent) of the maximum density.[21] The CBR method (Sec. 4.18) has shown that after the density of compacted materials has reached a certain value, its shearing strength starts to decrease.[12]

Compacted soils sometimes tend to swell or to consolidate further. The former occurs if the soil has pronounced expansive qualities, is *overcompacted*, or both. Too many passes of a roller in certain cohesive soils apparently may cause residual stresses in the earth mass and thus a tend-

* After R. R. Proctor of Los Angeles, California, who introduced this test.

ency to rebound.   Consolidation after compaction may occur if a soil
mass has been compacted on the dry side and at too low a moisture content.[22]   Chemical or, more appropriately, geochemical changes also may
occur.[23]

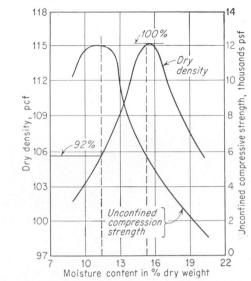

FIG. 16.14. Relationship between dry density and unconfined compressive strength of a
compacted material.[15]

16.12. Laboratory Control of Embankments.   *Before* the compaction
test is made, the designer should be assured that the foundation material
at the selected site will support the embankment safely, particularly without detrimental settlement.   Expansive soils, both *water-sensitive*, as discussed in Sec. 4.10, and *frost-sensitive*, as discussed in Sec. 10.3, should
not be used in the upper parts of the embankments supporting slabs (e.g.,
highway pavements) or light structures that may be heaved and damaged.
The test for expansion in suspicious cases should also precede the compaction test.   All materials chemically changed or changeable should be
carefully examined before using them in a compaction test.   For example, dams are rarely built from tailings (crushed rock) resulting from
mining or ore milling operations, though large quantities of tailings are
available in many potential dam locations, particularly in the Western
United States.   In fact the tailings may contain chemicals detrimental
to the density and permeability of the embankment.

After the compaction test is performed, the material compacted to the
proper density and moisture content should be tested for shear and probable settlement.   In the case of dams and other structures that are to
hold water, the compacted soil also should be tested for permeability.

Shear Tests. Highway and railroad embankments and small dams seldom require shear testing of the embankment materials. Careful examination of the foundation is recommended in all cases, however. For fills covering large areas (such as runways or housing projects), unconfined compression tests will suffice, and the foundation should be tested to a depth equal to 1½ to 2 times the thickness of the thickest portion of the fill. The sampling of the foundation material should be done at the periphery of the fill rather than at the center, since shear failures seldom occur at the center. For large dams, shear testing is done in triaxial compression devices.[24] If the specimens contain a considerable percentage of rock, direct shear tests previously have been used. However, triaxial compression devices of sufficient size now are available to perform tests on samples containing large percentages of rock. USBR investigators have noted that an admixture of 10 per cent rock does not affect the shear value of a soil.[22]

Consolidation Tests for Estimating Settlement. Laboratory consolidation tests for large dams generally are conducted in large ring consolidometers (e.g., 4¼ in. diameter by 1¼ in. high). For foundation soil tests, specimens are cut to fit the device; embankment soils are tested (when necessary) by compacting the material into the consolidometer at anticipated moisture content and density conditions.

Consolidation tests also are important for fills covering large areas. If the fill is to be placed on top of a thick, compressible deposit, borings should be made to firm material and samples taken of representative materials. The settlement at different sections of the area may be variable. Correlation between the consolidation laboratory and field data is not always favorable; field data often show smaller settlements.

Permeability Tests. Permeability tests should be performed on soils to be used in earth structures intended to hold water, such as earth dams, canals, and reservoirs. If the slopes of these structures are to be covered with lining, only the lining permeability may be tested. Impervious core material for earth dams also may be tested. Permeability of compacted material cannot be tested in the field until the actual structure is built (except, perhaps, in test sections, if any). Therefore laboratory testing only is used in this case.

16.13. Field Control of Embankments. Borrow material that is too wet for proper compaction should be placed in windrows (piles) and dried; if too dry, it should be moistened by sprinkling the placed lifts before compaction (Fig. 16.2) or sometimes sprinkling the borrow area or both. To assure that the fill is being placed in accordance with the laboratory compaction characteristics (dry density and moisture content), samples are taken from each lift and tested[25] (ASTM Designation D420–45). Figure 16.15 illustrates making a hole in the fill, extracting earth from it for weighing and determining the moisture content, and filling the hole with sand (or oil) from a calibrated flask.   For measuring moisture content in a fill, a special penetration-resistance Proctor needle may be used. The needle readings are calibrated against moisture contents of the earth material determined in a usual way (e.g., by oven drying).

FIG. 16.15. Field density test.   (USBR photograph.)

16.14. Free-draining Materials.   Well-graded mixtures of sand and gravel may be used in the downstream sections of an earth dam to provide a free-draining layer.   Concrete floor slabs and, often, large footings are placed on sand and gravel to permit more even distribution and spreading of high loads to underlying weak soils.   Sometimes buildings are founded on compacted sand piles (Sec. 13.6).   In all these cases it is necessary to compact the sand and gravel to a density sufficient to prevent detrimental consolidation.

For large embankments, such as dams, the sand-gravel portions of the fill can be compacted most efficiently by a combination of *vibration, weight application,* and *sluicing.*   Special equipment such as the Cedarapids compactor afford one method.[26]   Another successful approach is to place the free-draining material in 12- to 36-in. layers, sluice it thoroughly with streams of water from hoses (Fig. 16.16), and then make four or more passes over the whole fill with a heavy, caterpillar-tread tractor (40,000 lb or more).   Satisfactory relative densities (Sec. 4.8) of 70 per cent or higher can be obtained by this method.   Smaller areas can be compacted by hand-held vibrator compactors, such as the Triplex tamper, which can achieve 80 to 90 per cent relative density at rates of 22 and 10 cu yd/hr, respectively, and tamps at a rate of about 2,000 blows per minute.[27] When free-draining materials are compacted, the sluicing water should not be permitted to erode the embankment or tend to float the soil particles.

FIG. 16.16. Placing and sluicing sand fill, Trenton Dam, Nebraska.   (*USBR photograph.*)

EARTH DAMS

16.15. Design Criteria. An earth dam basically is a trapezoidal embankment or fill built in a valley to form a water reservoir. It has to be impervious enough to prevent *excessive* loss of water from the reservoir. The design also has to ensure stable slopes. The postconstruction settlement of the dam crest must not be so great as to reduce the freeboard to a dangerous point. The upstream slope of the dam must be protected from the destructive action of the waves, and the downstream slope must withstand rainfall erosion. There must be sufficient bond between the embankment and its foundation to prevent, as much as possible, the development of detrimental seepage paths (piping); excessive hydrostatic uplift must be controlled by proper drainage.

Where adequate earth material is available, construction of an earth dam often is preferable to that of a concrete dam. In fact, even high earth dams can be built on earth or *poor* rock foundations, a practically impossible foundation for a high concrete dam. It usually is more economical to construct an earth dam than a concrete one in order to block a *broad valley* (Fig. 16.17). The broad crest of an earth dam will efficiently accommodate a highway where it is necessary to route roads across the valley.   Earth dams are favored in cold climates because of their resistance to the damaging effects of freezing weather.   Construction of an earth dam may prove very expensive, however, if the broad base of the dam necessitates long diversion tunnels or conduits to handle the river during construction.[28b]

The *permeability of the fill* may be controlled by various means.   A reinforced-concrete core wall (Fig. 16.21a) constitutes an almost impervious diaphragm.   However, as concrete can withstand very little settlement, the wall must be placed on sound foundations.   If the wall cracks,

FIG. 16.17. Rolled fill dam and ungated overflow spillway, Medicine Creek Dam, Nebraska. (*USBR photograph.*)

water will force its way through the surrounding earth and create dangerous piping conditions (Sec. 16.19).   Diaphragms also may be made of interlocking steel or timber sheet piling.   These generally are driven normal to the stream and at the longitudinal axis of the dam or slightly upstream from that axis.   The most commonly used diaphragm is an *impervious-earth core* constructed of compacted or puddled fine-grained materials of low permeability.

16.16. Types of Earth Dams.   Earth dams often are classified according to method of construction.   In a *hydraulic-fill dam* (Sec. 16.20 and Fig. 16.18d) material of the borrow pit is transported to and placed in the fill by means of water.   In a *semihydraulic-fill dam*, earth material is excavated and transported to the damsite mechanically but is placed in the fill with water jets.   The *rolled-fill dam* is built with compaction

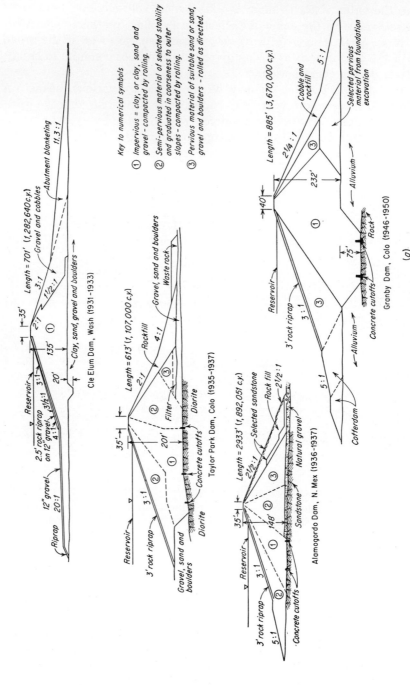

Key to numerical symbols

① Impervious = clay, or clay, sand and gravel - compacted by rolling.

② Semi-pervious material of selected stability and graduated in coarseness to outer slopes - compacted by rolling.

③ Pervious material of suitable sand or sand, gravel and boulders - rolled as directed.

Cle Elum Dam, Wash (1931-1933)

Taylor Park Dam, Colo (1935-1937)

Granby Dam, Colo (1946-1950)

Alamogordo Dam, N. Mex (1936-1937)

(a)

FIG. 16.18a. Zoned rolled-fill dams.

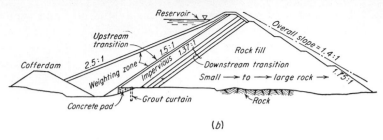

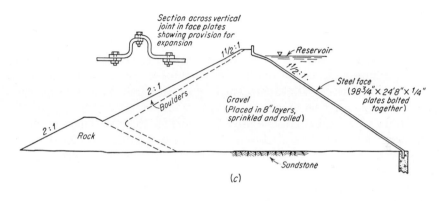

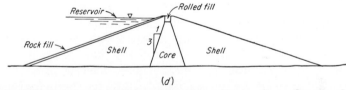

FIG. 16.18*b*. Rock-fill dam, compacted impervious face, Kenney Dam, Canada, 325 ft high.[32]

FIG. 16.18*c*. Steel-faced rock fill, El Vado Dam, New Mexico, 170 ft high, built in 1934–1936. (*USBR.*)

FIG. 16.18*d*. Hydraulic-fill dam. Note steep slopes of core.

equipment, and the earth material is hauled by earth movers. Hydraulic-fill-type dams seldom are built in modern practice owing to the difficulty in analyzing their potential stability and their past record of failures.

Earth dams also may be classified according to the characteristics of the fill. An earth fill may be zoned or homogeneous. A fill may be made exclusively from rock. A *zoned* dam is a rolled-fill dam composed of several layers, or *zones*, that increase in permeability from the core toward the outer slopes (Fig. 16.18*a*). The material of the core and the zones is placed and rolled simultaneously by the usual compaction methods.   The number of zones depends on the availability and type of borrow material.   The zones are numbered by symbols 1, 2, and 3, and the higher the number of the zone, the greater its permeability (see "Key to numerical symbols" in Fig. 16.18a).   If there is only one type of borrow material, a *homogeneous fill* may be constructed (Fig. 16.18a, Cle Elum Dam).   Stability of a zoned dam is mostly due to the weight of heavy outer zones.   To compensate for the absence of these favorable loads in a homogeneous fill, its slopes are *flattened*, which also contributes to the seepage control by decreasing the velocity of the percolating water (Sec. 16.19).   A *rock-fill dam* is built of coarse rock and is provided with an impervious upstream face, an impervious core, or a similar arrangement (Sec. 16.21).

16.17. Slope Protection.   Usually, waves are caused in the reservoir by wind action and, exceptionally, by earthquakes or massive slides of the reservoir slopes.   The upstream face of an earth dam is protected against this wave action by riprap, a layer of broken rock of specific grading underlain by graded gravels (Sec. 8.1 and Fig. 16.3c).   These prevent fine soils of the embankment core from being sucked out through the riprap by water action and also prevent the heavy riprap blocks from sinking into the core.

The downstream slope of an earth dam can be protected by covering it with graded gravel or crushed rock or by seeding with protective grasses.   The covering is intended to prevent erosion but must not permit water to accumulate ("pocket") and saturate the fill.

16.18. Slope Design.   In order to keep a uniform vertical pressure at all points of an earth dam, the slopes of the dam are gradually flattened from its top toward the foot of the face.   This is generally why the steepest slopes are at the top of a dam.*   Average slopes for the upstream face built of soil materials are $2\frac{1}{2}:1$ or $3:1$ below the maximum reservoir water level (water line) and $2:1$ above the water line.   The downstream faces generally are $2:1$ or flatter.   Generally, the downstream slope (and occasionally the upstream) includes one or more berms.   Usually these are spaced about 50 ft apart vertically and are provided with proper drainage for surface water.   When material is available, *rock-fill toes* are placed either at the downstream face only (Fig. 16.19a) or at both faces.   These toes tend to increase the stability of the dam and control the seepage.

The finer the particle sizes used in the fill, the flatter the slopes.   Thus, an embankment composed entirely of homogeneous silts may have slopes as flat as $4:1$ below the water line.   With a predominance of clays, dams have sometimes been built with $10:1$ slopes near their base.   The slope angles of a dam also are dependent upon the competency of the underly-

* Slope values in various dams are listed in pp. 778–781 of Ref. 29.

ing foundations. The less competent the foundation, the flatter the slope. By such means the heavy embankment load is more widely distributed over the underlying soft materials; this reduces foundation set-

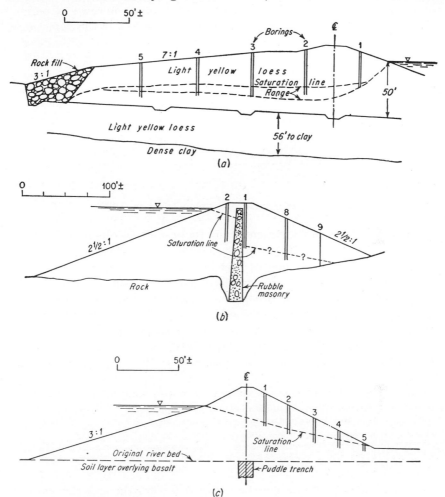

FIG. 16.19. Actual saturation lines in earth dams: (a) Boz-Su Dam, Russia; (b) Titicus Dam, Canal Zone; (c) Ashti Tank (dam), India. (After R. C. Haven. See "Additional Reading" at end of chapter.)

tlement and danger of sliding in the foundation. Stability of slopes and their design are discussed in Sec. 17.2.

16.19. Piping and Seepage. Internal erosion of the foundation or embankment caused by seepage is known as *piping*. Commonly, erosion starts at the downstream toe and works back toward the reservoir, forming channels, or "pipes," under the dam (Fig. 16.20a). The channels follow paths of maximum permeability and may not develop until many years after construction. The water first may emerge as a small spring, which gradually increases in size, and when muddy water appears at the toe, a failure may occur within hours. Sometimes, however, a permanent "boiling" of a sand-water suspension at the toe of the dam, i.e., a continuous movement of sand up and down, may not necessarily be serious but does require attention. When an embankment pipes (Fig. 16.20*b*), it is evidenced eventually by a progressive backward sloughing or raveling of the saturated downstream slope. Both types of piping may be combined.

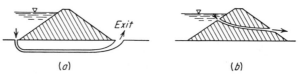

FIG. 16.20. Final stages of piping: (*a*) through foundation, (*b*) through fill.

Resistance of the embankment or its foundation to piping depends on the plasticity of the soil, its gradation, and degree of compactness. Compactness, in natural soil deposits such as the foundation, may be characterized by its dry density. Plastic clays with a PI value of over 15, both well and poorly compacted, are the most resistant materials. Minimum piping resistance is found in (1) poorly compacted, though well-graded, cohesionless materials with practically no binder and (2) very uniform, fine, cohesionless sand, even though well-compacted.[30] Settlement cracks, even in resistant materials, also may produce piping.

Apishapa Dam in Colorado was constructed of two different soils. The lower materials settled more than the upper, causing the upper soils to arch across the valley and form a horizontal crack between the two materials. When a heavy rain raised the reservoir level above the crack, piping through the dam started. In a few hours, the 112-ft-high dam was practically gone.[30] This arching phenomenon sometimes is termed "roofing."

Piping Control. Piping can be avoided by lengthening the path of percolation of the water within the dam and its foundations. This, in turn, decreases the hydraulic gradient of the water flow and hence its velocity. The path of percolation can be increased by cutoff walls, impervious cores, and impervious blankets extending upstream from the upstream face (Fig. 16.21*a*, *b*, *c*) and by widening the base of the dam, particularly where blanketing materials are scarce.

Seepage Control. There is seepage or continuous movement of water from the upstream face of the dam toward its downstream face. The upper surface bounding this continuous stream of percolating water is the *phreatic line.**   The pressure at the phreatic line is atmospheric, but below it, positive *pore pressures* exceeding the natural hydrostatic pressure in water may develop.   Above the phreatic line, there is a fringe of capillary water with negative pressures or tensions in it (Sec. 4.2) which *also moves* toward the downstream face.

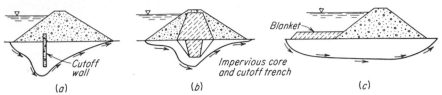

FIG. 16.21. Methods to increase the path of percolation of water seeping through a dam foundation.

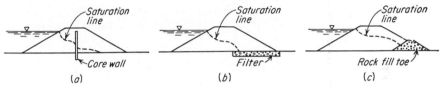

FIG. 16.22. Various types of seepage control.

The phreatic line can be kept at or below the downstream toe by properly designed cores or core walls (Fig. 16.22a).   Inverted filters of sand and gravel may be used to cover the area below the downstream toe. This provides an area for percolating water to escape without damage to the dam (Fig. 16.22b).   A rock toe similarly will divert the phreatic line (below the downstream toe, Fig. 16.22c).   Cutoff trenches, properly backfilled, not only may sharply depress the phreatic line but may stop practically all seepage into the downstream section of the dam.

If the rock foundation contains open seams, they may be sealed by grouting under pressure.   Blanket grouting, i.e., comparatively shallow, low-pressure grouting over the entire foundation area, sometimes is used. Earth dam foundation grouting is basically similar to that used for masonry dams (Secs. 15.19 to 15.25).[31]

16.20. Hydraulic-fill Dams.[29]   The material for these dams is obtained by directing high-pressure streams of water (100 to 150 psi) against the exposed face of the borrow pit.   The resulting suspension of soil and water is transported to the damsite by channels (sluiceways) and is discharged at the "sluice" (Fig. 16.23).   A temporary dike (small embankment in Fig. 16.23) prevents the suspension from overflowing down the slopes.   As the suspension flows toward a central pool, coarse particles are deposited on the "beaches" and finer particles reach the central pool to form the impervious central core.

*The term "saturation line" is also used.

A hydraulic-fill dam consists basically of a *core* and a *shell* or *shoulder* (Fig. 16.18*d*). The core serves as a more or less impervious membrane (as in masonry dams), and the outer shell assures the stability of the dam. During construction (Fig. 16.23), the outer shell has to resist the lateral pressure exerted on it by the still fluid core. Failures have occurred where the shell weight was insufficient to resist this pressure. As the core consolidates, this pressure gradually decreases owing to the dissipation of the pore water from the core into the pervious shell. Very careful design of the outer shell is required to prevent the development of high pore pressures in the semifluid core.

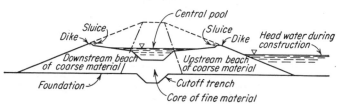

FIG. 16.23. Construction of hydraulic fill.

The suspension delivered to the fill must not contain too many fines. Otherwise, the core, which forms almost automatically in the central pool, will be disproportionately large. Excess fines can be removed by siphoning them from the pool. If, however, the borrow is deficient in fines, an additional pit with the required content of fines must be located. The design then is based on a mixture of the two pits. If this is not done, the core will be too pervious. The shoulders must contain a reasonable percentage of stones and cobbles.

Suitable and economical borrow material contains all grain sizes from clay to gravel with good gradation. Colloidal materials and an excess of materials finer than 0.1 mm should be avoided. Glacial deposits often are good borrow sources because of their erratic composition. Generally they contain all soil and rock sizes, although their grading may be poor. Some alluvium in mature valleys may be acceptable providing there is not too much fine material.

Figure 16.24 shows the sizes of core material in 14 hydraulic and semihydraulic fills. Curves 9 and 12 are Calaveras Dam, California, and Alexander Dam, Hawaii, respectively. Both dams failed during construction but were repaired. Curve 13 is for Cobble Mountain Dam, Massachusetts. Curve 14 is for Quabbin Dike, Massachusetts. The first and second largest hydraulic fills in the world are Fort Peck Dam, Montana, and Kingsley Dam, Nebraska, respectively.

16.21. Rock-fill Dams.[31-34] All rock-fill dams built prior to 1942 and the majority built since consist of three basic elements: (1) a loose rock-fill dump, which constitutes the bulk of the dam and resists the thrust of the reservoirs; (2) impervious facing of the upstream slope with concrete,

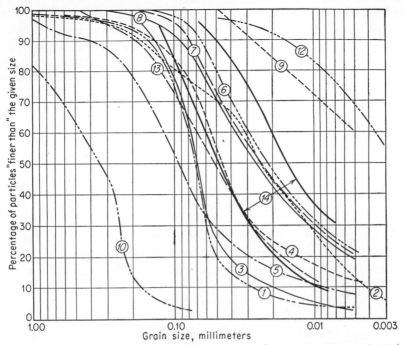

FIG. 16.24. Core material of several hydraulic fills. (*From ASCE Trans,* vol. 102.)

FIG. 16.25. Drew Dam, Oregon. Timber face on 8-in. concrete on rock fill. (*USBR photograph.*)

timber (Fig. 16.25), steel (Fig. 16.18c), or gunite; and (3) rubble masonry between (1) and (2) to act as a cushion for the membrane and resist destructive deflections. Composite dams are formed by replacing some of these elements with others having equivalent performance. For example, the rubble cushion may be replaced by a graded filter, and the impervious facing may be replaced by an impervious core.

Rock-fill dams may prove economical in localities where (1) concrete is expensive, (2) sufficient and adequate earth materials for a rolled or hydraulic fill are lacking, (3) foundations are not favorable for a masonry dam, (4) proper quality rock in sufficient quantity is available, (5) frost or permafrost would make the excavation of rolled- or hydraulic-fill materials very costly, or (6) there is considerable earthquake hazard.   The great number of earthquakes in California may partially account for the world's greatest concentration of rock-fill dams in that state—25.

Rock-fill dams have been designed with 2.5:1 to 3:1 slopes.   Recent practice, however, is to use considerably steeper slopes, such as 1.3:1 and 1.4:1.   Rock fill forming the main part of the dam must be of sound rock. Rocks that slake or severely weather such as shale should not be used. Rock should not be used that because of blasting or the transportation process shatters into small pieces or dust.

The disadvantage of an artificial impervious facing, such as concrete facing, is its relative inflexibility.   Consolidation of the main rock mass may tend to leave the face unsupported.   Resulting cracks can cause severe leakage.   One advantage claimed for impervious-faced rock fills is their purported ability to withstand overtopping by an unexpected flood during or after construction.[32]   In some dams built since 1942, a rolled-sloping impervious earth core (Fig. 16.18b) was used.   This construction has the advantage of a flexible core that does not crack despite a continuous settlement of the foundation and the embankment.   Also the impervious-earth section does not disintegrate or require repairs in the foreseeable future as is the case of steel- or concrete-faced dams.   A dam so constructed resembles the natural watertight dam formed in an old stream bed by a terminal moraine.   Such natural dams generally consist of coarse gravel that becomes progressively finer in an upstream direction and is sealed at the surface by glacial till.[33]

The Bear River Dam in California is a concrete-faced rock-fill dam made of large rocks carefully placed by a crane in a layer along the upstream face.   This formed a solid, mortarless masonry varying in thickness from 10 ft at the top to 22 ft at the bottom, underlying the concrete slab.   The composite dam in Fig. 16.18b is Kenney Dam, one of the principal structures of Alcan's British Columbia project.[34]

16.22. Earth-dam Failures.   The most frequent causes of earth-dam failures are (1) overtopping of the dam by water because of insufficient spillway capacity or excessive floods during construction when the flow exceeds the capacity of the diversion tunnels or conduits, (2) piping, (3) shear failures in the foundation, and (4) embankment slides.   Occasionally, inadequate protection to the upstream face will result in slides caused by wave action.

A common type of shear failure that occurs primarily during construction is shown in Fig. 16.26a.   The gradually increasing weight of the dam acts on the lower portion of the dam and its foundation in two mutually opposite ways, namely: (1) It consolidates the body of the dam and increases the pressure on the potential shearing surface, thus contributing to the stability of the dam, but simultaneously (2) causes additional shearing stresses along the potential shearing surface, thus increasing the chances of failure.   This is similar to the situation created by a person who with one hand presses down on an object on a table to keep the object in equilibrium but simultaneously with the other hand tries to push the object forward.   Since the consolidation of a foundation is a slow process but the shearing stresses appear immediately after application of the load, the shear failure may occur if the earthwork is done *too fast*.   An eccentric loading also may cause the same type of shear failure.

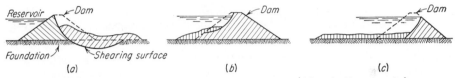

FIG. 16.26. Typical shear failures in earth dams.   (*After A. Casagrande.*)

Sudden drawdown caused by a rapid emptying of a reservoir may cause a shear failure (Fig. 16.26*b*).   In this case, the buoyancy effect is lost, and thus the fully saturated earth material weighs even more than during construction.   Shearing stresses are suddenly applied along the potential shearing surface that becomes an actual one, and failure occurs.   A failure similar in shape to that shown in Fig. 16.26*b* may occur also if the slope is made of expansive material (even though it is well-compacted) that swells in contact with water and loses all or part of its shearing strength.   This loss of strength is not necessarily instantaneous.   In both cases of failure the damage is usually not very serious and is easily repaired.[35]

Liquefaction and spreading of fine sand material (Fig. 16.26*c*) is another type of shear failure.   A. Casagrande[35] stated that:

Such failures develop chiefly on the upstream side of dams and may affect also foundations consisting of such materials.   This type of failure is a rare occurrence, but when it happens, it may be very serious; and it is treacherous because, in spite of extensive research, our understanding of its mechanism is meager. . . .

River deposits of fine, uniform sand are capable of liquefaction, but only if subjected to large strains such as would result from steepening of slopes or from yielding in underlying weak strata.   This conclusion is of particular importance in connection with the design of earth dams on foundations consisting of river sands.

Liquefaction due to vibrations (earthquakes, nearby pile driving, or blasting) has been observed in both natural and artificial deposits, particularly in thixotropic clays (Sec. 4.6).

Fort Peck Dam, on the Missouri River, has a maximum height of 242 ft and a total length of about 4 miles. The foundation is the Bearpaw shale, a dark, bluish-gray, clayey shale of marine origin. It is relatively firm when unweathered, but when exposed to air, it slacks and weathers to a fat clay within a short time. The firm shale is covered by weathered shale containing bentonite seams. During construction in 1938, a slide in the upstream portion of the dam near the east abutment lasted about 10 min and killed eight persons. A board of consultants concluded that the slide "was due to the fact that the shearing resistance of the weathered shale and bentonite seams was insufficient to withstand the shearing force to which the foundation was subjected. . . ."

The shear failure was accompanied by liquefaction of a large mass of sand, both in the foundation and in the saturated portion of the dam. This may have been caused by the strain in the underlying shale deposit. Apparently the true cause of failure was a combination of these two factors.[35] The Fort Peck failure has been described in detail in Refs. 36 and 37 and pages 797 to 803 of Ref. 29.

Rock-fill dams have failed or partially failed because of (1) cracking and shearing of a concrete core wall (Oued Kebir Dam, Africa, 1929), (2) overtopping (Lower Otay Dam, California, 1916), and (3) inadequate sluicing of the rock fill during construction, causing excessive settlement of the dam and face failure (San Gabriel No. 2, California, 1935).[29] In 1912, wave action resulted in the failure of the rock paving and underlying earth-gravel fill in the upstream embankment of Minidoka Dam, Idaho.[28a]

GEOTECHNICAL INVESTIGATIONS FOR EARTHWORK

16.23. Geotechnical Report for a Highway.[39,40] In the case of any earthwork on a large scale, *slopes* of deep cuts and *foundations* of large embankments may be critical and should be given special attention. Study of economic factors, as in any other engineering work, is of importance. The following questionnaire has been prepared for the case of a new important highway with large earthwork yardage. With pertinent changes it may be adapted for other similar earthwork cases.

1. What kind of material is to be excavated in cuts? What should be the steepness of the slopes considering the properties of the rocks and soils in which the slopes are excavated and the ground-water possibility?

2. Can the excavated material be used for adjacent fills, or will part of it be wasted because it is unsuitable? In the latter event, what part (in per cent, roughly)?

3. Are the strata in the embankment foundations capable of supporting the dead and live loads they have to carry? If the bearing power is adequate, is there a potential danger of its decrease due to some unfavorable factor such as ground water?

4. How extensive and deep are peat, muskeg, or similar swampy areas, if any? What is the nature of the material at the bottom?

5. If the excavated material in required cuts is unsuitable or insufficient, where should borrow pits be sunk? What are the length of haul and the probable supply of good material in these pits? How much unsuitable overburden should be stripped?

6. In the case of sidehill sections, what are the geological details, especially in rocky locations? Are natural slopes stable? Are there water-bearing strata, and is deep drainage necessary? Should the fill slopes be supported by retaining walls or otherwise? What are the foundation conditions for such walls?

7. What are the climatological, hydrological, and meteorological conditions at the site or sites? Is it possible to carry on the work during the time designated? Can the finished earthwork be damaged by periodic floods or periodic surface-water arrival? What are the approximate dates of floods and surface-water arrival at the given site if such information can be obtained? What are the intensity and frequency of earthquakes, if any?

For an actual highway project, a detailed outline of necessary geotechnical investigations should be prepared, considering the preceding considerations as a set of *suggestions* only. When the exploration program in the field and laboratory is finished, a geotechnical report should be prepared. Presumably such a report should include also information on river crossings and bridge foundations (Chap. 14).

16.24. Geotechnical Report for an Earth Dam.[38] The geotechnical reports and exploration programs are prepared more or less on the same basis as for masonry dams (Secs. 15.29 through 15.35). Earth dams can be founded on considerably less competent materials, however. Also, spillway studies for earth dams are more important than for masonry dams. In fact, the best design practice places a spillway adjacent to or at some distance from the earth dam rather than over the face.

In the geotechnical report the following items should be discussed and necessary measures proposed:

Foundations. 1. In earth dams founded on critical materials, tight contact between the embankment and the foundation and proper control of seepage along the plane of contact should be ensured.

2. Potentially unstable layers, such as bentonite seams, must be considered in the stability analysis of the embankment.[41]

3. If there are thick clay seams in the foundation, subject to excessive settlement, the embankment slopes should be flattened to reduce the per-square-foot (unit) load on the foundations.

4. In earth dams founded on alluvium and glacial deposits, it is necessary to determine the continuity of permeable strata and the possible direction of eventual seepage.

Cutoff Trench. 1. The characteristics of the materials through which the cutoff trench will be excavated should be given, and the proper method of draining the excavation indicated.

2. The slope ratio (Sec. 16.1) of the excavation slopes should be recommended.

3. It should be considered whether or not blasting will be needed for excavation, since this item may materially affect the cost or bid price of the dam.

16.25. Borrow Materials.   The methods of search for *pervious* materials, e.g., sands and gravels, and also for concrete aggregate are outlined in Sec. 8.10.   One method of searching for impervious materials is outlined hereafter:

A very suitable impervious material for an earth dam is a mixture of sand, silt, and clay with moderate to low plasticity.   It should be sufficiently cohesive to permit a small sample to be easily molded in the hand but should not be sticky or spongy. The most desirable borrow area contains a homogeneous distribution of the select material to a depth of at least 20 ft (a normal shovel cut for a dragline or power shovel) with a minimum amount of stripping required and no ground water.

As such homogeneity rarely occurs in nature, the borrow pit is *zoned*.   Auger holes are placed on a grid pattern (50 to 100 ft, or more, center to center, depending upon the size of the area and the heterogeneity of the soil).   The logs of these holes then are plotted on cross sections which determine the location of the various types of materials.   Earth material from a certain zone either is directly used in construction or is conveniently mixed with that from some other zone, following the construction engineer's orders.

The important factors to include in a borrow area report are (1) location of the area with respect to the dam; (2) accessibility relative to roads, lakes, topography, etc.; (3) depth and type of material available; (4) location of ground water and its fluctuations; (5) amount of stripping; (6) possibility of flooding by streams; (7) possibility of permafrost or deep frost; and (8) service history of materials if used locally.   Photographs and sack samples for laboratory tests should accompany the report.

CANALS

16.26. Terminology.   The two basic classes of canals are *navigation* and *irrigation*.   An irrigation canal is an excavation along a predetermined route and at a predetermined grade for water distribution.   Diversion and feeder canals, also termed "main" canals, obtain water directly from a dam.   *Laterals* are secondary irrigation canals that distribute the water from the main canal directly to each farmer's land.   If the material bounding the excavation is relatively impervious, the canal will be *unlined*.   Materials subject to high seepage rates are traversed by *lined* canals.   The lining may consist of compacted earth, concrete, gunite,[42] bituminous materials, mortared masonry blocks, earth and bentonite mixtures, or any combination of these.[43a]

The inside canal section generally is trapezoidal (Fig. 16.27). As in a tunnel, the canal floor is the *invert*. The portion of the inside of the canal in contact with canal water is the *wetted perimeter*. If the canal invert grade is above the natural ground surface and the canal thus is placed entirely in fill, it is said to be in *thorough fill*. When the entire excavation is in natural ground, the canal is in *thorough cut*.

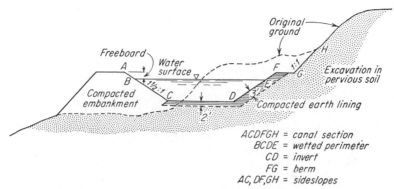

ACDFGH = canal section
BCDE = wetted perimeter
CD = invert
FG = berm
AC, DF, GH = sideslopes

FIG. 16.27. Typical cross section of canal with earth lining.

Canal construction involves numerous concrete or metal auxiliary structures. A concrete *drop* is placed where, owing to topography, it is necessary to increase the water gradient suddenly without producing erosion in the canal. A *check* is placed in the canal to raise the water level in order to divert water into laterals. A *turnout* is a small gate placed below water level in the side of the canal used to release water into laterals. Where it is necessary to cross a valley, a *siphon*, a concrete or steel pipe, carries the canal water down one slope and up the other slope of the valley. Siphon foundations must be particularly protected against settlement, as a change in the design grade can cause a siphon to cease operating and also may tear the siphon pipe loose from its inlet and outlet structures. A *flume*, usually an open metal or concrete trough, may be used to carry the water along a steep sidehill where excavation for the canal is impracticable. *Siphon spillways*, of concrete or steel, are placed in the canal embankment to permit excessive flood flows to be dumped *through* the canal walls without damage to the embankment. *Wasteways* are spillway-like structures which permit excessive water to be dumped *over* the top of the embankment without damage to the latter. *Weirs* are metal, concrete, or wood boxes of certain specified dimensions containing depth gauges that are placed in the canal to measure the flow. Other auxiliary structures may be boxes or bridges to carry the canal over highways, railroads, or other canals. In rare cases, canal water is carried over a deep draw by building a *detention dike* across the draw. This is an earth embankment which has a culvert to pass the water in the draw through the embankment, and the canal water flows along the top of the dike in a lined or unlined channel.

16.27. Tractive Force. "Stability" of the canal channel is the ability of the materials at its bottom and side slopes to withstand the scouring action of the flowing water without accepting objectionable sediments likely to be deposited in connection with the scour.   In some channels, scour is not accompanied by the deposition of sediments, however; in others there is both scour and sediment deposition; and finally, in some there is deposition of sediments without scour.[44]   The movement of cohesionless materials at the perimeter of a canal is caused by the *tractive force* of the flowing water.   If γ_w is the unit weight of water (62.4 pcf), d the depth of flow in feet, and s the slope of gradient of the flow (a dimensionless number), the tractive force in pounds per square foot would

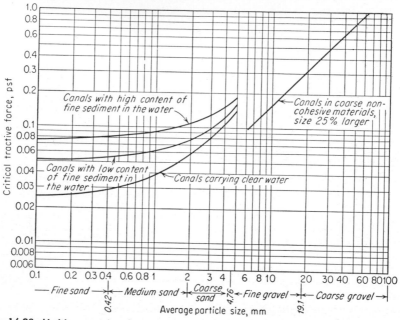

FIG. 16.28. Limiting tractive forces recommended for canals in noncohesive materials. (After Lane.[45])

be $\gamma_w ds$ for most irrigation canals.   Obviously the tractive force acts in the direction of the flow,[45] and because of the law of equality of action and reaction, the canal perimeter exerts an equal and opposite force on the flowing water, which, in a broad sense, is the cause of deposition of the sediments.   [Compare Eq. (12.1).]

To prevent erosion in the canal, the tractive force computed in pounds per square foot should be less than the critical value of tractive force for a given material traversed by the canal (see graph, Fig. 16.28).   For example, if the soil along the canal perimeter has a median size of 0.1 mm and clear water is flowing down the canal, the value of the tractive force as expressed in pounds per square foot should be less than 0.025.   Information on the erosion of cohesive materials in the canals may be found in Ref. 45.   The corresponding data are experimental in character.

16.28. Canal Linings.   It may be necessary to line a canal in order to (1) prevent excessive water loss by seepage, (2) prevent slides in unstable materials adjacent to the canal by properly controlling seepage, and (3) prevent excessive silting in the canal likely to be caused by erosion of natural materials in an unlined canal.

It has been estimated that about 25 per cent of the water entering unlined canals and laterals is lost by seepage before it reaches the farmer's field.[43a]   The cost of lining to prevent such losses may exceed the value of the lost water, and in the design these factors should be properly balanced.   In appraising the economic factor, it should also be considered

FIG. 16.29. Concrete canal lining destroyed by expansive soil. (*USBR photograph.*)

that the seepage from an unlined canal may raise the water table too high to grow satisfactory crops on adjacent farm lands.   From a construction and a stability standpoint, most canal slopes requiring lining should be excavated no steeper than $1\frac{1}{2}:1$.[43a]

Concrete Linings.   A concrete lining may be from $\frac{1}{2}$ to 4 in. thick. Generally, if properly placed on suitable foundations, it will have a service life of about 40 years.[43a]   Such a lining is not absolutely impervious; generally it is estimated that the loss from a canal with a properly maintained concrete lining will not exceed 0.05 cu ft per square foot of wetted perimeter in 24 hr.   The greatest source of leakage will be cracks due to expansion or settlement of the foundation.   Concrete is very sensitive to expansion pressure, as its weight is insufficient to withstand even moderate pressure (Fig. 16.29).   Hence, in critical cases, either expansive soils have to be removed or replaced by nonexpansive materials, or another type of lining used. The main advantages of concrete lining[43a] follow: (1) Weed growth is practically nil; (2) its high resistance to erosion permits a smaller canal section, as high velocities can be withstood; (3) burrowing animals cannot penetrate it; (4) livestock walking on it or canal cleaning equipment does not readily damage it; and (5) it may exert a stabilizing effect on otherwise unstable side slopes. The main disadvantages are (1) high cost, (2) subjection to temperature cracking, (3) relatively low resistance to external hydrostatic pressures due to ground water or from rapid drawdown of the canal water, and (4) possibility of serious damage by frost heave or expansive soils.

Siltation or silting is the simplest method of reducing water loss but is not always effective or durable. Fine-grained soil or mixtures of soil, bentonitic compound, and dispersing chemicals are mixed in the canal water by hydraulic or mechanical methods. The resulting mixture flows down the canal and deposits on the canal perimeter.

Compacted Earth. A thick, compacted-earth lining has proved to be one of the best types of canal lining.[46] A wide range of soils may be used for this purpose; the best results are obtained from well-graded gravel with sand-clay binder (a borderline GW-GC soil in the Unified Soil Classification). The clayey gravels are next in quality; they are followed by sand with clay binder and clayey sands, respectively. The earth lining generally is placed in 6-in. compacted lifts to an over-all thickness of 2 or 3 ft in the canal bottom and to a width of about 5 or 6 ft measured horizontally on the slopes. These dimensions point to one of the main disadvantages of earth lining—the considerable amount of overexcavation required. Another disadvantage is increased weed growth, which considerably reduces the carrying capacity of the canal. Mechanical removal of the weeds often destroys the lining and generally is expensive. Treatment of the compacted-earth lining with cement or certain chemicals may be helpful, however.

Asphaltic compounds of various types may be used for linings depending upon placing conditions and maintenance cost. Flexibility of such linings permits their use on settling or expansive soils; however, they do not have a satisfactory resistance to the pressures and shocks imposed by equipment and livestock. Selected fine-grained, erosion-resistant soils may be loosely placed to form the lining. The canal perimeter may be scarified and compacted by usual compaction methods, water being added if necessary. Soil-cement lining may be used. This is formed by mixing sandy soils with portland cement and compacting at optimum moisture content. Mixing bentonite with soil and compacting the mixture in the side slopes and bottom of the canal or stabilization of the soil in place with resins, plastics, or chemicals may also form a canal lining.

16.29. Canal Investigations. Engineering surveys for a canal are similar to those for a railroad or highway. Longitudinal and transverse profiles along the chosen alignment are prepared, and the depths of cuts and heights of embankments are determined. Geotechnical studies can proceed after these data are available. The following basic geotechnical requirements are to be met in the design of a canal: (1) There should be no detrimental settlement of the canal into the underlying material, (2) side slopes should be stable, and (3) the bottom and the side slopes should be sufficiently impervious with regard to permissible water losses.

A surface geologic map should be prepared. This map should encompass a strip about 200 ft wide, plus the width of the top of the proposed canal. The location of outcrops and contacts between rock formations, between soil and rock, and between different soil types should be shown. The accompanying report should describe the characteristics of the soils and rocks. Of particular importance is the description of the soil in the farm lands adjacent to the canal. If canal seepage is apt to occur, it is necessary to know whether these adjacent soils will *drain* properly or the farm land is likely to be "swamped out." Airphotos are invaluable for classifying soil types and geological mapping.

The surface geologic studies will determine the need for and the intensity of subsurface exploration by borings or geophysical methods. For main canals, test holes are placed about 1,000 ft apart along the canal-route center line. Additional holes are placed where geological or topographical conditions sharply change and at the locations of major auxiliary concrete structures. In the latter case, critical foundations such as those in loess may cause settlement and destruction of the structure; therefore, the respective test holes should be carried through the critical materials to underlying competent materials. All test holes should be carried to a minimum depth of 10 ft below invert grade. If the invert grade is deep and there is an absolute certainty that materials encountered at shallow depths prevail to invert grade, the drilling of the hole may be discontinued at shallow depths. Where the geology is well known, frequent test holes may be unnecessary, as, for example, in certain areas in Nebraska, where loess deposits of consistent character are known to extend to depths of 80 ft or more. The results of the subsurface studies are plotted on interpretative profiles parallel to and along the canal-route center line to aid the designer in estimating the type of excavation and the necessity for lining certain sections (or "reaches"), and in determining proper foundation designs for structures along the canal (Fig. 16.30).

The permeability of the foundation materials can be roughly estimated by performing the usual tests in drill holes (Secs. 5.14 and 5.15). More accurate determinations can be made by utilizing the constant-head permeameter (Sec. 5.13) providing such tests are performed in the field in materials similar to those which will occur in the wetted perimeter of the canal. Highly pervious layers can be successfully located by geophysical methods. For instance, a sand layer in silt can be traced by using electrical-resistivity equipment. Where it is necessary accurately to

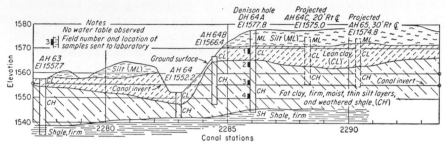

FIG. 16.30. Geologic profile of a portion of a canal.

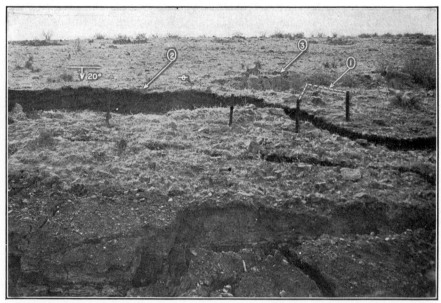

FIG. 16.31. Canal bank slide showing (1) surface displacements; (2) slip surface; (3) vertically dipping sandstone overlain by clay seams. Steep slopes caused failure of clay and motion of sandstone.

determine the bedrock profile, seismic methods as well as electrical-resistivity methods can be used, providing the seismic characteristics of the rock are considerably different from those of the overburden.

The possibility of landslides should be considered (Fig. 16.31). Existing slides should be mapped, and an estimate made of their influence on future canal construction and operation. It should be recalled that landslides often can be most easily detected from airphotos. Slide studies often give an idea as to the future stability of canal side slopes excavated in undisturbed materials.

In connection with the study of frost action the following data should be obtained: (1) the depth of frost penetration, (2) the location of and possible fluctuations of the water table, and (3) the capillary properties of local soils (Sec. 10.3). The behavior of local structures with respect to frost heave may be indicative.

16.30. Canal Drains. If a canal is located in rock, a careful estimate must be made of the permeability of the rock. This is necessary to determine whether or not ground water can or does flow through fissures in the rock. If the flow is restricted owing to the tightness of the natural rock openings, the geologist should indicate the most convenient location of *drains* for a lined canal to remove the excess of ground water. Failure to do so may result in the ground water causing frost heave or hydrostatic uplift of the lining.

REFERENCES

1. Catalogues: Caterpillar Tractor Co.; B. G. Le Tourneau-Westinghouse; Marion Co.; Bucyrus-Erie Co.; Allis-Chalmers; etc.
2. Advertisements: *Engineering News-Record, Construction Methods, Public Works, Roads and Streets, Civil Engineering,* etc.
3. *War Department Technical Manuals* TM 5-252, Use of Road and Airdrome Construction Equipment, 1945; TB 5-9720-1, Grading, Excavating, and Earth Moving Equipment, 1944; TM 5-255, Aviation Engineers, 1944.
4. Park, K. F., and F. C. Ruhloff: Modern Heavy Excavating Equipment, A Symposium, *Civil Eng.,* vol. 12, 1942.
5. Raggio, J. M.: Large Scraper Excavators, *Engineer,* vol. 177, 1944; trans. from *La Ingeniería,* Buenos Aires, May, 1942.
6. Hammond, Rolt: Soil Moving Equipment, *Engineer,* vol. 178, 1944.
7. Cleaves, A. B.: Sedimentation and Highway Engineering, "Applied Sedimentation," John Wiley & Sons, Inc., New York, 1950.
8. Terzaghi, K., and R. Peck: "Soil Mechanics in Engineering Practice," pp. 181–183, 358–360, John Wiley and Sons, Inc., New York, 1948.
9. "Field Manual of Soil Engineering," pp. 85–99, Michigan State Highway Department, Lansing, Mich., 1946.
10. Hennes, R. G.: Analysis of Stresses in Cuts and Embankments, and discussion by J. Feld, *Highway Research Board Proc.,* vol. 24, 1944.
11. Johnson, G. E.: Stabilization of Soil by Silt Injection Method, *Proc. ASCE Sep.* 323, vol. 79, 1953.
12. Foster, C. R.: Reduction in Soil Strength with Increase in Density, *Proc. ASCE Sep.* 228, vol. 79, 1953.
13. Stanton, T. E.: Sand Drains for Foundations, *Calif. Hwys. & Pub. Works,* vol. 27, nos. 3–4, 1948.
14. Anon.: Consolidation and Accelerating Settlement of Embankments, *Calif. Hwys. & Pub. Works,* vol. 27, 1948; also *Proc. 2d Conf. on Soil Mech. and Foundation Eng.,* vol. III, Rotterdam, 1948.

15. Porter, O. J., and L. C. Urquhart: Sand Drains Expedited Stabilization of Marsh Section of the New Jersey Turnpike, *Civil Eng.*, vol. 22, 1952.
16. Barron, R. H.: Consolidation of Fine-grained Soils by Drain Wells, *ASCE Trans.*, vol. 113, 1948.
17. Kjellman, W.: Accelerating Consolidation of Fine-grained Soils by Means of Cardboard Wicks, *Proc. 2d Conf. on Soil Mech. and Foundation Eng.*, vol. II, Rotterdam, 1948.
18. Johnson, S. J., and W. G. Shockley: Field Penetration Tests for Selection of Sheepsfoot Rollers, *Proc. ASCE Sep.* 363, vol. 79, 1953.
19. Holtz, W. G.: Determination of Limits for the Control of Placement Moisture in High Rolled-earth Dams, *ASTM Proc.*, vol. 48, 1948.
20. Progress Report of the Subcommittee on Consolidation of Materials in Earth Dams and Their Foundations, *Proc. ASCE Sep.* 48, 1950.
21. Glossop, R., and G. C. Wilson: Soil Stability Problems in Road Engineering, *Proc. Inst. Civil Eng.*, part 1, June, 1953.
22. Walker, F. C., and W. G. Holtz: Control of Embankment Material by Laboratory Testing, *Proc. ASCE Sep.* 108, 1951.
23. Forbes, H.: Geochemistry of Earthwork, *ASCE Trans.*, vol. 116, 1951.
24. Noell, O. A.: Triaxial Shear Equipment for Gravelly Soils, *Proc. 3d Conf. on Soil Mech. and Foundation Eng.*, vol. I, Zurich, 1953.
25. "Earth Manual" and revisions, United States Bureau Reclamation, Denver, Colo., 1951–1956.
26. Catalogue (Cedarapids compactor), Iowa Mfg. Co., Cedar Rapids, Iowa.
27. "Compaction Handbook," Gunderson-Taylor Machinery Co., Denver, Colo., 1954.
28. "Treatise on Dams," United States Bureau of Reclamation, Denver, Colo.: (*a*) Chap. 1, History, February, 1949; (*b*) Chap. 4, Basic Considerations, October, 1948; (*c*) Chap. 8, Earth Dams, May, 1948.
29. Creager, W. P., J. D. Justin, and J. Hinds: "Engineering for Dams," 3 vols., but mainly vol. III, John Wiley & Sons, Inc., New York, 1945.
30. Sherard, J. L.: Influence of Soil Properties and Construction Methods on the Performance of Homogeneous Earth Dams, *USBR Tech. Memo.* 645, 1953.   A comprehensive treatise on earth dam failures of the United States.
31. Thompson, T. F.: Foundation Treatment for Earth Dams on Rock, *Proc. ASCE Sep.* 548, vol. 80, 1954.
32. Bleifuss, D. J., and J. P. Hawke: Rock-fill Dams Design and Construction Problems, *Proc. ASCE Sep.* 514, vol. 80, 1954.
33. Wise, L. L.: Renaissance in Rockfill Dams, *Eng. News-Record*, vol. 150, 1953.
34. Anon.: Kenney Dam Notes, *Eng. News-Record*, Nov. 29, 1951, Sept. 25, 1952. Further data on Kenney Dam in Refs. 32 and 33.
35. Casagrande, A.: Notes on the Design of Earth Dams, *J. Boston Soc. Civil Engrs.*, vol. 37, 1950.
36. Report on the Slide of a Portion of the Upstream Face of the Fort Peck Dam, Corps of Engineers, U.S. Army, Montana, July, 1939.
37. Middlebrooks, T. A.: Fort Peck Slide, *ASCE Trans.*, vol. 107, 1952.
38. Simpson, Brian: The Geological Aspects of Dam and Reservoir Construction, *Water & Water Eng. (England)*, December, 1953.
39. Maddalena, Leo: Géologie appliquée au autostrades modernes, *Congr. intern. mines, mét. et géol. appl.*, Paris, vol. II, pp. 569–573, 1935; trans. in *USGS Trans.* 4.
40. Bean, E. F.: Engineering Geology of Highway Location, Construction, and Materials, "Berkey Volume."

41. Feil, L. G.: Unusual Design Problems, Harlan County Dam, *ASCE Trans.*, pp. 321–332, 1949.
42. Womack, D. E.: Gunite Lining to Stop Seepage, *Western Construction*, May, 1952,
43a. "Lower Cost Canal Linings," U.S. Bureau of Reclamation, June, 1948.
43b. Operation and Maintenance Equipment and Procedures, *USBR Release* 10, October, November, December, 1954.
44. Thomas, A. R.: 1945 Annual Report of the Central Board of Irrigation of India.
45. Lane, E. W.: Progress Report on Results of Studies on Design of Stable Channels, *USBR Hydraulic Lab. Rept.* Hyd-352, 1952.
46. Holtz, W. G.: Construction of Compacted Soil Linings for Canals, *Proc. 3d Conf. on Soil Mech. and Foundation Eng.*, vol. I.   Zurich, 1953.

Additional Reading

Excavating Equipment

Nichols, Herbert L., Jr.: "Moving the Earth," North Castle Books, Greenwich, Conn., 1955.

Soft Soil Foundations

Lehmann, H. L.: Highway Embankment on Yielding Subsoil, *ASCE Preprint* 3, New Orleans Convention, 1952.

Dams

"Low Dams," National Resources Committee, Washington, D.C., 1938.
Haven, R. C.: Saturation of Existing Earth Dams, *USBR Tech. Memo.* 389, June, 1934, and 493, Sept., 1935, Denver, Colo.
Middlebrooks, T. A.: Earth Dams, "Applied Sedimentation," John Wiley & Sons, Inc., New York, 1950.

Railway Embankments

Concrete Supported Railway Track, *PCA*, Chicago, Ill.
Stabilizing Railroad Track by Pressure Grouting, *PCA Concrete Inform. Bull.* ST 60, Structural Bureau, April, 1943.
Grouting Subballast for Track Bases, *PCA Concrete Inform. Bull.* ST 49, Structural Bureau, March, 1942.
"AREA Manual," American Railway Engineering Association, Chicago, Ill.   In loose-leaf form and revised annually.

Canals

Quantity Estimates for Irrigation Structures, *USBR Spec. Release*, September, 1955. Excellent drawings of typical canal structures.

CHAPTER 17

LANDSLIDES AND OTHER
CRUSTAL DISPLACEMENTS

17.1. Terminology. The term "displacement," as used hereafter, involves change in position vertically, horizontally, or obliquely of certain sections of the earth crust, in some cases together with the engineering structures connected with these sections. Such crustal displacements as faults and folds are discussed in Chap. 2. One group of crustal displacements discussed in this chapter is caused by the weight of huge earth and rock masses, by the influence of ground and surface water, and by other factors that do not depend or depend slightly on the weight of the engineering structures supported by those masses or connected with them. This group of displacements consists of *landslides* (or simply "slides"), *creep* and *flow* of earth masses, and *subsidences* of certain areas. Another group of crustal displacements discussed in this chapter is connected with *settlements*. This term as used in engineering generally refers to the vertical displacements of the structure itself under the action of its weight alone or in combination with other forces. A building or an embankment *settles*, whereas a loaded pavement *deflects*. There may be horizontal movements of structures, usually combined with settlements.

SLIDES

17.2. Stability of Slopes. The theoretical principles of designing higher slopes are as follows. Separation of a *wedge* bounded by a continuous rupture surface is assumed, and stability of the wedge checked from conditions of equilibrium. The shape and the position of the rupture surface (termed generally "failure surface" or "slip surface") are influenced by the distribution of pore pressure and the variations of the shearing strength within the earth mass. A simplified shape is therefore adopted; usually this is a *circular arc*. The slope itself is assumed to be very long and identical at all cross sections. The shearing stress and the shearing strength are assumed to be uniformly distributed along the slip surface.

The two basic theoretical cases of wedge separation are shown in Fig. 17.1a and b, showing "slope failure" and "base failure," respectively. In both figures the underlying "rigid boundary" may be rock, very stiff clay, or other firm material of greater shearing strength than the overlying soil. Note that the slip surface is always "in the soil" and at its limiting position just touches the underlying firm boundary. Failure of the mass occurs by cracking along the slip surface and rotating about center O of the arc. The force causing failure (driving force) is primarily the weight of the wedge; the "resisting force" is primarily the shearing strength across the slip surface. The ratio of the resisting force to the driving force for various positions of the slip surface is computed, and the minimum value of this ratio is the *safety factor* of the slope against failure. For no position of the slip surface should the safety factor be below 1. If the safety factor is exactly 1, the slope is in the state of

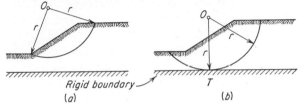

FIG. 17.1. Two theoretical cases of slope failure: (a) slope failure, (b) base failure.

limit equilibrium. It is worth while to note (Fig. 17.1b) that in the case of a base failure the center of rotation O is located on the vertical passing through point T, the tangency of the slip surface and rigid boundary.

There are no slip surfaces in dry, clean, cohesionless materials (sand, gravel, and, to a certain extent, silt). The slopes in these materials cannot make with the horizon an angle steeper than their respective angle of repose (Sec. 4.16).

In soft (i.e., nonstiff) and nonconsolidated clays the stability of a slope may be analyzed by using the $\phi = 0$ method (Sec. 4.18). In this case the value of the unit cohesion c may be determined either from a quick triaxial test or simply as a half of the unconfined shearing strength obtained in an unconfined compression test. (Sec. 4.18). In a more general case of a cohesive earth mass the shear characteristics of the material c and ϕ and its unit weight γ should be determined. The critical, or maximum possible, height of a slope may be determined from Eq. (17.1).

$$H_{cr} = \frac{1}{SN} \frac{c}{\gamma} \tag{17.1}$$

where SN is the *stability number* variable for different values of the angle of shearing resistance ϕ. For a simple case when there is no pore pressure and the soil material is reasonably moist, the graph of Fig. 17.2 may be used.

Example. The critical height of a slope $1\frac{1}{2}:1$ in a clay with $\phi = 5°$, $c = 600$ psf, and $\gamma = 120$ pcf, the stability factor from Fig. 17.2 is about $SN = 0.12$. Hence the critical height

$$H_{cr} = \frac{1}{0.12} \times \frac{600}{120} = 42 \text{ ft}$$

or using a factor of safety of 1.5, the admissible height of the slope to be constructed is about 28 ft.

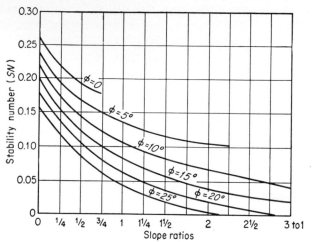

FIG. 17.2. Graph for determining the steepness of slopes.

17.3. Sliding and Slides.

Landslides occur on sloping terrain in all kinds of earth, earth-rock, and rock materials. Generally a slide may be defined as a downward and outward movement of a portion of the soil or soil-rock mass, sometimes conventionally termed a "wedge" with respect to that portion remaining in place. This movement is accompanied by a breaking of the bond* connecting the separating and remaining portion of the mass across the slip surface. The separation usually starts at some weak point, e.g., an old tensile crack or cracks at the slope itself or the adjacent rather horizontal ground surface, and consists first of more or less slow movements along the potential slip surface followed by a more rapid movement of the separated portion. Slides often exhibit a characteristic topography. There is a typical area near the upper end of the slide from which the material has been excavated and carried away and an area of deposition near the lower end of the slide. The slip surface is bounded by a continuous perimeter crack, usually well defined. If the slide occurs on a slope resembling a smooth plane, the perimeter crack will resemble a smooth parabolic curve. The crack bounding the slip surface is a *characteristic feature* distinguishing a slide from other earth-crust displacements such as creep or flow in which a continuous

---

* The term "bond" as used here is a counterpart of the term "shearing strength" of homogeneous materials.

crack is often absent.   The mass that has separated from the rest of the slope is inevitably headed for the topographic bottom of the area but is impeded in its motion by various factors such as friction, occasional reversed gradients, and the dense consistency of the mass itself.   If the excavated material is reasonably fluid, it will be spread on the adjacent terrain, mostly in thin layers.   The configuration of this *area of deposi-tion* thus formed depends essentially on the local topography.   Figure 17.3 is a schematic plan view of a slide on a slope roughly resembling a plane.   If the exca-vated material is but slightly fluid, it will be squeezed out and remain next to the exca-vated area in the form of a *bulge*.   Sometimes the excavated material moves just several inches along the slip surface.   A slide that is prevented from fully developing by the presence of a structure (e.g., a wall)

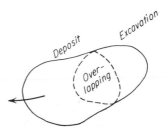

FIG. 17.3. Schematic plan view of a slide.

at its toe exerts an upheaving force on that structure and possibly damages or destroys it.

The study of slides leads to the conclusion that the slide characteris-tics within a given region should depend on the geology, topography, and climate of that region.   In fact, often certain slide characteristics are reported either from within a large typical area or from two areas that are similar in some respects.   For example, the tremendous Colorado landslides often consist primarily of large boulders.   In older glaciated zones both rock and soils are often remarkably stable as in some New England regions, and vice versa, in the regions of the so-called "young geology," e.g., in some portions of the San Francisco Bay area, slide scars on natural slopes are so abundant that they really should be considered as characteristic landforms of the region.   Serious consideration should be given, therefore, to the "regional concept" of landslide classification. According to this concept the slides within a "geomorphic (or physio-graphic) province" should possess similar characteristics.   A geomorphic province may be defined as an area within which the method of deposi-tion of rocks and soils is approximately the same, landforms are similar, and the climate is approximately identical.[1]   This regional concept is accepted by some of the workers interested in landslides, but much evi-dence is still needed.

17.4. Types of Slides.   In rocks and soils as in any other material, a failure follows the pattern (or path) of *least resistance*.   This means that the actual slip surface along which an actual slide has taken place offered less resistance to the separation of the "wedge" than any other possible slip surface.   In a more or less homogeneous cohesive material, e.g., in some clays, a slope would fail primarily by shear, and the slip surface, as already explained in Sec. 17.2, would be approximately circular (or, more accurately, circular-cylindrical). Thus *rotational* slides are produced. Like any other sliding (Sec. 17.3) the rotational sliding takes place first by little jerks, thus gradually destroying the bond along the slip surface until the separating mass slumps down. Hence the term "slump" is used sometimes for this kind of slide, especially in geological literature.[2]

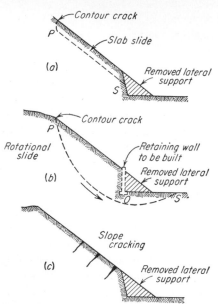

If the shearing strength of the earth material is less in the horizontal direction than in the vertical, the circular arc may flatten out. Conversely, cases of steepening of the circular arc may also be observed.

If the soil or rock deposit is stratified, the upper strata may slide down with respect to the lower strata along the boundary, which in this case becomes the slip surface. Slides formed in this way and generally slides characterized by the translation of the separated material rather than by rotation are termed *translational slides* hereafter. A common case of translational slides are *slab slides*, in which

FIG. 17.4. Sliding caused by the removal of lateral support.

the slip surface is roughly parallel to the ground surface (Fig. 17.4a). Whereas rotational slides occur on steeper slopes and are of a relatively limited length, the slab slides may occur on gentler slopes and are usually long, i.e., several hundred feet or longer.

Rock displacements are discussed in Sec. 17.9. Most of them are *translational slides*, since irregular crack and fissure pattern in rock masses is not favorable for the development of smooth, circular slip surfaces (except in very soft rocks). A separate class of rock displacements are *rock falls*, also discussed in Sec. 17.9.

Minor Soil and Rock Displacements. There are several types of displacements that deserve to be noted from the purely geological point of view but, as a rule, are of little importance to the engineer. Such are, for instance, *debris slides* and *debris avalanches*. These are displacements of unconsolidated foreign earth and rock materials accumulated at the slope that tend to move in a slab-slide manner. There are *terracettes*, or small steplike discontinuities on the surface of a slope. In this category also belong small *slipouts* and *popouts* of the slopes. Formation of *talus* at the foot of a slope has already been mentioned (Sec. 3.2).

Deep and Shallow Slides. All slides may be subdivided into two large classes, namely, *deep* slides and *shallow* slides, according to the depth of the excavations produced. Though there is no numerical criterion for such subdivision, rotational slides often belong to the deep class whereas slab slides are usually relatively shallow. Other translational slides may be either deep or shallow. Corrective measures (Sec. 17.11) may vary according to the depth of the slide, hence the importance of the classification.

17.5. Causes of Slides. In the analysis of slopes, both natural and artificial (i.e., made by man), the element of *probability* prevails rather than that of certainty. Even if accurate engineering computations of stability are made and observable geologic factors are considered, it is more correct to interpret the expression "this slope is stable" in the sense *"probably* is stable." An old slope that has withstood a number of rainy seasons without visible damage is *probably* stable if it is not overloaded.[8]

The most common driving force tending to destroy a slope is *gravity*, generally the weight of the slope material and that of superimposed loads, and an increase in this weight decreases the stability of the slope. The most common resisting force, the shearing strength of the material, may be decreased by *excess moisture*. Excess of free water may even convert the material into a suspension totally or almost deprived of shearing strength. Most slides occur during or soon after rainstorms or rainy spells; dry slides have been reported as a very rare exception. It should be noted, however, that the shearing strength of a rock or soil material may be decreased also by chemical changes.[3,4]

Increase in gravity and decrease in shearing strength acting together or separately generally constitute the *cause* of the slide. Besides these two *common components* of the cause of a slide, there may be one or several events or combinations of events that facilitate the occurrence of a slide. These are *contributing factors*.

The weight of the slope may be increased by *saturation* during rainy spells. This increase is relatively small in comparison with heavy loads that are sometimes applied at the upper edge of the slope by storing heavy materials or passing heavy vehicles. The load may be applied even at the middle of the slope as discussed in Sec. 17.6.

Generally speaking, *excavations close to the toe of a slope* cause the same effect as the loading of the slope. The *removal of the lateral support*, particularly the removal of the toe of the slope for emplacement of a building or a highway, is one of the very common causes of a slide (Fig. 17.4). In this case the normal pressure on the potential slip surface (*PS* in the figure) is decreased, and the tensile and shearing stresses in the unsupported earth or rock body are increased. It should be understood that each of the causes discussed may produce either a rotational (shear) slide or a translational slide. As already stated, a rotational (shear) slide will probably take place if the material is homogeneous. A translational slide will probably take place if the material has a tendency to slide along a plane or some other definite surface. In the cases shown in Fig. 17.4a and b even if no slide occurs because of the high shearing strength of the material or firm bond, tensile cracks may appear on the slope (Fig. 17.4c). Such cracks eventually may develop into a slide, especially in expansive soils. In fact, a supported slope tends to expand vertically only[6] but expands laterally if the support is removed.

Immediate Cause of a Slide. An engineer or a geologist working on a slide may be asked to give to the court where a landslide case is being heard a clear-cut answer as to the "cause" of the slide, and it is advisable in this connection to distinguish sharply between the *real* and the *immediate* cause of the slide. The immediate cause may be an increase of driving forces, a decrease in resisting forces, or such factors as earthquakes, shocks, or vibrations. The immediate cause may be sudden ("trigger action") or may last for a certain time.

The slide in the Department of Maritime Alps, France,[3] occurred on a steep slope crowned by a terrace of old alluvium. The terrace is underlain by clayey marl and anhydrite. The ground water under pressure from the adjacent mountains alters anhydrite into soluble gypsum with formation of voids and channels. A week-long heavy rainstorm caused the failure of the slope. The heavy rain was the immediate cause of the slide, and its real cause was the gradual transformation of the anhydrite into gypsum.

17.6. Slab Slides.

Slab slides occur in upper strata that either are loosely bound to the underlying materials or are structurally different from them as in stratified rocks. Examples follow.

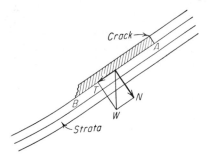

FIG. 17.5. Slab slide in stratified rock.

1. *Slab Slides in Stratified Rocks.* In the case of Fig. 17.5 there is an old tension crack in the upper stratum, and the portion of the stratum that tends to slide down along the potential sliding surface AB (cross-hatched in figure) is held in place by the bond along that surface. In this case the bond is due to the shearing strength of the cementing material at the potential sliding surface AB and to the adhesion of the cementing material to the rock. The weight W of the portion tending to slide may be broken into a force N normal (perpendicular) to the plane sliding surface and a force T along the sliding surface. Force T is the *component*

of gravity pushing the sliding portion downhill, whereas friction fN, together with the bond, tends to keep it in place.   The symbol f means the coefficient of static friction between the portion tending to slide and the underlying material.   The driving and the resisting forces are equal if

$$T = fN + \text{bond resistance} \tag{17.2}$$

When the cementing material has been gradually weakened by the rains of the preceding years, a heavy rainfall may cause the slide if friction

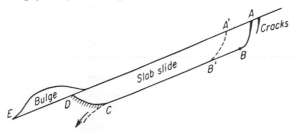

FIG. 17.6. Typical slab slide.

alone cannot keep the rock mass in equilibrium.   In such a case the immediate cause of the slide is the last rainfall, and the real cause the gradual destruction of the cementing material in the previous years. When stability of rock massifs such as in Fig. 17.5 are investigated, due attention should be given to the properties and the condition of the cementing material.

2. *Shallow Slab Slides in Earth Materials.*   In regions with predominantly shaly bedrock, like West Virginia,[5] certain parts of California,[6,7] and others, soil deposits developed or located over the bedrock are often impervious clays with a top layer of 4 to 10 ft of more pervious, sometimes weathered material.   Numerous slides occur in these upper soil strata, these slides being sometimes classified as creep or flows of the soil mantle.   In reality most of these displacements are slab slides, i.e., fast bodily translations of portions of the upper soil layer.   Such shallow slab slides may occur, of course, on any slope the upper few feet of which are structurally different from the underlying materials.   Figure 17.6 shows a typical section of a slab slide.   The slip surface often starts with a roughly circular arc or other curved surface at the top of the slide and continues along plane BC if the bond resistance along this plane is weak. Very often the slab thus separated moves bodily so that arc AB gradually assumes position $A'B'$.   If at some point C further movement is prevented, e.g., by the presence of a strongly resisting material CD, the excavated material forms a *bulge* (as anticipated in Sec. 17.3).   Conversely, if at point C the ground surface bends down (as shown by the dotted line), the deposition area will be as in Fig. 17.3.

3. *Retrogressive Slides.*   The concept of retrogressive slides has been introduced by Scandinavian engineers.   These are translational slides,

mostly of slab type. The terms "successive" and "progressive" slides are also used.[9] A retrogressive slide develops upslope and not in the downslope direction as in the usual translational slides.

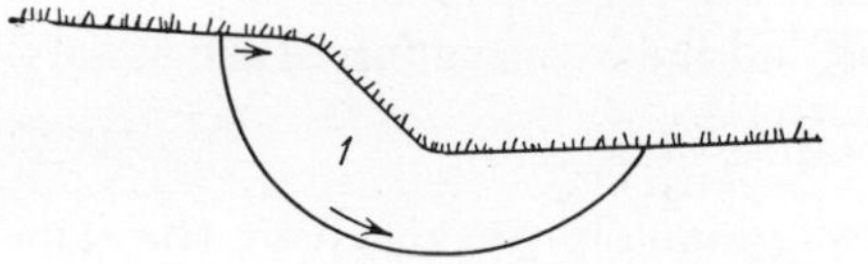

According to the explanation of the Geotechnical Commission of the Swedish State Railways (Ref. 10 and Fig. 17.7), a retrogressive slide is composed of a series of simple rotational slides following each other in succession. Each simple slide is supposed to affect the stability of the ground behind it and thus cause a new slide. Though the appearance of the ridges observed in such slides (see figure) is thus explained, this hypothesis cannot be applied to numerous slablike failures on mild slopes, some of which are practically horizontal terrains.

A more realistic hypothesis has been advanced by the Swedish Geotechnical Institute.[9] Assume that in a slope (Fig. 17.8) portion ACC' is in the state of limit equilibrium whereas portion $CDD'C'$ has a considerable margin of safety. A decrease of the bond value in portion ABB', e.g., by pore pressure, would increase the pressure on BB' by some value ΔF. This would cause an equal and opposite reaction ΔF of the firm portion of the slope and compression of portion $BB'C'C$. As a result portion ABB' would slide down somewhat. Practice has shown that sliding even along a few inches causes remolding and loss of a considerable percentage of the shearing strength of the clay material. Thus clay beyond point A will be deprived of the lateral support and free to slide.

FIG. 17.7. Mechanics of retrogressive slides according to the Geotechnical Commission of Swedish State Railways. (*From Géotechnique, vol. 5, no. 1, 1955.*)

Many Scandinavian slides, though of a progressive type, were formed in the usual downslope direction.[11] These include the huge Surte slide in Sweden[12] that started at the moment when a commuters' train was leaving the Surte depot, so that the commuters were able to see their own houses were coming down, many of them practically without tilting. The slide started at a higher elevation and was translating toward Göta River, finally covering an area of about 110 acres.

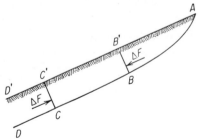

FIG. 17.8. Mechanics of retrogressive slab slides. (*After Kjellman.*[9])

Figure 17.9 represents a retrogressive slide observed in San Francisco. In this case an abandoned, several-year-old fill caused a slab slide during an exceedingly heavy rain period. The rain was thus the immediate cause of this slide; its real cause was

the weight of the fill.   The portions of the fill labeled M and L in Fig. 17.9 broke the
retaining wall RW.   The upper part of the slab U thus lost its lateral support and
joined the slide.

4. *Slab Slides at a Slope Break.*   Such slides occur in long slopes of
cohesive earth material the upper layer of which is pervious and underlain by impervious material, e.g., a stiff or hard clay deposit.   At a break

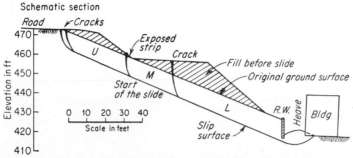

FIG. 17.9. Sliding caused by the overloading of a slope, San Francisco, California, 1953.

B of the slope the upper layer consists of an upstream (steeper) portion
U and a downstream (flatter) portion D, as shown in Fig. 17.10.   During
a rainstorm water flows on the ground surface and into the pervious
layer, thus reaching the top of the impervious deposit.   As soon as the
upper few inches of this deposit become saturated (crosshatched strip in
Fig. 17.10), it constitutes an obstacle or a curtain for the downward passage of the still arriving rain water.   Water is then obliged to flow downslope, along both the ground surface and the top of the impervious deposit,
and as the rain proceeds, the whole upper soil layer becomes saturated.
Since the downstream portion D is flatter than the upstream portion U,
its discharging capacity is smaller
and it becomes saturated first.
Water still arriving from uphill cannot pass through the downstream
portion D, and it either breaks
the ground as shown in Fig.
17.10 or produces a swamping
condition around the break B.
Such a hazardous condition may

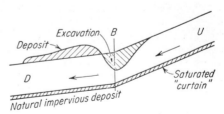

FIG. 17.10. Slide at the break of a slope.

take place either at the foot of the hill or anywhere at a break of the
hill slopes.

It is obvious that the rain water flowing inside the upper ground layer is *gravitational
water* in its own right and should be classified as such.   In this way, a *temporary
aquifer* is being formed next to the ground surface.

17.7. Rotational Slides.   *Checking Stability of a Slope.*   In important
cases of slopes in homogeneous materials at which a rotational slide may

be expected, conditions of stability of a planned (or existing) slope are studied by assuming several possible slip surfaces and by estimating the values of the driving and resisting forces for each of them.   The smallest value of the ratio of the resisting forces to the driving forces is the *factor of safety* of the slope.   If the factor of safety is less than 1, the slope should be considered unsafe.   If a slope is standing up with the safety factor under 1, this possibly means that the factor of safety has been

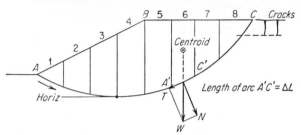

FIG. 17.11. Slice method of checking stability of slopes.

computed incorrectly, perhaps on the basis of wrong assumptions.   To check the stability of a slope the values of the angle of shearing resistance ϕ and the unit cohesion c should be known.

In the *slice method* (Fig. 17.11) the slope is subdivided into several slices and the stability of each slide investigated (see slice 6 in the figure).   The slices are supposed to be 1 ft thick.   The weight W of a slice is broken into a tangential force T, which is the driving force, and a normal (perpendicular) force N.   The resisting forces are (1) friction along slip surface $N \tan \phi$ ($\tan \phi$ is the coefficient of friction along the slip surface), and (2) cohesion along arc $A'C'$ equal to $c(\Delta L)$.   The driving and the resisting forces for all slices are summed up. The total driving force should not be larger than the total resisting force divided by the safety factor (e.g., 2 or 3).   Note that the weight of slices 1, 2, 3 in Fig. 17.11 is favorable for the stability of the slope because the tangential forces T of these slices are resisting forces.   There are also other methods of checking stability of slopes.

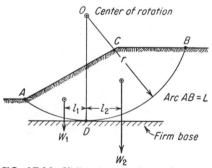

FIG. 17.12. Sliding in cohesive soil underlain by a firm base.

Figure 17.12 shows the method of checking stability of a cohesive slope underlain by a firm base (or "rigid boundary").   The center O of the arc AB, that is, the center of the circular slip surface in this case, is located in the vertical DO.   Rotation about center O is considered.   The driving moment is the product of the weight W_2 of the portion of the slide right (on the figure) of vertical OD times the arm l_2.   The resisting moment equals (1) the product $W_1 l_1$, where W_1 is the weight of the portion of the slope left (on the figure) of vertical OD, *plus* (2) the product of the unit cohesion times the length of arc AB times its radius. The factor of safety should be considered as in Fig. 17.11. All lengths (including those of arcs) are measured usually in feet.

The upper part of the slip surface from point C to level of the bottom of the crack, if any (horizontal dotted line, Fig. 17.11), is not considered in computations.

If the wedge ABC is in the state of limit equilibrium (Fig. 17.13) with a safety factor of 1, any geometrically similar small wedge as $A'B'C$ has a larger safety factor that may be proved by applying the slice method to both wedges. This means that there are no detrimental shears *inside* the wedge, which explains the practically *intact* condition of the wedge sometimes observed.

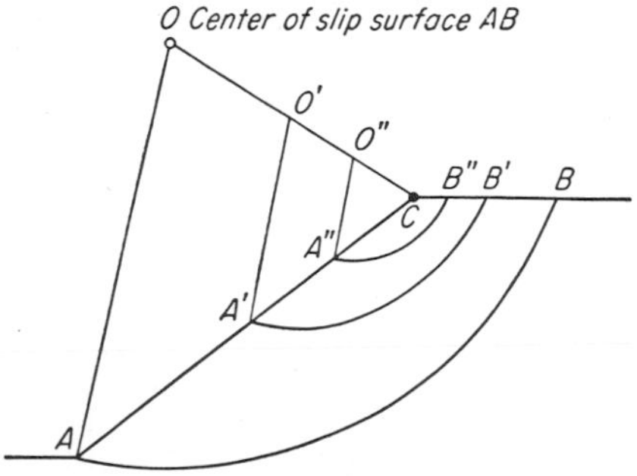

O Center of slip surface AB

FIG. 17.13. Critical shears on the slip surface AB; no critical shears inside wedge ABC.

Slides in Quasi-homogeneous Materials. Rotational slides occur not only in purely homogeneous materials but also in materials that may be termed "quasi-homogeneous" such as varved clays, clays with abundant small gravel, or fissured and jointed materials. Figure 17.14 shows a displacement presented in Ref. 2. This is a typical rotational slip with a bulge. The terminology of the original has been conserved in the figure; particularly "earth flow" is what is termed "bulge" in this book. Vertically fissured and jointed siltstones and claystones are very sensitive to the presence of

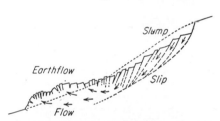

FIG. 17.14. Rotational slide (slump) of a fissured slope. (After Sharpe.[2])

water in their fissures or joints. Failures may occur in cuts made of similar material covered in nature by an impervious clay layer if the latter is removed in construction. The exposed material easily slumps after the first good rain following the construction. In this connection a slide occurs very similar to that shown in Fig. 17.14. The immediate cause of such a slide is rain water filling the fissures, softening the base of the chunks shown in figure, and eroding the cementing material. The real cause of this slide is, of course, the removal of the protective cover. In nature, displacements of this kind are sometimes formed at the edge of clay terraces by the water running toward the thalweg of the valley.

If the soil material is stratified and the shearing strength of individual strata is approximately the same, the slip surface will be a slightly broken curve that, without considerable error, still may be considered as approaching an arc of circle.

Additional Comments.   If the upper stratum is stronger than the underlying material, the slip surface may develop in the underlying material. For example, to investigate the possibility of failure of a harbor wall on a pile foundation (Fig. 17.15) the slip surface clearing the tips of the piles should be considered.   Weak piles may be broken by a landslide, however.

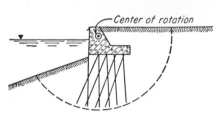

FIG. 17.15. Slip surface clearing the tips of the piles.

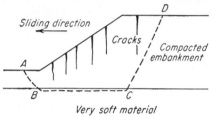

FIG. 17.16. Failure of a compacted slope built on underlying soft material.

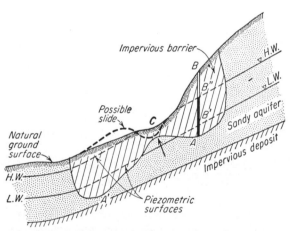

FIG. 17.17. A "break-through" caused by pore pressure.

Figure 17.16 illustrates the case of a compacted earth slope built on underlying soft material, a procedure which is generally not advisable (compare Sec. 16.7).   Instead of following a possible theoretical slip surface *ABCD*, the slope in such cases fails in tension.   It cracks and may spread forward.

17.8. Sliding Caused by Pore Pressure.   Under the conditions shown in Fig. 17.17 the ground-water flow is constricted in the sandy aquifer below points *A* and *A'* of an impervious deposit.   If a boring at point *B* is made to reach point *A*, water in the boring will rise to some level *B'* when the ground-water level is low because of seasonal and other changes.

It would rise to a higher level B'' when the ground-water level is high. If borings are made in several points of the impervious deposits, points such as B' and B'' will be found in each of them.   By plotting these points on the drawing (Fig. 17.17) and joining them, one may trace the *piezometric surfaces* of the ground water (dotted lines in the figure).   It can be concluded from the figures that at the time of the high ground-water level the soil material at point B'' (before the boring was made at that point) was acted upon by a force equal to the distance from A to B'' in feet multiplied by the unit weight of water (62.4 pcf).   If at any point such as B'' there is a danger of sliding and a potential slip surface, the danger will be aggravated each time the ground water rises.   If the impervious layer at some point C is thin, the water in the underlying aquifer will either reach the ground surface by capillarity or simply break through the impervious layer and cause a small slide.   Such occurrences are sometimes observed in new excavations at the foot of hill ranges where ground water under pressure is common.

Figure 17.18 refers to a large slide that took place on July 24, 1941, in a flood-control dike under construction along the Connecticut River in Hartford, Connecticut.   At the time of the failure the dike was practically finished (dotted lines in Fig. 17.18) and the space between this dike and the old Colt Dike (at the left, Fig. 17.18) was being filled by the sluicing method for highway fill.   The free-water level was at elevation 40.   Water was draining through the dike fill, down over the top of the sheet piling at the foot of the dike, and thence to the river.   At the time of the failure there was a high pore pressure in the silt layers of the varved clay.

The material at the site was by no means homogeneous, and the slide was not typically rotational.   There was a depression in the clay caused by the weight of the new dike and bulging close to the river.   Silt and fill material settled down and caused the movement of the piles supporting the sewer.   It is also possible that because of the resulting shock, part of the material of the new fill may have been temporarily liquefied; at least the position of the pipes 4, 3, 2, and 1 after the slide suggests this.   A remarkable feature of this slide was the translation of the sheet piling wall (1,200 ft long) along a distance of 52 ft without losing verticality and with no damage whatsoever.   The real cause of the slide is difficult to establish.   The pore pressure was an important contributing factor and was probably the immediate cause of the slide.

17.9. Rock Falls and Rock Slides.   A *rock fall* may be defined as a free fall of rock fragments of various sizes detached from a slope and may be differentiated from a *rock slide* by the absence of the slip surface.   The fall may be combined with rolling and leaping of fragments, which may be broken into pieces in the process.   Fallen-rock fragments may remain at the foot of the slope, forming talus deposits, or may be taken away by gravity.   One major cause of rock falls is the force developed by alternate freezing and thawing in joints and fractures in the springtime.

At Niagara Falls, hard limestone overlies a thick bed of less resistant sandstone and shale.   Churning water containing rock fragments erodes the less resistant lower rocks, causing overhanging ledges and resultant

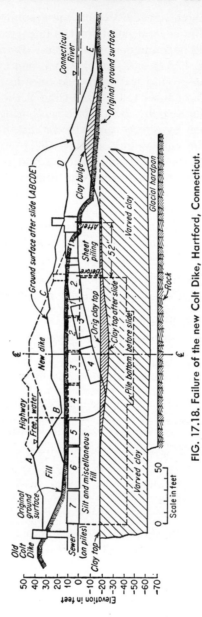

FIG. 17.18. Failure of the new Colt Dike, Hartford, Connecticut.

rock falls.   The rock fall of July 28, 1954, involved about 185,000 tons of rock.

The protective measures against rock falls depend on the nature of the rock and the degree and character of its fissure pattern.[13]   Stratified calcareous rocks are easily cracked, but only in the direction normal to bedding.   In sandstone, besides the fracture in the direction normal to bedding, other fractures, some of them dangerous, occur easily.   Crystalline and marmoreal limestones often split according to the crystal planes of the calcite.   Igneous rocks cooled at the surface (eruptive rocks) with contraction joints are subject to complex fissuring and abundant rock falls.   The rocks of the granitic type generally do not present such dangers.   However, in the case of rocks that have taken part in orogenic processes the possibilities of fracture and falling rock are high.

The following types of protective structures against rock falls are used by railroads: (1) *covered galleries* for badly fissured igneous rocks—porphyries, trachytes, basalts—and for coarsely fractured rocks of all kinds; (2) *retaining walls and cribs* (bottomless boxes) made of rails and timber ties for eruptive igneous and finely broken sedimentary rocks; (3) *tie bars, underpinning, and roof bolts* (*Sec.* 9.5) for sedimentary rocks with one plane of fissuration and igneous rocks with one or two planes of fissuration; (4) *protective facing or gunite application* for all easily weathered rocks.   Trees, bushes, and even stumps left on the slope give effective protection against rock falls.   Power plants are protected against rock falls by masonry or concrete deflector walls (Sec. 13.17) to divert rocks from the plants.   Numerous strong supporting columns are added to the design.   Heavy roof constructions are used and the roofs covered with protective sand and gravel layers from 3 to 6 ft thick.

Rock Slides.   Rock slides from natural causes are frequent in rugged areas like the Rocky Mountains.   They are generally caused by the gradual wearing out of the bond interconnecting rock massifs and are mostly translational in character.

The great Turtle Mountain[2,7,8] slide of 1903 at Frank, Alberta, was a rock slide of the combined slide-fall type, the real cause of which was gravity or the weight of heavy limestone deposits (at the left, Fig. 17.19) prevented from sliding toward a lake by softer sandstones and shales (fine crosshatching at the right, Fig. 17.19).   The sandstones and shales were creeping and could not withstand the thrust exerted by the limestone massif.

Figure 17.20 shows a slide caused by undercutting the foot of the slope.   Measured vertically from the water level at the foot of the slide, the latter was over 200 ft high.   The mighty perimeter crack is clearly seen in the photograph.   The slide involved a separation of a relatively thin mantle or surface layer from the rest of the rock massif and should be classified as a shallow, translational slide very similar to the slablike type.

Figure 17.21 depicts a slide of sandstone beds on interbedded clays.   The beds in limit equilibrium were brought into motion by excavation.   The slide is similar to that shown in Fig. 17.4 and is essentially a slab slide.

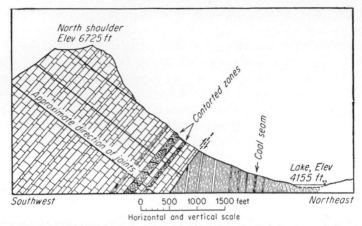

FIG. 17.19. Structure of north shoulder of Turtle Mountain, Frank, Alberta.   (*After Sharpe.*[2])

FIG. 17.20. Panoramic view of a rock slide at the Grand Valley Project, Colorado. (*USBR photograph.*)

The reader's attention is called to the fact that a considerable number of slides, both in soil and in rock, occur on the slopes of river valleys, with partial obstruction or complete damming of the river.   Figure 17.22 is a typical example of such slides.

17.10. Prevention of Landslides.   In a general way it is difficult to forecast a slide unless there is definite evidence of a horizontal motion of sloping ground, such as cracking or relative motion of some points with respect to the others.   In each particular case when the possibility of a slide is suspected, it should be investigated to determine whether or not the two most important slide-producing agencies (gravity and subsurface water) are acting detrimentally and in what form.

In the case of rocks the gravity force may be a cause of sliding if massive heavy beds are underlain or prevented from sliding along a plane by weaker rocks (as in the case of Turtle Mountain, Sec. 17.9); if competent and incompetent rock strata are alternated, especially if some are argillaceous (clayey); if the dip of a rock stratum is too great and the stratum

FIG. 17.21. Slide of sandstone beds on interbedded clays (Flatiron power plant excavation, Colorado).

may slide over a bedding, foliation, cleavage, joint, or fault plane; or if the rock is badly fissured. These and similar rock defects should be investigated by a competent geologist.

In the cases of both rock and earth slopes the action of gravity may be intensified by the construction of fills for highways, railroads, and housing projects, or the dumping of waste materials from quarries, mines, and excavations. Therefore, a suspicious slope should be investigated to determine if there are loads threatening the stability of the slope and whether the lateral support has been or is going to be removed. Excavation at the foot of a slope, demolition of an existing structure at the foot of a slope, deep dredging of a channel if there are structures (for instance, tanks on spread foundations) next to the shore—all these are examples of the removal of the lateral support.

It should be noted that detrimental subsurface water may reach a given locality in some *indirect way*, e.g., because of the removal of vegetation with consequent increase in percolation of the surface water, interception of the subsurface water flow somewhere downhill of the investigated locality with consequent rising of the ground-water level, or leakage

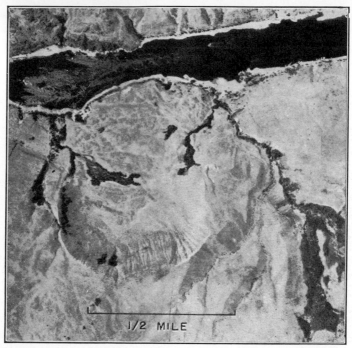

FIG. 17.22. Landslide on Columbia River, Washington. Note partial obstruction of the river by the slide. (*USDA-PMA airphoto.*)

from reservoirs, aqueducts, sewers, canals, or irrigation systems. Expansion of water freezing in rock fissures and joints also increases the sliding risk.

Preventive measures against sliding depend, of course, on the findings of the preliminary investigation described. Detrimental loads should be prevented from being placed and should be removed if already placed. Weak material at the foot of an earth slope may be replaced by a better one with proper compaction. Water should not be permitted to come to the slope in excessive quantities. Proper design of *preventive drainage* in an unknown locality is a difficult problem, however, since considerable time and careful observations are required to ascertain correctly the critical amount of water to come. In fact, the magnitude of the subsurface water flow changes from year to year. The methods of preventive drainage are basically similar to those of *corrective drainage* discussed in the next section.

17.11. Rehabilitation (Correction) of Slides. After a slide has occurred, the survey and investigation of the cause (analysis of the slide) should be made. A hand-level survey is sufficient in many cases. For larger slides airphotos are of great help. An airphoto facilitates the ground survey and often makes it wholly or partly unnecessary. Usually the cost of aerial photography is not so high as is often believed. Borings are done for locating the slip surface, determining the position of the natural ground surface, and studying ground-water and soil conditions. If rock is involved, borings may be helpful for determining the stratification and nature of the rocks. The investigation of the slide should furnish all data required for establishing its cause and studying the possibilities of its correction on the comparative basis as far as the cost of rehabilitation is concerned:

Drainage. A slide occurs or recurs as a result of the combined action of the cause and the contributing factors. Even if the cause of the slide still persists but the contributing factors are sufficiently reduced in value or in efficiency or both, a poorly corrected slide might not recur. Since water is practically always a contributing factor to a slide, thorough drainage may be helpful as a preventive measure against the recurrence of a slide in the majority of cases. If the cause of the slide is a high pore pressure, good drainage should lower the water table, actual or piezometric, to such a level as to overcome this danger. In this case not all ground water should be removed, but only the detrimental portion. It is very difficult, however, to give rules or numerical data for the design of drainage to correct a slide. This is the reason why the actual drainage often proves insufficient or otherwise unsatisfactory in such cases and has to be gradually improved. This should be made clear to the client as early as possible.

There are two basic ways of draining a slide which will be explained hereafter in the example of an embankment in which a circular slide has occurred. Either the arriving ground water may be intercepted by a covered drain and taken to a sink or place of discharge, or the drainage area (i.e., cross section) of the coming ground-water flow decreased by inserting tubes in it and taking water away.

A more complicated case is shown in Fig. 17.23, which shows a sloping locality on which three consecutive slides occurred, all practically at the same place. Eleven wells 24 in. in diameter were dug and filled with riprap, and from a point a fan of oblique holes was driven to reach those wells. Metallic slotted tubes 2 to 4 in. in diameter, the so-called "horizontal drains" (sometimes termed "hydraugers"), were inserted into the oblique holes, and ground water was taken to the adjacent brook. The ground-water level under the berm of an adjacent pond was lowered. Incidentally, the existence of consecutive slides at the same place may be indicative of a condition as at point C, Fig. 17.17.   Hydrauger hole drainage is very popular on the California highways.[14]   It was also successfully used in stopping sliding ground along relocated highways and railroads at Big Dam in central Oregon,[15] and was applied in the construction of Oahe Dam in South Dakota.   The idea of a "longitudinal drain" for a slide correction is discussed in Ref. 16.

The principle of drainage by intercepting the ground-water flow may be used in the form of deep galleries or toe trenches perpendicular to the direction of the flow and filled with pervious material.   In all cases of interception of the water flow (both surface and subsurface) water should be conveyed to a low place (sink). Other methods of drainage varying according to the ingenuity of the engineer or geologist may be used.

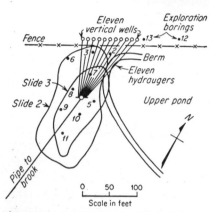

FIG. 17.23. Fanlike "horizontal drain" system or hydrauger drainage.[6]

Removal of Loads and Restoration of the Lateral Support.   Loads that have caused a slide should be removed, and if this is impossible (as, for instance, of a railroad or highway embankment that cannot be relocated), a good lateral support should be furnished.   *Retaining walls* in this case may be of great help, but they are of little or no value if the slide has been caused by water.   *Cribs* made of timber, concrete, or other suitable material and filled with coarse rocks or boulders are a special case of retaining walls.   A heavy *toe buttress* ("key") of compacted earth or rock also offers a lateral support to a slope.

Driving sufficiently strong piles through an earth slope to a sufficient depth in a reliable stratum may be considered as both a preventive and correction measure.

It goes without saying that in the case of a slide, removal and sometimes reshaping of the slide material should be done.   New material may be added.   Bulldozers are often used for these earthwork purposes.   In the case of large rock slides correction is difficult, and relocation of the structure involved (often a highway or a railroad) is perhaps the sole remedy.

A discussion of landslide corrective measures may be found also in Refs. 5, 8, 14, and 17.   Reference 18 contains a vast bibliography on slides, including their correction.   Reference 19 is a comprehensive discussion of the stability analysis of actual slides.*

* A comprehensive treatise on slides by the Landslide Committee of the Highway Research Board was almost ready for publication early in 1956.

CREEP AND FLOW

17.12. Characteristics of Creep Movements.   Creep is a *slow motion* of the upper strata of a generally *unloaded* soil, soil-rock, or rock mass with respect to the underlying strata.   It usually affects the first few feet below the ground surface in both homogeneous and heterogeneous (nonhomogeneous) masses.

In a laboratory direct shear test (Sec. 4.18) clay starts to move along a shear surface when the gradually increasing shearing stress constitutes *only a fraction* of the shearing strength of the clay.   In a like manner the

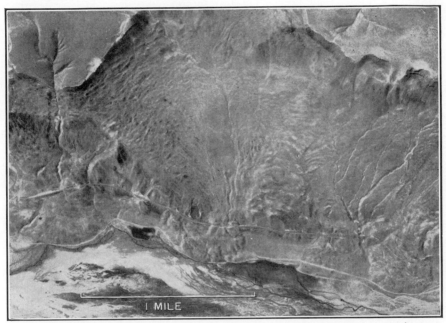

FIG. 17.24. Airphoto of soil creep, Washington.   (*USDA-PMA photograph.*)

upper layers of a homogeneous thick clay deposit may creep under the action of the shearing stresses, which overcome only a fraction of the shearing strength.   This fraction is termed "fundamental strength" by some[8,20] and "creep strength" by others.   Since a creeplike landslide or any other displacement in the earth crust follows the pattern of least resistance, the creep in *heterogeneous* masses will tend to separate a stratum from the underlying one if the *bond resistance* between the two strata is weaker than the fundamental strength of either of the strata in contact.   Creep in heterogeneous masses essentially consists in the motion of the upper, softer portion of a mass over the sloping, flat, or curved surface of a firmer underlying deposit.   The creeping portion of the mass may move as a unit or may consist of two or more portions of variable firmness or consistency which may (but not necessarily do) creep with respect to each other.[5]

Since, in order to overcome the fundamental strength of the clay or break the bond resistance, shearing stresses of limited value only are needed, the creep may take place on mildly sloping terrain. If the terrain is loaded or unloaded along a line, the creep will be perpendicular to that line. For example, there may be creep toward an erosion valley on a very mild slope. In hitting an obstacle (or descending to a valley) the creeping strata may bulge, fold, or break. Tension cracks perpendicular to the direction of the creep also may be formed.

Following the same principle that governs landslides, namely, heading toward the topographic bottom of the area, creep may be directed toward a river.

The intensity of the fundamental strength or the bond between two different strata may be modified by atmospheric agents. A change in the moisture content in a portion of a stratum may shift its center of gravity and thus cause a motion. Freezing and thawing or swelling and shrinkage may on different occasions accelerate, decelerate, or stop the creep. Soil creep may cease in dry weather. If the creeping mass reaches a steep slope, the bond between the strata may be totally broken, in which case the creep degenerates into a landslide or even a fall, particularly in the case of rock creep.

17.13. Creep Types. One of the creep types, namely, *solifluction*, has already been discussed in Chap. 10 in connection with permafrost. It should be noted, however, that solifluction is not exclusively confined to arctic zones but may be observed in all cold climates. The same is true in all cases of rock and soil creep with ice inclusions or ice creep with rock and soil inclusions ("rock-glacier creep").

A thorough discussion of creep phenomena may be found in Ref. 2. There may be *rock creep, talus creep,* and *soil creep.* From the engineering point of view the last type is the most important. When the stability of an earth slope is being investigated, attention should be paid, among other things, to tilted trees, which generally are indicative of creep. (A tree tilted in its lower portion only and vertical in the rest of its height generally indicates an old landslide, however, and sometimes by counting the rings in the tree section, it is possible to estimate the date of an old landslide.) Tilted fence posts, especially at the upper edge of a slope; broken or displaced retaining walls; and other structures also indicate creep. If the wall is low and very stable, however, the creeping material may overflow it.[7] Figure 17.24 depicts creep on a huge scale.

Engineering structures should not be erected in the creep zones unless their foundations reach strata in which no potential danger of creep exists.   Reservoirs of the dams constructed in valleys with distorted and broken creep strata are likely to leak.

17.14. Earth Flow.   A soil or a soil-rock mass may flow as a liquid if it possesses a certain degree of *fluidity*.   A solid earth mass may become fluid (1) by simple addition of water, (2) because of a shock or a series of shocks as occurs to thixotropic clays during earthquakes or pile driving in the neighborhood, and (3) because of remolding as in the case of very sensitive, so-called quick clays.   In some cases of apparently stable, saturated silt masses a sudden liquefaction may occur because of excessive pore pressure.   The agencies turning a solid earth mass into a fluid may act separately or in combinations.

Mudflow.   The degree of fluidity of an earth mass ranges from very dense, doughlike mixtures to quite fluid dilute suspensions such as muddy waters of a stream eroding its valley (the Colorado River, for example). Examples of fluid soil suspensions are also *mudflows*, generally produced when a rapidly moving stream of storm water washes out a suitable abundant load in a valley or a depression.   A stream of flowing mud may move heavy objects; on the other hand, cases have been reported when water made bogs burst and a mud stream thus produced carried large masses of peat.   Since mudflows often recur along the same channels, such channels should be under close observation and measures against possible damage of nearby structures should be taken.

Displacements of Semifluid Masses.   A conventional term "semifluid mass" is used here to designate a mass that translates not so fast as the mud flow, for example, but differs from the creep by greater mobility and greater velocity of the moving material.   Such flow may end by turning into creep.

In Montier-Court Gorges, Switzerland,[21] abundant marl deposits in an anticline covered with a thick layer of debris were saturated by heavy precipitation.   The semifluid mass from 15 to 20 m thick was translating with a decreasing velocity, from 1 m per day in April, 1937, to 1 mm in 15 days in February, 1938.   The flow was directed toward La Birse River and dammed it.

Flow of extrasensitive (quick) glacial clays oversaturated to the "stiffening limit" (Sec. 4.9) has been reported from Norway and Canada. It is generally difficult to establish the immediate cause of a practically instantaneous transformation of an apparently solid clay substance into a semifluid mass.   The clay deposits in this case are generally *very thick*, and the flow leaves a depression with practically *vertical walls*, or cliffs, in the remaining clay deposit.   The flow (sometimes classified as a "landslip") is generally *retrogressive;* i.e., the transformation of clay into a semifluid mass proceeds upstream, i.e., against the flow.

In dealing with high clay slopes in the regions where highly sensitive clays exist, it is convenient at least to test the clay in the field by remolding it between the fingers and subjecting it to vibrations. Simple comparative tests on undisturbed and remolded samples are recommended in doubtful cases.

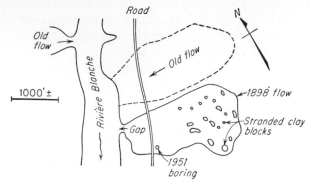

FIG. 17.25. Sketch map of clay-flow area, St. Thuribe, Quebec, Canada. (After Peck et al.)

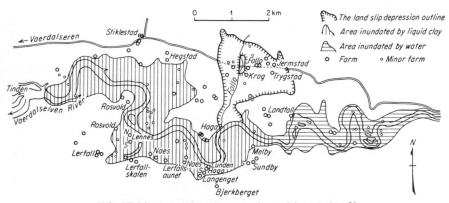

FIG. 17.26. Vaerdalen, Norway, flow of liquid clay.[24]

In the case of St. Thuribe flow, 40 miles west of Quebec, Canada, about 3,500,000 cu yd of extrasensitive clays together with overlying cultivated land were moved. The material rushed forward through a 200-ft gap to the Rivière Blanche (a tributary of the St. Lawrence River) which was dammed to a height of about 25 ft over a length of about 2 miles. The depression left in the clay deposit was about 30 ft deep with practically vertical walls (Fig. 17.25). The bottom of the depression was flat, and so was the surrounding terrain.[2,22,23]

In 1951 (i.e., over 50 years after the flow) a boring within the depression next to its slope was made for research purposes (Ref. 22 and Fig. 17.25). Some soil samples taken were relatively brittle when undisturbed, but when remolded, they were transformed into viscous liquid which, if held in the hand, gradually flowed out and between the fingers in droplets, which accounts for a high sensitivity of the material. Other tests have shown that the tested material could be identified as clay of moderate plasticity (LL = 33 per cent, PL = 21 per cent) and that it contained about 36 per cent of clay-size particles, the rest being practically all silt size. The average moisture content was about 44 per cent.

The great Vaerdalen, Norway, flow took place in 1893. A freehand sketch of the present state of the depression caused by the flow is shown in Fig. 17.27. The flow lasted about 30 min, and the flow material (clay with lenses of quick clay) dammed the adjacent river. The slopes of the depression were vertical, as may be seen from the sketch, and the plan of the depression was similar to that of the St. Thuribe depression. Farms and entire fields joined the churning stream of clay and floated away. Some inhabitants of a farm were rescued after having sailed about 4 miles down the clay stream on the roof of their own house.[24]

FIG. 17.27. Freehand sketch of the Vaerdalen, Norway, clay flow. (By H. Reusch.[24])

A flow in Japan was set up by an earthquake.[25] The depression left in the original deposit had "sheer" walls, and the flow material consisted of weathered volcanic ash saturated with water. In another flow in Japan the depression was 20 ft deep and its walls were "vertical on both sides." Flows of this kind moved out great boulders "some of the size of a room."[25]

SETTLEMENTS AND SUBSIDENCES

Settlement is defined as a downward movement of a structure as caused by its dead and live load and other forces. Subsidence may be defined as a downward movement of the natural ground surface on or close to which no structures have been built that could produce a settlement of that magnitude.

17.15. Settlement Observations. If considerable settlements of a structure or of an earth body are expected or noticed, settlement observations should be organized. In a general case both horizontal and vertical movements of the structure or the ground surface are observed. All observations should be recorded in plan and the longitudinal and transverse profiles. Two intersecting base lines should be traced in the proximity of the structure, and these lines should be oriented on remote stationary objects so that they can be retraced at any time. Horizontal motion, if any, should be tied to these axes. Good bench marks (symbol B.M. on maps) should be established. Their elevations should be tied to the existing bench marks of a well-known organization such as USGS or Army Engineers. In many cases a conventional elevation of the bench mark, such as 100 or 10, will suffice for observation purposes. For small or temporary projects bench marks are fixed on points not subject to disturbance, e.g., stumps or walls of large buildings. Bench marks in permafrost areas are described in Chap. 10.

Setting up of substantial bench marks is described in Ref. 26. These should be founded on bedrock if possible. Outcrops, if any, may be used to great advantage in this connection. When the soil mantle is thick, a large pipe, e.g., 4 in. in diameter, is driven through all compressible strata. Soil material is cleaned out by using water or air jet or some other means, and the hole extended to bed rock or at least to a firm soil stratum, such as glacial hardpan or other similar material. A 2-in. pipe to serve as a bench mark is placed into the casing formed by the 4-in. pipe and securely grouted, for instance with cement mortar. The simplest type of reference point *within a building* is a bolt embedded in masonry or concrete.

In settlement surveys the engineer's level is generally used. The frequency of leveling operations depends on the purpose of observations. For long-term observations elevations should be taken at least once a year.

By definition a downward movement of an embankment is its settlement. If the embankment is well compacted, its settlement is mostly due to the displacement of the foundation material. Movement of metallic or concrete slabs placed at the surface of an embankment or monuments embedded in it should be observed if embankment settlement records are required.

17.16. Field Time-settlement Curves. Various types of time-settlement curves are shown in Fig. 17.28. To draw such curves observed values of settlement are plotted against time, during and after construction. In Fig. 17.28 the figures on the horizontal axis are years, the duration of the construction period being assumed to be 1 year. In the case of *structures on sand* (curve *a*) there is usually settlement during the construction period, after which, as a rule, the structure remains stable. In the case of *clay foundations* the settlement may last for a long time, sometimes for decades and even centuries, while the rate of settlement decreases (curve *b*). If there is plastic flow from underneath a structure, the time-settlement curve degenerates into a straight line *c*, which shows *leakage of plastic material*. Sudden (spontaneous) change in the properties of the foundation material is illustrated by a break in the time-settlement curve (curve *d* in Fig. 17.28).

17.17. Causes of Settlement. Some causes of settlement other than dead and live loads of the structure are discussed hereafter.

Removal of Lateral Support.   If an excavation is made next to a structure (Fig. 17.29), the structure tends to move downward at point *D* because of the lateral yield of slope *AB* and moves somewhat along the potential sliding surface *BC* (shown in dotted lines in Fig. 17.29).   In

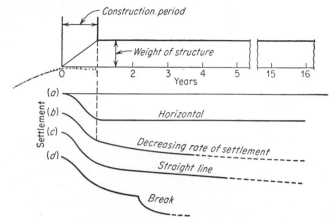

FIG. 17.28. Field time-settlement curves.

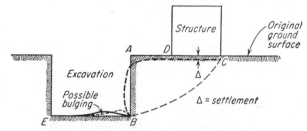

FIG. 17.29. Settlement caused by excavation.

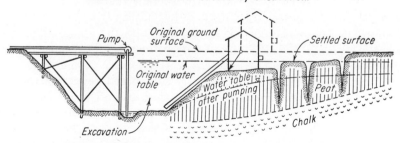

FIG. 17.30. Settlement and cracking due to pumping.[27]

many similar cases the bottom of the excavation *BE* bulges up.   If a deep excavation is filled with water, pumping decreases the hydrostatic pressure on its slopes, and this removes the lateral support to the adjacent structure.

Figure 17.30 shows the effects of pumping on a peat deposit.   The settlement in this case was accompanied by wide tensile cracks about 10 ft deep.[27]

Settlement Caused by a Shear Failure.   A shear failure may cause *tipping* of the structure and its irregular settlement.

An example of tipping is the Leaning Tower of Pisa (Fig. 17.31) constructed gradually between 1174 and 1350.   Tipping was observed when about 35 ft of masonry was laid, but it apparently stopped in 1186.   Vertical settlement proceeded, however. Since 1932 more than 1,000 tons of high-strength cement in the form of grout has been injected into the soil through 361 holes, 2 in. in diameter.   During 1934 and 1935 a study of the movements of the tower with a specially designed inclinometer has shown that the tower not only has a north-south cyclic and recuperative tendency but also moves persistently eastward (Fig. 17.31c).   The cause of the settlement has not been established with certainty.

Fine sand masses, if confined, represent a good foundation material, but if a saturated fine sand mass is permitted to flow laterally (which is a shear failure), it may flow along together with the structure it supports. In this connection the structure moves both vertically and horizontally. Bridge piers have moved in this way both in the United States and abroad.   Different methods have been used to stop such movements. Particularly, soil stabilization by grouting with cement or by injecting special stabilizers may help.   Such stabilization of *fine* sandy materials may or may not be successful, however.   Long piles reaching reliable material may be driven, and a moving pier tied to them with metallic or concrete ties.

Settlements Caused by the Fluctuations of the Water Table.   The weight of that portion of the earth mass below the water table is decreased by buoyancy or hydrostatic pressure on the solid phase of the submerged mass.   The same is true of that part of the structure below the water table.   Should the water table drop, either naturally or by pumping, the buoyancy in the dewatered region vanishes and the apparent weight of the earth mass and the structure, if any, increases.   This causes an additional settlement of the structure.   A rise of the ground-water table causes an opposite effect, and the newly submerged portion of the structure and earth mass moves up.   Removal of ground water from underneath a structure that is still settling may cause its spontaneous settlement and a break in its time-settlement curve (Fig. 17.28d).

In the case of buried structures such as subaqueous tunnels the rise of the water level in the river over the tunnel constitutes an additional load on the tunnel and hence causes an additional settlement.[28,29]

17.18. Causes of Subsidences.   Subsidences may be instantaneous and violent or gradual and may be caused by natural forces or by the activities of man.   The general cause of a subsidence is the removal of some solid or fluid phase of the earth crust, sometimes, but not

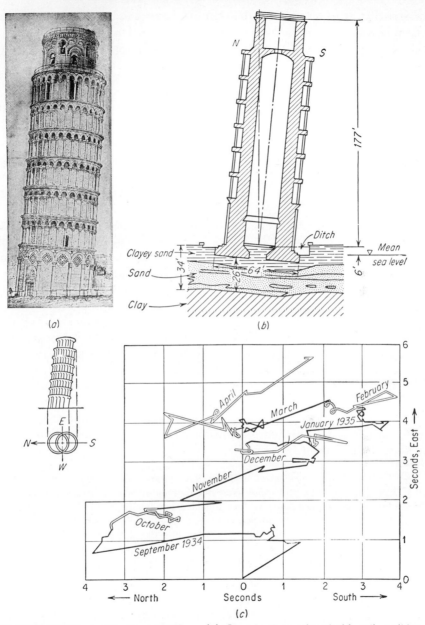

FIG. 17.31. (a) View of the Tower of Pisa. (b) Cross section and probable soil conditions (c) Movement observed with precise instruments.

necessarily, at a considerable depth under the ground surface. Erosional effect of water is often a cause of subsidence. Such is, for example, downfall of large portions of river banks undermined by the repeated rise and fall of the water level in the river.

Subsidences Caused by Pumping. The difference between a subsidence and a settlement caused by pumping consists mainly in the part played by the dead and live load of the structure, which in the case of a subsidence is negligible. Subsidences generally affect large portions of the ground surface, but not necessarily. For instance, unreasonable pumping from a well under construction may cause only a limited subsidence of the ground surface around that well. Subsidences may be caused by pumping of either ground water or oil.

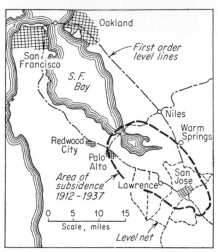

FIG. 17.32. Subsiding area in Santa Clara Valley, California. (*Eng. News-Record,* Apr. 1, 1937.)

The mechanism of a subsidence caused by pumping water from a sand deposit is as follows: (1) Buoyancy acting on the sand grains is relieved; (2) hence pressure on the underlying layers is increased; and (3) water from the overlying clay layers, if any, is gradually removed. This causes additional compression and consolidation and hence additional slow downward motion of the surface of the deposit.

Santa Clara Valley Subsidence in California. (Started about 1920.) This was due to pumping water from the upper several hundreds of feet for irrigation purposes. In the first 12 years the ground surface subsided 1 in. in Redwood City, 6 in. at Palo Alto, and over 4 ft in the city of San Jose (Fig. 17.32). The total subsidence at San Jose was measured as 5.7 ft, the total area affected being about 200 sq miles. Apparent rising of well casings and pumps (as in Fig. 13.25) was common.

Terminal Island Subsidence.[30] Terminal Island in the Long Beach, California, harbor area was built up from a small spit at the ocean level as it was in 1900 to a large island rising about 13 ft above that level. It is generally believed, however, that the weight of the fill is not a major factor in the subsidence of this island.

The sediments in the upper 7,000 ft of the area are shales and sands of both ancient and geologically recent origin. Ancient strata bent into a gentle anticline form the structure of the largely oil-producing Wilmington field. These strata are all highly faulted, the faults being traced to an unconformity about 2,000 ft below surface (Fig. 17.33).

The upper 1,800 ft of the deposit contains fresh water. This was intensively pumped out during the dewatering of the site for a dry dock, which caused the original subsidence around many oil wells.   When pumping for the dry dock stopped, water in the adjacent wells rose to the original level (about the ocean level).   There was a small rebound, but otherwise, the subsidence remained.

There was also considerable pumping of petroleum from the Wilmington oil field. In the case of such pumping there is a local decrease in pressure in the liquid due to suction, a gradient toward the pumping well from all sides, and a compensating increase in the intergranular (effective) pressures in the materials within the affected stratum and compression of such materials.   It is believed by some observers that the decrease in the thickness of the compressible shale strata is predominantly responsible for the subsidence.[31]   This point of view is far from being universal, however.

During high tide the pavements of Terminal Island facing the ocean are covered with water and the industrial plants there have been obliged to take measures against the water invasion.   For instance, Edison Co.'s steam plant subsided about 9 ft. The total subsiding area is roughly elliptical in shape, being miles long in either direction.

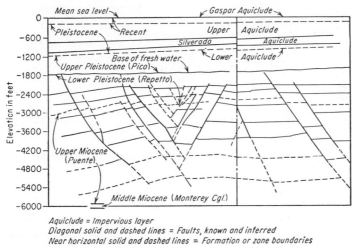

FIG. 17.33. Section through Wilmington oil field.   (*After ASCE Trans., vol. 113, p. 382, Fig. 5, 1948.*)

Shore Subsidences.   The foot of the shore slope is in water, and the part of the slope above water is moist because of the capillary rise which decreases the shearing strength of the material.   River currents and lake waves, respectively, erode the lower part of the slope.   Downfalls of undermined shores have already been mentioned (Sec. 17.16).   The water-level changes in the case of lake shores act similarly to the "sudden drawdown" in the earth dams, and as in a dam the shore slope may slump and even liquefy (compare also Fig. 17.30).   Problems of shore subsidences are closely connected to those of coastal engineering in general (Chap. 11).

The shore subsidence at Lake Gerzen, canton Bern, Switzerland,[32] took place in a flat-lying succession of loam, marl, and peat.   The upper part of this formation was frozen.

The collapse of the shore started upon a drop of the water level in the lake, followed by tree felling and root blasting.   The layer covering the marl was broken into blocks which, in combination with the shock from the felled trees and loss of support from the roots, destroyed the structure of the marl and caused its liquefaction followed by its squeezing out into water.   The soil remaining in place was fissured parallel to the shore line.

Subsidences Caused by Mining and Tunneling.   Damage that can be done to buildings by subsidences in mining districts is illustrated in Fig. 17.34, which shows a wide strip of coal mined out and the consequences

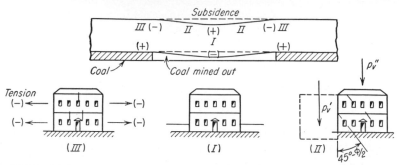

FIG. 17.34. Slablike subsidence due to mining. (*From D. P. Krynine, "Soil Mechanics," 2d ed., McGraw-Hill Book Company, Inc., New York, 1947.*)

of such mining.   If the rock mass above this strip is able to resist stresses, it acts as an unsupported slab and its top surface will subside.   Tension and compression in this slab are marked in Fig. 17.34 with signs "minus" and "plus," respectively.   In the central part of the subsidence there is, close to the surface, a compression that may break pipes and culverts and warp roads and sidewalks.   Buildings in this zone settle down.   In the outside zone (III) there are tensile stresses at the top of the mass. These may cause cracks in the earth mass and also vertical fissures in the buildings.   These cracks obviously appear first at the weakest parts of the building, namely, close to doors and windows.   The case of the intermediate zone II is very similar to that of the differential settlement, and the cracks are oblique.   Cases of sand flow into a tunnel under construction are discussed in Chap. 9.   In conjunction with such a flow a subsidence of the ground surface may be produced.

Figure 17.35 represents such a subsidence over a tunnel being driven in sandstone.[33] According to the shape of fissures the sandstone deposit was folded and fractured. "Open" fissures correspond to tension at the bottom of the deposit, whereas the "filled" fissures have been formed by tension at its top during the process of folding. This fact explains why the overlying sand and fine gravel pour through the funnel-like (in cross section) fissures into the tunnel as soon as the tunnel hit the fissures.

The photograph of Fig. 17.36 represents the fall of a chunk of concrete from the tunnel arch.   The latter was damaged by a huge slide over the tunnel shown in Fig. 17.20.

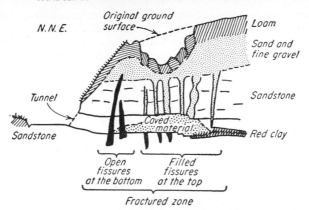

FIG. 17.35. Subsidence over a tunnel in sandstone (Dreisbach, Saar).

FIG. 17.36. Fall of a chunk of concrete from the tunnel arch damaged by a slide, Grand Valley Project, Colorado. (*USBR photograph.*)

Secular Subsidences. Perhaps a unique record of the changes of level of the earth crust is represented by the ruins of the so-called "Temple of Serapis" in the city of Pozzuoli, near Naples, Italy,[34] in the neighborhood of Mount Vesuvius.* The ruins

* The building known as the Temple of Serapis, constructed at Pozzuoli in the second century B.C., was not a temple at all but a commercial building with a 3-ft-high statue of the god in a niche.

of this building contain among other things the remnants of three marble columns about 42 ft in height. The lower 12 ft of the columns is intact, whereas a zone of about 8 ft immediately above is worn out by the action of marine mollusks. The top part of the columns is intact. There is no generally accepted explanation of this phenomenon.

REFERENCES

1. Fenneman, N. M.: (a) "Physiography of Western United States," (b) "Physiography of Eastern United States," McGraw-Hill Book Company, Inc., New York, 1931 and 1938, respectively.
2. Sharpe, C. F. Stewart: "Landslides and Related Phenomena," Columbia University Press, New York, 1938.
3. Messines, J.: Les éboulements dans les gypses, *Compt. rend. acad. sci. Paris*, vol. 226, 1948; trans. in *USGS Trans.* 14, 1948.
4. Proix-Noé, M.: Etude d'un glissement de terrain dû à la présence de glauconie, *Comp. rend. acad. sci. Paris*, vol. 222, 1946; trans. in *USGS Trans.* 6, 1947.
5. Baker, R. F.: Analysis of Corrective Actions for Highway Landslides, *ASCE Trans.*, vol. 119, 1954.
6. Woodward, R. J., *et al.*: Experiences with Subsurface Water, *Highway Research Board Proc.*, vol. 32, 1953; discussion by E. Buckingham.
7. Forbes, Hyde: Landslide Investigations and Corrections, *ASCE Trans.*, vol. 112, 1947; discussion by E. Buckingham.
8. Terzaghi, K.: Mechanics of Landslides, "Berkey Volume," 1950.
9. Kjellman, W.: Mechanics of Large Swedish Landslides, *Géotechnique*, vol. 5, no. 1, March, 1955.
10. Statens Järnväger Geotekniska Kommissionen 1914–1922, *Slutbetänkande*, Stockholm (*cf.* Fig. 40 in Ref. 9).
11. The Proceedings of the European Conference on the Stability of Earth Slopes, *Géotechnique*, vol. 5, no. 1, March, no. 2, June, 1955.
12. Jacobson, B.: The Landslide at Surte on the Göta River, *Proc. Roy. Swedish Geotech. Inst.*, vol. 5, 1952.
13. Maddalena, L.: Protection des voies contre la chute des rochers, *Proc. congr. intern. mines, met. et géol. appl.*, Paris, vol. II, 1935; trans. in *USGS Trans.* 4.
14. Root, A. W.: California Experience in Correction of Landslides and Slipouts, ASCE San Francisco Convention, *Sep.* 81, 1953.
15. Editorial: Huge Slides Halted by Hydrauger Holes, *Contractor's Engrs. Monthly*, vol. 49, 1952.
16. Marivoet, L.: Control of Stability of a Sliding Slope in a Railway Cut near Wetteren (Belgium), *Proc. 2d Conf. on Soil Mech. and Foundation Eng.*, vol. II, Rotterdam, paper IV, p 3, 1948.
17. Ladd, G. E.: Landslides, Subsidences and Rock Falls, *ARE Assoc. Proc.*, vol. 36, 1935.
18. Tompkin, J. M., and S. H. Britt: "Landslides," Highway Research Board Publication, Washington, D.C., 1951; a selected annotated bibliography.
19. Berger, Louis: "Stability Analysis of Actual Slides," Northwestern University dissertation (unpublished), Evanston, Ill., 1951.
20. Griggs, D. T.: Deformations of Rocks under High Confining Pressure, *J. Geol.*, vol. 44, 1936.
21. Peter, A.: Glissement de terrain dans les gorges de Montier-Court, *La Route et la circulation moderne*, Solothurn, Switzerland, 1938; trans. in *USGS Trans.* 7, 1947.

22. Peck, R. B., *et al.:* Earth Flow, St. Thuribe, Quebec, *Univ. Illinois Civil Eng. Studies,* 1951.

23. Dawson, G. M.: Remarkable Landslip in the Portneuf County, Quebec, *Bull. GSA,* vol. 10, 1899.

24. Holmsen, P.: Landslips in Norwegian Quick Clays, *Géotechnique,* vol. 3, 1953.

25. Lowdermilk, W. C.: Problems in Reducing Geological Erosion in Japan, *Proc. assoc. intern. d'hydrol. sci.,* vol. 2, Brussels, Belgium, 1951.

26. Terzaghi, K.: Settlement of Structures in Europe, *ASCE Trans.,* vol. 103, 1938.

27. Haefeli, R., and W. Schaad: Setzungen infolge Senkung, Schwankung und Stroemung des Grundwassers, *Schweiz. Bauzt.,* vol. 123, 1944.

28. Jacobs, C. M.: The New York Tunnel Extension of the Pennsylvania Railroad, *ASCE Trans.,* vol. 68, 1910.

29. Les affaissements de la gare transatlantique du Hâvre, *Centre d'études et de recherches géotech. Bull.* 3, Paris, 1935.

30. Gilluly, J., and U. S. Grant: Subsidence in the Long Beach Harbor Area, *Bull. GSA,* vol. 60, 1949.

31. Harris, F. R., and E. H. Harlow: Subsidence of the Terminal Island, Long Beach Area, California, *ASCE Trans.,* vol. 113, 1948; discussion by Hyde Forbes.

32. von Moos, A., and R. T. Rutsch: Ueber einen durch Gefugestoerung verursachten Seeufereinbruch, *Eclogae Geol. Helv.,* vol. 37, 1944.

33. Sobotha, E.: Erdfalle im Gebiet des Mittleren Buntsandsteins, *Z. deut. geol. Ges.,* vol. 93, 1947; trans. in *USGS Trans.* 8, 1947.

34. Günther, R. T.: Changes in the Level of the City of Naples, *Geograph. J.,* vol. 24, 1904; also *Nature,* vol. 69, 1904. Both are London publications. There is also a considerable amount of Italian literature on the subject.

CHAPTER 18

EARTHQUAKES AND ASEISMIC* DESIGN

Earthquakes have been reported from all parts of the globe. Many have taken place in the zones close to the great mountain chains, and a number of earthquakes have been registered in the circum–Pacific Ocean belt. Of known United States earthquakes the severest were in New Madrid, Missouri, 1811–1812; Charleston, South Carolina, 1886; and San Francisco, California, 1906. The West Coast of the United States is chronically affected by earthquakes of variable intensity. Knowledge of the physical nature of earthquakes and their influence on engineering structures is essential for engineers and engineering geologists working in the zones where the occurrence of earthquakes is probable.

18.1. Engineering Problems Related to Earthquakes. Engineering problems related to earthquakes may be grouped into two large categories. *First*, it is necessary to establish what the nature of the *risk* would be if the planned structure is built in the given region. This breaks down to a consideration of (1) *seismicity* of the region, i.e., the probability of earthquake occurrence and their intensity and magnitude, and (2) the possible nature and consequences of earthquake *damage* to the particular structure. The earthquakes are of variable intensity, and there are classifications of earthquakes, or *intensity scales*, discussed in Sec. 18.2 of this chapter. As soon as the intensity detrimental to the planned structure is established, the *probability* of the earthquake of this intensity in the given region should be estimated. Statistical data on observed earthquakes and knowledge of the local geology are of help in this connection.

Protection of life and major property—for which the existing building codes offer only minimum standards of design—should be considered by the engineer. If the designer feels that the requirements of the building code are insufficient, he must decide in each particular case if more stringent design criteria are warranted.

The second category of engineering problems related to earthquakes is to *establish certain bases of structural design* in seismic regions. These

---

* The more correct term "aseismatic" is not used by engineers.

result from theoretical studies and actual observations of the ground motion and damage during an earthquake.   Considerable work has been done along these lines since the San Francisco earthquake of 1906 and particularly since the Long Beach, California, earthquake of 1933.

18.2. Earthquake Intensity Scales.[1-4]   It is necessary to make clear the difference between the *intensity* and the *magnitude* of an earthquake. Whereas the intensity is characterized by earthquake effects and of necessity is a qualitative concept, the magnitude is an instrumentally measured quantity related to the total energy released during the earthquake.

The earthquake intensity scales depend on the *perceptibility* (or sensibility) and *destructivity* of an earthquake.   In other words, the earthquakes are classified on the basis of the answers to the following two questions: (1) Where and how was the earthquake felt, and (2) what damage was done by the earthquake?   Whereas the degree of damage may be estimated correctly and objectively, the perceptibility of an earthquake depends on the location of the observer and on what is termed his "personal equation."   Observers at rest or located at a high story of a building or on loose ground feel slight earthquakes which may not be perceived by others.   Quite exaggerated estimates of the height of ground waves or of displacement of objects are common.   The weakness of existing intensity scales is that lack of information makes difficult the correlation of observed effects with instrumentally recorded earthquake motion. Intensity scales in general use are the Rossi-Forel and the modified Mercalli.   These should not be confused with the Gutenberg-Richter scale, which relates to magnitude and energy.

MODIFIED MERCALLI (M-M) EARTHQUAKE INTENSITY SCALE (abridged)

I. Not felt except by a very few under especially favorable circumstances.

II. Felt only by a few persons at rest, especially on upper floors of buildings. Suspended objects may swing.

III. Felt quite noticeably indoors, especially on upper floors of buildings.   Standing motor cars may rock slightly.

IV. During the day felt indoors by many, outdoors by few.   At night some awakened.   Dishes, windows, doors disturbed.   Standing motor cars rocked noticeably.

V. Felt by nearly everyone, many awakened.   Some dishes, windows, etc., broken; unstable objects overturned.   Pendulum clocks may stop.

VI. Felt by all; many frightened and run outdoors.

VII. Everybody runs outdoors.   Damage negligible in buildings of good design and construction, however.   Shock noticed by persons driving motor cars.

VIII. Damage slight in specially designed structures; considerable in ordinary substantial buildings; great in poorly built structures.   Fall of chimneys, stacks, columns.   Persons driving motor cars disturbed.

IX. Damage considerable even in specially designed structures; well-designed frame structures thrown out of plumb.   Buildings shifted off foundations.   Ground cracked conspicuously.

X. Some well-built wooden structures destroyed; ground badly cracked; rails bent. Landslides and shifting of sand and mud.

XI. Few if any (masonry) structures remain standing. Broad fissures in ground.

XII. Damage total. Waves seen on ground surfaces.

18.3. Mechanism of Earthquakes.[4,5] An earthquake is a vibration, or oscillation, of the ground surface by a transient disturbance of the elastic or gravitational equilibrium of the rocks at or beneath the surface. The earthquake may be artificial if the disturbance was caused by man, as through a blast of explosives done for subsurface exploration purposes (Chap. 6). In the case of a natural earthquake as discussed hereafter the disturbance and the consequent movements which give rise to elastic impulses, or *waves*, are caused by natural processes in the earth. Natural earthquakes are classified as *tectonic, plutonic,* or *volcanic* accord-

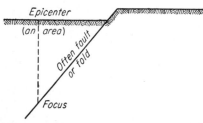

FIG. 18.1. Focus and epicenter of an earthquake (rough sketch).

ing to whether the stresses which cause the movement at the *source* are, respectively, structural in origin or proceed from deep-seated changes or from volcanic activity. The earthquakes in the continental United States are tectonic with exception of volcanic earthquakes in the region of Mt. Lassen in the northern part of California. The earthquakes in Hawaii as caused by Mauna Loa volcano and numerous Japanese earthquakes are volcanic.

The place in the earth crust where an earthquake originates is its *focus*. The projection of the focus on the ground surface or a limited area of the ground surface is the epicenter (or epicentrum) of the earthquake. The "elastic rebound theory" of tectonic earthquakes[5] explains the occurrence of an earthquake by the gradual accumulation of strain in a given zone and hence a gradual increase in the amount of elastic forces stored. A sufficiently large strain ruptures the earth crust with the formation of a fault (Fig. 18.1). In the 1906 San Francisco earthquake the San Andreas fault broke for a length of 270 miles. This fault is roughly parallel to the Pacific Ocean shore and in parts is hidden under water.

Friction between the moving faces of the fault generates the elastic waves. Another opinion is that the waves are generated by the vibrations of the strained mass. The energy stored in the rock before the earthquake is released in producing those waves and is partly used up in heat. As already stated, the amount of energy released is the measure of the magnitude of an earthquake.

18.4. Ground Vibrations. Earthquakes and the earth waves are studied by the science of seismology, one of the earth sciences. The theoretical approach to this study presupposes that the earth is an elastic medium, possessing a certain density and certain values of the modulus of elasticity and Poisson's ratio.   In the theoretical study of waves the concept of *simple* (or free) *harmonic motion* is used.   If a particle M travels counterclockwise with a constant linear velocity about a circle of radius r (Fig. 18.2a), its projection M_1 on the X axis (obtained by dropping the perpendicular MM_1) travels along this axis back and forth. The *amplitude* of this oscillation equals the radius of the circle.   If particle M starts its motion at point X, the right end of diameter X_1X describes the whole circle and returns to point X, and its projection M_1 would travel horizontally from X to X_1 and back.   At each moment the

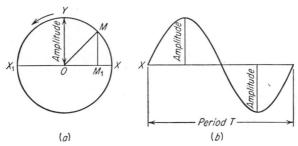

FIG. 18.2. Simple (or free) harmonic vibration.

displacement of point M_1 is measured by its distance from the center of the circle.   The time required for projection M_1 to complete one such to-and-fro oscillation is the *period* of oscillation (symbol T).   The frequency of oscillation f is the number of cycles the particle M described in 1 sec, or the number of periods per second.   For example, if the period is $\frac{1}{30}$ sec, frequency f is 30.

It should be made clear that the seismologists do not believe that earthquake waves follow exactly the law of simple harmonic motion. This concept is brought in to define period and amplitude and get a rough relationship between them and acceleration.   In a simple harmonic motion, rock or soil particles located along a certain line XX_1 (Fig. 18.2a) before the earthquake, under the influence of the impulses from the focus of the earthquake and from elsewhere in the rock mass, start to move in a way similar to that shown in Fig. 18.2b.   In reality, the amplitudes and periods of oscillations are irregular.   It should also be clearly understood that though the rock or soil particles *oscillate* about their original positions, they do not advance with the wave.   The wave pattern is advancing, however.

A rock particle starts to oscillate or vibrate when the *wave front* reaches it.   When a period of oscillation of the given particle is completed, the wave front has advanced a certain distance, which is the *wave length*.

18.5. Vibrations of Structures.

Vibrations of points within a medium such as rock or vibrations of elastic systems such as structures are *free* if they are caused by *one impulse* and no other factors interfere with the vibration.

If during a free vibration some other impulses influence the structure, this would be a case of *forced vibrations*. If there are forces tending to stop or reduce vibrations, these forces are termed *damping forces*. A very common damping force is that of friction.

Generally an elastic system, such as a structure, can perform vibrations of different shapes, or *modes*. If there is only one mode of vibrations, the system is said to have one degree of freedom. There may be a *fundamental* mode with the longest period, the first, the second mode, etc., with gradually decreasing periods and increasing frequency. The Golden Gate suspension bridge in San Francisco, for instance, during a storm, has the largest *amplitude* in a higher mode, i.e., with shorter period of vibration. If translational vibration of the system is considered, the fundamental mode in translation is called *natural frequency* in engineering. The reciprocal of the natural frequency is the *natural period*. In the case of a structure the natural frequency depends on the properties of the structure itself and those of its foundation.

18.6. Earth Waves and Seismographs.[4-7]

In the interior of a vibrating elastic medium including the earth globe, there are *body waves* which travel through the interior of the elastic medium and *surface waves* bound to the free surfaces of that medium. The two kinds of body waves are *longitudinal, compressional*, or P waves, and *transverse*, shear, or S waves. Longitudinal waves in which particles oscillate in the direction of propagation cause alternate compression or tension (rarefaction) of the medium (push or pull). In the case of transverse or shear waves, particles oscillate perpendicularly to the direction of propagation. P waves arrive at a given point, including seismograph, before S waves.

A seismic sea wave (*tsunami* or *tunami*, the Japanese term) is due to a sudden rising or dropping of the ocean floor. In some areas, as the northeastern coast of Japan, they are frequent.

Microseisms are minute waves that are continuously moving through the rocks at the globe surface (compare Sec. 2.6). They may have periods from about 4 to 7 min and amplitudes such as 0.001 in. or less. Apparently they are of meteorological origin as are water waves.

The movement at a given point of the ground surface as caused by an earthquake (or by any other impulse) may be resolved into three translations parallel to three mutually perpendicular axes (for instance, vertical, north, and east from the given point). There are also three rotations about those axes which, being small, may be neglected in the case of an earthquake. Translations (or displacements) are measured by *seismo-*

graphs. The seismograph records of an explosion or dropping weight are very little different from the records of a small earthquake. For natural earthquakes seismographs of various types are used. Some seismographs measure vertical motion only; others, horizontal. Actual translations are *magnified* in recording, and usually a recording station is equipped with instruments having a large range of magnification.

18.7. Engineering Use of Accelerograms. From the engineering point of view the *strong-motion* earthquakes are the most important, since such earthquakes bring damage and destruction to engineering structures. Records of shocks produced by such earthquakes are obtained by use of

TABLE 18.1. Data on Ground Motion during Some Strong-motion Earthquakes

Refer-ence	Locality and date	Maximum acceleration								Maximum horizontal velocity		Maximum horizontal displace-ment	
		Horizontal				Vertical							
		a, cm/sec²	$\frac{a}{g}$	Peri-od, sec	Dis-place-ment, cm	a, cm/sec²	$\frac{a}{g}$	Peri-od, sec	Dis-place-ment, cm	cm/sec	Peri-od, sec	cm	Peri-od, sec
(1)	(2)	(3)	(4)	(5)	(6)	(7)	(8)	(9)	(10)	(11)	(12)	(13)	(14)
8	Long Beach, Calif., 1933	230	0.24	0.30	0.50	250	0.26	0.11	0.07	25	1.50	6.0	1.5
9	Helena, Mont., 1935	115	0.12	0.13	0.05	78	0.08	0.125	0.03	96	1.20	1.8	1.2
10	El Centro, Calif., 1940	320	0.33	0.31	1.30	220	0.23	0.10	0.07	30	4.00	19.0	4.0
11	Olympia, Wash., 1949	321	0.33	0.34	0.94	107	0.11	0.10	0.03	25	0.34	0.9	0.34
12	Taft, Calif., 1952	128	0.13	0.24	0.19	99	0.10	0.54	0.73	32	0.84	2.1	0.84

ruggedly constructed seismographs (called also accelerometers), and the useful approximations of accelerations or displacements are plotted against time. Three mutually perpendicular components of motion are usually recorded. Table 18.1 summarizes some of the key data obtained from the accelerograph records for several strong-motion earthquakes. The accelerations measured instrumentally are designated with symbol a, and the ratio a/g is computed, g being the acceleration of gravity.

The wide use of strong-motion accelerographs is relatively recent. Only in 1932 the USCGS initiated a strong-motion earthquake program in order to obtain instrumental records of ground motion in destructive areas. This is the reason why Table 18.1 contains but a limited number of data. Even so, some conclusions from the table are obvious. Horizontal accelerations are, as a rule, considerably stronger than the vertical. Vertical displacements are very small in comparison with the horizontal. The accelerations, velocities, and displacements shown in the table are maxima and do not necessarily correspond in time one with the other. Horizontal velocities and displacements are great if the accelerations during a given earthquake are great. According to El Centro records, acceleration periods from 0.1 to 0.3 sec could be considered predominant in the case of strong-motion earthquakes but not exclusive because of the random nature of the ground motion.

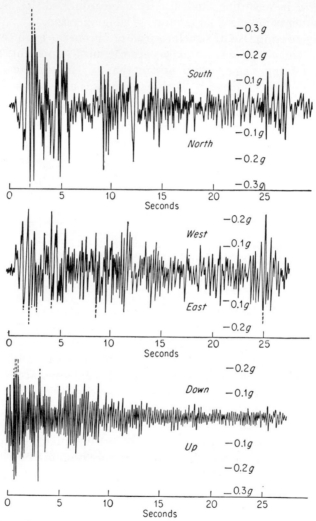

FIG. 18.3. Example of an accelerogram, El Centro, Imperial Valley, California, 1940. (*From ASCE Trans., vol. 117, p. 722, 1952.*)

Other minor and moderate earthquakes recorded seem to have peak acceleration periods within that range. The ground motion as recorded at El Centro (Fig. 18.3) exhibits an irregular pattern lacking a simple harmonic motion. Information obtained in these measurements, however, has permitted engineers to lay a purely engineering background for the aseismic design of structures.

18.8. Spectrum of an Earthquake, Base Shear. A careful examination of the accelerogram of an earthquake permits one to distinguish various vibration periods and measure the corresponding acceleration. By plotting accelerations against the period values, one may obtain an irregular curve called the *spectrum* of a given earthquake. If the spectra of a large number of strong-motion earthquakes could be thus constructed, an average spectrum indicating the relationship between the period and acceleration values in strong-motion earthquakes could be ascertained. Insufficiency of statistical data (as those in Table 18.1) does not permit such a construction, however.

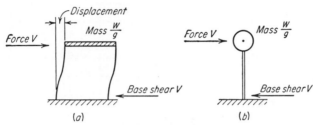

FIG. 18.4. Spring-mass (or one-mass) model of a structure.

The behavior of an actual structure during a shock depends on its natural frequency and damping forces. In order to establish the earthquake *design forces*, researchers replace actual structures by simplified models consisting, for example, of springs with heavy masses on top. Models are vibrated at different vibration periods, and the corresponding accelerations are determined. Thus an idealized spectrum may be obtained applicable to any structure and any earthquake.

If the weight of the experimental spring mass (Fig. 18.4) is W lb, its mass is W/g, where g is acceleration of gravity ($g = 32.2$ ft/sec²). If W is expressed in grams, $g = 981$ cm/sec². If a shock is simulated, the structure vibrates, and the force V making it vibrate is its mass times acceleration a. This force tends to shear off the structure from its base. Hence the *base shear* value V is

$$V = \frac{W}{g} a = \frac{a}{g} W \tag{18.1}$$

Mathematical procedures required for the computation of the acceleration a are far from simple. This value has been determined, however, by using various auxiliary devices, such as mechanical analogy, electric analogue computer, or computing machines. If the maximum acceleration during a shock is designated by A, the maximum or *spectral base shear* will be, by analogy with Eq. (18.1),

$$\text{Max } V = \frac{A}{g}\, W \tag{18.2}$$

On the basis of numerous experiments by various investigators, M. A. Biot constructed the average response curve known as "standard acceleration spectrum" that is shown in Fig. 18.5. According to this curve, if a spring-mass system with a 0.2-sec natural period develops a base shear of 100 kips, a one-mass structure of the same weight but with a natural period of 0.73 sec would develop a base shear of only 30 kips under the influence of the same earthquake pattern.

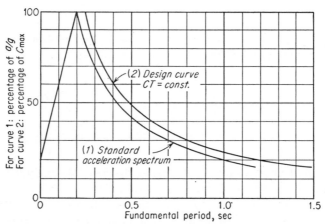

FIG. 18.5. Standard acceleration spectrum and the aseismic design curve. (*After Biot.*)

18.9. Lateral Forces of an Earthquake. The Joint Committee of the San Francisco engineering organizations,* upon a study of the earthquake spectra and considering the design and construction practice of its members in seismic regions, adopted the following decisions:[7]

The lateral force acting on a structure during an earthquake or, which is the same, the base shear V should be expressed by Eq. (18.3):

$$V = CW \tag{18.3}$$

where W is the weight of the structure and C is the base shear coefficient.

The value of C is at a maximum when the length of the period T is 0.25 sec, which is close to the period value $T = 0.2$ sec given by the standard acceleration spectrum (Fig. 18.5). As the length of the period T increases, the value of the base shear coefficient C decreases. The

---

* The San Francisco Section of ASCE and the Association of Structural Engineers of Northern California, *cf.* Ref. 7.

product CT remains constant for all values of T less than 0.25 sec, which is graphically represented by the "design curve" in Fig. 18.5.

If a structure is designed by using Eq. (18.2), the design will be very much on the safe side.   The theoretical value A/g is entirely too high for

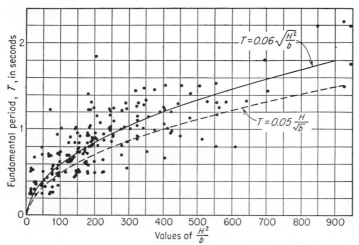

FIG. 18.6. Natural vibration period of buildings.   (*From ASCE Trans., vol.* 117, *p.* 732, 1952.)

practical use, since the damping, tilting, and slipping of structures during a shock tend to reduce the shear transmitted to a structure from the ground.   Instead, the Joint Committee recommends the following base shear coefficient C for buildings:

$$C = \frac{0.015}{T} \tag{18.4}$$

where T is the natural period of the structure.   The range of the C values varies from $C = 0.06$ for $T = 0.25$ sec or less to a minimum of $C = 0.02$.

Natural periods of buildings were measured by the USCGS and plotted against the value of H^2/b, where H is the height and b the width of the building in the direction considered (Fig. 18.6).   In the absence of better criteria for the determination of the length of the period T, a conservative formula (18.5) from that figure may be used:

$$T = 0.05 \frac{H}{\sqrt{b}} \tag{18.5}$$

H and b should be expressed in feet.

For structures other than buildings which generally have little damping and few members that might essentially contribute to their earth-quake resistivity, a larger coefficient C is applied, namely,

$$C = \frac{0.025}{T} \tag{18.6}$$

with a maximum of T of 0.10 and a minimum of 0.03. As to the weight W [Eq. (18.4)] this is the estimated weight of the structure during the earthquake. It equals the dead load (DL) plus a reasonable portion of the live load (LL), according to the service of the structure.

In the case of a multistory building a portion of the shear V is applied at each floor. If the weight of a story assumed concentrated at the floor level is w_x and the height of that floor above ground is h_x, the lateral earthquake force ΔV at that level is

$$\Delta V = V \frac{w_x h_x}{\Sigma wh} \tag{18.7}$$

where Σwh is the sum of the products wh for all stories of the building.

The method of aseismic design described in this chapter may be called *dynamic*, since it is based on actual observations of the ground motion. In the other method in use, the analysis of structures is made under the assumption that during the earthquake a horizontal acceleration, usually of one-tenth gravity, is applied to the structure. In other words, besides the forces used in the static design, an additional horizontal body force of 10 per cent of weight is considered. In an actual design, consultation of local building codes is essential. Horizontal acceleration used by the USBR in the aseismic design of dams is $0.1g$, which in certain cases may be increased to $0.3g$.[20,21]

18.10. Earthquake Damage.[14-18] *Buildings.* Steel-framed buildings in which the frame supported all walls and floor loads behaved satisfactorily in San Francisco's 1906 earthquake, though damage of varying degrees occurred to walls, facings, and utilities. In Tokyo (1923 earthquake) out of 16 large steel-framed buildings, six were absolutely undamaged and the rest suffered damage of varying degrees; one building was even on the edge of collapse. All reinforced-concrete buildings in the Santa Barbara, California, earthquake of 1923 fared satisfactorily, although there were cracks in walls and piers. Masonry-bearing walls with wood interior construction prove to be very vulnerable to earthquake damage unless the whole building including its foundation is properly tied into a rigid unit to resist the shock. Lime-cement mortar apparently proved to be the best material for such masonry. Good workmanship contributes to stability as demonstrated by the example of many excellent masonry walls that survived the Charleston earthquake in 1886.

Most earthquake hazards in the cities are due not to major structural failures but rather to falling debris, in the first place, parapets, filler walls, veneer, and other non-essential appendages, and to disruption of utilities (electricity, gas, water). Fire is an important contributing factor in earthquake disasters (short circuits, kitchen fires). The ground motion in itself is not hazardous except in the vicinity of the actual faulting.

Foundations. Deep foundations generally are not hazardous in an earthquake. Shallow foundations are sensitive to the vertical displacement component, especially if the structure is light and the water table near. In this connection it should be noted that apparently vibration is most easily transmitted through sand saturated with ground water. The common general opinion is that soft soils of limited cohesion are more dangerous than compact soils. Loose alluvial soils like those close to a beach are considered most hazardous, though exceptions exist.

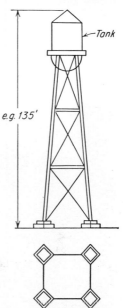

Elevated Tanks. Tank towers of the rod-braced type have been extremely vulnerable to earthquake damage. One of many types of such tanks is shown in Fig. 18.7. Apparently the X bracing (see figure) fails first, after which the remaining frame deflects and collapses. In the case of the Long Beach earthquake of 1933, a steel water tank 150 ft above the ground surface collapsed completely, whereas a short distance away another identical tank withstood the shock with minor damage. The former tank was on spread footings, and the latter on poured-in-place concrete footings, tied up and resting on piles. This fact, though very significant, cannot be considered as an unquestionable proof of the better service of pile foundations, however. In interpreting such cases the presumable but unpredictable occurrence of nodal and antinodal points should be considered. At such points surface vibrations either emphasize or cancel one another (interference phenomenon).

FIG. 18.7. Sketch of an elevated tank.

Slopes and Embankments. Vibrational effects are especially strong in the case of thixotropic materials. In the New Zealand earthquakes of 1929 and 1931, a section of the highway turned fluid, but a couple of hours after the shock the resident engineer was able to walk on it.

At another section of the highway a 22-ft-high, 6-year-old embankment decreased in height by 3 ft, but the vegetation on its slopes was not disrupted.

Dams. During the San Francisco earthquake of 1906, the San Andreas fault zone passed through a knoll that formed the left abutment of the main San Andreas Dam. No offset was produced in the dam itself, but a tunnel from the overflow weir was cut in two, partly crushed, and offset about 5 ft.

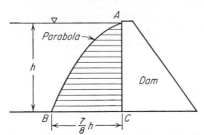

FIG. 18.8. Pressure on dam during earthquake. *ABC* is a water body moving with the dam. (*After Westergaard.*)

South of the San Andreas Dam in the same fault-line valley is the Upper Crystal Springs earth dam, about 90 ft high. The fault movement of 1906 sheared the structure diagonally and offset it about 8 ft.

During a shock the body of a dam moves back and forth. When the dam moves away from the water in the reservoir, hydrostatic pressure on the dam decreases, since the water does not follow it closely. Conversely, if the dam moves toward the water, pressure increases and there is some rise of water level. In a study based on the assumption of simple harmonic motion, Westergaard came to the conclusion that for design purposes a portion of water in the reservoir, parabolic in shape (Fig. 18.8), may be assumed as moving together with the dam.[19]   Information on the hydrodynamic pressure on the face of a dam due to horizontal earthquake shock also may be found in Ref. 20.

Tunnels.   The railroad tunnel near Wright Station in the Santa Cruz Mountains, California, was intersected by the San Andreas fault in 1906.   In this 6,200-ft tunnel a fracture was observed at 400 ft from the northeast portal, making an angle of 80° with the longitudinal axis of the tunnel.   The maximum offset of 5 ft was at the fracture, gradually dying out at a distance of about 5,100 ft.

FIG. 18.9. Failure of a tunnel during the Kern County, California, earthquake of 1952. Note rail shifted out.   (*Southern Pacific Company.*)

During the Kern County, California, earthquake of July 21, 1952, tunnel 3 of the Southern Pacific Railroad (near Bealville) was damaged.   After the earthquake one wall of the tunnel was found straddling the distorted track (Fig. 18.9).   The rail, though badly distorted, remained firmly fixed to the ties.

REFERENCES

1. Byerly, P.: "Seismology," Prentice-Hall, Inc., New York, 1942.
2. Gutenberg, Beno, and C. F. Richter.: Seismicity of the Earth, *GSA Spec. Paper* 34, 1941; also published by the Princeton University Press.
3. Leet, L. D.: "Practical Seismology and Seismic Prospecting," Appleton-Century-Crofts, Inc., New York, 1938.
4. Macelwane, J. B., S. J.: Definition and Classification of Earthquakes and other articles in "Physics of the Earth," vol. VI, Seismology, Bulletin 90, National Research Council, 1933.

5. Reid, H. D.: Mechanics of Earthquakes.   Rebound Theory.   Regional Strain, Bulletin 90, National Research Council; also Bulletin, vol. 6, Department of Geology, University of California, Berkeley, 1911.
6. Leet, L. D.: "Earth Waves," Harvard University Press, Cambridge, Mass., 1950.
7. Joint Committee of the San Francisco, Calif., section ASCE, and the Structural Engineers Association of Northern California, John E. Rinne, chairman: Lateral Forces of Earthquake and Wind, *ASCE Trans.*, vol. 117, 1952; contains comprehensive bibliography.
8. U.S. Department of Commerce, U.S. Coast and Geodetic Survey: "United States Earthquakes 1933," Serial 579, by Frank Neumann.
9. "United States Earthquakes 1935," Serial 600.
10. "United States Earthquakes 1940," Serial 647.
11. "United States Earthquakes 1949," Serial 748.
12. "United States Earthquakes 1952," Serial 773.
13. Duke, C. M., and M. Feigen (eds.): Proceedings of the Symposium on Earthquakes and Blast, University of California, Los Angeles, 1952.   Contains 12 papers on earthquakes, particularly (*a*) G. W. Housner, Spectrum Intensities of Strong Motion Earthquakes, and (*b*) J. E. Rinne, Building Code Provisions for Aseismic Design.
14. Hudson, D. E., and G. W. Housner: Structural Vibrations Produced by Ground Motion, *Proc. ASCE Sep.* 816, October, 1955.
15. Freeman, J. R.: "Earthquake Damage and Earthquake Insurance," McGraw-Hill Book Company, Inc., New York, 1932.
16. Martel, R. R.: Effect of Foundations on Earthquake Motion, *Civil Eng.*, vol. 10, 1940.
17. Ruge, A. C.: Earthquake Resistance of Elevated Water Tanks, *ASCE Trans.*, vol. 103, 1938.
18. Furkert, F. W.: The Effect of Earthquakes on Engineering Structures, *Proc. Inst. Civil Engrs. (London)*, vol. 23, 1932.
19. Westergaard, H. M.: Water Pressure on Dams during Earthquakes, *ASCE Trans.*, vol. 98, 1938.
20. Kirn, F. D.: Design Criteria for Concrete gravity and Arch Dams, *USBR Eng. Monogr.* 19, Denver, Colorado, 1953.
21. Esmiol, E. E.: Seismic Stability of Earth Dams, *USBR Tech. Memo* 641, Denver, Colorado, 1951.

CHAPTER 19

SOME LEGAL ASPECTS OF GEOTECHNICS

The person in responsible charge of foundation construction is bound by certain legal liabilities that are inherent in excavating a hole into the earth's surface. For engineering works of any magnitude, legal advice should be sought and followed from the time exploratory drill holes are put into the ground until the signing of the final payment contract. In this chapter are presented some of the basic principles of law applicable to foundation excavation.

19.1. Basic Legal Principles. As it is not possible within the scope of this book to present the full extent of the legalities involved in foundation excavation, only a few out of many hundreds of such cases are discussed. These cases are cited to indicate some of the many difficulties that can be encountered and are not to be construed as presenting the absolute law for any general type of case. Cases will differ even though they seem to involve similar facts; thus the courts have to rule on the individual merits of each case and find the basic legal principles which may apply.

The engineer or geologist should be acquainted with the fact that certain laws, statutes, or ordinances may affect the legal prosecution of his work.

A *law** is "The binding custom or practice of a community; rules or mode of conduct made obligatory by some sanction which is imposed and enforced for their violation by a controlling authority; . . . In highly developed societies, a law becomes chiefly of special enactment or of statutory origin, and the authority imposing the sanction is the sovereign state."[1]

From a legal standpoint, a *statute* is generally defined as "the written will of the legislature, solemnly expressed according to the forms necessary to constitute the law of the state" (82 C.J.S., Statutes, §1a).

An *ordinance** as customarily defined in the United States is "a local law or regulation enacted by a municipal government, as a common council, board of aldermen, or the like."[1] The building code of a city is an example of an ordinance.

*** By permission. From "Webster's New International Dictionary" Second Edition, copyright, 1934, 1939, 1945, 1950, 1953, 1954, by G. & C. Merriam Co.

Prior to the construction of any structure, the city building code, if one exists, always should be consulted. The code usually is in conformance with other ordinances, statutes, and laws governing the community. There are various legalities concerned in the contract for the work, the method in which the excavation is performed, any expert opinions rendered by the engineer or geologist, or the material appearing in the geologic or engineering report.

Although there are many basic legal principles applicable to foundation work, the *laws of contracts*, the *laws concerning negligence*, and the *laws concerning adjoining landowners* are of considerable importance.

A *contract* is "an agreement which creates an obligation. Its essentials are competent parties, subject matter, a legal consideration, mutuality of agreement, and mutuality of obligation" (17 C.J.S., Contracts, §1a).

The laws involved in *adjoining landownership* generally agree that "the mutual and reciprocal duties of adjoining landowners require that each use his own land in a reasonable manner, with due regard for the rights and interests of the other; no liability attaches for consequential injuries from a reasonable use" (2 C.J.S., Adjoining Landowners, §1).

The word *negligence* has no simple legal definition; usually it is subject to specific definition by the courts. Generally in a colloquial sense, negligence means "a mere absence of care or the want of ordinary care," and in a legal sense, negligence is said to be "actionable" (65 C.J.S., Negligence, §1a).

Cases pertinent to the above principles are discussed in succeeding paragraphs. For more details, the reader should consult "Corpus Juris," "Corpus Juris Secundum," and "American Jurisprudence," which present the law in general terms and cite various cases in support of the existing interpretation of the law. The reader is cautioned, however, that although these references give a general idea of the laws applicable to his situation, he should always consult local legal talent. Just as a geologic cross section is subject to several interpretations, so is every law subject to different interpretations by different courts. Only a competent lawyer can ascertain the most recent and most valid interpretation of the applicable law, and in many cases, court action may be required to establish the correct interpretation.

In the cases cited in this chapter, the *plaintiff* is the party who starts the suit and is listed first and the suit is defended by the *defendant*. Thus, if the citation reads "Jones v. (versus) Smith," Jones is the plaintiff and Smith is the defendant. The abbreviations used in the citations are explained in the references at the end of the chapter.

CONTRACTS

19.2. Scope of Contracts. The old saying that you should not sign anything without reading it first is very applicable in the case of contracts. Any document to which you append your name may constitute a legal contract, in which case you can be held liable for anything that is stated in the contract, regardless of whether you have read the document or not. Many contractors have become sadly aware of this fact. Some contracts, for example, may require that the builder erect a structure according to the specifications (which are considered or made part of the contract) and, furthermore, that the structure shall be ready for use and occupation and delivered so finished and ready at the end of the said contract. Such a contract has been termed a "turn-key" contract —all the owner has to do is "turn the key" in the door and open it, and the building is ready for his use. This type of contract is illustrated in the case of Ingle v. Jones (69 U.S. 762).

In this case, the building was constructed, but before completion, owing to defects in the soil, the foundation sank, causing the building to crack and start to fail. The owner had to remove and rebuild the foundation and rebuild part of the structure. Unfortunately (for the contractor) the contract provided that

[the building] shall be executed, finished, and ready for use and occupation; and delivered over, so finished and ready, . . . and that he [the contractor] shall and will forthwith prosecute and carry on the said work with all diligence and practical expedition.

The court ruled that it was the contractor's duty to fulfill the contract and that he could have guarded against hardship by a *provision* in the contract. However, since he did not do so, it was not in the power of the court to relieve him, for he did not make the building *"fit for use and occupation."* The court stated that unforeseen difficulties, no matter how great, could not excuse him (the contractor). The principle on which this decision is based is the *sanctity of contracts*, wherein all parties are required to do what they have agreed to do, and if the parties make no provision for dispensations, the rule of law gives them none. A party must perform his contract unless performance is excused by agreement or certain rules of law.

However, even with turn-key contracts, if it becomes impossible for the contractor to fulfill his obligation because of faulty specifications, the contractor may under some circumstances not be held liable if the structure is not built. In Sickels v. U.S. (1 Ct. Cl. 214; 53 A.L.R. 112) the contractor could not complete construction of a lighthouse on a site selected by the United States government. The site was underlain by quicksand, *although* the plans had stated that hard sand was present. The court ruled that

. . . if the owner selects both the plan and the exact site of the location of the structure, the contractor is excused from the performance if it develops (*contrary to repre-*

sentations) that the subsoil is of such character as to render it impossible to erect the structure under the plans specified.

In this case, the contractor recovered for the work he had performed and the materials he had furnished up to the time it was discovered that the soil was defective.   Other cases concerning turn-key contracts and foundation defects are in 20 Minn. 494, Gil 448, and 27 N.J.L. 573.

19.3. Specifications and Owner Responsibility.   The owner has the responsibility of presenting *all* the known *facts* in his specifications.   If he fails to do so, through either carelessness or willfulness, he may be held liable for any difficulties encountered by the contractor as result of such omissions.   In M. H. Sobel v. U.S. (88 Ct. Cl. 149), a faulty contract resulted in a damage claim.   The plaintiff sued because when he began to excavate for hangars at an airfield, he found that the soil was unsuitable to support the foundations.   When the matter was called to the attention of the government, the contractor was ordered to place the foundation in question on piles.   The contractor then filed a claim for the additional expenses which he incurred by reason of the change order. The court ruled in favor of the contractor, as the change made necessary by the soil conditions was unknown when the contract was made and therefore the change was not contemplated by the contract.   The basis for this ruling was the case of Rust Engineering Co. v. U.S. (86 Ct. Cl. 461); the construction specifications stated

Should the contractor encounter, or the Government discover, during the progress of the work, subsurface and (or) latent conditions at the site materially different from those shown on the drawings or indicated in the specifications . . . the contracting officer shall thereupon promptly investigate the conditions, and if he finds that they materially differ from those shown on the drawings, or as indicated in the specifications, he shall at once . . . make such changes in the drawings and (or) specifications as he may find necessary, and may increase or decrease all costs and (or) difference in time resulting from such changes shall be adjusted. . . .

The drawings did show and specify the depth to which the footing excavations were to extend.   However, at this specified depth, the contractor deemed that the subsoil conditions were unsuitable for placement of the footings.   The government agreed and directed the contractor to conduct exploration and place the footings deeper.   In the ensuing claim by the contractor for the extra work, the court held that the plaintiff was entitled to recover the extra costs caused by delays in beginning of the superstructure, delays due to the unknown conditions encountered in excavating.   They also stated

The changes made necessary by reason of the conditions encountered in excavating for the foundation of the building were not reasonable changes within the scope of the drawings and specifications as contemplated in . . . the contract, but represented important changes based upon the conditions which were unknown and materially different from those shown on the drawings or indicated in the specifications.

If ground conditions materially change between the time specifications are prepared and the job is undertaken, the contractor cannot be held liable for damages caused by such changes. In Oklahoma City v. Derr (235 Pac. 218), a sewer system was constructed some years after the plans had been drawn. During the interim between design and construction, some water lines had been installed and the grades over part of the proposed sewer excavation had been considerably changed. These changes in conditions were not shown in the specifications. During construction, the hidden water lines were damaged, and the changes in grade required unforeseen excavation; as a result the contractor was forced to bear unforeseen costs. The court ruled that

. . . where the contractor must build and complete a structure according to the plans and specifications of the owner. The contractor will not be required to bear extra expense resulting from the performance of the contract on account of defects in the plans and specifications prepared and submitted by the owner.

Although the contractor has a legal responsibility to complete a job according to the contract, the owner has the responsibility to prepare adequate drawings and specifications. This is well-illustrated in U.S. v. Spearin, and Spearin v. U.S. (248 U.S. 132; 86 Ct. Cl. 461, 475). During construction of a dry dock and relocation of a sewer, a flood overloaded the sewer, causing it to break and flood the working area. The government insisted that it was the contractor's responsibility to complete the job according to the specifications. The contract finally was annulled, and the contractor sued for the amount he had expended and the profit he would have earned if allowed to complete the contract. The court held that

The contractor is bound to build according to plans and specifications prepared by the owner, [but] the contractor will not be responsible for the consequences and defects in the plans and specifications. *This responsibility of the owner is not overcome by the usual clauses requiring builders to visit the site* [the authors' italics], to check the plans, and to inform themselves of the requirements of the work, as is shown by Christie v. U.S. (237 U.S. 234) where it was held that the contractor should be relieved if he was misled by erroneous statements in the specifications.

In the case at bar, the sewer . . . must be built in accordance with the plans and specifications. . . . The insertion of the articles describing the character, dimensions, and location of the sewer imported a warranty that if the specifications were complied with, the sewer would be adequate. . . . The obligation [of the contractor] to examine the site did not impose upon him the duty of making a diligent inquiry into the history of the locality, with a view to determining, at his peril, whether the sewer specifically prescribed by the Government would prove adequate. The duty to check plans did not impose the obligation to pass upon the adequacy to accomplish the purpose in view. The provision concerning contractor's responsibility cannot be construed as abridging rights arising under specific provisions of the contract.

In a somewhat analogous situation, that of Woods v. Amulco Products (235 P.2d 273), the courts ruled that the construction contractor is not liable to the owner for loss or damage which results solely from defective or insufficient plans or specifications furnished by the owner or his architect or engineer and followed by the contractor providing that the contractor has not been negligent or has not given the owner an express warranty that the plans or specifications are free from defects.

Statutes concerning the liability of the owner under the aforementioned circumstances differ in various states. In Texas, for example, the courts have held that the owner does not warrant that the plans and specifications submitted by him are sufficient to produce the desired results, and therefore the builder must know for himself that they are sufficient. For this reason the builder may not recover extra costs incurred by reason of faulty specifications (17 C.J.S., Contracts, §371c). The necessity for accurate specifications is summarized in the following statement from 17 C.J.S., Contracts, §371c:

Where the necessity for extra work results from the acts, errors, and mistakes of the architect or engineer of the owner under whose supervision the work is to be done, the loss should fall on the owner, and the builder may recover additional compensation, and this rule applies if the extra work is outside the original contract, although it [the contract] provides that the contractor will not charge for extra labor. Additional compensation may be recovered for extra work which becomes necessary because the building cannot be constructed according to the plans and specifications furnished . . . such as for alterations rendered necessary by defective plans; or for additional work or expense which is rendered necessary by the owner's negligence or failure to perform his part of the contract. . . .

19.4. Contractor's Responsibility. Obviously, the contractor is not always on the winning side, as he has certain obligations when performing a contract. For instance, in Brent v. Head *et al.* [16 L.R.A. (n.s.) 801], the court held that a contractor undertaking the erection of a building in freezing weather and attempting to do a first-class job is not relieved from this responsibility just because the foundation settles owing to the weather. The contractor should have foreseen the difficulties likely to ensue because of weather conditions and based his bid accordingly.

In the often-cited Grier-Lourance Construction Co. v. U.S. (98 Ct. Cl. 434) case, the court ruled that the specifications for construction of a bridge over the Potomac River had warned the contractor of abrupt variations in the bedrock in the area. It was the contractor's responsibility to reflect these variabilities in his bid price. In other words, the government had already paid the contractor for the risk and therefore did not have to pay again. In the similar case of Simpson v. U.S. (34 Ct. Cl. 539; 172 U.S. 372), the plans upon which the contractor had bid did not specify the character of the underlying ground. When the foundation material turned out to be quicksand, the contractor's claim for damages was not allowed, as the court ruled that

. . . a warranty that the ground was free from quicksand or was of the character shown by the profile could not be implied from the terms of the contract, and hence the plaintiffs were not entitled to recover damages sustained by the presence of quick-sand in the formation at the location selected.

Further information on the contractor's responsibility to perform a job according to specifications even though they may be faulty is given in 248 U.S. 132 and 235 P.2d 273.

19.5. Geological Specification Drawings. The placing of interpreted geological information on specification drawings may result in serious legal consequences; therefore, the geologist preparing the drawings should take care that he shows *only* such conditions as are actually encountered in the exploration hole. If he attempts to extend the bore hole information by interpreting the contacts between formations and depths to bedrock, etc., and his interpretations prove to be incorrect, the contractor may have a valid basis for a claim. An example of the latter occurred in the case of H. M. Baruch v. U.S. (93 Ct. Cl. 107), which involved the construction of several buildings. During excavation, it was found that a rock dike extended upward into what was to have been the subbasement of one of the buildings. The borings (and logs), however, had stated that this rock was only a boulder submerged in blue clay; thus the building had been designed to rest on hard blue clay. The court ruled that

. . . the changes made necessary by reason of the conditions encountered in excavating for the foundation . . . were not reasonable changes within the scope of the terms of the contract and specifications, but were important changes that were not within the contemplation of either party of the contract at the time it was made. Under the previous decisions of the court, the plaintiffs may recover the actual costs thus incurred.

While, ordinarily, interpretative information would be of value to the contractor, if it is incorrect, the contractor has no recourse but to sue for the damages incurred as a result of its use. The most "foolproof" method of presenting the results of subsurface exploratory data in specifications would be to include merely exact replicas of the bore hole logs. Usually, however, for ease in using the drawing, the bore holes are plotted on a profile. Occasionally some geological interpretation is presented for the convenience of the contractor; for example, the "assumed" bedrock surface may be indicated by connecting the points at the top of bedrock in each bore hole with a straight or curved line. However, the use of the word "assumed" does not relieve the preparer of the drawings of responsibility if the interpretation is faulty. If the contractor should base his bid for excavation on the "assumed" bedrock surface and then find that during excavation the bedrock line between bore holes was actually several feet higher or lower than "assumed," the owner may be sued for the increase in excavation costs.

Correct presentation of water-table (or ground-water) data on a specification drawing is also of importance.   Adverse water conditions can be detrimental and costly to a contractor if they are unforeseen or miscalculated owing to erroneous information on the drawings.   Therefore, it is advisable to place the water level on each bore hole log and beside it give the date upon which it was measured.   Then if, at the time of excavation, the water table is higher than indicated on the drawing and this fact causes damages to the contractor, he has to prove that the change is due, not to seasonal fluctuations, but rather to negligence or carelessness on the contractee's part.   A further difficulty may result if on some of the logs the water table is given but on others no mention of water conditions is made.   The contractor may draw the logical assumption that water was not encountered in the latter bore holes, whereas actually, owing to erroneous data or carelessness, the water-table level and date were omitted.   Thus, if the contractor encounters water in the excavation where the logs did not show water (due to the aforementioned circumstances), he will have a basis upon which to file a claim for any extra costs that he might incur.   Therefore, if water is not encountered in a bore hole, this fact should be so stated on each log to which it applies, and the date should be given upon which the absence of water was noted.

The factual results of *all* exploratory work should be given on the geologic specification drawing.   If some logs are omitted, either purposely or carelessly, and the contractor encounters construction difficulties in those areas where logs were not shown (although holes had been drilled in those areas), he may have a valid claim.   (This matter of omission of data is closely allied to "negligence" and "misrepresentation" problems and is discussed in those sections.)   For the same reasons, the *locations* of all exploratory work should be carefully checked and shown correctly on the drawings.

For further information on contracts and their relation to the engineer, the reader should consult 17 C.J.S., Contracts, §1 through 11 ("Contracts—Definitions, Kinds, and Distinctions"), and 17 C.J.S., Contracts, §360 through 372 ("Building and Construction Contracts").

NEGLIGENCE

19.6. Definition: Personal Injuries.   The term "negligence" does not have a specific legal definition; however, in many instances the courts have held that negligence is "a failure to act with the care and forethought usually exercised by reasonable and prudent men" (65 C.J.S., Negligence, §1).   Generally, the judge defines the term to the jury, and the definition depends upon the particular case being tried.

Negligence suits (involving engineering projects) may involve injury to the person of other individuals.   A simple example is that of an excavation which undermines an adjacent sidewalk, injuring the persons on the walk.   If the engineer had used proper geological knowledge, he could have controlled his excavation slopes by shoring or other means.   The courts have held that

> The owner of land abutting on a highway generally owes a duty to travelers on the way not unreasonably to endanger them by excavations on his land (65 C.J.S., Negligence, §78a).

In such cases, the term "highway" includes the "sidewalk."   Under certain circumstances an owner has an obligation to protect an excavation (test pits, basements, etc.) on his land.   If the excavation is in close proximity to a thoroughfare, it generally is advisable to protect the excavation with fencing.   The law has held that the owner is negligent if a person who has exercised ordinary care has fallen into the excavation because of its close proximity to a thoroughfare (65 C.J.S., Negligence, §85a).   If excavations are at some distance from generally traveled thoroughfares, the owner is not necessarily required by law to protect them unless they may create a peril to children, livestock, or others who should not be subjected to dangers (because they cannot judge for themselves whether dangers exist).

The contractor generally is permitted to rely on plans and specifications which appear to be of average quality.   However, unless these plans and specifications are

> . . . so apparently and obviously defective as to put a contractor of average skill and ordinary prudence on notice that the work is dangerous and likely to cause injury and that he should not attempt construction according to the plans and specifications. . . .

there would be no negligence on his part if he did follow them (65 C.J.S., Negligence, §95a).   The law generally holds that

> As a general rule, an independent contractor is liable for injuries caused by his own negligence in the performance of the work.

The term "contractor" as used in this ruling also can apply to an architect or engineer in charge of construction, as it has been ruled that

> A supervising engineer in complete charge of the work is liable for injury due to defects in plans which he has prepared or approved, arising from his negligence (65 C.J.S., Negligence, §95c).

19.7. The Consultant's Liability.   The engineer or geologist can be sued for "actionable" negligence if he prepares a faulty report, gives faulty testimony when acting as an expert witness, or performs his work in a faulty manner, providing certain circumstances exist.   The status of the consultant in such cases was summarized in the case of Staloch **v.** Hohm[9 L.R.A. (n.s.) 712]:

The distinction between an error of judgment and negligence is not easily determined. It would seem, however, that, if one, assuming a responsibility as an expert, possesses a knowledge of the facts and circumstances connected with the duty he is about to perform, and bringing to bear *all his professed experience and skill* [authors' italics], weighs these facts and circumstances, and decides upon a course of action which he faithfully attempts to carry out, then want of success, if due to such course of action, would be due to error of judgment, and not to negligence. But, if he omits to inform himself as to the facts and circumstances, or does not possess the knowledge, experience, or skill which he professes, then a failure, if caused thereby, would be negligence.

The above judgment was further verified in the ruling given in the case of L. B. Laboratories v. Mitchell (237 P.2d 84; 244 P.2d 385):

. . . a member of the learned professions and anyone who takes employment because of his exceptional skill, impliedly represents that he possesses and will employ the degree of learning and skill usually possessed by those in good standing practicing their specialities in the same locality, and impliedly agrees to use his best judgment but does not guarantee results. Where a person undertakes to employ a special skill, there is no breach of duty if such skill and operator's best judgment are put forth, and under such circumstances, mere errors of judgment are not actionable. While the failure to achieve a guaranteed result is actionable, independently of negligence, this is not true where there is no guarantee of the result.

It is apparent from these rulings that a consultant or an expert witness is required to exercise his *best* skill and judgment; when he does so, the court will not find him liable for negligence if he makes an error in such judgment. Conversely, if it could be proved that the consultant did not exercise his best skill or judgment, it is possible that he would be liable for negligence.

19.8. Misrepresentation. Although closely allied to negligence, misrepresentation has no established legal definition. The facts may be misrepresented, either purposely or carelessly. This situation can develop in the preparation of specification drawings, particularly those that have bore hole data. For example, in the case of G. B. Christie v. U.S. (237 U.S. 234), contractual difficulties due to foundation conditions during the construction of locks and dams were legally considered to result from a misrepresentation of conditions. In this case, the contractors relied upon the government boring sheets which showed gravel, sand, and clay and no other materials. The written part of the specifications stated that the material to be excavated was (as shown on the drawings) correct "as far as known"; however, "the bidders must inform and satisfy themselver as to the nature of the material." During excavation, the contractor encountered stumps, cemented sand and gravel, and a sandstone con-glomerate. This material was more difficult and expensive to excavate then could be inferred from the specification drawings, and thus the contractor filed a claim.

In the ensuing testimony, it was learned that when the borings were made, the drill had met "obstructions which, from the particles broken off and floating to the surface, would indicate they might be logs." This note, however, was not shown on the drawings. Also, it was learned that the drillers, whenever they met obstructions that hampered their drilling, would move the drill to places where the drill would penetrate; the final boring logs, however, indicated that the hole had been drilled at its original location. It was further brought out that the specification drawings showed only the record of the completed borings and did not show the cemented sands and gravels in the partially completed drill holes (from which the drillers had moved due to "obstructions"). It was stated that the resident engineer apparently did not consider it of enough importance to note these partially completed drill hole logs. The court expressed the opinion that there was a deceptive representation of the material and thus the contractor was misled, and it made no difference to the legal aspects of the case that the omissions of the boring results had no sinister purpose. "There were representations made which were relied upon by the claimants, and properly relied upon by them, as they were positive." The final decision (after an appeal) was that the contractor's claims should be allowed.

Misrepresentation also occurred in the case of the U.S. v. the Atlantic Dredging Co. (253 U.S. 1). A channel was to be dredged into the Delaware River, and the specifications stated that the material to be removed were believed to be mainly mud or mud with an admixture of fine sand. It was further stated that

Bidders were expected to examine the work, however, and decide for themselves as to its character, and to make their bids accordingly, as the U.S. did not guarantee the accuracy of this description. . . . A number of test borings had been made in all the areas where dredging is to be done . . . and the results, therefore, may be seen by impending bidders on the maps on file in this office. No guarantee is given as to the correctness of these borings, in representing the character of the bottom over the entire vicinity in which they were taken, although the general information given thereby is believed to be trustworthy.

The results of all the borings had been recorded in the logs or in field notes, but the map shown to the bidders did not show the field notes. When dredging was started, material was encountered that differed from that shown on the map. Some of it was a compacted sand and gravel with a small percentage of cobbles which required the contractor to bring in heavier type of equipment.   In the ensuing claim by the contractor, the court held that

The case comes within the ruling of the U.S. v. Spearin (248 U.S. 132) where it is stated that the direction to contractors to visit the site and inform themselves of the actual conditions of a proposed undertaking will not relieve from defects in the plans or specifications.   It is held (in certain cases) "that the contractor ought to be relieved if he was misled by erroneous statements in the specifications."   The present case is certainly within the principle expressed. . . . The company did not know of the concealment of the actual test borings. . . . It did not know, at that time, of the manner in which the test borings had been made.   Upon learning that they had been made by the probe method, it then elected to go no further with the work; that is, upon discovering that the belief expressed was not justified and was, in fact, a deception.

The latter points in this ruling should be noted.   In effect, it is necessary that the specification drawings indicate the method by which the exploratory work was done—core drills of certain sizes, wash borings, churn drills, test pits, hand or power augers, etc.   The effect of these rulings can be summarized as follows: If the owner makes subsurface investigations and then willfully or otherwise withholds some of the data when the specifications are issued, thus misleading the contractor as to the true nature of foundation conditions, the owner probably will be held liable (also see Sec. 19.3).

ADJOINING LANDOWNERS

19.9. Lateral Support of Land.   The various legal rulings on this subject are very complex and subject to individual court interpretations.   Of most interest to the foundation engineer are the rulings concerning the lateral support of the excavation.   Under common law, the landowner has the *absolute right to the lateral support of his land in its natural state*. However, when a building is erected upon the adjacent land, the law may be subject to different interpretations by local ordinances and statutes (2 C.J.S., Adjoining Landowners, §5).   For example, in one case (1 Am. Jur., Adjoining Landowners, §25), the court ruled that

It is now well settled that the owner of land is entitled to have it supported and protected in its natural condition by the land of his adjoining proprietor, and that if such adjoining owner removes such natural support, whereby the soil of the former is disturbed or falls away, he is legally liable for all damages so occasioned . . . and the fact that the falling of the soil on the adjoining land was due to the action of the elements does not constitute a defense.
. . . a person who is excavating adjacent to another person's property whereon a house is constructed, the excavator can lawfully carry the foundation of his house lower than the foundation of a neighboring house and is not liable for damage, pro-

viding he uses due care and diligence in preventing any injury to the other house. . . .
However, he must give notice to the owner of the adjoining building so that the owner
can take the necessary precautions to protect his property after the excavator has
exercised due care.

These same interpretations also apply to the removal of *subjacent* support. For example, in construction of a sewer trench, quicksand flowed into the trench from underneath the adjoining land, and the adjoining land subsided because its support was removed. The contractor on this job was held liable for negligence. Similarly if subsurface water is withdrawn from under the adjacent land and causes it to subside, the user of the water will be liable for negligence [5 L.R.A. (n.s.) 1086]. It is necessary when excavating next to land with buildings on it that the excavator conduct the work skillfully and prudently in order to avoid liability for injury to the adjoining owner's structures (2 C.J.S., Adjoining Landowners, §12).

19.10. Notice of Excavation. Notice to the adjoining landowner in the event of excavation is necessary unless he has full knowledge of the work. However, this full knowledge must include the proximity of the excavation as well as its intended depth. Unless local ordinances or statutes provide otherwise, if the above conditions are complied with, the excavator has no liability for any injury that might result to the adjacent building, providing he uses due care in the performance of his work (1 Am. Jur., Adjoining Landowners, §28). This holds true even if the adjoining owner fails to protect his building adequately after receiving notice in sufficient time to perform such protective measures. It is interesting to note that if the adjoining building is in such poor condition that it falls, even though the excavation was conducted with prudence and skill, the contractor has no liability unless it has been so imposed by statute (2 C.J.S., Adjoining Landowners, §18c).

19.11. Blasting. If, during the excavation, blasting operations are required and such blasting causes damage to the adjacent property, the laws in various states differ with regard to the amount of liability to the excavator. Some hold that such damage results in liability unless there is " . . . no element of negligence or unskillfulness present." Others hold, however, that "there is liability regardless of negligence or lack of skill" (2 C.J.S., Adjoining Landowners, §45c).

19.12. General Advice. It cannot be too strongly emphasized that laws, statutes, and ordinances vary in different states, and therefore competent legal advice should be sought if excavation is to be undertaken. Occasionally certain circumstances will arise whereby existing interpretations of law become questionable, and therefore a new interpretation has to be sought in the courts.

REFERENCES

Legal terms and phrases are explained in the following:

1. "Webster's New International Dictionary," 2d ed., G. & C. Merriam Company, Springfield, Mass., 1950.
2. "Black's Law Dictionary," West Publishing Co., St. Paul, Minn.
3. "Words and Phrases," West Publishing Company, St. Paul, Minn.
4. *Cases cited* in Chap. 19 were checked, as far as possible, in August to October, 1954, issues of *Shephard's Federal Reporter Citator* or *Shephard's Pacific Reporter Citator* published at 420 No. Cascade Ave., Colorado Springs, Colo., to assure that there were no decisions prior to the publication of those books that reversed the court rulings given in the citation.
5. *Standard abbreviations* for law books referred to in this Chapter:

A.L.R.	"American Law Reports Annotated," Lawyers Cooperative Publishing Co., Rochester, N.Y., and Bancroft-Whitney Co., San Francisco, Calif.
Am. Jur.	"American Jurisprudence," Lawyers Cooperative Publishing Co. and Bancroft-Whitney Co.
C.J.	"Corpus Juris," American Law Book Co., Brooklyn, N.Y.
C.J.S.	"Corpus Juris Secundum," American Law Book Co.
Ct. Cl.	"Cases Decided in the Court of Claims of the United States," Government Printing Office, Washington, D.C.
Fed.	"The Federal Reporter," West Publishing Company, St. Paul, Minn
L.R.A.	"The Lawyers Reports Annotated," Lawyers Cooperative Publishing Co.
L.R.A. (n.s.)	"The Lawyers Reports Annotated, New Series," Lawyers Cooperative Publishing Co.
Pac.	"Pacific Reporter," West Publishing Company.
P.2d	"Pacific Reporter, Second Series," West Publishing Company.
U.S.	"U.S. Supreme Court Reports," Lawyers Cooperative Publishing Co.

Cases are cited according to volume number and page number of a reference; e.g., 235Pac. 218 means Vol. 235 of the "Pacific Reporter," page 218; or by volume and section number, e.g. 82 C.J.S., Statutes, §1*a*, means Vol. 82 of "Corpus Juris Secundum," §1*a* of the title "Statutes."

6. Readers of this text unfamiliar with the importance of law in engineering should inform themselves about this subject in the corresponding books, e.g., McCullough and McCullough, "The Engineer at Law," Iowa State College Press, Ames, Iowa, 1946, and Werbin's "Legal Cases for Contractors, Architects, and Engineers," McGraw-Hill Book Company, Inc., New York, 1955.
7. The authors acknowledge the publishers' permission to quote passages from "American Jurisprudence" and "American Law Reports" and from other series of the Annotated Reports System. Similar permissions were received from the publishers of the "Pacific Reporter" and "Pacific Reporter, Second Series"; and "Corpus Juris" and "Corpus Juris Secundum."

INDEX OF ROCKS, MINERALS, AND SOILS

701

GENERAL INDEX